ANALOG
Communication

V. CHANDRA SEKAR
Professor and Head
Department of Electronics and Communication Engineering
SASTRA University, Kumbakonam

OXFORD
UNIVERSITY PRESS

OXFORD
UNIVERSITY PRESS

Oxford University Press is a department of the University of Oxford.
It furthers the University's objective of excellence in research, scholarship,
and education by publishing worldwide. Oxford is a registered trademark of
Oxford University Press in the UK and in certain other countries

Published in India by
Oxford University Press

22 Workspace, 2nd Floor, 1/22 Asaf Ali Road, New Delhi 110002, India

© Oxford University Press 2010

The moral rights of the author have been asserted

First published 2010
Fourth impression 2013

ISBN-13: 978-0-19-806185-4
ISBN-10: 0-19-806185-4

Typeset in Times Roman
by Anvi Composers, New Delhi 110 063

Printed in India by Manipal Technologies Limited, Manipal

Third-party website addresses mentioned in this book are provided
by Oxford University Press in good faith and for information only.
Oxford University Press disclaims any responsibility for the material contained therein.

To
my beloved mother
Late Smt. V. Muthulakshmi
Had it not been for her sacrifices at a young age,
I would not be what I am today.

Preface

The subject of analog communication dates back to the period when radio was invented. Since radio waves can be transmitted from one place to another, a method was to be found to encode information onto these waves. This problem was solved by what is known as modulation. Modulation is the process of changing a physical characteristic of a wave to make it carry the information we need.

In an electromagnetic wave, there are three characteristics that can be modified, namely, amplitude, frequency, and phase. If the amplitude characteristic of a wave is changed, the process is called amplitude modulation (AM). This was first proposed and demonstrated by Fessenden and Lee de Forest in the year 1906. Frequency modulation (FM), where the frequency characteristic of a wave is changed, was patented in the year 1933 by its inventor Edwin H. Armstrong. It offered many benefits (like less immunity to noise) over AM, which dominated radio at that time. Similarly, phase modulation takes place when the phase characteristic of a wave is changed. Analog communication entails the study of these different modulation methods.

Rapid expansion of digital communication, particularly in the field of radio (software defined radio), TV (high-definition TV), and mobile phones does not override the need for a clear understanding of analog communication. Moreover, analog technology still plays a crucial role in the field of communication and will continue to do so in the years to come. Additionally, to understand clearly the concept of modulation/demodulation, a complete course on analog communication is a must for the students. With this background, the students will find it much easier to follow the concept of digital and data communication.

About the Book

Analog Communication is mainly written as a textbook for undergraduate students of electronics and communication engineering of all the major universities and engineering colleges in India. The book talks at length about the concepts, principles, and applications of analog communication and also provides an overview of digital communication concepts.

Students taking this course are expected to have a background in calculus, linear algebra, basic electronic circuits, linear system theory, and signals and system. A few topics such as phase locked loop, Costas loop, and digital data synthesis and a few case studies are included to make the book more application oriented. This book also introduces the concept of digital communication. Although in most of the universities and engineering colleges in India, there is a separate course on digital communication, a brief introduction can go a long way for the students to get an insight into this topic and they will find it easier to follow a separate course on the subject. This book can also be used as a reference book for practising engineers in the form of a self-learning tool.

Every effort has been made to help the students gain an understanding of the subject, apart from making them understand the theoretical concepts. The text is designed in such a way that it presents the basics, circuits, and building blocks of communication system. Many worked-out examples are provided throughout the book to help the students appreciate the concepts presented in the text. A large number of problems of different levels of difficulty are provided at the end of each chapter. Most of the chapters also include worked-out examples using MATLAB.

The main features of the book are as follows:

1. Case Studies on software defined radio (SDR) in Chapter 3 and direct digital synthesis (DDS) and frequency synthesizers using PLL in Chapter 4
2. Emphasis on important concepts such as phase locked loop, Costas loop, and digital data synthesis
3. Additional MATLAB exercises for thorough understanding of the concepts as an appendix

Content and Structure

The book comprises nine chapters. A brief outline of each chapter is given below.

Chapter 1 gives a brief introduction to communication theory, describing the meaning of different types of modulation and their properties. The chapter also gives a brief account of the digital communication system of a typical transmitter and receiver with the help of block diagrams. The concept of multiplexing of signals is also introduced.

Chapter 2 provides an introduction to the concept of signals and discusses typical signals and their properties. Linear and non-linear, time variant and invariant, discrete time and continuous time, and causal and non-causal systems are explained in the chapter. It also discusses Fourier series and transforms, Laplace transform, and z-transform as well as their properties. The relationship between energy and power with respect to frequency is also examined.

Chapter 3 is entirely devoted to the study of amplitude modulation (AM). It starts with baseband communication and theory of AM followed by the study of AM frequency spectrum and bandwidth, amplitude modulation index, and methods to calculate the average power for sinusoidal modulation. Theory and principles of double sideband (DSB), single sideband (SSB), independent sideband (ISB), vestigial sideband (VSB), and quadrature amplitude modulation (QAM) are also explained in detail along with the. generation and detection of various types of modulation. Block diagrams of transmitter and receivers are included. A detailed discussion on Costas loop, which is used for carrier phase recovery in double sideband suppressed carrier system, is also provided.

Chapter 4 extensively covers angle modulation. The introduction is followed by the concept of instantaneous frequency. The chapter highlights the differences between phase modulated (PM) and frequency modulated (FM) signals and elaborates the theory of FM (sinusoidal narrowband and wideband) and PM (sinusoidal and digital). Various methods for FM generation and FM detection are also discussed. Working of a typical transmitter and receiver is explained. Phase locked loop (PLL) and direct digital synthesis (DDS) are discussed in detail. At the end, a comparison between angle modulation and amplitude modulation is given.

Chapter 5 on pulse modulation discusses how the advent of computer has brought about a revolution in communication field. Now data is not transmitted as an analog or a continuous signal but is transmitted in digital form, which helps in processing the data more efficiently. After introducing the sampling theorem, various types of modulation, such as pulse amplitude modulation (PAM), pulse position modulation (PPM), pulse width modulation (PWM), pulse code modulation (PCM), pulse frequency modulation (PFM), and pulse time modulation (PTM), are discussed. The chapter introduces the concept of delta and adaptive delta modulation and explains in detail frequency division multiplexing (FDM) and time division multiplexing (TDM).

Chapter 6 presents the study of noise. The chapter includes different types of noise and their description as well as noise figure, noise temperature, and their measurements. Effects of noise on various types of modulation are discussed extensively. It emphasizes the need for pre-emphasis and de-emphasis circuits and covers threshold effects in angle modulation, mathematical representation of noise, quadrature component of noise, representation of noise using orthogonal representation, and narrowband noise. A brief discussion on frequency modulation feedback (FMFB) is also included.

Chapter 7 introduces the concepts of digital communication with a discussion on digital amplitude modulation, I/Q modulation, frequency shift keying (FSK), pulse shift keying (PSK), minimum shift keying (MSK), and quadrature amplitude modulation (QAM). The chapter analyses the bandwidths of various digital modulation techniques and discusses spectral efficiency against power consumption. It also deals with time and frequency domain view of digitally modulated signal and explains block diagrams of digital transmitters and receivers.

Chapter 8 introduces information theory. The topics covered are measure of information, average information, entropy (joint and conditional entropy and differential entropy), mutual information, source coding theorem, algorithm, modelling of communication channels, Shannon's theorem, channel capacity, and information capacity of coloured noisy channel. The chapter presents an introduction to coding discusses various types of codes. It also touches upon automatic repeat request (ARQ), communication over noisy channel, and application of information theory.

Chapter 9 provides an introduction to probability. It discusses the elementary set theory, various types of probabilities, random variable, statistical averages, random process (both continuous and discrete), and Gaussian process and its properties.

Latest advances in analog communication like SDR, DDS, and frequency synthesizers using PLL are discussed in the book as Case Studies in Chapters 3, 4, and 6, respectively.

Appendices A, B, and *C* provide MATLAB exercises, important mathematical relations/formulae, and Fourier series representation and its properties, respectively. *Appendix D* provides convolution table, Laplace transform and its properties, *z*-transform pairs, ITU voiceband (telephone line) modem standards, some useful constants, and recommended unit prefixes.

To conclude, this book is written to make learning pleasurable by presenting the subject in a clear, understandable manner. The flow of chapters is designed in such a way that learning later chapters becomes easier.

Acknowledgements

I wish to thank first of all Prof. R. Sethuraman, Vice Chancellor of SASTRA University, for encouraging me to write this textbook. Thanks are due to Oxford University Press (India) for coming forward in publishing this textbook. I also thank Mr K. Venkataramanan, Mr S. Krishanan, Mr V. Prabhu, and Mr K. Suryanarayanan for helping me in preparing figures for the book. I would like to express my sincere thanks to Mr S. Swaminathan and Mr V. Balaji for helping me in the preparation of MATLAB exercises. Finally, I thank my wife, Mrs B. Rukmani, for converting the handwritten manuscript into a presentable MS Word format, and my family for their tolerance, patience, encouragement, and support throughout this very fruitful but time-consuming effort.

Throughout the preparation of this book, care has been taken to maintain technical accuracy as far as possible and present an error-free text. However, suggestions and comments to improve it further will be appreciated.

V. Chandra Sekar

Contents

1

Introduction

1.1 WHAT IS COMMUNICATION?

It is the study of the fundamental concept and principles of transferring information from one place to another. This involves the process of transmission, reception, and processing of information between locations. The source can be in a continuous form as in the case of analog signals or in a digital form.

As in the case of discrete signals, all forms of information, however, should be converted into an electrical signal before being sent via some medium. The medium can be a wire, a coaxial cable, a waveguide, an optical fibre, or atmosphere as in the case of radio and TV broadcasting. The medium is sometimes called a channel.

The first communication system was telegraphy followed by telephony and then the wireless system, which was used to broadcast radio programmes.

Invention of transistors and later integrated circuits, LSI, and VLSI has made the design and development of low-power, small-size, lightweight, high-speed, and reliable communication systems possible. Introduction of fibre optic cable as a medium resulted in providing an extremely high bandwidth and making possible transmission of voice, data, and picture over the same channel. The world is witnessing a significant growth in the field of communication in the form of cellular or mobile phones and high-speed communication networks with the help of powerful and faster computers. Today the world has become smaller, thanks to the modern advancement in communication engineering.

Initial communication systems were analog but present-day communication systems are mostly digital.

1.2 MODULATION AND ITS TYPES

The original information is mostly not in the form that is suitable for transmission. If the distance is quite small, this problem never arises. In this case, we call the transmission as baseband transmission. However, for a long distance, original information has to be transformed into some other form so that it is most suitable for transmission. The process of impressing such information onto a high-frequency component, called carrier, is known as *modulation*.

1.2.1 Need for Modulation

Suppose you are on the 36th floor of a building and your friend is standing down on the ground floor. Now you want to convey some information to him. (Assume that no mobile phone is available with you or him.) If you write this information on a piece of paper and drop it down to him through the balcony or window, chances are that it may not reach him. This is due to the fact that this piece of paper containing the information is so light that it will float in the air and drift away and will never reach your friend. To ensure that the message reaches him, just wrap this piece of paper around a small stone and drop it. Due to the weight of the stone and the gravity, the stone just drops down straight and your friend can pick it up. He takes the piece of paper containing the information and throws the stone. Precisely the same method is followed when we transmit a signal over a long distance. The original low-frequency signal is impressed onto a high-frequency signal called carrier (since this carries the low-frequency information) and transmitted over a long distance. On the receiver end, this signal is received and the carrier is removed and discarded and the low-frequency signal containing the information is retained.

We can summarize the need for modulation as follows.

- To translate the frequency of a low-pass signal to a higher band so that the spectrum of the transmitted bandpass signal matches the bandpass characteristics of the channel.
- For efficient transmission, it has been found that the antenna dimension has to be of the same order of magnitude as the wavelength of the signal being transmitted. Since $C = \lambda f$ for a typical low-frequency signal of 2 kHz, the wavelength works out to be 150 km. Even assuming the height of the antenna half the wavelength, the height works out to be 75 km, which is impracticable.
- To enable transmission of a signal from several message sources simultaneously through a single channel employing frequency division multiplexing.
- To improve noise and interference immunity in transmission over a noise channel by expanding the bandwidth of the transmitted signal.

1.2.2 Frequency Translation

We have seen that the modulation process shifts the modulating frequency to a higher frequency, which in turn depends on the carrier frequency, thus producing upper and lower sidebands. Hence, signals are upconverted from low frequencies to high frequencies and downconverted from high frequencies to low frequencies in the receiver. The process of converting a frequency or a band of frequencies to another location in the frequency spectrum is called *frequency translation*.

1.2.3 Types of Modulation

Depending on whether the amplitude, frequency, or phase of the carrier is varied in accordance with the modulation signal, we classify the modulation as amplitude modulation, frequency modulation, or phase modulation. The method of converting information into pulse form and then transmitting it over a long distance is called *pulse modulation*.

1.3 TRANSMITTER

The message as it arrives may not be suitable for direct transmission. It may be voice signal, music, picture, or data. The signals, which are not of electrical nature, have to be converted into electrical signals. Hence the need for transducer arises. Examples are microphone for speech and camera for pictures. The electrical signals thus generated are called modulating signals. These signals modulate a carrier and this modulated carrier is transmitted. The type of modulation depends on systems. They may be of high level or low level. They can also be any variation or a combination of these. Figure 1.1 shows a typical transmitter.

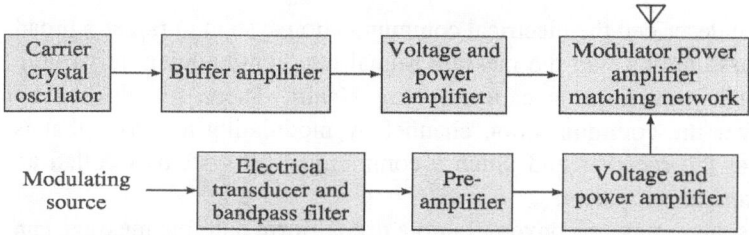

Fig. 1.1   Block diagram of a typical transmitter

The information to be transmitted comes out as an electrical signal from the transducer. This signal is bandlimited through a bandpass filter and is connected to a preamplifier, then to a voltage and power amplifier and finally is given as one of the inputs to the modulator. The other input to the modulator is the carrier, which is generated normally from a crystal oscillator and is then connected to a buffer amplifier and a voltage and power amplifier before connecting to the modulating input. The output of the modulator is connected to a power amplifier and this signal is coupled to the antenna through a matching network to avoid reflection, etc. The power of the transmitter depends on the range of the transmission.

1.4 RECEIVER

Many types of receivers are available in communication systems. A typical receiver is shown in Fig. 1.2. The type of receiver depends on the type of modulation, carrier frequency, the strength of signal received, etc. Most of the modern-day receivers are of superheterodyne type. The received signal from the antenna is fed to an RF amplifier and is given as one of the inputs to a mixer. The other input is the local oscillator, which can be tuned to different frequencies. The output of the mixer is the intermediate frequency, which is fixed irrespective of the frequency of the received signal. This is fed to an intermediate frequency amplifier and to a demodulator. The detector output is given to an audio/video amplifier depending on the original information and is fed to a loudspeaker or a video display unit as the case may be.

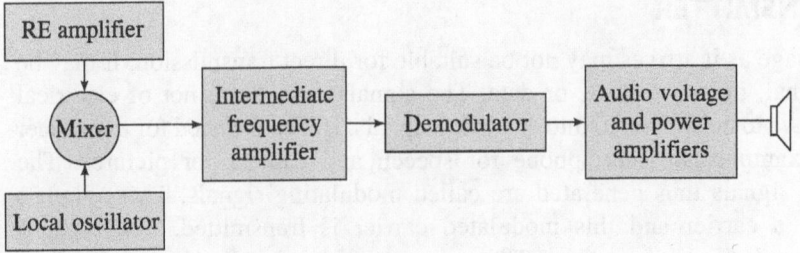

Fig. 1.2 Block diagram of a typical receiver

1.5 DIGITAL COMMUNICATION SYSTEM

So far, we have described the electrical communication system in rather a broad sense on the assumption that the message signal is a continuous time varying waveform. Such waveform is called *analog signal*. These signals can be transmitted over the communication channel by modulating a carrier that is demodulated at the receiver end. Such a communication system is called an *analog communication system*.

An analog source may be converted into a digital form and this message can be transmitted as digital data. At the receiver, these digital data are converted back into analog signals. There are numerous advantages with this type of transmission. Signal fidelity is better controlled. Digital transmission allows us to regenerate the digital signal in long-distance transmission, thus eliminating the effects of noise at each regeneration point. But in the case of an analog transmission, the noise added is amplified along with the signal. Another advantage in digital transmission is removal of redundancy, which is inherent in analog systems. In digital systems, redundancy is removed prior to the modulation, which results in conserving bandwidth. They are also cheaper to implement. Figure 1.3 gives the block diagram of a basic digital communication system transmitter.

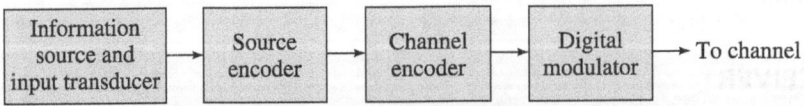

Fig. 1.3 Block diagram of a digital communication transmitter

The analog input is converted into a sequence of binary digits by a source encoder, which is generally an analog-to-digital converter. We normally represent the message signal with as few binary digits as possible. This helps obtain the output with little or no redundancy. The process of efficiently converting the output of either an analog or a digital source into a sequence of binary digits is called *source encoding* or *data compression*. The source encoded outputs, which are a sequence of binary digits, are called *information sequence*. This is passed on to the channel encoder. The channel encoder is introduced in a controlled manner. Some redundancy in the binary information sequence can be used at the receiver to overcome the effects of noise and interference encountered in the transmission of signal through the channel. Thus, the added redundancy serves to increase the reliability of the received data and improves the fidelity of the received signal. The redundancy in the information sequence aids the receiver in decoding the desired information sequence. The binary sequence at the output of the channel encoder is passed through the digital modulator, which serves as the interface to the communication channel.

At the receiving end, the digital demodulator processes the received waveform and passes it onto a channel decoder. The channel decoder output is connected to the source decoder, which is generally a digital-to-analog converter, and the original analog signal is obtained. Figure 1.4 gives the receiver block diagram.

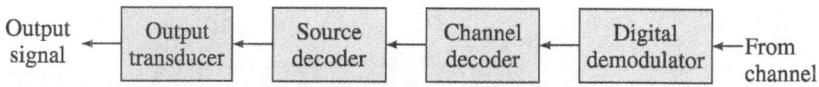

Fig. 1.4 Block diagram of a digital communication receiver

It has to be kept in mind that in all communication systems, the transmitter and receiver must be in agreement with the modulation method used.

1.6 MULTIPLEXING OF SIGNALS

When it is required to transmit more signals on the same channel, baseband transmission fails, as in the case of audio signals being broadcast from different stations on the same channel. The reason for this is the interference between each audio signal due to their frequencies being more or less the same. To avoid this, either frequency division multiplexing or time division multiplexing is employed.

1.6.1 Frequency Division Multiplexing

In this method, various carrier frequencies, which are quite apart, are chosen and these carriers get modulated by different baseband signals. Thus, the modulated carriers are transmitted over the same channel. At the receiver, tunable bandpass filters are used to separate each modulated carrier and then demodulate it to recover the baseband signal. This method of transmitting several channels simultaneously is known as *frequency division multiplexing* (FDM).

Here the bandwidth of the channel is shared by various signals without any overlapping.

1.6.2 Time Division Multiplexing

In this method, several signals are transmitted over a time interval. Each signal is allotted a time slot and it gets repeated cyclically. The only difference compared to FDM is that the signals are to be sampled before sending. Hence, the signals will be in the form of pulse trains. At the receiver, there will be a synchronizer to recover each signal.

2

Signals: An Introduction

2.1 BASIC CONCEPTS

A signal is defined as any function that carries information. It is also a description of how one parameter varies with another parameter. Examples include a voltage changing over time and brightness varying with distance as in an image. But we will consider only the signals with time as an independent variable. Hence, we can call signals as something that change with time, which is taken as an independent variable, such as audio signals like speech and music, video signals, and data signals. Signals are represented mathematically as per their types. A continuous signal is represented as $x(t)$, $f(t)$, or $y(t)$, that is using parentheses, whereas discrete signals are represented using brackets, e.g. $x[n]$ and $y[n]$. We also note that signals use lower case letters. This is particularly so when they are represented in a time domain. Upper case letters are reserved for a frequency domain. Also the name given to a signal is usually descriptive of the parameters it represents. For example, a voltage dependent on time is represented as $v(t)$ or the price of a commodity that changes with time is denoted as $p(t)$. If more descriptive name for a signal is not available, then as in algebra, the input signal to a continuous system is called $x(t)$ and the output signal is called $y(t)$. For a discrete system, the input is represented as $x[n]$ and the output as $y[n]$. Another class of signal is speech signal, which consists of different frequencies.

2.2 CLASSIFICATION OF SIGNALS

Signals can be classified in a variety of ways—continuous time signals and discrete time signals, periodic and non-periodic signals, causal and non-causal signals, even and odd signals, deterministic and random signals, real and complex signals, and energy-type and power-type signals.

2.2.1 Continuous and Discrete Time Signals

A continuous time signal is a signal $x(t)$ for which the independent variable 't' takes real numbers.

Example of continuous signal is a sinusoidal signal $x(t) = A \cos (2\pi f_0 t + \theta)$. It is shown in Fig. 2.1. It can be observed that the signal has a value at every instant of time.

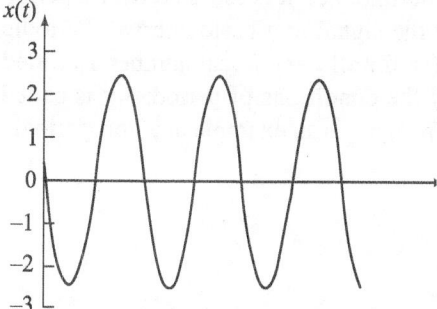

Fig. 2.1   Sinusoidal signal

A discrete time signal represented as $x[n]$ is a signal for which the independent variable n takes its value as a multiple integer. Compared to a continuous signal, this type of signal has values only at regular intervals of time, which depends on the sampling interval. An example of such signal is $x[n] = A\cos(2\pi f_0 n + \theta)$. The waveform of a discrete time signal is shown in Fig. 2.2.

Example 2.1   For what value of f_0, the discrete sinusoidal signal $x[n] = A$
$\cos (2\pi f_0 n + \theta)$ is periodic?

Solution

For this signal to be periodic, the condition

$$2\pi f_0 (n + N_0) + \theta = 2\pi f_0 n + \theta + 2m\pi$$

must be satisfied for all integers n and some positive integer m.
  Hence,

$$2\pi f_0 N_0 = 2\pi m$$

i.e.,
$$f_0 = \frac{m}{N_0}$$

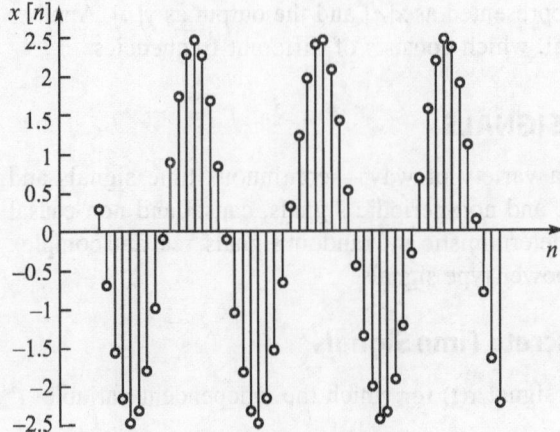

Fig. 2.2 Discrete time sinusoidal signal

2.2.2 Periodic and Non-periodic Signals

A periodic signal repeats itself regularly in time, i.e., it is repeated over a period of time. Hence, it is sufficient to specify the signal in a basic interval. Periodic signals satisfy the property $x(t + T_0) = x(t)$, for all t and a real number T_0 called the period. A signal that does not satisfy the conditions of periodicity is called non-periodic or aperiodic signal. Unit-step signal is an example of a non-periodic signal, which is shown in Fig. 2.3.

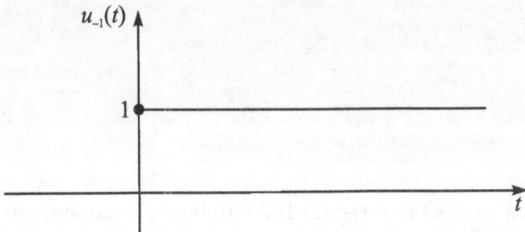

Fig. 2.3 The unit-step function

2.2.3 Causal and Non-causal Signals

A signal $x(t)$ is called causal if for all $t < 0$, $x(t) = 0$, otherwise the signal is non-causal.

Similarly, a discrete time signal is causal if it is identically equal to zero for $n < 0$. An example of causal signal is shown in Fig. 2.4. A non-causal signal is defined as a signal whose time inverse is causal, i.e., a non-causal signal is identically equal to zero for $t > 0$.

2.2.4 Even and Odd Signals

A signal $x(t)$ is even if it has mirror symmetry w.r.t. the vertical axis. A signal is odd if it is symmetric w.r.t. the origin. A signal is even if and only if for all t: $x(t) = x(-t)$ and is odd if and only if for all t: $x(-t) = -x(t)$. Figure 2.5 shows even and odd signals.

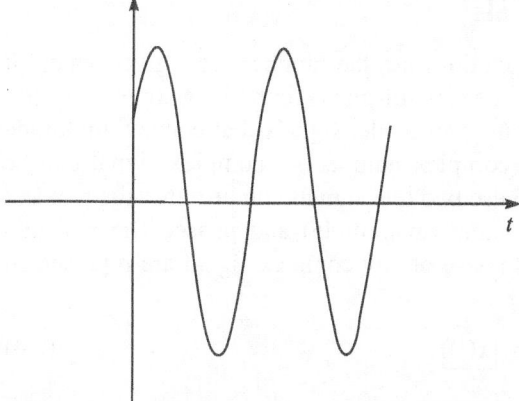

Fig. 2.4 Causal signal

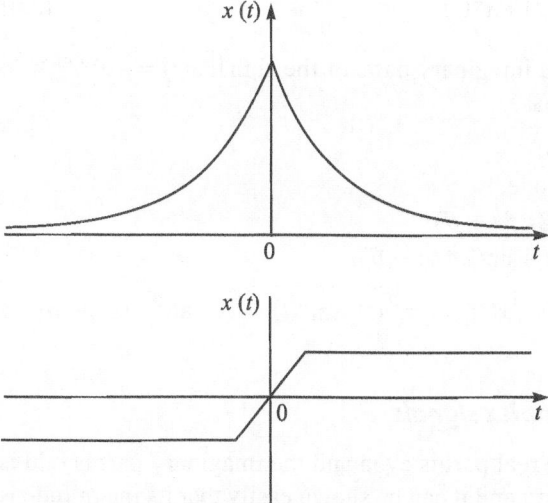

Fig. 2.5 · Even and odd signals

In general, any signal $x(t)$ can be written as the sum of its even and odd parts as

$$x(t) = x_e(t) + x_0(t) \tag{2.1}$$

$$x_e(t) = \frac{x(t) + x(-t)}{2} \tag{2.2}$$

$$x_0(t) = \frac{x(t) - x(-t)}{2} \tag{2.3}$$

2.2.5 Deterministic and Random Signals

A deterministic signal is defined at any instant of time t for which the value of $x(t)$ is given as a real or complex number. In a random or stochastic signal, at any given time instant t, $x(t)$ is a random variable and it is defined by a probability density function.

2.2.6 Real and Complex Signals

Since signals are functions and functions are the numbers at a given value of their independent variable, they can be real, imaginary, or both. In communication engineering, common signals are used to model signals that convey amplitude and phase information. Just like complex numbers, a complex signal can be represented by two real signals. These two real signals can be either the real and imaginary parts or the absolute value (magnitude) and phase. The real and complex component modules and phase of any complex signal are represented by the following relations:

$$x_r(t) = |x(t)| \cos(\angle x(t)) \tag{2.4}$$

$$x_i(t) = |x(t)| \sin(\angle x(t)) \tag{2.5}$$

$$|x(t)| = \sqrt{x_r^2(t) + x_i^2(t)} \tag{2.6}$$

Example 2.2 Find the real and imaginary parts of the signal $x(t) = Ae^{j(2\pi f_0 t + \theta)}$. Also find its magnitude and phase.

Solution
The given signal is a complex one.
  Its real part is $x_r(t) = A\cos(2\pi f_0 t + \theta)$.
  The imaginary part is $x_i(t) = A\sin(2\pi f_0 t + \theta)$.

  Magnitude of the signal is $|x(t)| = \sqrt{x_r^2(t) + x_i^2(t)} = |A|$ and its phase is $\angle x(t) = 2\pi f_0 t + \theta$.

Hermitian symmetry for complex signals

A complex signal in which the real part is even and the imaginary part is odd is said to have Hermitian symmetry and it can be shown easily that its magnitude is even and the phase is odd. The signal $x(t) = Ae^{j2\pi f_0 t}$ is an example of Hermitian signal.

2.2.7 Energy-Type and Power-Type Signals

This type of classification deals with the energy content and the power content of the signals. For any signal $x(t)$ the energy content of the signal is defined by

$$E_x = \int_{-\infty}^{\infty} |x(t)|^2 \, dt = \lim_{T \to \infty} \int_{-T/2}^{T/2} |x(t)|^2 dt \tag{2.7}$$

The power content is given by $P_x = \lim_{T \to \infty} \dfrac{1}{T} \int_{-T/2}^{T/2} |x(t)|^2 \, dt \tag{2.8}$

  For a real signal, $|x(t)|^2$ is replaced by $x^2(t)$. A signal $x(t)$ is an energy-type signal if and only if E_x is finite and power type and P_x satisfies $0 < P_x < \infty$.

Example 2.3 Find the energy content of the signal $x(t) = 2, \quad |x| < 2$
$$= 0, \quad \text{else}$$

Solution

Energy of the signal is given by $E_x = \int_{-\infty}^{+\infty} |x(t)|^2 \, dt = \int_{-2}^{2} 4 \, dt = 16$

Example 2.4 Find the energy content and power content of the signal $x(t) = A \cos(2\pi f_0 t + \theta)$.

Solution

Energy of the signal is given by $E_x = \lim_{T \to \infty} \int_{-T/2}^{T/2} A^2 \cos^2(2\pi f_0 t + \theta) \, dt = \infty$

Hence, this signal is not an energy signal. However, the power of the signal is

$$P_x = \lim_{T \to \infty} \frac{1}{T} \int_{-T/2}^{T/2} A^2 \cos^2(2\pi f_0 t + \theta) \, dt$$

$$= \lim_{T \to \infty} \frac{1}{T} \int_{-T/2}^{T/2} \frac{A^2}{2} [1 + \cos(4\pi f_0 t + 2\theta)] \, dt$$

$$= \lim_{T \to \infty} \left[\frac{A}{2} + \left\{ \frac{A^2}{8\pi f_0 T} \sin(4\pi f_0 t + 2\theta) \right\} - T/2 \right]$$

$$= \frac{A^2}{2} < \infty.$$

Hence, $x(t)$ is a power-type signal and its power is $\dfrac{A^2}{2} < \infty$.

2.3 TYPICAL SIGNALS AND THEIR PROPERTIES

In this section, we will deal with some typical signals and their properties.

2.3.1 Sinusoidal Signal

A sinusoidal signal is defined by $x(t) = A \cos(2\pi f_0 t + \theta)$, where A is the amplitude, f_0 is the frequency, and θ is the phase of the signal. A sinusoidal signal is periodic with the period $T_0 = 1/f_0$. Figure 2.6 shows a sinusoidal signal.

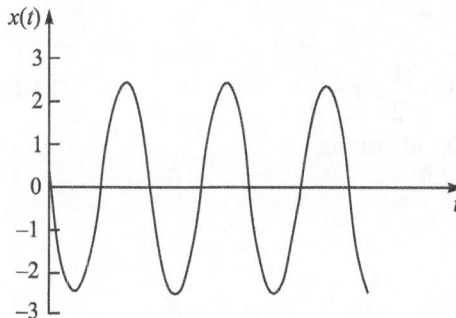

Fig. 2.6   Sinusoidal signal

2.3.2 Complex Exponential Signal

This is defined by $x(t) = A e^{j(2\pi f_0 t + \theta)}$, where again A is the amplitude, f_0 is the frequency, and θ is the phase angle. The signal is shown in Fig. 2.7.

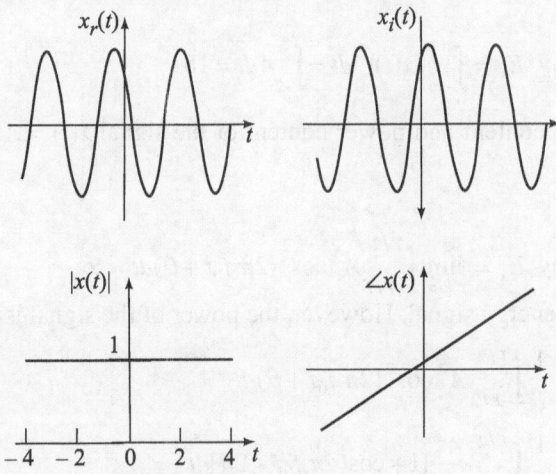

Fig. 2.7 Complex exponential signal

2.3.3 Unit-Step Signal

A unit-step signal is another frequently encountered signal. Any signal that gets multiplied by this signal results in the signal becoming causal. A unit-step signal is shown in Fig. 2.8.

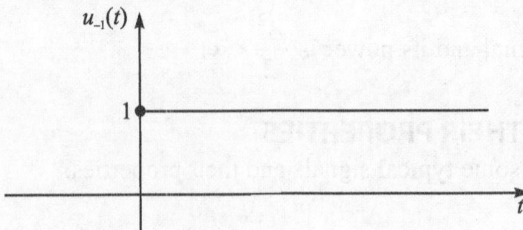

Fig. 2.8 Unit-step signal

The unit-step function is defined as follows: $U(t) = 1$, for $t > 0$ (2.9)

$$= 0, \text{ for } t < 0$$

2.3.4 Rectangular Pulse

A rectangular pulse is defined as $x(t) = 1$, $-\dfrac{1}{2} \le t \le \dfrac{1}{2}$ (2.10)

$$= 0, \text{ otherwise}$$

A rectangular pulse is shown in Fig. 2.9.

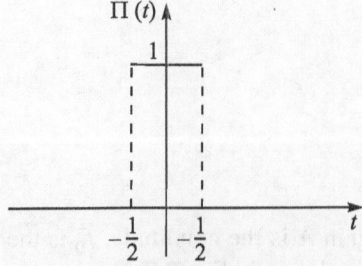

Fig. 2.9 Rectangular pulse

2.3.5 Triangular Signal

This signal (shown in Fig. 2.10) is given by the function

$$f(t) = t + 1, \quad -1 \le t \le 0$$
$$= -t + 1, \quad 0 \le t \le 1 \tag{2.11}$$

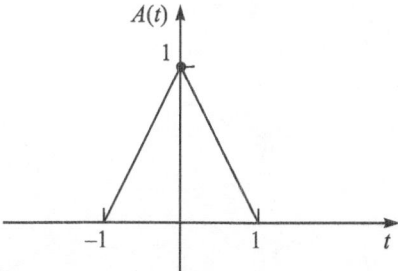

Fig. 2.10 Triangular signal

2.3.6 The Sinc Signal

It is defined as

$$\text{sinc}\,(t) = \frac{\sin(\pi t)}{\pi t}, \quad t \ne 0$$
$$= 1, \qquad t = 0 \tag{2.12}$$

The waveform of sinc signal is shown in Fig. 2.11. The sinc signal achieves its maximum of 1 at $t = 0$ and zero at $t = \pm1, \pm2, \pm3, \ldots,$ etc.

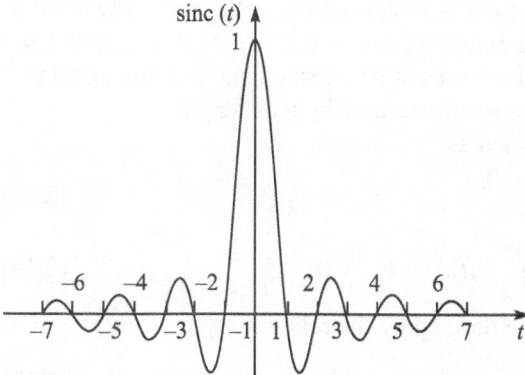

Fig. 2.11 The sinc signal

2.3.7 Sign or Signum Signal

This signal denotes the sign of the independent variable t, which is defined by

$$\text{sgn}(t) = 1, \quad t > 0$$
$$= -1, \quad t < 0 \tag{2.13}$$
$$= 0, \quad t = 0$$

The signal is shown in Fig. 2.12. The signum signal can be expressed as the limit of the signal $x_n(t)$, which is defined by

$$x_n(t) = e^{-\frac{t}{n}}, \quad t > 0$$

$$= -e^{\frac{t}{n}}, \quad t < 0 \tag{2.14}$$

$$= 0, \quad t = 0$$

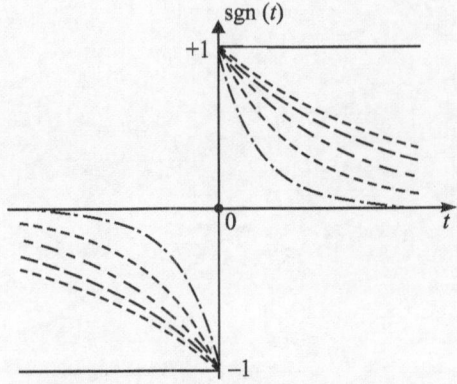

Fig. 2.12 The signum signal as the limit of $X_n(t)$

2.3.8 Impulse or Delta Signal

This signal is used to represent physical phenomena that occur in a very small infinitesimal time duration and they are too small to be measured by any instrument. This duration can be assumed to be equal to zero.

An example of such a phenomenon is a very narrow voltage or current pulse. The impulse signal $\delta(t)$ is not a function, but is a distributed or generalized function. A distribution is defined in terms of its effect on another function under the integral sign. An impulse is generally defined by its strength.

A unit impulse function is defined as

$$d(t) = \infty \text{ for } t = 0$$
$$= 0 \text{ for } t \neq 0 \tag{2.15}$$

i.e., $\quad \displaystyle\int_{-\infty}^{\infty} \delta(t)dt = 1 \text{ and } \int_{-\infty}^{\infty} v(t)\delta(t)dt = v(0) \tag{2.16}$

The impulse distribution can be defined by the relation

$$\int_{-\infty}^{\infty} \phi(t)\delta(t) = \phi(0) \tag{2.17}$$

This expression gives the effects of impulse distribution on the test function $\phi(t)$. It may be noted that it is not defined in terms of different values of t. The impulse signal can be visualized as the limit of certain known signals. The first example is a rectangular pulse, i.e.,

$$\delta(t) = \lim_{\Delta t \to 0} \frac{1}{\Delta t} f\left(\frac{t}{\Delta t}\right) \tag{2.18}$$

This can be interpreted as follows. As we reduce the width of the pulse, the amplitude increases (to keep the area constant). As the width approaches zero, the amplitude approaches infinity.

For a sinc signal, $\delta(t) = \lim\limits_{\Delta t \to 0} \dfrac{1}{\Delta t} \operatorname{sinc}\left(\dfrac{t}{\Delta t}\right)$ (2.19)

Figure 2.13 shows the waveform of these signals.

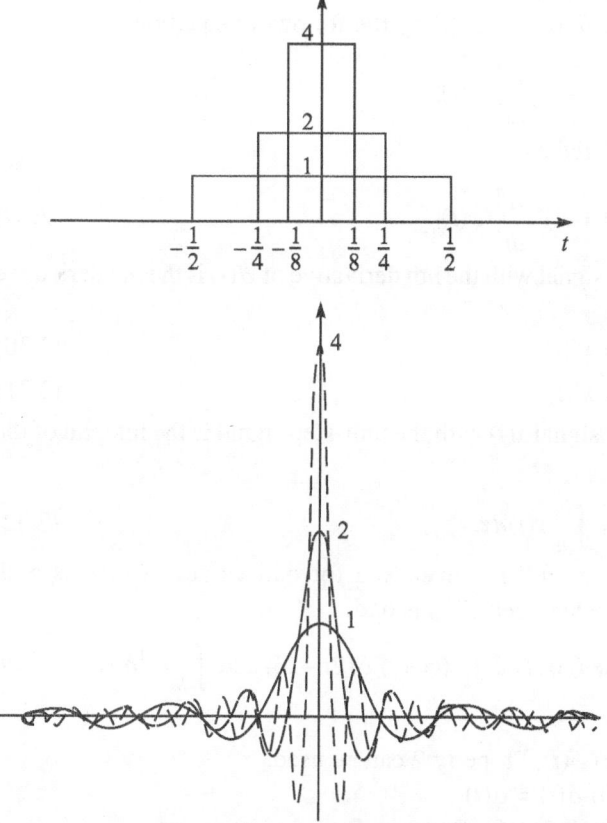

Fig. 2.13 The impulse signal formed from a rectangular pulse as the width of the pulse approaches zero

Properties of impulse signals

1. $\delta(t) = 0$ for all t not equal to zero and $\delta(0) = \infty$ (2.20)
2. $x(t)\,\delta(t - t_0) = x(t_0)\,\delta(t - t_0)$ (2.21)

3. For all a not equal to zero $\delta(at) = \dfrac{1}{|a|}\delta(t)$ (2.22)

4. For any $\phi(t)$ continuous at t_0, $\displaystyle\int_{-\infty}^{\infty} \phi(t)\delta(t - t_0)dt = \phi(t_0)$ (2.23)

5. For any t continuous at t_0, $\displaystyle\int_{-\infty}^{\infty} \phi(t + t_0)\delta(t)dt = \phi(t_0)$ (2.24)

6. Convolution of any signal with the impulse is the signal itself, i.e.,
   $$x(t) * \delta(t) = x(t)$$ (2.25)
   Also $x(t) * \delta(t - t_0) = x(t - t_0)$ (2.26)

7. A unit-step signal is the integral of the impulse signal, and the impulse signal is the generalized derivative of the unit-step signal, i.e.,

$$u_{-1}(t) = \int_{-\infty}^{\infty} \delta(\tau)d\tau \tag{2.27}$$

and $$\delta(t) = \frac{d}{dt}u_{-1}(t) \tag{2.28}$$

8. We can define $\delta(t)$, $\delta''(t)$, ..., $\delta^n(t)$ by the following equation:

$$\int_{-\infty}^{\infty} \delta^n(t)\phi(t)dt = (-1)^n \frac{d^n}{dt^n}\phi(t)\Big|_{t=0}$$

This can be generalized as

$$\int_{-\infty}^{\infty} \delta^n(t-t_0)\phi(t)dt = (-1)^n \frac{d^n}{dt^n}\phi(t)\Big|_{t=t_0} \tag{2.29}$$

9. Convolution of any signal with the nth derivative of $\delta(t)$ is the nth derivative of $x(t)$, i.e.,

$$x(t)*\delta^n(t) = x^n(t) \tag{2.30}$$

Hence, $$x(t)*\delta'(t) = x'(t) \tag{2.31}$$

10. Convolution of any signal $x(t)$ with the unit-step signal is the integral of the signal $x(t)$, i.e.,

$$x(t)*u_{-1}(t) = \int_{-\infty}^{t} x(t)d\tau \tag{2.32}$$

11. For even values of n, $\delta^n(t)$ is even and for odd values of n, it is odd. However, if $\delta(t)$ is even, then $\delta'(t)$ is odd.

Example 2.5 Determine $(\cos t)\,\delta(t)$, $(\cos t)\,\delta(2t-5)$, and $\int_{-\infty}^{\infty} e^{-1}\delta'(t)(t-3)dt$.

Solution

(a) To determine $(\cos t)\,\delta(t)$, Property 2 can be used:

$(\cos t)\,\delta(t) = (\cos 0)\,\delta(t) = \delta(t)$

(b) To determine $(\cos t)\,\delta(2t-5)$, Property 5 can be used:

$$\delta(2t-5) = \frac{1}{2}\delta(t-\frac{5}{2})$$

Then from Property 1, we have

$$(\cos t)\,\delta(2t-5) = \frac{1}{2}\cos(t)\,\delta(t-\frac{5}{2}) = \cos\frac{2.5}{2}\,\delta(t-\frac{5}{2}) \approx 0.99976(t-\frac{5}{2})$$

(c) To determine $\int_{-\infty}^{\infty} e^{-1}\delta'(t)(t-3)dt$, use Property 8 to obtain

$$\int_{-\infty}^{\infty} e^{-1}\delta'(t)(t-3)dt = (-1)\frac{d}{dt}e^{-3} = 0.04978$$

2.3.9 Singular Function

The unit-step and impulse functions are called singular functions. The property of singular functions is that they do not have finite derivatives everywhere.

2.3.10 Shifting, Inversion, Scaling, and Convolution of Signal

These properties are related to the independent variable time t.

Shifting

This is an operation by which a signal is time shifted. If $f(t)$ is the original signal, then $s(t) = f(t + T)$ is the time-shifted version of $f(t)$. If T is negative, then the signal is shifted towards the right and if T is positive, it is shifted towards the left, i.e., whatever occurs for $f(t)$ in time t occurs for $s(t)$ in time $t + T$. For δ function, $\delta(\tau - T)$ will be a shifted version of $\delta(\tau)$ along the τ axis by $+ T$, i.e.,

$$\int_{-\infty}^{\infty} f(\tau)\delta(t - T)d\tau = f(T) \tag{2.33}$$

Time inversion

This operation inverts the signal along the time axis to get the mirror image of the signal. If $f(t)$ is the original signal, then $s(t) = f(-t)$ is the time inverted version of $f(t)$, i.e., what occurs at t for $f(t)$, the same occurs for $s(t)$ at $-t$.

Time scaling

This operation compresses or expands the signal along the time axis. If $f(t)$ is the original signal, then $s(t) = x(at)$, is the time scaled version of $x(t)$ for $a > 0$. Negative a signifies both time scaling and time inversion.

The inversion comes from negative sign. If $a < 1$, then $s(t)$ is an expanded version of $x(t)$ and if $a \geq 1$, then $s(t)$ is a compressed version of $x(t)$. Figure 2.14 shows all these operations.

Convolution

In this process, one signal is time reversed, shifted, and multiplied with another signal and then its integral is calculated to generate the third signal. Generally, the output of a system is described by the convolving input signal and its unit impulse response. The convolution of a signal with another signal is represented as

$$y(t) = f(t)*h(t) = \int_{-\infty}^{\infty} f(t)h(t - \tau)d\tau \tag{2.34}$$

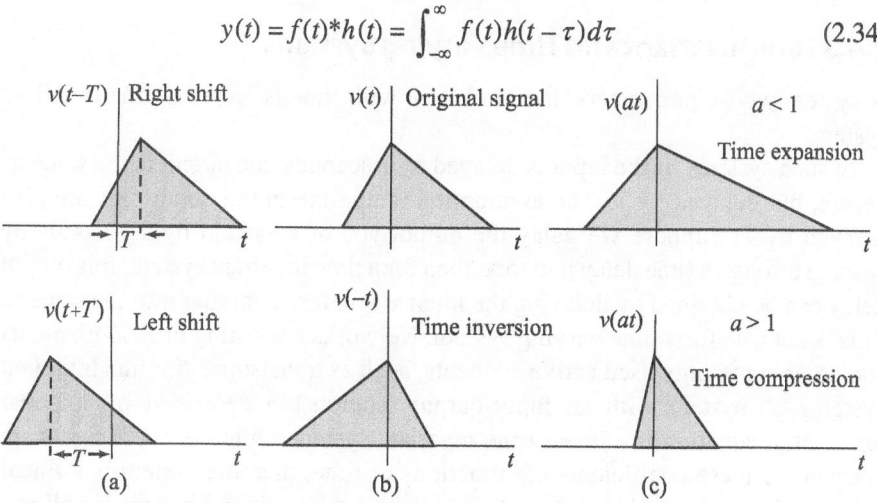

Fig. 2.14   (a) Time shifting, (b) time inversion, and (c) time scaling

2.4 CLASSIFICATION OF SYSTEMS

A system can be classified as an interconnection of various elements or devices. In communication systems, a system is an entity that is exited by an input signal, which results in an output. A system has to give a unique output for any legitimate input, i.e., $y(t) = x(t)\ h(t)$, where $x(t)$ is the input to the system, $y(t)$ is the output from the system, and $h(t)$ is the transfer function of the system.

2.4.1 Discrete Time and Continuous Time Systems

A discrete time system accepts a discrete time signal as the input and produces a discrete time signal at the output, whereas in a continuous time system, the input is a continuous time signal and the output is also a continuous time signal.

2.4.2 Linear and Non-Linear Systems

A system whose output is proportional to its input is an example of a linear system. Linearity implies additive and scaling properties. If several inputs are acting on a system, then the total effect on the system due to all these inputs can be determined by considering one input at a time while assuming all the other inputs to be zero. The total effect is then the sum of all outputs, i.e., for a linear system if an input x_1 acting alone has an output y_1, and another input x_2 also acting alone has an output y_2, then with both inputs acting simultaneously on the system, the total output will be $y_1 + y_2$. This is the additive property. The scaling property states that, for an arbitrary real or imaginary number S, if an input is increased s fold, the output also increases s fold. Thus, if $x \rightarrow y$, $sx \rightarrow sy$. Thus, linearity implies both additive and scaling properties. These two properties are combined into a single property called *superposition property*, which is expressed as follows: if $x_1 \rightarrow y_1$ and $x_2 \rightarrow y_2$, then for all values of s_1 and s_2: $s_1 x_1 + s_2 x_2 = s_1 y_1 + s_2 y_2$. A system that does not satisfy this relationship is called a non-linear system.

2.4.3 Time Invariant and Time Varying Systems

A system whose parameters do not change with time is called a time invariant system.

In such systems, if the input is delayed by T seconds, the output is the same as before, but delayed by T. The assumption is that the initial conditions are also delayed by T. Suppose we delay the output $y(t)$ of a system by T seconds by passing through a time delay network, then for a time invariant system, this output delay can be obtained by delaying the input $x(t)$ before applying it to the system. This is not true for a time varying system. Networks consisting of RLC elements and other commonly used active elements, such as transistors, are time invariant systems. A system with an input-output relationship described by a linear differential equation is a linear time invariant system while the coefficients are constant. If these coefficients are functions of time, then the system is a linear time varying system. Hence, a system is *time invariant* if and only if for all $x(t)$ and all values of t_0, its response to $x(t - t_0)$ is $y(t - t_0)$, where $y(t)$ is the response of the system to $x(t)$. The response to the time invariant system can be derived simply by finding the convolution of the input and impulse response of the system. A time invariant system is shown in Fig. 2.15.

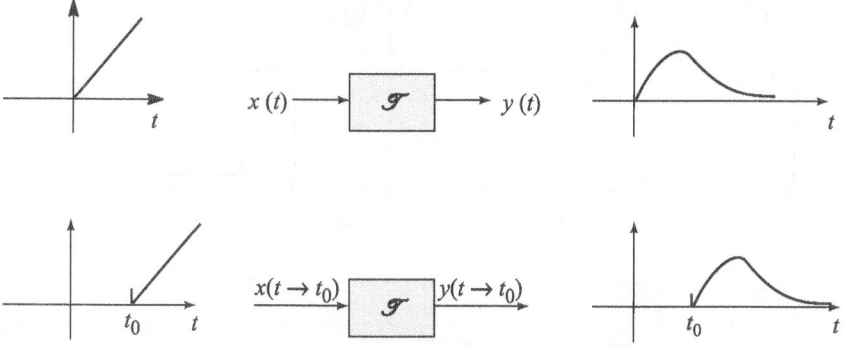

Fig. 2.15 A time invariant system

2.4.4 Causal and Non-causal Systems

Causality deals with the physical realizability of the system. No physical system can predict the values that its input signal will assume in the future, i.e., a causal system, also called a physical or non-anticipative system, is one for which the output at any instant of time t_0 depends only on the value of the input $x(t)$ for $t \le 0$.

This means that the value of the output at the present instant depends only on the past and present values of the input $x(t)$, but not on its future value. For a causal system, the output cannot be present before the input is applied. A system that violates the condition of causality is called a non-causal system. Any practical system that operates in real time must necessarily be causal. For a non-causal system, if we apply an input starting at $t = 0$, the output begins even before $t = 0$. For example, consider the system specified by $y(t) = x(t - 2) + x(t + 2)$. This equation shows that $y(t)$, the output at t, is given by the sum of the input values 2's before and after t, i.e., at $(t - 2)$ and $(t + 2)$, respectively. But if the system is operated in real time at t, we cannot predict the value of the input 2's later. Hence, it is impossible to implement this system in real time. This is the reason for such systems to be unrealizable in real time. However, we can realize a non-causal system satisfactorily approximated in real time by using a causal system with a delay, i.e.,

$$y(t) = y(t - 2) \tag{2.35}$$

$$\hat{y}(t) = x(t - 4) + x(t) \tag{2.36}$$

The output $y(t)$ at any instant of time is the sum of the values of the input $x(t)$ at t and at 4 seconds earlier. This system can be considered as causal since the output does not depend on the future value of the input. A non-causal system and its realization by a delayed causal system are shown in Fig. 2.16.

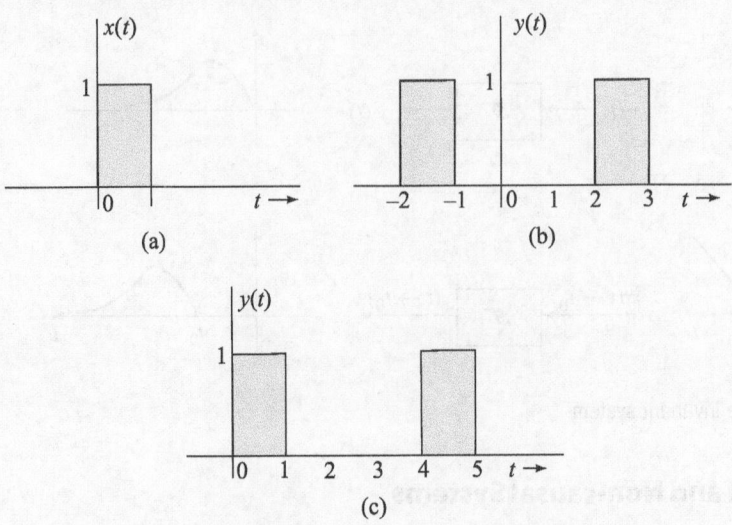

Fig. 2.16 A non-causal system and its realization by a delayed causal system

2.4.5 Instantaneous and Dynamic Systems

We have seen that in a causal system, the output depends on the past and present inputs. But there are systems in which the output depends only on the present input, i.e., the output at any instant t depends only on the strength of its input at the same instant t and not on any past or future values of the input. Such systems are called *instantaneous systems* or *memoryless systems*.

If the system output depends on its past and present inputs, then it is called a *dynamic system* or a system with memory. A system whose response at t is completely determined by the input signal over past Ts, i.e., in the interval $(t - T)$ to t is called a *finite memory system* with a memory of T s. A network containing an inductor and a capacitor generally has an infinite memory since the response of such a network at any instant of time t is determined by its inputs over the entire period $(-\infty, t)$.

2.4.6 Stable and Unstable Systems

Systems can be classified as stable or unstable. If every bounded input applied at the input results in a bounded output, the system: is said to be stable externally. This is also called a BIBO (bounded input and bounded output) system. If the system output does not stabilize for a bounded input, then it is called an unstable system.

2.5 DELTA FUNCTION AND CONVOLUTION

We will first deal with the delta function.

2.5.1 Delta Function

Continuous signals can be decomposed into scaled and shifted delta functions. This function is explained as follows. Consider a system consisting of linear components. If the input to the system consists of various shapes of short pulses, the resultant output is the waveform due to each of these individual inputs. It can be observed that the shape of the input pulse does not affect the shape of the output signal. Figure 2.17 shows various shapes of short input pulses, which produce exactly the same shape of the output pulses. The shape of the output waveform is entirely determined by the characteristic of the system.

The amplitude of the output pulse is directly proportional to the area of the input pulse, i.e., the output will have the same amplitude for an input of 0.1 V for 1 μs, 1 V for 0.1 μs, and 10 V for 1 ns, etc.

This relationship allows for the input pulses with a negative area. If there is a combination of two input signals, one 3 V pulse lasting for 2 μs being quickly followed by –1 V pulse lasting for 6 μs, then the output will be zero, since the total area of the input signal is zero.

An input signal that has the properties explained above is called impulse. We can define an impulse signal as one that is entirely zero except for a very short duration of time. An impulse has strength but does not have a voltage. Mathematically, it is defined as a signal that is infinitesimally narrow. The continuous delta function is a normalized version of an impulse. Continuous delta function is mathematically defined by three idealized characteristics:

1. The signal must be infinitesimally narrow.
2. The signal must occur at time zero.
3. The pulse must have a unit area.

As the delta function is defined to be infinitesimally narrow and has a fixed area, its amplitude is considered to be infinite. The continuous delta function is defined by $\delta(t)$. The output of a continuous system in response to a delta function is called the *impulse response* and is often denoted by $h(t)$. Impulses are represented as a vertical line with an upward arrow with the length of the arrow indicating the area of the impulse.

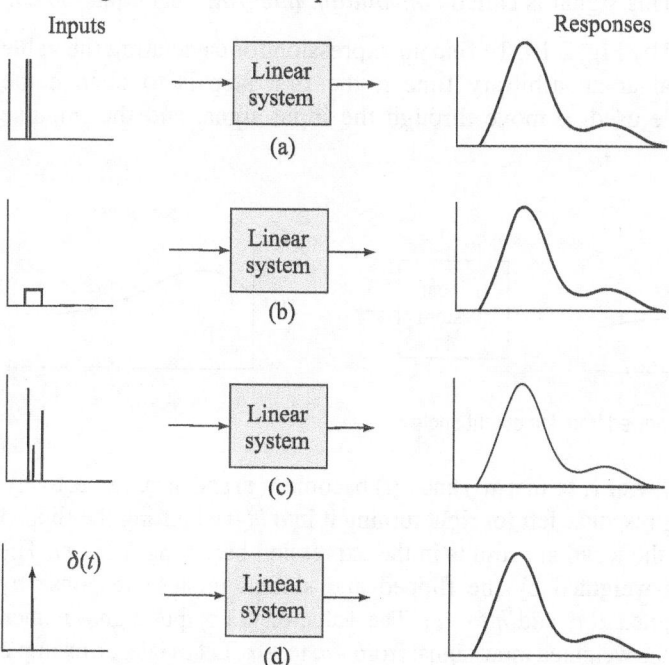

Fig. 2.17 Short input pulses

2.5.2 Convolution

Convolution of a continuous signal can be viewed from the input signal or the output signal. Convolution viewed from the input side can be interpreted as follows: An input signal $x(t)$ is passed through a system characterized by an impulse response $h(t)$ to produce an output $y(t)$. The mathematical equation for convolution is $y(t) = x(t)*h(t)$. The input signal is divided into a set of impulses, i.e., the input signal is decomposed into an infinite number of scaled and shifted delta functions. Each of these impulses produces a scaled and shifted version of the impulse response in the output signal. The final output is then equal to the sum of all the individual responses. Thus, we see that how a single point or a narrow region in the input signal affects a larger portion of the output signal. Figure 2.18 shows this concept. From the output viewpoint, it can be seen how a single point in the output signal is determined by the various values from the input signal.

Fig. 2.18 Convolution viewed from the input angle

Each instantaneous value in the output signal is affected by a section of the input signal weighted by the impulse response flipped left or right. The signals are multiplied and integrated. In the equation form, it is shown as

$\int_{-\infty}^{\infty} x(t)h(t - \tau)d\tau$. This signal is called *convolution integral*. This equation can be better understood by Fig. 2.19. To find an expression for calculating the value of the output signal at an arbitrary time t, the first step is to change the independent variable used to move through the input signal and the impulse response.

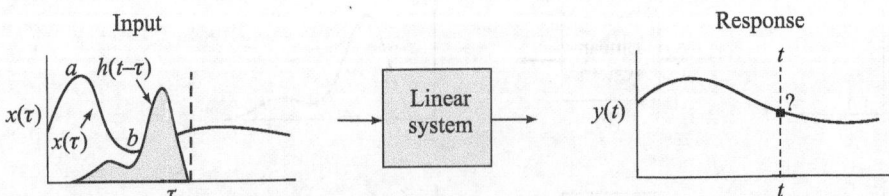

Fig. 2.19 Convolution viewed from the output angle

To do so, replace t with τ, so that $x(t)$ and $h(t)$ become $x(\tau)$ and $h(\tau)$, respectively. Next flip the impulse response left for right turning it into $h(-t)$. Shifting the flipped impulse response to the location t results in the expression becoming $h(t - \tau)$. The input signal is then weighted by the flipped and shifted impulse response by multiplying the two, i.e., $x(\tau)$ and $h(t - \tau)$. The value of the output signal is then found by integrating the weighted input signal from $-\infty$ to $+\infty$. Let us take an example of a continuous convolution. Consider an RC circuit as shown in Fig. 2.20.

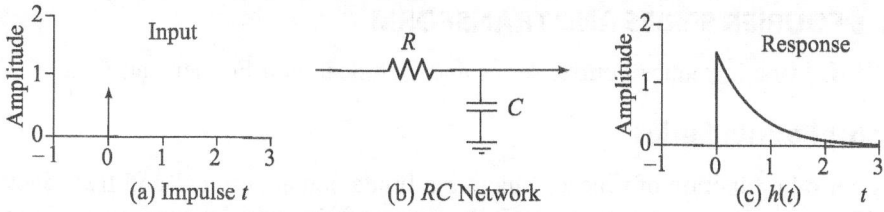

Fig. 2.20 (a) Input impulse, (b) *RC* network, and (c) impulse response of the system

This circuit is a low-pass filter. If an impulse is given to this network as an input, the output waveform is as shown in the figure. It can be observed that the output quickly gets some value and then exponentially decays towards zero, i.e., the impulse response of this circuit is one-sided exponential. Mathematically, we can express the output as

$$h(t) = 0 \text{ for } t < 0 \tag{2.37}$$

$$h(t) = \frac{1}{RC} e^{-\frac{t}{RC}} \tag{2.38}$$

for $t \geq 0$ where $T = 1/RC$ is the time constant of the circuit.

The continuous impulse response contains the complete information about the system, i.e., how it reacts to all possible signals.

Example 2.6 Find the response to a complex exponential input signal $x(t) = Ae^{j(2\pi f_0 t + \theta)}$ from a linear time invariant system whose impulse response is $h(t)$.

Solution

The response $y(t) = \displaystyle\int_{-\infty}^{+\infty} h(\tau) A e^{j[2\pi f_0 (t-\tau)+\theta]} dt$

$$= A e^{j\theta} e^{j2\pi f_0 t} \int_{-\infty}^{+\infty} h(\tau) e^{-j2\pi f_0 \tau} dt$$

$$= A \left| H(f_0) \right| e^{j(2\pi f_0 t + \theta + \angle H(f_0))}$$

where $\quad H(f_0) = \left| H(f_0) \right| e^{j\angle H(f_0)} = \displaystyle\int_{-\infty}^{+\infty} h(\tau) e^{-j2\pi f_0 \tau} d\tau$

This shows that the response of an LTI system to the complex exponential with frequency f_0 is a complex exponential with the same frequency. By multiplying the amplitude of the input by $|H(f_0)|$, the amplitude of the response can be obtained and by adding $\lfloor H(f_0)$ to the input phase, its phase is obtained. It should be noted that $H(f_0)$ is a function of the impulse response and the input frequency. Because of this property, complex exponentials are called *eigen functions* of the class of linear time invariant system. The eigen function of a system is the set of inputs for which the output is a scaling of the input. Because of this important property for complex exponential, finding the response of an LTI system is simple.

2.6 FOURIER SERIES AND TRANSFORM

We will first introduce Fourier series and then deal with Fourier transform.

2.6.1 Fourier Series

A periodic function of time $x(t)$ having a fundamental period T and frequency $1/T$ can be represented as an infinite series of sinusoidal waveforms whose frequency spectrum consists of the fundamental frequency and its harmonics. This summation is called *Fourier series*. The time domain signal used in the Fourier series is periodic and continuous. The first harmonic, i.e., the frequency that the time domain repeats itself is also called *fundamental frequency*. We can view the frequency spectrum in two ways:

(i) The frequency spectrum is continuous, but zero at all frequencies except at the harmonic frequencies.

(ii) The frequency spectrum is discrete and only defined at the harmonic frequencies, i.e., the frequencies between the harmonics can be thought of as having a value zero or not existing and do not contribute to the formation of the time domain signal.

The Fourier series creates a continuous periodic signal with a fundamental frequency f and scaled cosine and sine wave frequencies f, $2f$, $3f$, $4f$, $5f$, etc. The amplitudes of the cosine waves are a_1, a_2, a_3, a_4, a_5, etc. and the amplitudes of the sine waves are b_1, b_2, b_3, b_4, b_5, etc. These are called coefficients. The coefficients a and b are the real and imaginary parts of the frequency spectrum, respectively. In addition, the coefficient a_0 is used to hold the DC value of the time domain waveform. This can be viewed as the amplitude of a cosine waveform with zero frequency, which is a constant value. In some cases, the coefficient is grouped with the other a coefficient, but it is often handled separately because it requires special calculations. There is no b_0 coefficient since the value of a sine wave with zero frequency is zero. Hence, the function $x(t)$ is represented as a Fourier series:

$$x(t) = a_0 + \sum_{n=1}^{\infty} a_n \cos(2\pi f t n) + \sum_{n=1}^{\infty} b_n \sin(2\pi f t n) \qquad (2.39)$$

The corresponding Fourier series equations are usually written in terms of the period of the waveform T, rather than the fundamental frequency. Since the time domain signal is periodic and the sine and cosine waves need to be evaluated over a single period, i.e., $-\dfrac{T}{2}$ to $\dfrac{T}{2}$ or 0 to T, $-T$ to 0, etc.

The Fourier series analysis equations are

$$a_0 = \frac{1}{T} \int_{-T/2}^{T/2} x(t) dt \qquad (2.40a)$$

$$a_n = \frac{2}{T} \int_{-T/2}^{T/2} x(t) \cos \frac{2\pi t n}{T} dt \qquad (2.40b)$$

$$b_n = \frac{2}{T} \int_{-T/2}^{T/2} x(t) \sin \frac{2\pi t n}{T} dt \qquad (2.40c)$$

The a and b coefficients will change if the time domain waveform is shifted left or right. If the waveform is even, i.e., symmetrical around $t = 0$, it will be composed of only even sine waves, (cosine waves). This makes all of the b coefficients zero. In the same way, if the waveform is odd, i.e., symmetrical but opposite in sign around $t = 0$, it will be composed of odd sinusoids (sine waves) and the a coefficients will be zero. If the coefficients are converted into polar notation R and ϕ, a shift in the time domain leaves the magnitude unchanged, but adds a linear component to the phase. An alternative form of the Fourier series, known as compact form, is

$$x(t) = C_0 + \sum_{n=1}^{\infty} C_n \cos\left(\frac{2\pi nt}{T} - \phi_n\right) \qquad (2.41)$$

where C_0, C_n, and ϕ_n are related to a_0, a_n, and b_n by the following equations:

$$C_0 = a_0 \qquad (2.42a)$$

$$C_n = \sqrt{a_n^{\,2} + b_n^{\,2}} \qquad (2.42b)$$

$$\phi_n = \arctan\left(\frac{-b_n}{a_n}\right) \qquad (2.42c)$$

The coefficient C_n is called spectral amplitude, i.e., it is the amplitude of the spectral component $C_n \cos(2\pi n f_0 t - \phi_n)$ at the frequency $n f_0$. A typical amplitude spectrum of a periodic waveform is shown in Fig. 2.21. The amplitude of the harmonic at each harmonic frequency is shown by a vertical line.

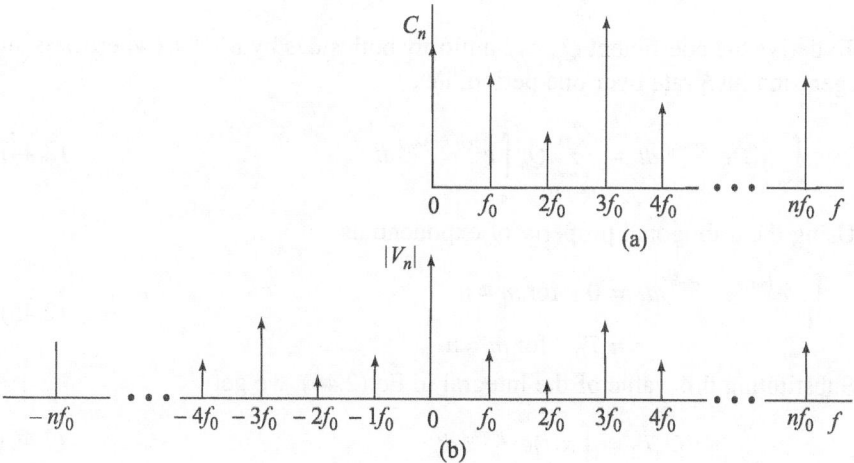

Fig. 2.21   Plot of the spectral amplitude of a periodic waveform: (a) one sided and (b) two sided

Example 2.7 Find the Fourier coefficient of the function given by

$$f(t) = \sum_{k=-\infty}^{\infty} \delta(t - kT_0)$$ and hence find the Fourier series of the given function.

Solution

The coefficients are given by

$$a_0 = \frac{1}{T_0} \int_{-T_0/2}^{T_0/2} \delta(t)\, dt = \frac{1}{T_0}$$

$$a_n = \frac{1}{T_0} \int_{-T_0/2}^{T_0/2} \delta(t) \cos\frac{2\pi nt}{T_0}\, dt = \frac{2}{T_0}$$

$$b_n = \frac{1}{T_0} \int_{-T_0/2}^{T_0/2} \delta(t) \sin\frac{2\pi nt}{T_0}\, dt = 0$$

Hence $f(t)$ may be written in the form $f(t) = \dfrac{1}{T_0} + \dfrac{2}{T_0} \displaystyle\sum_{n=1}^{\infty} \cos\frac{2\pi nt}{T_0}$

Exponential form of the Fourier series

This form of Fourier series finds extensive applications in communication theory. By using Euler's identity, we can express $\cos\omega_0 t$ and $\sin\omega_0 t$ in terms of the exponentials $e^{jn\omega_0 t}$ and $e^{-jn\omega_0 t}$. Hence, we will be able to represent the trigonometric Fourier series in terms of exponentials of the form $e^{jn\omega_0 t}$ with index n taking all values from $-\infty$ to ∞ including zero.

The exponentials series for a periodic signal $x(t)$ can be expressed as

$$x(t) = \sum_{n=-\infty}^{\infty} Q_n e^{jn\omega_0 t} \tag{2.43}$$

To derive the coefficient Q_n, we multiply both sides by $e^{-jm\omega_0 t}$ (where m is an integer) and integrate over one period, i.e.,

$$\int_{T_0} x(t) e^{-jm\omega_0 t}\, dt = \sum_{n=-\infty}^{\infty} Q_n \int_{T_0} e^{j(n-m)\omega_0 t}\, dt \tag{2.44}$$

Using the orthogonal property of exponentials,

$$\int_{T0} e^{jn\omega_0 t} e^{-jm\omega_0 t}\, dt = 0 \quad \text{for } m \neq n$$

$$= T_0 \quad \text{for } m = n \tag{2.45}$$

Substituting this value of the integral in Eq.(2.44), we get

$$Q_n T_0 = \int_{T_0} x(t) e^{-jm\omega_0 t}\, dt \tag{2.46}$$

Hence the exponential series can be expressed as

$$x(t) = \sum_{n=-\infty}^{\infty} Q_n e^{jn\omega_0 t}\, dt \quad \text{where } Q_n = \frac{1}{T_0} \int_{T_0} x(t) e^{-jm\omega_0 t}\, dt \tag{2.47}$$

The above two equations show the compactness of the expressions, compared to the corresponding trigonometric Fourier series. The exponential form is more compact. Q_n can be related to the trigonometric series coefficients a_n, $b_n^{'}$ by setting $n = 0$ in the equation above. We thus obtain $Q_0 = a_0$.

For $n \neq 0$:

$$Q_n = \frac{1}{T_0} \int_{T_0} x(t) \cos(n\omega_0 t) dt - \frac{j}{T_0} \int_{T_0} x(t) \sin(n\omega_0 t) dt = \frac{1}{2}(a_n - jb_n)$$

(2.48)

and $\qquad Q_{-n} = \frac{1}{T_0} \int_{T_0} x(t) \cos(n\omega_0 t) dt + \frac{j}{T_0} \int_{T_0} x(t) \sin(n\omega_0 t) dt = \frac{1}{2}(a_n + b_n)$

(2.49)

These results are generally valid for $x(t)$, real or complex. When $x(t)$ is real, Q_n and Q_{-n} are conjugates, i.e., $Q_{-n} = Q^*_n$.

Also since $\qquad a_n - jb_n = \sqrt{a_n^2 + b_n^2}\, e^{j \arctan(b_n/a_n)} = C_n e^{j\theta_n}$ $\qquad\qquad$ (2.50)

Hence, $\qquad\qquad Q_0 = a_0 = C_0$

and $\qquad\qquad Q_n = \frac{1}{2} C_n e^{j\theta_n}, \ Q_{-n} = \frac{1}{2} C_n e^{-j\theta_n}$ $\qquad\qquad$ (2.51)

Therefore, $\qquad\qquad |Q_n| = |Q_{-n}| = \frac{1}{2} C_n \quad \text{for } n \neq 0$ $\qquad\qquad$ (2.52)

and $\qquad\qquad \lfloor Q_n = \theta_n \text{ and } \lfloor Q_{-n} = -\theta_n$ $\qquad\qquad$ (2.53)

Note that $|Q_n|$ are the amplitudes and $\lfloor Q_n$ are the angles of various components. From the above equations, it can be seen that when $x(t)$ is real, the amplitude spectrum ($|Q_n|$ versus ω) is an odd function of ω. For a complex $x(t)$, Q_n and Q_{-n} are not generally conjugates. For the exponential form of a Fourier series, we can observe the following.

(a) The coefficients Q_n are called the Fourier series coefficients of the signal $x(t)$. Also these are generally complex signals even when $x(t)$ is real.

(b) The limits of integral are arbitrary, generally chosen as 0 or $\frac{-T_0}{2}$.

(c) The frequency $f_0 = 1/T_0$ is called the fundamental frequency of the signal $x(t)$. The frequencies of the complex signals are multiples of this fundamental frequency. The nth multiple of f_0 is called the 'nth harmonic'.

(d) The above-mentioned statement states that the periodic signal $x(t)$ can be described by the period T_0, i.e., the fundamental frequency f_0 and a sequence of complex numbers (q_n). Thus, to describe $x(t)$, we may specify a countable number set of complex numbers. This reduces the complexity of describing $x(t)$, since to define $x(t)$ for all values of t, we must have to specify its value on an uncountable set of points.

(e) Fourier series expansion can be expressed in terms of the angular frequency $\omega_0 = 2\pi f_0$ by

$$Q_n = \frac{\omega_0}{2\pi} \int_{k}^{k + \frac{2\pi}{\omega_0}} x(t) e^{-jn\omega_0 t} dt$$

(2.54a)

$$x(t) = \sum_{n=-\infty}^{\infty} Q_n e^{jn\omega_0 t} \tag{2.54b}$$

In general, $Q_n = |Q_n| e^{j\underline{Q_n}}$. Thus $|Q_n|$ gives the magnitude of the nth harmonic and $\underline{Q_n}$ gives its phase. Figure 2.22 shows a graph of the magnitude and phase of various harmonics in $x(t)$. This type of graph is called discrete spectrum of the periodic signal $x(t)$.

(f) The Dirichlet conditions (discussed below) are only sufficient conditions for the existence of a Fourier series expansion. Even for signals that do not satisfy these conditions, we can still find a Fourier series expansion.

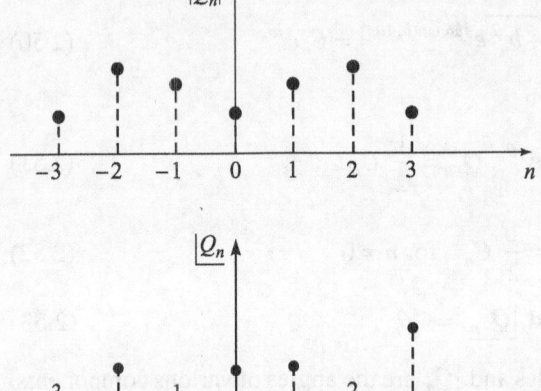

Fig. 2.22 Discrete spectrum of $x(t)$

Example 2.8 Find the Fourier series representation of an impulse train given by the expression $x(t) = \sum_{n=-\infty}^{+\infty} \delta(t - nT_0)$.

Solution

$$Q_n = \frac{1}{T_0} \int_{-T_0/2}^{+T_0/2} x(t) e^{-j2\pi \frac{n}{T_0} t} dt$$

$$= \frac{1}{T_0} \int_{-T_0/2}^{+T_0/2} \delta(t) e^{-j2\pi \frac{n}{T_0} t} dt$$

$$= \frac{1}{T_0}$$

Hence, the Fourier series expansion for the given function is

$$\sum_{n=-\infty}^{+\infty} \delta(t - nT_0) = \frac{1}{T_0} \sum_{n=-\infty}^{\infty} e^{j2\pi \frac{n}{T_0} t}$$

Positive and negative frequencies

From the Fourier series expansion of a periodic signal for

$$x(t) = \sum_{n=-\infty}^{\infty} Q_n e^{j2\pi \frac{n}{T_0} t}$$

it can be observed that both positive and negative frequencies are present. A positive frequency corresponds to a term of the form $e^{j\omega t}$ for a positive ω and a negative frequency corresponds to $e^{-j\omega t}$. The term $e^{j\omega t}$ corresponds to a phasor rotating anticlockwise at an angular frequency ω, and $e^{-j\omega t}$ corresponds to a phasor rotating clockwise at the same angular frequency. However, if two signals $e^{j\omega t}$ and $e^{-j\omega t}$ are added, their sum is $2\cos\omega t$ which is a real signal with two frequency components at $\pm \omega/2\pi$. Hence, we can conclude that in real signals, frequencies appear in positive and negative pairs with amplitudes that are conjugates.

Dirichlet conditions

If $x(t)$ satisfies certain conditions, its Fourier series is guaranteed to converge pointwise at all points where $x(t)$ is continuous. Also at the point of discontinuities, $x(t)$ converges to the value midway between the two values of $x(t)$. On the either side of discontinuity, these conditions are as follows.

(i) The function $x(t)$ must be absolutely integrable, i.e., it must satisfy the condition

$$\int_{T_0} |x(t)| dt < \infty \tag{2.55}$$

(ii) The function $x(t)$ must have only a finite number of discontinuities in one period.

(iii) The function $x(t)$ must contain only a finite number of maxima and minima in one period.

2.6.2 Fourier Transform

We have seen earlier that any periodic signal can be represented by a Fourier series. Now we will discuss how a non-periodic (aperiodic) signal is represented. It can be seen that it is possible to expand an aperiodic signal in terms of complex exponentials.

However, the resulting spectrum will no longer be discrete, but continuous, i.e., it covers a range of frequencies. This is the well-known *Fourier transform*. Sometimes, it is also called Fourier integral. For this signal $x(t)$ must satisfy the following. Dirichlet conditions.

(i) The function $x(t)$ is absolutely integrable on the real time, i.e.,

$$\int_{-\infty}^{\infty} |x(t)| dt < \infty \tag{2.56}$$

in one period.

(ii) The number of maxima and minima of $x(t)$ in any finite interval on the real time is finite.

(iii) The number of discontinuities of $x(t)$ in any finite interval on the real time is finite.

Then the Fourier transform or the integral of $x(t)$ is defined as

$$X(f) = \int_{-\infty}^{\infty} x(t)e^{-j2\pi ft} dt \qquad (2.57)$$

and the original signal can be obtained from

$$x(t) = \int_{-\infty}^{\infty} X(f)e^{j2\pi ft} dt \qquad (2.58)$$

The function $X(f)$ is generally a complex function. Its magnitude $|X(f)|$ and phase $|X(f)|$ represent the amplitude and phase of various frequency components in $x(t)$. It is sometimes referred to as the spectrum of signal $x(t)$.

To denote that $X(f)$ is the Fourier transform of $x(t)$, the following notation is employed:

$$X(f) = \mathcal{F}[x(t)] \qquad (2.59)$$

or

$$x(t) = \mathcal{F}^{-1}[X(f)] \qquad (2.60)$$

i.e., $x(t)$ is the inverse Fourier transform of $X(f)$. The relation between both of them is shown as

$$x(t) \Leftrightarrow X(f) \qquad (2.61)$$

In terms of ω, these equations become

$$X(\omega) = \int_{-\infty}^{\infty} x(t)e^{-j\omega t} dt \qquad (2.62)$$

and

$$x(t) = \frac{1}{2\pi} \int_{-\infty}^{\infty} X(\omega)e^{j\omega t} d\omega \qquad (2.63)$$

The relation between the Fourier transform and the inverse Fourier transform can be written as

$$x(t) = \int_{-\infty}^{\infty} \left[\int_{-\infty}^{\infty} x(\tau)e^{-j2\pi f\tau} d\tau \right] e^{j2\pi ft} df$$

where

$$X(f) = \int_{-\infty}^{\infty} x(\tau)e^{-2\pi\tau} d\tau \qquad (2.64)$$

and changing the order of integration, we get

$$x(t) = \int_{-\infty}^{\infty} \left[\int_{-\infty}^{\infty} e^{j2\pi f(t-\tau)} df \right] x(\tau) d\tau \qquad (2.65)$$

Since we know that $x(t) = \int_{-\infty}^{\infty} \delta(t-\tau)x(\tau)d\tau \qquad (2.66)$

Comparing the above two equations, we get

$$\delta(t-\tau) = \int_{-\infty}^{\infty} e^{-j2\pi f(t-\tau)} df \qquad (2.67)$$

or

$$\delta(t) = \int_{-\infty}^{\infty} e^{j2\pi ft} df \qquad (2.68)$$

Example 2.9 Determine the Fourier transform of signal sgn(t).

Solution

Signal sgn(t) is defined as a limit of exponential and is given by

$$x_n(t) = e^{-t/n}, \quad t > 0$$

$$= -e^{t/n}, \quad t < 0$$
$$= 0, \quad t = 0$$

The Fourier transform of this signal is

$$X_n(f) = F[x_n(t)]$$

$$= \int_{-\infty}^{0} (-e^{t/n})e^{-j2\pi ft}\, dt + \int_{-\infty}^{+\infty} e^{-t/n} e^{-j2\pi t}\, dt$$

$$= \int_{-\infty}^{0} e^{t\left(\frac{1}{n}-j2\pi f\right)}\, dt + \int_{0}^{+\infty} e^{-t\left(\frac{1}{n}+j2\pi f\right)}\, dt$$

$$= \frac{1}{\dfrac{1}{n} - j2\pi f} + \frac{1}{\dfrac{1}{n} + j2\pi f}$$

$$= \frac{-j4\pi f}{\dfrac{1}{n^2} + 4\pi^2 f^2}$$

Now letting $n \to \infty$, we get

$$F[\text{sgn}(t)] = \lim_{n\to\infty} X_n(f)$$

$$= \lim_{n\to\infty} \frac{-j4\pi f}{\dfrac{1}{n^2} + 4\pi^2 f^2}$$

$$= \frac{1}{j\pi f}$$

Fourier transform of real, even, and odd signals

The Fourier transform is generally written as

$$F[x(t)] = \int_{-\infty}^{+\infty} x(t)e^{-j2\pi ft}\, dt$$

$$= \int_{-\infty}^{+\infty} x(t)\cos(2\pi ft)\, dt - j\int_{\infty} x(t)\sin(2\pi ft)\, dt \qquad (2.69)$$

For a real $x(t)$, both integrals are real. Hence, they denote the real and imaginary parts of $X(f)$, respectively. Also we know that for a real $x(t)$, the real part of $X(f)$ is an even function of f and the imaginary part is an odd function of f. Hence, in general for a real $x(t)$, the transform $X(f)$ is a Hermitian function.

$$X(-f) = X*(f) \tag{2.70}$$

$$\text{Re }[X(-f)] = \text{Re }[X(f)] \tag{2.71}$$

$$\text{Im }[X(-f)] = -\text{Im }[X(f)] \tag{2.72}$$

$$|X(-f)| = |X(f)| \tag{2.73}$$

$$\underline{X(-f)} = \underline{-X(f)} \tag{2.74}$$

Typical plots of $|X(f)|$ and $\underline{X(f)}$ for a real $x(t)$ are shown in Fig. 2.23. If in addition to being real, $x(t)$ is an even signal, then the integral $\int_{-\infty}^{\infty} x(t)\sin(2\pi ft)dt$ vanishes since the integral is a product of even and odd signals and it is odd. Hence, the Fourier transform $X(f)$ will be real and even.

Similarly, if $x(t)$ is real and odd, the real part of its Fourier transform vanishes and $X(f)$ will be imaginary and odd.

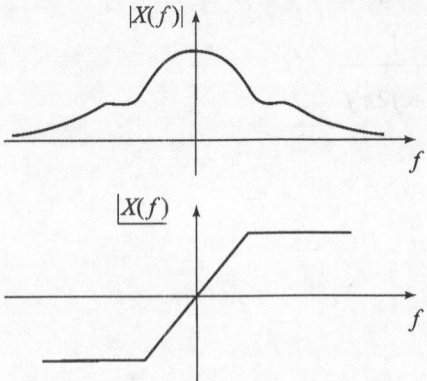

Fig. 2.23 Spectrum of a real signal: (a) magnitude and (b) phase

2.7 LAPLACE TRANSFORM

Laplace transform is a mathematical technique for solving a differential equation. It changes one signal into another according to some fixed set of rules. The Laplace transform changes a signal in a time domain into a signal in a frequency domain or in an s-domain called the s-plane. The time domain signal is continuous, extends to both positive and negative infinity, and may be either periodic or aperiodic. The Laplace transform allows the time domain to be complex.

The s-domain is a complex plane, the distance along the real axis is expressed by the variable σ and the imaginary axis by a variable ω, the natural frequency. Any location in the s-plane is represented by a complex variable $s = \sigma + j\omega$. As with the Fourier transform, signals in the s-plane are represented by capital letters.

A time domain signal $x(t)$ is transformed into an s-domain signal. $X(s)$ or $X(\sigma, \omega)$. The s-plane is continuous and extends to ∞ in all four directions. As the location on the s-plane is defined by a complex number, each point in the s-domain has a value that is a complex number. In other words, each location in the s-plane has real and imaginary parts. As with all complex numbers, the real and imaginary parts can alternatively be expressed as the magnitude and phase.

The Laplace transform also analyses the signals in terms of sinusoids and exponential. One can view the Fourier transform as a subset of the Laplace transform. The Laplace transform can be obtained from the Fourier transform as follows:

$$X(\omega) = \int_{-\infty}^{\infty} x(t)e^{-j\omega t}dt \tag{2.75}$$

To get the Laplace transform, multiply the above time domain signal by the exponential term $e^{-\sigma t}$, i.e.,

$$X(\sigma,\ \omega) = \int_{-\infty}^{\infty} [x(t)e^{-\sigma t}]e^{-j\omega t}dt \tag{2.76}$$

The two exponential terms can be combined to give

$$X(\sigma,\ \omega) = \int_{-\infty}^{\infty} x(t)e^{-(\sigma+j\omega)t}dt \tag{2.77}$$

Substituting for $s = \sigma + j\omega$, where s is the complex variable that represents the location in the complex plane, the above equation reduces to

$$X(s) = \int_{-\infty}^{\infty} x(t)e^{-st}dt \tag{2.78}$$

This is called the final form of the Laplace transform and e^{-st} is called the complex exponential. This complex exponential is a compact way of representing both sinusoids and exponentials in a single expression. Although the Laplace transform was explained as a two-stage process, multiplication by an exponential curve followed by the Fourier transform, it is a single equation relating $x(t)$ and $X(s)$. The equation for $X(s)$ given above describes how to calculate each point in the s-plane (given by σ and ω) based on the values of σ, ω and the time domain signal $x(t)$. Using the Fourier transform to simultaneously calculate all the points along a vertical line is just a convenience but not a requirement. The values in the s-plane along the y-axis (i.e., $\sigma = 0$) are exactly equal to the Fourier transform.

Let us now look at several individual points in the s-domain and examine how the values at these locations are related to the time domain signal. Each point in the frequency domain identified by a specific value of ω corresponds to two sinusoids $\cos(\omega t)$ and $\sin(\omega t)$. The real part is found by multiplying the time domain signal by the cosine wave, and then integrating from $-\infty$ to ∞. The imaginary part is found in the same way except that the sine wave is used. When we deal with a complex Fourier transform, the value at the corresponding negative frequency $-\omega$ will be the complex conjugate (having the same real part, but negative imaginary part) of the values at ω. The Laplace transform is just an extension of these same concepts.

Figure 2.24 shows three pairs of points in the s-plane: A and A', B and B', and C and C'. As in the complex frequency spectrum, the points at A, B, and C (the positive frequencies) are the complex conjugates of the points at A', B', and C' (the negative frequencies). The top half of the s-plane is a mirror image of the lower half, and both halves are needed to correspond to a real time domain signal. Since in each of these pairs, s has specific values for σ and $\pm\omega$, there are two waveforms associated with each pair: $\cos(\omega t)\,e^{-\sigma t}$ and $\sin(\omega t)\,e^{-\sigma t}$. In Fig. 2.24, points C and C' are at locations $\sigma = 1.5$ and $\omega = \pm 30$ and, therefore, correspond to the waveform $\cos(30t)e^{-1.5t}$ and $\sin(30t)e^{-1.5t}$. Depending on the value of σ, these sin waves remain with the same constant amplitude when $\sigma = 0$, with an exponentially increasing amplitude when $\sigma = $ negative, or with an exponentially decreasing amplitude when $\sigma = $ positive.

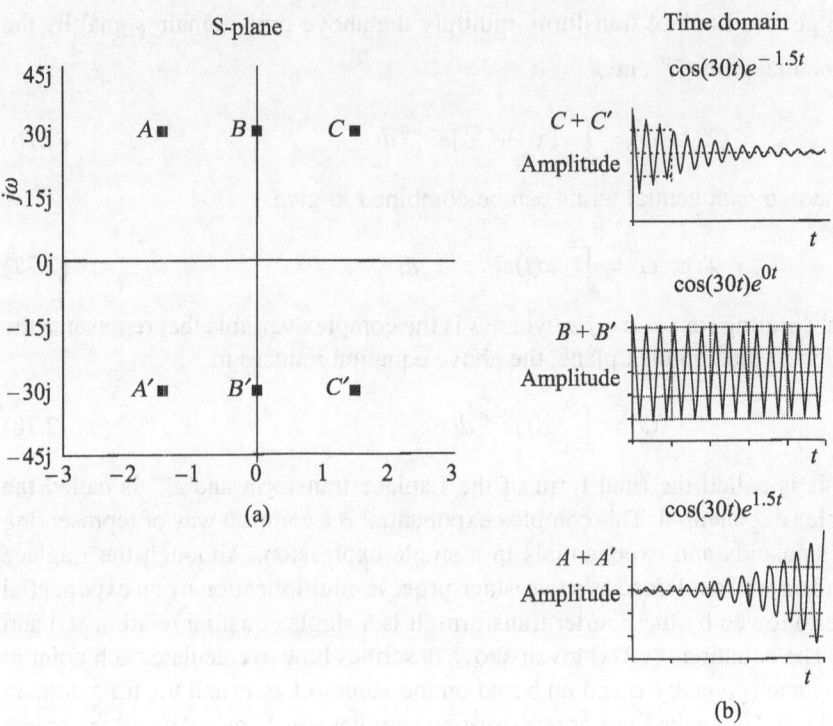

Fig. 2.24 (a) The s-plane and (b) waveforms in a time domain

The value at each location in the s-plane consists of a real part and an imaginary part. The real part is found by multiplying the time domain signal by the exponentially weighted cosine wave and then integrated from $-\infty$ to ∞. The imaginary part is found in the same way except the exponentially weighted sine wave is used instead. The equation formed using the real part of C and C'.

$$\text{Re } X(\sigma = 1.5,\ \omega = \pm\ 30) = \int_{-\infty}^{\infty} x(t)\cos(30t)e^{-1.5t}\,dt \qquad (2.79)$$

Figure 2.25 shows an example of a time domain waveform, its frequency spectrum, and its s-domain representation. Let us take the example of a rectangular pulse of width 2 and height 1 in the time domain. As shown, the complex Fourier transform of this signal is a sinc function in the real part and an entirely zero signal in the imaginary part. The s-domain is an undulating two-dimensional signal, displayed as a topographic surface of real and imaginary parts. The Laplace transform of this signal is given by

$$X(s) = \int_{-\infty}^{\infty} x(t)e^{-st}\,dt = \int_{-1}^{1} 1e^{-st}\,dt \qquad (2.80)$$

$$X(s) = \frac{e^s - e^{-s}}{s} \qquad (2.81)$$

Substituting for s, $s = \sigma + j\omega$ and separating the real and imaginary parts,

$$\text{Re } X(\sigma, \omega) = \frac{\sigma \cos\omega(e^\sigma - e^{-\sigma}) + \omega \sin\omega(e^\sigma + e^{-\sigma})}{\sigma^2 + \omega^2} \qquad (2.82)$$

$$\text{Im } X(\sigma, \omega) = \frac{\sigma \sin\omega(e^\sigma - e^{-\sigma}) - \omega \cos\omega(e^\sigma - e^{-\sigma})}{\sigma^2 + \omega^2} \qquad (2.83)$$

The topographical surfaces in Fig. 2.25 are graphics of these equations. These equations are quite long and tedious and to find out whether they are correct, one method to quickly verify is that these equations reduce to the Fourier transform along the y-axis. This we can get by setting $\sigma = 0$ in the equations and simplifying

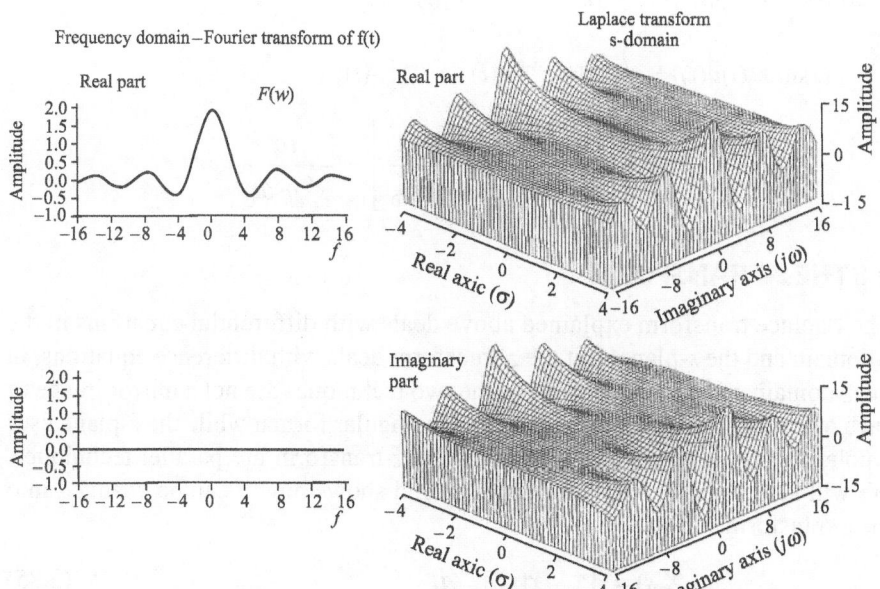

Fig. 2.25   Time domain signal (rectangular pulse) transformed into a frequency domain using the Fourier transform, then to s-domain using the Laplace transform

$$\text{Re } X(\sigma, \omega) \Big|_{\sigma=0} = \frac{2 \sin \omega}{\omega}, \quad \text{Im } X(\sigma,\omega) \Big|_{\sigma=0} = 0 \tag{2.84a}$$

These are the correct frequency domain signals for the sinc function.

Inverse Laplace transform

The inverse Laplace transform $x(t)$ is defined as

$$x(t) = \frac{1}{2\pi j} \int_{-\infty}^{+\infty} X(s)e^{st} ds \tag{2.84b}$$

Example 2.10 Find the Laplace transform of the following:
(a) $\delta(t)$
(b) $u(t)$
(c) $\sin \omega_0 t\, u(t)$

Solution

(a) $L[\delta(t)] = \int_{0^-}^{\yen} \delta(t)e^{-st} dt = 1$ for all s, i.e.,

$\delta(t) \Leftrightarrow 1$ for all s

(b) Since $u(t) = 1$ for $t \geq 0$,

$$L[u(t)] = \int_{0^-}^{\infty} u(t)e^{-st} dt = \int_{0^-}^{\yen} e^{-st} dt = -\frac{1}{s}e^{-st} \Big|_{0^-}^{\infty} = \frac{1}{s}$$

(c) As $\sin \omega_0 t u(t) = \frac{1}{2}\, [e^{j\omega_0 t} - e^{-j\omega_0 t}](t)$

$$L[\sin\omega_0(t)u(t)] = \frac{1}{2}\, L[e^{j\omega_0 t} u(t) - e^{-j\omega_0 t} u(t)]$$

$$= \frac{1}{2}\left[\frac{1}{s - j\omega_0} - \frac{1}{s + j\omega_0} \right] = \frac{\omega_0}{s^2 + \omega_0^2}$$

2.8 THE z-TRANSFORM

The Laplace transform explained above deals with differential equations in the s-domain and the s-plane. But the z-transform deals with difference equations, in the z-domain and z-plane. However, the two techniques are not a mirror image of each other. The s-plane is arranged in a rectangular format while the z-plane uses a polar format. The Laplace transform and z-transform are parallel techniques. We will start with the Laplace transform and shows how it can be changed into the z-transform.

$$X(s) = \int_{t=-\infty}^{\infty} x(t)e^{-st} dt \tag{2.85}$$

where $x(t)$ and $X(s)$ are the time domain and the s-domain representations of the signal, respectively. This equation analyses the time domain signal in terms of sine and cosine waves, which has an exponentially changing amplitude. This can be better understood by replacing the complex variable s with its equivalent expression $s = \sigma + j\omega$. Now the Laplace transform

$$X(\sigma, \omega) = \int_{t=-\infty}^{\infty} x(t)e^{-\sigma t}e^{-j\omega t} dt \qquad (2.86)$$

The Laplace transform can be changed into the z-transform in three steps.

1. The first step is most obvious: change from a continuous signal to a discrete signal. This is done by replacing the time variable t with the sample number n and changing the integral to summation.

$$X(\sigma, \omega) = \sum_{n=-\infty}^{\infty} x[n]e^{-\sigma n}e^{-j\omega n} \qquad (2.87)$$

It can be observed that $X(\sigma, \omega)$ is continuous (hence uses parentheses) not discrete. Although we are dealing with a discrete time domain signal $x[n]$, the parameters σ and ω can still take on a continuous range of values.

2. The second step is to rewrite the exponential term. Since we can represent an exponential signal as

$$y[n] = e^{-\sigma n} \quad \text{or} \quad y[n] = r^n \qquad (2.88)$$
$$r^n = [e^{\ln(r)}]^n = e^{n \ln(r)} e^{-\sigma n} \qquad (2.89)$$

where $\sigma = -\ln(r)$

The second step of converting the Laplace transform into the z-transform is completed by using the other exponential form

$$X(r, \omega) = \sum_{n=-\infty}^{\infty} x[n]r^n e^{-j\omega n} \qquad (2.90)$$

This equation is not in the most compact form for complex notation. This problem was overcome in the Laplace transform by introducing a new complex variable s defined to be $s = \sigma + j\omega$. In the same way, we will define a new variable for the z-transform, $z = re^{-j\omega}$. This defines the complex variable z as the polar notation combination of the two real variables r and ω.

3. The third step in deriving the z-transform is to replace r and ω with z. Thus, we get the standard form of the z-transform.

$$X(z) = \sum_{n=-\infty}^{\infty} x[n]z^{-n} \qquad (2.91a)$$

As in the case of Laplace transforms, the inverse z-transform is given by

$$x[n] = Z^{-1}\{X[z]\} \quad \text{or} \quad x[n] \Leftrightarrow X[z] \qquad (2.91b)$$

Figure 2.26 shows the difference between the Laplace transform's s-plane and the z-transform's z-plane. Locations in the s-planes are identified by two parameters σ, the exponential delay variable along the horizontal axis, and ω, the frequency variable along the vertical axis, i.e., these two real parameters are arranged in a rectangular coordinate system, i.e., each point in the s-plane is represented by $s = \sigma + j\omega$.

In comparison, the z-domain uses the variables r and ω arranged in the polar coordinate. The distance from the origin r is the value of the exponential delay.

The angular distance measured from the positive horizontal axis ω is the frequency, i.e., $z = re^{-j\omega}$, the complex variable representing position in the z-plane is formed by combining the two real parameters in a polar form.

These differences result in vertical lines in the s-plane matching circles in the z-plane. For example, the s-plane in Fig. 2.26 shows a pole-zero pattern where all poles and zeros lie on vertical lines. The equivalent poles and zeros in the z-plane lie on circles concentric with the origin. This follows from the fact that $\sigma = -\ln(r)$. For instance, the s-plane's vertical axis ($\sigma = 0$) corresponds to the z-plane's unit circle, i.e., $r = 1$. Vertical lines in the left half of the s-plane correspond to circles inside the z-plane's unit circle. Similarly, vertical lines in the right half of the s-plane match circles that are outside the z-plane's unit circle. Hence, we can infer that left and right sides of the 's' plane correspond to the interior and the exterior of the unit circle, respectively. A continuous system is unstable when poles occupy the right half of the s-plane. In the case of a discrete system when poles lie outside the unit circle in the z-plane, the system is unstable. When the time domain signal is completely real, the upper and lower halves of the z-plane are mirror images of each other as with the s-domain.

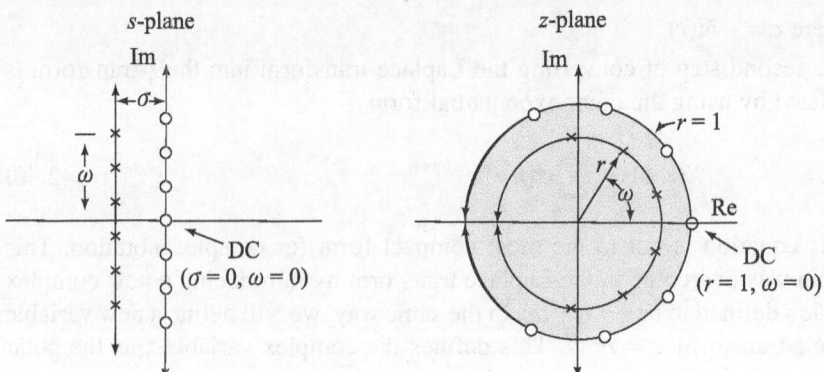

Fig. 2.26 Relationship between the s-plane and z-plane

The frequency variable ω can have any value between zero (i.e., DC) and infinity for a continuous sinusoid. This means that the s-plane must allow ω to run from negative to positive infinity. In comparison, a discrete sinusoid can only have a frequency between DC and one-half the sampling rate, i.e., the frequency must be between 0 and 0.5 when expressed as a fraction of the sampling rate or between 0 and τ when expressed as a natural frequency (i.e., $\omega = 2\pi f$). Hence, in the z-plane, positive frequencies correspond to angles of 0 to π rad and the negative frequencies correspond to 0 to $-\pi$ rad. Sometimes, the letter Ω (an upper case omega) is used to represent frequency in the z-plane and ω (a lower case omega) for frequency in the s-plane.

In the s-plane, the values that lie along the vertical axis are equal to the frequency response of the system, i.e., the Laplace transform evaluated at $\sigma = 0$ is equal to the Fourier transform. Similarly, the frequency response in the z-domain is found along the unit circle. This can be seen by evaluating the z-tansform at $r = 1$, resulting in the equation reducing to the DTFT (discrete time

Fourier transform). This increase the value of the frequency from 0 to 1 on the horizontal axis in the s-plane. The spectrum positive frequencies are positioned in a counterclockwise pattern from this DC position occupying the upper semi-circle. Likewise, the negative frequencies are arranged from the DC position along the clockwise path forming the lower semi-circle. The positive and negative frequencies in the spectrum meet at the common points $\omega = \pi$ and $\omega = -\pi$. This circular geometry also corresponds to the frequency spectrum of the discrete signal being periodic, i.e., when the frequency angle is increased beyond π, the same values are encountered in between 0 and π.

Example 2.11 Find the z-transform of
  (a) $\delta(n)$
  (b) $u(n)$
Solution

We have
$$X(z) = \sum_{n=0}^{\infty} x[n]z^{-n}$$

$$= x[0] + \frac{x[1]}{z} + \frac{x[2]}{z^2} + \frac{x[3]}{z^3} + \cdots$$

(a) For $x[n] = \delta(n)$, $x[0] = 1$, and $x[1] = x[2] = x[3] = \cdots = 0$
  Therefore, $\delta(n) \Leftrightarrow 1$ for all z
(b) For $x[n] = u[n]$, $x[0] = x[1] = x[2] = x[3] = \cdots = 1$

Therefore, $X[z] = 1 + \frac{1}{z} + \frac{1}{z^2} + \frac{1}{z^3} + \cdots$

$$= \frac{1}{1 - \frac{1}{z}} \text{ for } \left|\frac{1}{z}\right| < 1$$

$$= \frac{z}{z-1} \text{ for } \left|\frac{z}{z-1}\right| > 1$$

2.9 SIGNAL ENERGY AND ENERGY SPECTRAL DENSITY

The energy E of a signal $x(t)$ is defined as the area under $|x(t)|^2$. Every non-zero periodic continuous time signal (CT) has a infinite total energy. Let $x(t)$ be a voltage signal and if it is applied to a $1\,\Omega$ resistor, then the power dissipated by it is $x(t) \cdot x(t)$. Thus, the total energy provided by $x(t)$ from $-\infty$ to ∞, assuming $x(t)$ to be real is

$$E = \int_{-\infty}^{\infty} x^2(t)dt \tag{2.92}$$

If $x(t)$ is complex, then the total energy is

$$E_T = \int_{-\infty}^{\infty} x(t)x^*(t) = \int_{-\infty}^{\infty} |x(t)|^2 dt \tag{2.93}$$

From the above relation, it is clear that every non-zero periodic signal has $E = \infty$. Hence, it is meaningless to discuss total energy for periodic signals; instead, we discuss its average power defined by

$$P_{av} = \lim_{L \to \infty} \frac{1}{2L} \int_{-L}^{L} |x(t)|^2 \, dt \tag{2.94}$$

For a periodic signal with period P, this equation reduces to

$$P_{av} = \frac{1}{P} \int_{0}^{P} x(t) x^*(t) dt = \frac{1}{P} \int_{0}^{P} |x(t)|^2 dt \tag{2.95}$$

This average power can be computed directly in a time domain. We can also compute in frequency domain from its Fourier series. Now

$$x(t) = \sum_{n=-\infty}^{\infty} C_n e^{jn\omega_0 t}$$

with

$$\omega_0 = \frac{2\pi}{P} \tag{2.96}$$

Substituting this value of $x(t)$ in the expression for P_{av}, we get

$$P_{av} = \frac{1}{P} \int_{0}^{P} \left(\sum_{-\infty}^{\infty} C_n e^{jn\omega_0 t} \right) x^*(t) dt \tag{2.97}$$

$$= \sum_{-\infty}^{\infty} C_n \left(\frac{1}{P} \int_{0}^{P} x^*(t) e^{jn\omega_0 t} dt \right) \tag{2.98a}$$

The term in the large parentheses is C_n^*. Hence,

$$P_{av} = \frac{1}{P} \int_{0}^{P} |x(t)|^2 dt = \sum_{n=-\infty}^{\infty} C_n C_n^* = \sum_{n=-\infty}^{\infty} (C_n)^2 \tag{2.98b}$$

This is called Parseval's theorem. It states that the average power equals the sum of the squared magnitude of C_n. Since $|C_n|^2$ is the power at the frequency $n\omega_0$, from C_n, we can see the distribution of power over frequencies. It can be seen that the average power depends only on the magnitude of C_n and is independent of the phase of C_n.

Are phase important? Yes, but only when we consider a video signal or image processing. But for an audio signal, it is not important.

Example 2.12 Verify Parseval's theorem for the signal $e^{-at}u(t)$ for $a > 0$.

Solution
Since the signal energy is given by

$$E_g = \int_{-\infty}^{\infty} x^2(t) dt = \int_{0}^{\infty} e^{-2at} dt = \frac{1}{2a}$$

Now let us determine the energy from the signal spectrum $X(\omega)$, which is

$$X(\omega) = \frac{1}{j\omega + a}$$

$$E_g = \frac{1}{2\pi} \int_{-\infty}^{\infty} |X(\omega)|^2 d\omega = \frac{1}{2\pi} \int_{-\infty}^{\infty} \frac{1}{\omega^2 + a^2} d\omega = \frac{1}{2\pi a} \tan^{-1} \frac{\omega}{a} \Big|_{-\infty}^{\infty} = \frac{1}{2a}$$

This verifies Parseval's theorem.

2.10 ENERGY SPECTRAL DENSITY

From Parseval's theorem, it is seen that the energy of a signal $x(t)$ is the result of the energies contributed by all the spectral components of the signal $x(t)$. The contribution of a spectral component of frequency ω is proportional to $|X(\omega)|^2$. Let us consider a signal $x(t)$. Using Parseval's formula, the total energy of $x(t)$ can also be computed from its magnitude spectrum. The magnitude spectrum reveals the distribution of energy in frequencies. The energy contained in the frequency range $[\omega_1, \omega_2]$ with $\omega_1 < \omega_2$ is given by

$$\frac{1}{2}\int_{\omega_1}^{\omega_2}|X(\omega)|^2 d\omega \tag{2.99}$$

If $x(t)$ has a finite energy, then its spectrum contains no impulses and

$$\int_{\omega_0^-}^{\omega_0^+}|X(\omega)|^2 d\omega = 0 \tag{2.100}$$

Hence, it is meaningless to talk about energy at a discrete or isolated frequency. For this reason, $X(\omega)$ is called *spectral density* of $x(t)$. If $x(t)$ is real, then

$$|X(-\omega)| = |X(\omega)| \text{ and} \tag{2.101}$$

and
$$E = \frac{1}{2\pi}\int_{-\infty}^{\infty}|X(\omega)|^2\,d\omega \tag{2.102}$$

$$= \frac{1}{\pi}\int_0^{\infty}|X(\omega)|^2 d\omega \tag{2.103}$$

Then the total energy of a real valued signal can be computed over positive frequencies. The quantity $\dfrac{|X(\omega)|^2}{\pi}$ may be called energy spectral density (ESD). It is independent of its phase spectrum. Time shifting of a signal will not affect its total energy.

We can conclude that the total energy of every non-zero CT periodic signal is indefinite and its average power equals

$$P_{av} = \sum_{n=-\infty}^{\infty}|C_n|^2 \tag{2.104}$$

Non-zero power appears only at the discrete frequency $n\omega_0$. The total energy of a bounded and absolutely integrable signal equals

$$E = \frac{1}{2\pi}\int_{-\infty}^{\infty}|X(\omega)|^2 d\omega \tag{2.105}$$

The energy at any isolated frequency is zero. Therefore, its energy distribution is discussed over non-zero frequency intervals. Its average power is defined by

$$P_{av} = \lim_{L\to\infty}\frac{1}{2L}\int_{t=-L}^{L}|(xt)|^2 dt \tag{2.106}$$

2.11 ESSENTIAL BANDWIDTH OF A SIGNAL

The spectra of most signals extend to infinity. Since the energy of a practical signal is finite, the signal spectrum must approach zero as $\omega \to \infty$. Most of the signal energy is contained within a certain band of B Hz, and the energy content of the component of frequencies greater than B Hz is negligible. Hence, the signal can be bandlimited to B Hz without much sacrificing the shape and energy of the signal. The bandwidth B is called the *essential bandwidth* of signal and the criteria for selecting B depend on the error tolerance in a particular application. Suppression of all the spectral components of $f(t)$ beyond the essential bandwidth results $\hat{x}(t)$ in a signal that is close to $\hat{x}(t)$.

Example 2.13 Find the essential bandwidth B of the signal $e^{-at}u(t)$ if the required bandwidth has to contain 85% of the signal energy.

Solution

Since $X(\omega) = \dfrac{1}{j\omega + a}$, the energy spectral density (ESD) is

$$|X(\omega)|^2 = \frac{1}{\omega^2 + a^2}$$

The energy of this signal found in Example 2.12 is $E_g = \dfrac{1}{2a}$.

Let B be the essential bandwidth, which contains 85% of the total signal energy E_g, i.e.,

$$\frac{0.85}{2a} = \frac{1}{2\pi}\int_{-B}^{B}\frac{d\omega}{\omega^2 + a^2} = \frac{1}{2\pi a}\tan^{-1}\frac{B}{a}\Big|_{-B}^{B} = \frac{1}{\pi a}\tan^{-1}\frac{B}{a}$$

or $$\frac{0.85\pi}{2} = \tan^{-1}\frac{B}{a}$$

Hence, $$B = 4.16529a \text{ radians per second}$$

Thus, it can be inferred that the spectral components of $x(t)$ in the band from zero (DC) to 4.16529 radians per second (238.65 Hz) contribute 85% of the total signal energy. The remaining spectral components in the band from 238.65 Hz to ∞ contribute 15% of the signal energy.

2.12 ENERGY OF MODULATED SIGNAL

Let $m(t)$ be the baseband signal bandlimited to B Hz. The amplitude modulated signal is $f(t) = m(t)\cos\omega_0 t$ and its frequency spectrum is

$$F(\omega) = (1/2)\,[M(\omega + \omega_0) + M(\omega - \omega_0)] \tag{2.107}$$

The energy spectral density of the modulated signal $f(t)$ is $|F(\omega)|^2$, i.e.,

$$|F(\omega)|^2 = (1/4)\,[M(\omega + \omega_0) + M(\omega - \omega_0)]^2 \tag{2.108}$$

If $\omega_0 \geq 2\pi B$, then $M(\omega + \omega_0)$ and $M(\omega - \omega_0)$ are nonoverlapping and

$$|F(\omega)|^2 = (1/4)\,[|M(\omega + \omega_0)|^2 + |M(\omega - \omega_0)|^2] \tag{2.109}$$

$$= (1/4)\,[F_g(\omega + \omega_0) + F_g(\omega - \omega_0)] \tag{2.110}$$

where $F_g(\omega)$ is the ESD of $m(t)$. The ESDs of both $m(t)$ and the modulated signal $f(t)$ are shown in Fig. 2.27. It can be seen that the modulation shifts the ESD of $m(t)$ by $\pm \omega_0$. Also we can observe that the area under $|F(\omega)|^2$ is half the area under $|F_g(\omega)|$.

Because the energy of a signal is proportional to the area under its ESD, it follows that the energy of $f(t)$ is half the energy of $m(t)$, i.e.,

$$E^x = (1/2) \, E_m, \quad \omega_0 \geq 2\pi B \tag{2.111}$$

The energy of a signal is proportional to the square of its amplitude and a higher amplitude contributes more energy signal. The signal $m(t)$ remains most of the time at a higher amplitude level. On the other hand, $f(t)$, because of the factor $\cos \omega_0 t$, crosses zero amplitude many times and reduces its energy.

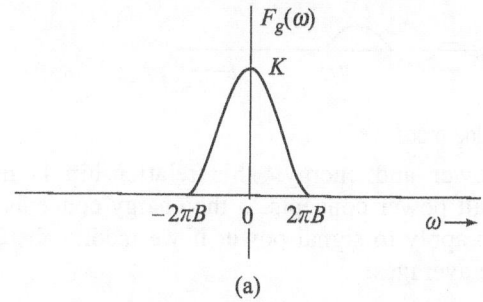

(a)

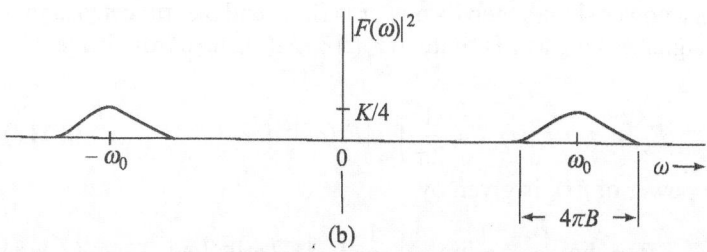

(b)

Fig. 2.27   Energy spectral densities of (a) modulating signal and (b) modulated signal

2.13 SIGNAL POWER AND POWER SPECTRAL DENSITY

For a power signal, the measure of its size is its power as the time average of the signal energy averaged over the infinite time interval. The power P_g of a real signal $f(t)$ is

$$P_g = \lim_{T \to \infty} \frac{1}{T} \int_{-T/2}^{T/2} f^2(t) dt \tag{2.112}$$

We can define a truncated signal $f_\tau(t)$ as

$$f_\tau(t) = f(t), \quad |t| \leq T/2 \tag{2.113}$$

$$= 0, \quad |t| > T/2 \tag{2.114}$$

The truncated signal is shown in Fig. 2.28. The energy of the truncated signal $f_\tau(t)$ is

$$\frac{1}{T} \int_{-T/2}^{T/2} f^2(t) dt \tag{2.115}$$

Hence,

$$P_g = \lim_{T \to \infty} \frac{E_{g\tau}}{T} \qquad (2.116)$$

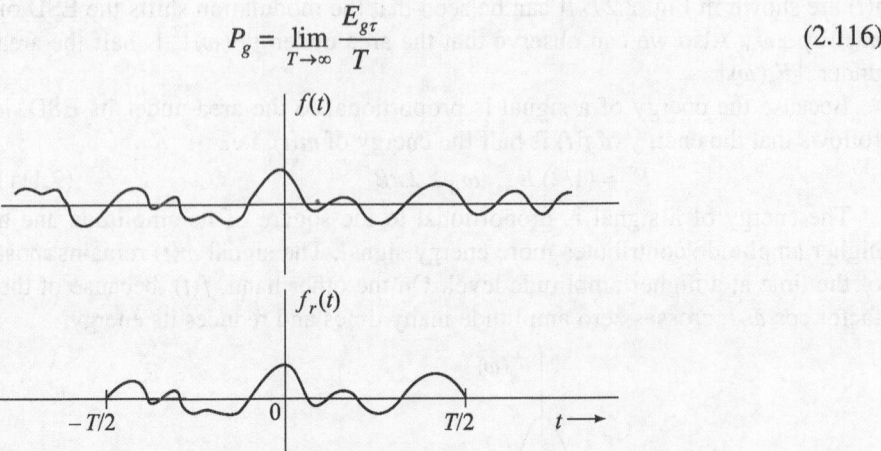

Fig. 2.28 PSD derivation by limiting process

This equation relates power and energy. This relationship is useful in understanding and relating all power concepts to the energy concepts and the results of signal energy also apply to signal power if we modify the concepts properly by taking their time averages.

2.13.1 Power Spectral Density (PSD)

If the signal $f(t)$ is a power signal, then its power is finite and the truncated signal $f_\tau(t)$ is an energy signal as long as T is finite. If $f_\tau(t) \Leftrightarrow F_\tau(\omega)$, then from Parseval's theorem,

$$E_{g\tau} = \int_{-\infty}^{\infty} f_\tau^2(t)dt = \frac{1}{2\pi} \int_{-\infty}^{\infty} |F_\tau(\omega)|^2 d\omega \qquad (2.117)$$

Hence, P_g, the power of $f(t)$, is given by

$$P_g = \lim_{\tau \to \infty} \frac{E_{g\tau}}{T} = \lim_{\tau \to \infty} \frac{1}{T} \left[\frac{1}{2\pi} \int_{-\infty}^{\infty} |F_\tau(\omega)|^2 d\omega \right] \qquad (2.118)$$

As T increases, the duration of $f_\tau(t)$ increases and its energy $E_{g\tau}$ also increases proportionately. This means that $|F_\tau(\omega)|^2$ also increases with T and as $T \to \infty$ $|F_\tau(\omega)|^2$ also approaches ∞ at the same rate as T because for a power signal, the above-mentioned equation must converge. This convergence permits us to interchange the order of the limiting process and integration. Then the equation can be rewritten as

$$P_g = \frac{1}{2\pi} \int_{-\infty}^{\infty} \lim_{\tau \to \infty} \frac{|F_\tau(\omega)|^2}{T} d\omega \qquad (2.119)$$

where we define $\lim_{\tau \to \infty} \dfrac{|F_\tau|^2}{T}$ as $S_g(\omega)$, the power spectral density (PSD).

Hence,

$$P_g = \frac{1}{2\pi} \int_{-\infty}^{\infty} S_g(\omega)d\omega \qquad (2.120)$$

or

$$P_g = \frac{1}{\pi} \int_{0}^{\infty} S_g(\omega)d\omega \qquad (2.121)$$

This is similar to the results for an energy signal. The power is $1/2\pi$ times the area under the PSD. The PSD is the time average of the ESD of $F_\tau(t)$. The PSD is positive real and even a function of ω. If $f(t)$ is a voltage signal, the units of the PSD are volts squared per hertz (V^2/Hz). In terms of frequency, the equation becomes

$$P_g = \int_{-\infty}^{\infty} S_g(\omega)df = 2\int_0^{\infty} S_g(\omega)df \qquad (2.122)$$

SUMMARY

This chapter begins with the basic concept of signals followed by classifications of the signals. Differences between the continuous and discrete signals are discussed. Systems are classified depending on their properties. The impulse response of a system is explained in detail and the advantage of finding it is discussed. The concept of convolution is explained. An important signal, called unit-step functions is introduced. This signal is very useful in representing causal signals and signals with different mathematical descriptions over different intervals.

The Fourier transform and its application in analysing any periodic signal is discussed in detail with examples. Another form of the Fourier transform called exponential transform, is explained. To analyse aperiodic signals, the Fourier integral is employed. The concept of positive and negative frequencies is explained. To analyse the linear time invariant system, another transform, called Laplace transform, is introduced. With the help of the Laplace transform, the signal $f(t)$ is broken into impulse components and then analysed. For dealing with a discrete system, z-transforms are introduced.

IMPORTANT FORMULAE

- Continuous signal: $x(t) = A\cos(2\pi f_o t + \theta)$
- Discrete time signal: $x[n] = A\cos(2\pi f_o n + \theta)$
- Even signal: $x_e(t) = \dfrac{x(t) + x(-t)}{2}$
- Odd signal: $x_o(t) = \dfrac{x(t) - x(-t)}{2}$
- Real and complex signals: $x_r(t) = |x(t)|\cos(\underline{x(t)})$

$$x_o(t) = |x(t)|\sin(\underline{x(t)})$$

$$|x(t)| = \sqrt{x_r^2(t) + x_i^2(t)}$$

- Energy-type and power-type signals: $E_X = \displaystyle\int_{-\infty}^{\infty} |x(t)|^2\, dt = \lim_{T\to\infty} \int_{-T/2}^{T/2} |x(t)|^2 dt$

$$P_x = \lim_{T \to \infty} \frac{1}{T} \int_{-T/2}^{T/2} |x(t)|^2 \, dt$$

- Unit step signal: $U(t) = 1$ for $t > 0$
$$= 0 \text{ for } t < 0$$

- The sinc signal: $\text{sinc}(t) = \dfrac{\sin(\pi t)}{\pi t}, \quad t \neq 0$
$$= 1, \quad t = 0$$

- The sign or the signum signal: $\text{sgn}(t) = 1, \quad t > 0$
$$= -1, \quad t < 0$$
$$= 0, \quad t = 0$$

- The impulse or delta signal: $\delta(t) = \infty$ for $t = 0$
$$= 0 \text{ for } t \neq 0$$

- The mathematical equation for convolution is $y(t) = x(t) * h(t) = \displaystyle\int_{-\infty}^{\infty} x(t) h \, (t - \tau) d\tau$

- Fourier series of a function: $x(t) = a_0 + \displaystyle\sum_{n=1}^{\infty} a_n \cos(2\pi ftn) + \displaystyle\sum_{n+1}^{\infty} b_n \sin(2\pi ftn)$

- Fourier series analysis equations:

$$a_0 = \frac{1}{T} \int_{\frac{-T}{2}}^{\frac{T}{2}} x(t) \, dt$$

$$a_n = \frac{2}{T} \int_{\frac{-T}{2}}^{\frac{T}{2}} x(t) \cos(\frac{2\pi tn}{T}) \, dt$$

$$b_n = \frac{2}{T} \int_{\frac{-T}{2}}^{\frac{T}{2}} x(t) \sin(\frac{2\pi tn}{T}) \, dt$$

- Fourier series in a compact form:

$$x(t) = C_0 + \sum_{n=1}^{\infty} C_n \cos(\frac{2\pi nt}{T} - \phi_n)$$

where C_0, C_n, and ϕ_n are related to a_0, a_n, b_n by the equations

$$C_0 = a_0$$

$$C_n = \sqrt{a_n^2 + b_n^2}$$

$$\phi_n = \arctan(\frac{-b_n}{a_n})$$

- Exponential form of Fourier series:

$$x(t) = \sum_{n=-\infty}^{\infty} Q_n e^{jn\omega_0 t} dt$$

where

$$Q_n = \frac{1}{T_0} \int_{T_0} x(t) e^{-jm\omega_0 t} dt$$

- Fourier transform of $x(t)$ is given by

$$X(f) = \int_{-\infty}^{\infty} x(t) e^{-j2\pi ft} dt$$

and the original signal can be obtained from

$$x(t) = \int_{-\infty}^{\infty} X(f) e^{j2\pi ft} dt$$

- Laplace transform: $X(s) = \int_{-\infty}^{\infty} x(t) e^{-st} dt$

- Inverse Laplace transform: $x(t) = \dfrac{1}{2\pi j} \int_{-\infty}^{+\infty} X(s) e^{st} ds$

- z-transform: $X(z) = \sum_{n=-\infty}^{\infty} x[n] z^{-n}$

- Inverse z-transform: $x[n] = Z^{-1}\{X[z]\}$ or $x[n] \Leftrightarrow X[z]$

ADDITIONAL EXAMPLES

1. Check the periodicity of the following signals:

(a) $x(t) = 5\sin\left(6t + \dfrac{\pi}{4}\right)$

(b) $x(t) = e^{j3t}$

(c) $x(t) = \cot(3t + \theta)$

Solution

(a) $x(t) = 5\sin\left(6t + \dfrac{\pi}{4}\right)$

$\omega_0 = 6$ rad/s, i.e., $T_0 = \dfrac{2\pi}{\omega_0} = \dfrac{2\pi}{6}$ sec

Hence, the signal is periodic with the fundamental period $T_0 = \dfrac{2\pi}{6}$ sec

(b) $x(t) = e^{j3t}$

$\omega_0 = 3$ rad/s, i.e., $T_0 = \dfrac{2\pi}{\omega_0} = \dfrac{2\pi}{3}$ sec

Hence, the signal is periodic with the fundamental period $T_0 = \dfrac{2\pi}{3}$ sec.

(c) $x(t) = \cot(3t + \theta)$

$x(t + T_0) = \cot\{3(t + T_0) + \theta\}$

The cotangent function repeats every π radian.

So, $3T_0 = \pi$

$$T_0 = \frac{\pi}{5} \text{ sec}$$

2. Prove that the following function $x(t) = t^6 + 2t^4 + 3t^2 + 4$ is even. Also show that the odd part is zero.

Solution

We have

$$x(-t) = (-t)^6 + 2(-t)^4 + 3(-t)^2 + 4$$

$$= x(t) = t^6 + 2t^4 + 3t^2 + 4$$

The function is even.

The odd part $= x_o(t) = \dfrac{1}{2}(x(t) - x(-t))$

$$= \frac{1}{2}(t^6 + 2t^4 + 3t^2 + 4 - t^6 - 2t^4 - 3t^2 - 4) = 0$$

3. Find the power and RMS value for the following signal:

$$x(t) = 5u(t)$$

Solution

(a) $x(t) = 5u(t)$

$$P = \lim_{T \to \infty} \frac{1}{2T} \int_{-t}^{T} 25 dt$$

$$= \lim_{T \to \infty} \frac{1}{2T} 25 \int_{0}^{T} dt = \lim_{T \to \infty} 25 \frac{T}{2T} = \frac{25}{2} = 12.5 \text{ watts}$$

RMS value of power $= \dfrac{5}{\sqrt{2}}$

4. Find the Fourier series in trigonometric form for the periodic function $x(t)$ $= e^{-t}$ with a period $T_o = 1$ second.

Solution

Since $T_o = 1$ second, $\omega_0 = \dfrac{2\pi}{T_o} = 2\pi$

$$a_0 = \frac{1}{T_0} = \int_{0}^{T_0} x(t) dt = \int_{0}^{1} e^{-t} dt = -[e^{-t}]_0^1 = 0.632$$

$$a_n = \frac{2}{T_0} \int_{0}^{T_0} x(t) \cos \omega_0 nt \, dt = 2\int_{0}^{1} e^{-t} \cos 2\pi nt \, dt$$

$$a_n = \frac{2}{(1+4\pi^2n^2)}\left[-\cos 2\pi nte^{-t} + e^{-t}2\pi n\sin 2\pi n\pi\right]_0^1$$

$$= \frac{2}{(1+4\pi^2n^2)}\left[e^{-t}(-\cos 2\pi nt + 2\pi n\sin 2\pi n\pi)+1\right]$$

$$= \frac{2}{(1+4\pi^2n^2)}(1-e^{-1}) = \frac{1.264}{(1+4\pi^2n^2)}$$

$$b_n = \frac{2}{T_0}\int_0^{T_0} x(t)\sin \omega_0 ntdt = 2\int_0^1 e^{-t}\sin 2\pi ntdt$$

$$= \frac{2}{(1+4\pi^2n^2)}\left[e^{-t}\{-\sin 2\pi nt - 2\pi n\cos 2\pi nt\}\right]_0^1$$

$$= \frac{2}{(1+4\pi^2n^2)}\left[-e^{-1}\{\sin 2\pi n - 2\pi n\cos 2\pi n + 2\pi n\}\right] = \frac{4\pi n}{(1+4\pi^2n^2)}(1-e^{-1})$$

$$\therefore b_n = \frac{2.53\pi n}{1+4\pi^2n^2}$$

Hence, $$x(t) = 0.632 + 1.264\sum_{n=1}^{\infty}\frac{n}{(1+4\pi^2n^2)}\cos 2\pi nt$$

$$+ 2.53\pi\sum_{n=1}^{\infty}\frac{n}{(1+4\pi^2n^2)}\sin 2\pi nt\ \sin 2\pi nt$$

5. Find the Fourier transform of the function $x(t) = 4t$.

Solution

$$X(j\omega) = \int_{-1}^{1} 4te^{-j\omega t}dt$$

Let $u = 4t$, $du = 4dt$, and $dv = \int e^{-j\omega t}dt; v = \frac{-1}{j\omega}e^{-j\omega t}$

$$X(j\omega) = uv - \int vdu$$

$$= \left[\frac{-4t}{j\omega}e^{-j\omega t}\right]_{-1}^{1} + j\frac{4}{\omega}\int_{-1}^{1}e^{-j\omega t}dt$$

$$= \left[\frac{-4t}{j\omega}e^{-j\omega t} + \frac{4}{\omega^2}e^{-j\omega t}\right]_{-1}^{1}$$

$$= 4\left[\frac{-e^{-j\omega}}{j\omega} + \frac{1}{\omega^2}e^{-j\omega} - \frac{1}{j\omega}e^{j\omega} - \frac{1}{\omega^2}e^{j\omega}\right]$$

$$= 4\left[-\frac{1}{j\omega}\left(e^{j\omega} + e^{-j\omega}\right) - \frac{1}{\omega^2}\left(e^{j\omega} - e^{-j\omega}\right)\right]$$

$$= 8\left[-\frac{1}{j\omega}\cos\omega + \frac{1}{j\omega}\frac{\sin\omega}{\omega}\right]$$

Hence, $X(j\omega) = \frac{8}{j\omega}(\sin\omega - \cos\omega)$

6. Find the Laplace transform of the function $x(t) = \dfrac{1}{2}at^2u(t)$, $t \geq 0$.

Solution

Taking the Laplace transform of the given function, we get

$$L\left[\frac{1}{2}at^2\right] = \int_0^\infty \frac{1}{2}at^2e^{-st}dt$$

Let $u = \dfrac{1}{2}at^2$ and $du = at$

$$dv = \int e^{-st}dt \text{ and } v = \frac{e^{-st}}{(-s)}$$

So, $\quad L\left[\dfrac{1}{2}at^2\right] = uv - \displaystyle\int_0^\infty vdu = \left[\dfrac{1}{2}at^2\dfrac{e^{-st}}{(-s)}\right] - \displaystyle\int_0^\infty \dfrac{ate^{-st}}{(-s)}dt$

$$= 0 + 0 + \frac{a}{s}\int_0^\infty te^{-st}dt$$

The integration in the right hand side of the above equation is nothing but a

ramp signal whose Laplace transform is $\dfrac{1}{s^2}$.

Hence, $L\left[\dfrac{1}{2}at^2\right] = \dfrac{a}{s^3}$

The region of convergence is the entire right half plane except the origin of the s-plane.

7. Find the inverse Laplace transform of $x(s) = \dfrac{10e^{-4s}}{(s+2)(s-2)}$.

Solution

Let $\quad X_1(s) = \dfrac{10}{(s-2)(s+2)}$

The function can be represented as

$$X_1(s) = \frac{A}{(s+2)} + \frac{B}{(s-2)}$$

Solving this partial fraction, we get the coefficients as $A = 2.5$ and $B = 2.5$.

$\therefore \quad X_1(s) = 2.5\left[\dfrac{1}{(s+2)} - \dfrac{1}{(s-2)}\right]$

Taking Inverse transform, we get

$$x_1(t) = 2.5\left(e^{-2t} - e^{2t}\right)u(t)$$

Since $X(s)$ is $X_1(s)$ multiplied by e^{-4s}, which results in time domain a shifting of 4,

$$x(t) = 2.5\left(e^{-2t} - e^{2t}\right)u(t-4)$$

8. Find $X[n]$ if $X[Z] = \dfrac{(7z - 23)}{(z - 3)(z - 4)}$.

Solution

Dividing both sides by z, we get

$$\frac{X(z)}{z} = \frac{(7z - 23)}{z(z - 3)(z - 4)} = \frac{A}{z} + \frac{B}{(z - 3)} + \frac{C}{(z - 4)}$$

or $(7z - 23) = A(z - 3)(z - 4) + Bz(z - 3)z - 4) + Cz(z - 3)$

Solving for A, B, and C, we get

$$A = -\frac{23}{12}, A = B = \frac{2}{3}, \text{ and } C = \frac{5}{4}$$

$$X(z) = -\frac{23}{12} + \frac{2}{3}\frac{z}{z - 3} + \frac{5}{4}\left(\frac{z}{z - 4}\right)$$

So, $X(z) = \left[-\dfrac{23}{12}\delta(n) + \dfrac{2}{3}(3)^n + \dfrac{5}{4(4)^n} \right]u(n)$

REVIEW QUESTIONS

1. Give the two advantages of the Laplace transform method of solving linear differential equations.
2. Define deterministic and random signals.
3. State the sampling theorem.
4. What is meant by an even signal?
5. State Parseval's relation.
6. List any two properties of a convolution integral.
7. What is the Fourier transform of a sequence $x(n)$?
8. State Parseval's relation for discrete time periodic signals.
9. What is the property of the Fourier spectrum of a discrete time aperiodic sequence?
10. Express the periodic sequence $x(n)$ with period N as discrete Fourier series.
11. What is the condition for the existence of the Fourier transform of a signal $x(t)$?
12. Find the inverse Fourier transform of $\delta(W)$.
13. Define stable and casual systems.
14. What are the classifications of systems?
15. Define energy and power signals.
16. What is aliasing error?

17. Define linear systems.
18. What is meant by aliasing?
19. What are the Dirichlet conditions under which a periodic signal can be represented by a Fourier series?
20. What is an anti-aliasing filter?
21. If $u(n)$ is the impulse of a system, what is its step response?
22. Define a linear time invariant system.
23. Verify linearity, causality, and time invariance of the system $y(n + 2) = ax(n + 1) + bx(n + 3)$.
24. Find the energy of the signal $X(t) = e^{-2t} u(t)$.
25. Explain casual, inverse, and linear systems through examples.
26. Explain the causality property of a continuous time system through an example.
27. Explain through examples power and energy signals.
28. Explain the following properties of a system: (i) memorylessness and (ii) non-linearity.
29. Write a short note on impulse function $\delta(t)$ and its properties.
30. Explain how the impulse response of a system explains everything about that system?

PROBLEMS

1. Plot the signal $u(t - \tau)$ against τ.
2. Plot the signal $u(t) - u(t - \tau)$, where τ is a positive constant.
3. Consider the periodic pulse train shown in Fig. 2.29 and find its Fourier coefficients. Hence represent it as (i) a trigonometric Fourier series and (ii) an exponential Fourier series.

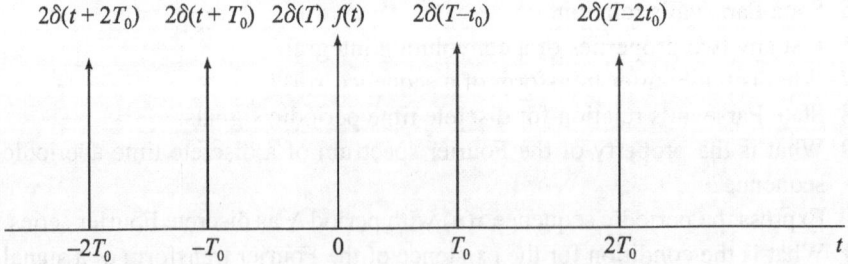

Fig. 2.29

4. Find the Fourier coefficients of the periodic square waveform shown in Fig. 2.30. Hence find its Fourier series representation.

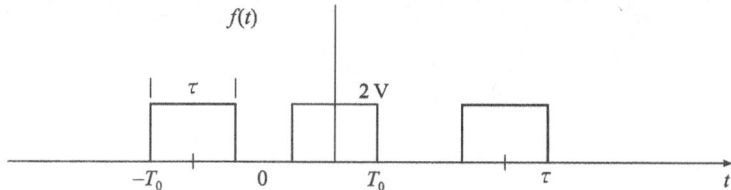

Fig. 2.30

5. A signal $m(t)$ is multiplied by a sinusoidal waveform f_c. The product signal is

$$f(t) = m(t) \sin 2\pi f_c t$$

If the Fourier transform of $m(t)$ is $M(f)$, i.e., $M(f) = \int_{-\infty}^{\infty} m(t)e^{-j2\pi ft} dt$, find the Fourier transform of $f(t)$.

6. A pulse of amplitude 5 V extends from $t = -\tau/2$ to $t = +\tau/2$. Find its Fourier transform $F(t)$. Consider also the Fourier series for a periodic sequence of such pulses separated by intervals T_0. Compare the Fourier series coefficients A_n with the transform in the limit $T_0 \to \infty$.

7. Find the Laplace transform of $t^n u(t)$.

8. Find the Laplace transform of $\cos bt \, u(t)$.

9. Find the z-transform of (i) $n^2 u(n)$ and (ii) $\cos \beta n \, u(n)$.

10. Show that $(t^3 + 2)\, \delta(t) = 2\delta(t)$.

11. Show that $\left[\sin\left(t^2 - \dfrac{\pi}{4} \right) \right] \delta(t) = -\dfrac{1}{\sqrt{2}} \delta(t)$.

12. Find the Fourier transform of the signal $f(t) = e^{-at}\, u(t)$, where $u(t)$ is a unit-step function.

13. Find the Fourier transform of the signal $\operatorname{sgn}(t)$.

14. Compute the convolution integral $e^{-2t}u(t)*(1-e^{-t})u(t)$.

15. Show that the Fourier transform of $\dfrac{1}{2}\delta(t+\dfrac{1}{2}) + \dfrac{1}{2}\delta(t-\dfrac{1}{2})$ is $\cos(\pi f)$.

Answers to problems

1.

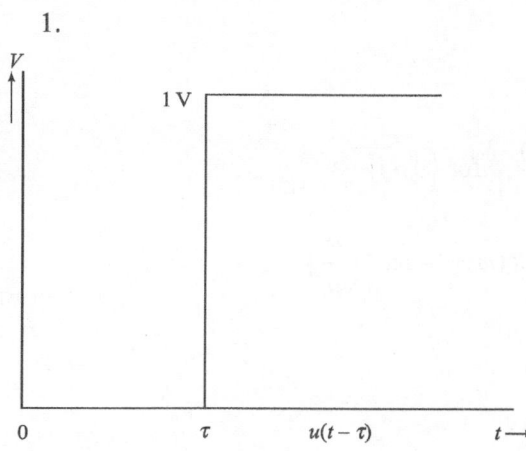

2.

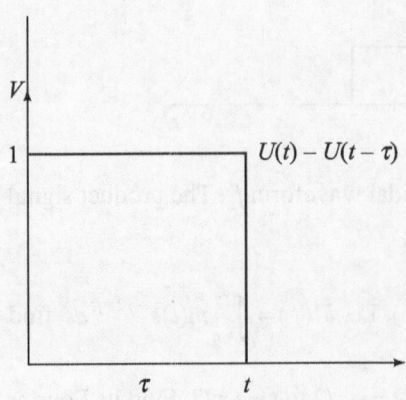

$U(t) - U(t - \tau)$

3. $f(t) = \dfrac{2}{T_0} + \dfrac{4}{T_0} \displaystyle\sum_{n=1}^{\infty} \cos \dfrac{2\pi n t}{T_0}$ and $\dfrac{2}{T_0} \displaystyle\sum_{n=-\infty}^{\infty} e^{\frac{j2\pi n t}{\omega T_0}}$

4. $A_0 = \dfrac{2\tau}{T_0}$

$A_n = \dfrac{4\tau}{T_0}\left[\dfrac{\sin(n\pi\tau / T_0)}{n\pi\tau / T_0}\right], \theta_n = B_n = 0$

$f(t) = \dfrac{2\tau}{T_0} + \dfrac{4\tau}{T_0} \displaystyle\sum_{n=1}^{\infty} \dfrac{\sin(n\pi\tau / T_0)}{n\pi\tau / T_0} \cos \dfrac{2\pi n t}{T_0}$

5. $F(f) = \dfrac{1}{2}M(f - f_c) - \dfrac{1}{2}M(f + f_c)$

6. $F(f) = 5\tau \dfrac{\sin \pi f \tau}{\pi f \tau}$, $A_n = \dfrac{5\tau}{T_0} \dfrac{\sin(\tau f \tau / T_0)}{nf\tau / T_0}$

$\displaystyle\lim_{\Delta f \to 0} A_n = 5\tau \dfrac{\sin \pi f t}{\pi f t}$

7. $\dfrac{\lfloor n}{s^{n+1}}$

8. $\dfrac{s}{s^2 + b^2}$

9. $\dfrac{z(z+1)}{(z-1)^3}$ and $\dfrac{z(z - \cos \beta)}{z^2 - 2z\cos\beta + 1}$ for $|z| > 1$

12. $|X(\omega)| = \dfrac{1}{\sqrt{a^2 + \omega^2}}$ and $\angle X(\omega) = -\tan^{-1}\left(\dfrac{\omega}{a}\right)$

13. $\dfrac{1}{j\pi f}$

14. $\left(\dfrac{1}{2} - e^{-t} + \dfrac{1}{2}e^{-2t}\right) u(t)$

MATLAB EXAMPLES

1.
```
%program to generate sine wave%
clc;
clear all;
close all;
f=20;
t1=1/f;
t2=2*t1;
t=0:0.000001:t2;
y=sin(2*pi*f*t);
plot(t,y);
xlabel('time in seconds');
ylabel('amplitude in volts');
title('SINE WAVE');
grid on;
```

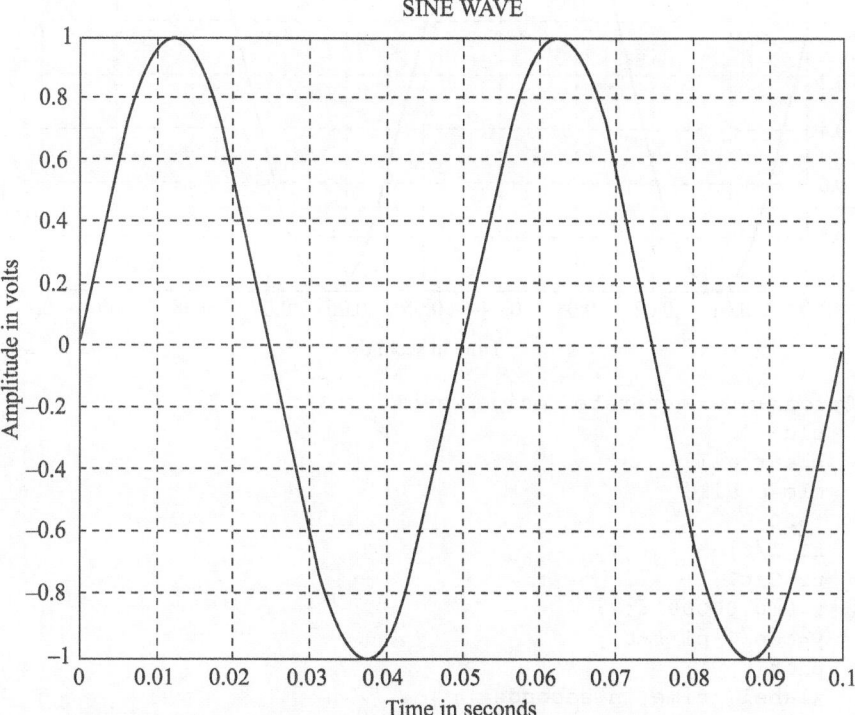

2.
```
%program to generate sine wave with phase shift%
clc;
clear all;
close all;
f=20;
m=180;
n=((pi*m)/180);
t1=1/f;
t2=2*t1;
```

```
t=0:0.000001:t2;
y=sin((2*pi*f*t)+n);
plot(t,y);
xlabel('time in seconds');
ylabel('amplitude in volts');
title('SINE WAVE WITH PHASE SHIFT');
grid on;
```

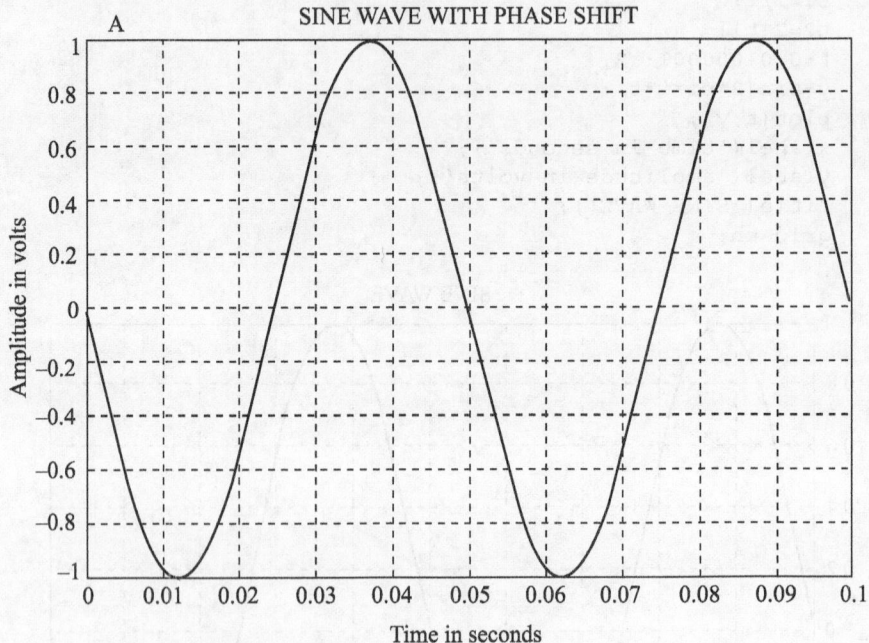

3. ```
   %program to generate cosine wave%
   clc;
   clear all;
   close all;
   f=20;
   t1=1/f;
   t2=2*t1;
   t=0:0.000001:t2;
   y=cos(2*pi*f*t);
   plot(t,y);
   xlabel('time in seconds');
   ylabel('amplitude in volts');
   title('COSINE WAVE');
   grid on;
   ```
   ```

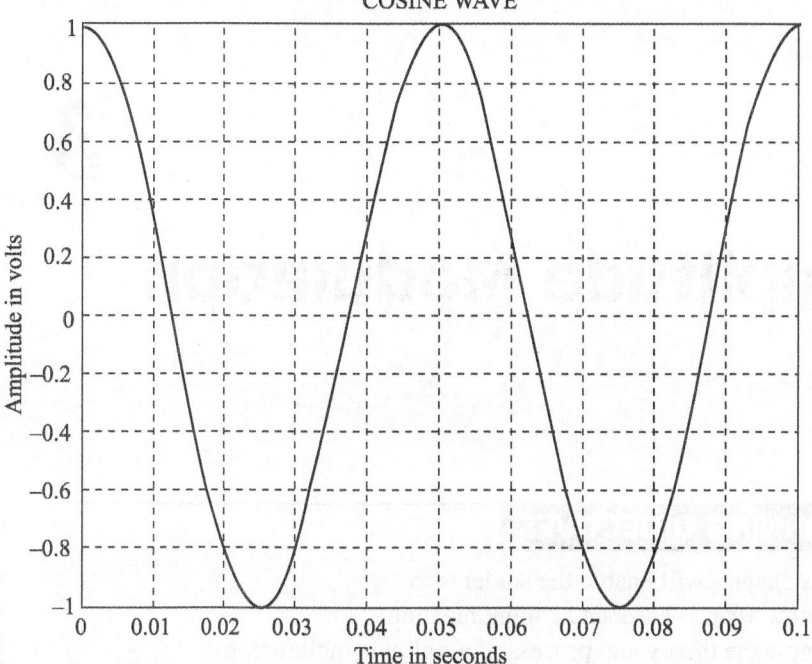

Amplitude Modulation

LEARNING OBJECTIVES

This chapter will enable the students to
- know what is baseband communication
- know the theory and process of amplitude modulation
- observe the frequency spectrum of sinusoidal amplitude modulation (AM)
- compute the modulation index and percentage of modulation
- understand the difference between overmodulation and undermodulation
- calculate the power distribution between the carrier and sidebands
- appreciate that AM does not affect the carrier signal
- know about double sideband (DSB) and its advantage over AM double sideband with full carrier (DSBFC)
- know the concept of single sideband (SSB), its types, and power saving due to SSB transmission
- know about independent sideband (ISB) amplitude modulation, vestigial sideband (VSB), and quadrature amplitude modulation (QAM)
- study various types of AM modulations
- know different types of SSB and ISB generation
- study various types of AM demodulation, SSB reception, demodulation of VSB signals, and detection of ISB signals
- know AM transmitters and receivers, SSB transmitters, and receivers.
- solve some numerical problems, which will make them to understand the theoretical concepts better

3.1 BASEBAND COMMUNICATION

Before we describe amplitude modulation (AM), let us briefly discuss transmission without any modulation. This type of communication is called *baseband communication*. In this method, the information after being converted into electrical signals is transmitted as it is. The term *baseband* is used to indicate the band of frequencies of the signal delivered by the source or input transducer. For example, in TV signals the baseband occupies 0 to 4.3 MHz, for voice band it occupies 0 to 3 KHz, and for digital data using bipolar signalling, the baseband is 0 to N Hz, where N is the bipolar signalling rate at N pulses per second. Baseband signals, which are transmitted without modulation, are generally sent over a pair of wires, coaxial cables, or optical fibres. Examples of baseband communication are local telephone communication, short hand pulse code modulation between exchanges, and long-distance PCM over optical fibres.

3.2 THEORY OF AM

The process of changing the amplitude of a carrier signal in proportion with the instantaneous value of the modulating signal is called *amplitude modulation* (AM). In AM, the carrier frequency should be relatively higher than the modulating frequency. This type of modulation is generally used in commercial broadcasting of both video and audio information. It also finds applications in two-way mobile radio communications such as in Citizens' Band radio.

AM modulators are non-linear devices having two inputs and one output. One of the inputs is the high-frequency carrier and the other is the low-frequency information signal, which can be of a single frequency or a band of frequencies as in the case of speech signals.

In the case of speech signals, this band of frequencies lies from 300 Hz to 3 kHz.

The AM envelope

The modulated output waveform for an AM modulator is called an AM envelope. A conventional AM waveform is shown in Fig. 3.1.

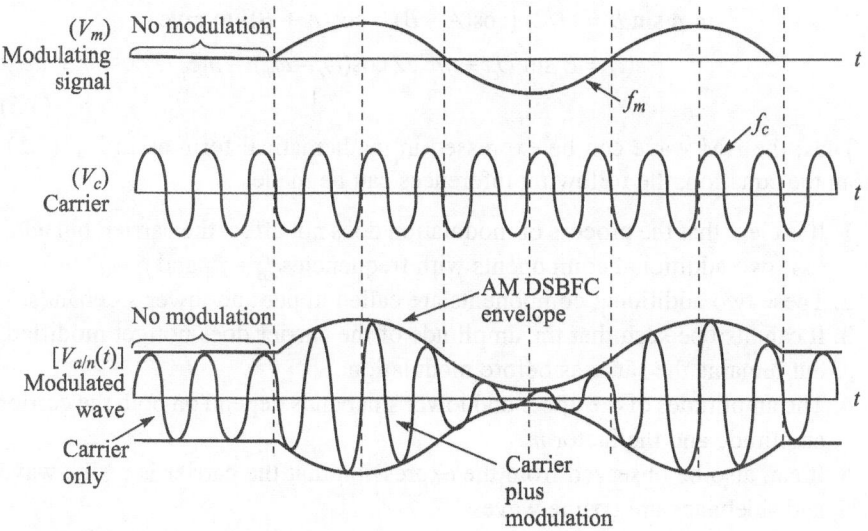

Fig. 3.1 Conventional AM waveform

3.3 FREQUENCY SPECTRUM OF SINUSOIDAL AM

Since an AM modulator is a sinusoidal device, the output is a complete wave and will be shown later to consist of the carrier frequency f_c and the sum $(f_c + f_m)$ and difference $(f_c - f_m)$ of the frequencies. The sum and carrier frequencies are displaced from the carrier by an amount equal to the modulating frequency. This can be shown mathematically as follows.

Let the carrier frequency be $E_c \sin \omega_c t$ and the modulator frequency be $E_m \sin \omega_m t$.

Observe that the fixed phase ϕ_c of the carrier and ϕ_m of the modulating signal are not brought into the picture since the modulation results are independent of these phase angles. By the definition of amplitude modulation, the equation for the sinusoidally modulated wave is

$$e(t) = [\, E_c + E_m \sin \omega_m t\,]\, \sin \omega_c t \tag{3.1}$$

$$= E_c \sin \omega_c t + E_m \sin \omega_m t \sin \omega_c t \tag{3.2}$$

Equation (3.2) can be rearranged as

$$e(t) = E_c \left[1 + \frac{E_m}{E_C} \sin \omega_m t \right] \sin \omega_c t \tag{3.3}$$

Let $$\frac{E_m}{E_c} = m$$

Then $$e(t) = E_c(1 + m \sin \omega_m t) \sin \omega_c t$$

$$= E_c \sin \omega_c t + m E_c \sin \omega_m t \sin \omega_c t \tag{3.4}$$

The term $m \sin \omega_m t \sin \omega_c t$ can be expanded in terms of the trigonometric identity

$$\sin A \sin B = (1/2)\,[\cos(A - B) - \cos(A + B)] \text{ to give}$$

$$e(t) = E_c \sin \omega_c t + m E_c/2 \cos(\omega_c - \omega_m)t - m E_c/2 \cos(\omega_c + \omega_m)t \tag{3.5}$$

Thus, the AM wave can be expressed in mathematical form as in Eq. (3.5). From the equation, the following inferences can be made.

1. It is clear that the process of modulation does not affect the carrier, but adds to it two additional components with frequencies $f_c + f_m$ and $f_c - f_m$.
2. These two additional components are called upper and lower sidebands.
3. It can also be seen that the amplitude of the carrier does not get modified, but remains the same as before modulation.
4. The amplitudes of the upper and lower sidebands depend on both the carrier amplitude and the factor m.
5. It can also be observed from the expression that the carrier is a sine wave and sidebands are cosine waves.
6. At the beginning of each cycle of the envelope, the carrier is 90° out of phase with both sidebands and they themselves are 180° out of phase with each other.

These phase relationships are shown in Fig. 3.2.

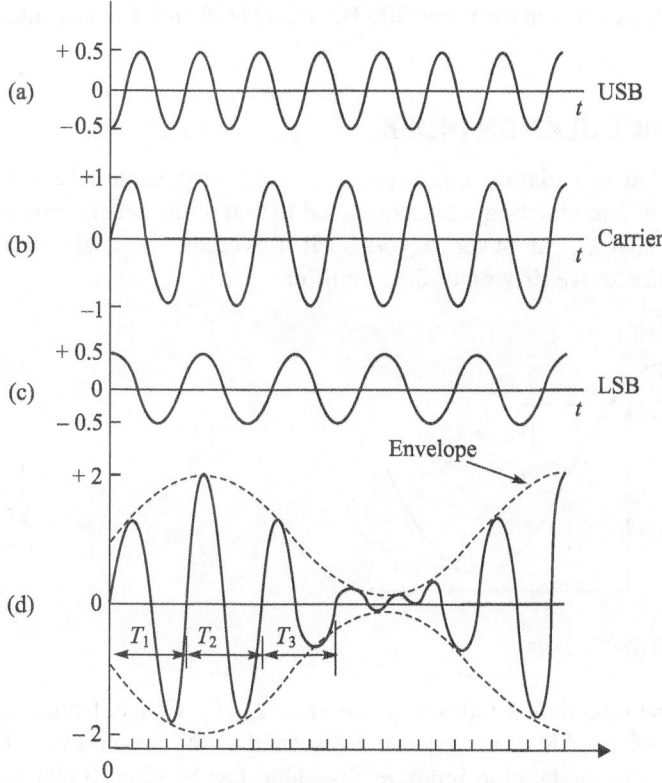

Fig. 3.2 Phase relationship between carrier, USB, and LSB

Frequency spectrum of AM wave

From the expression for an AM wave given by Eq. (3.5), we can plot its frequency spectrum. It can be seen from Fig. 3.3 that the two sidebands are equally spaced on both sides of the carrier. While the carrier amplitude remains the same, the amplitudes of the sideband depend on the value of m. When $m = 1$, the amplitudes of the sidebands equal $E_c/2$, i.e., half that of the carrier. The bandwidth required for AM is twice that of the modulating frequency.

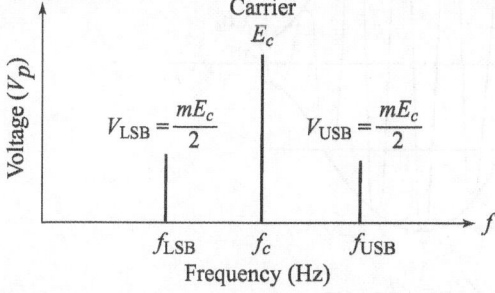

Fig. 3.3 Voltage spectrum: AM DSBFC wave

If the modulation is due to several sine waves as in the case of speech signals, the bandwidth will then be equal to twice the highest modulating frequency. For example, if the speech signal is in the range 300 Hz to 3 kHz, then the bandwidth required will be 6 kHz.

3.4 AMPLITUDE MODULATION INDEX

Before we define what modulation index is, let us see what should be the maximum amplitude of a modulating wave compared to that of the carrier. From Fig. 3.4, it is evident that E_m can never exceed E_c. If it exceeds, there will be a distortion, which is due to overdriving of the amplifier.

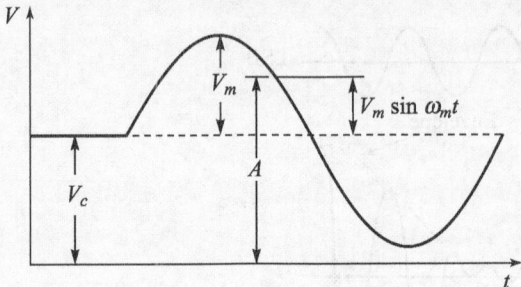

Fig. 3.4 Amplitude of AM DSBFC wave

Thus, we define the modulation index m as the ratio E_m/E_c. Hence, from the frequency spectrum of an AM wave, it can be seen that the amplitudes of sidebands depend on the modulation index m. Its value lies between 0 and 1. When $m = 0$, there is no modulation and only the carrier will be present. When $m = 1$, full modulation takes place. If $m > 1$, then it is called overmodulated. Normally, the modulation index m is expressed as a percentage. Thus, 0% represents no modulation and 100% represents full modulation. We can calculate the modulation index from the waveform as shown in Fig. 3.5.

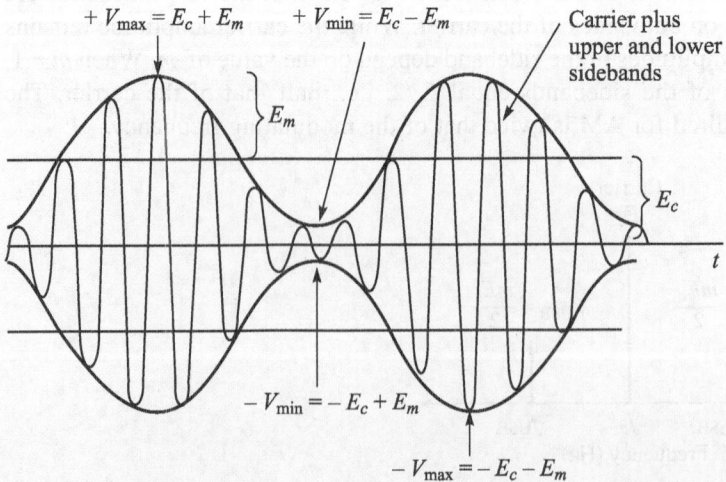

Fig. 3.5 Modulation index

The figure shows an AM signal for a pure sine wave modulated with a single-frequency sine wave. The modulation is symmetric, i.e., both positive and negative excursions of the envelope's amplitude are equal. Since by definition $m = E_m/E_c$, from the figure, it can be seen that

$$E_m = (1/2)(V_{\max} - V_{\min}) \qquad (3.6)$$

$$E_c = (1/2)(V_{\max} + V_{\min}) \qquad (3.7)$$

Dividing Eq. (3.6) by (3.7), we get

$$m = \frac{V_{\max} - V_{\min}}{V_{\max} + V_{\min}} \qquad (3.8)$$

Percentage of modulation m is given by

$$M = m \times 100\% = \frac{V_{\max} - V_{\min}}{V_{\max} + V_{\min}} \times 100\% \qquad (3.9)$$

From the above equation, it can be seen that for 100% modulation, $V_{\min} = 0$ or $E_m = E_c$ and for 50% modulation, $V_{\min} = V_{\max}/3$ or $E_m = E_c/2$. Figure 3.6 shows these cases.

(a)

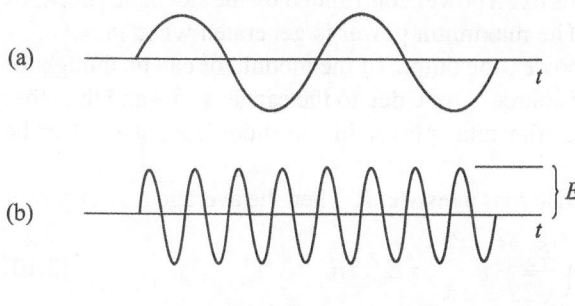

(b)

(c)

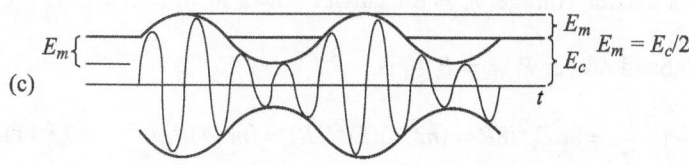

(d)

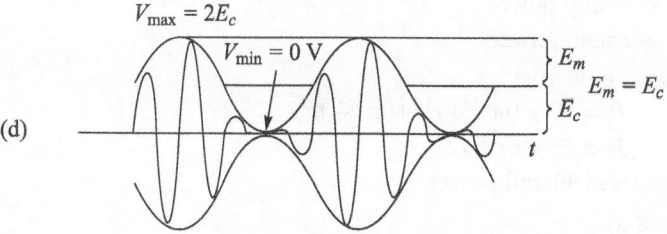

Fig. 3.6 Percent modulation of AM DSBFC envelope: (a) modulating signal, (b) carrier, (c) 50% modulated wave, and (d) 100% modulated wave

Example 3.1 A carrier frequency of 5 MHz and peak value of 5 V is amplitude modulated by a 4 kHz sine wave of amplitude 3 V. Determine the modulation index, the upper and lower sideband frequencies, and their amplitudes.

Solution

Modulation index, $\quad m = \dfrac{3\,V}{5\,V} = 0.6$

     Upper sideband frequency = 5.004 MHz
     Lower sideband frequency = 4.996 MHz

     Amplitude of each sideband = $0.6 \times \dfrac{5\,V}{2} = 1.5\,V$

3.5 AVERAGE POWER FOR SINUSOIDAL AM

The total average power in an amplitude modulated wave is the contribution due to the carrier and the two sidebands. Since the carrier amplitude does not get affected due to modulation, the power due to the carrier remains the same. However, the addition of two sidebands to the carrier results in an extra energy, and hence more power. But this extra power contributed by the sidebands depends on the modulation index m. The maximum power is generated when $m = 1$.

To find the total average power, the output of the modulator can be thought of as consisting of three voltage sources—one due to the carrier wave and the other due to the sidebands. Hence, the total power in the modulated wave can be calculated as follows.

Let the power be developed across a resister R. Then the average carrier power

$$P_c = \left(\frac{E_c}{\sqrt{2}}\right)^2 \times \frac{1}{R} = E_c^2/2R \tag{3.10}$$

where E_c is the peak carrier voltage, P_c is the carrier power in watt, and R is the load resistance in Ω.

The average sideband power $P_{LSB} = P_{USB}$ is

$$\left(\frac{(m/2)E_c}{\sqrt{2}}\right)^2 \frac{1}{R} = m^2 E_c^2/8R = (m^2/4)(E_c^2/2R) = (m^2/4)P_c \tag{3.11}$$

where P_{USB} = upper sideband power

     P_{LSB} = lower sideband power

Hence, the total average power is

$$P_t = P_c + (m^2/4)\,P_c + (m^2/4\,)P_c$$

or $\qquad\qquad\qquad P_t = P_c + m^2 P_c/2 \tag{3.12}$

where $m^2 P_c/2$ is the total sideband power.

P_t can be simplified as

$$P_t = P_c(1 + m^2/2) \tag{3.13}$$

Figure 3.7 shows the power spectrum for an AM wave. The figure shows that the carrier power remains the same with or without modulation, but the sideband power depends on m, the modulation index. The maximum sideband power $P_c/4$ occurs when $m = 1$. This is only 25% of the carrier power. This is one of the main disadvantages of AM, since most of the power is wasted in the carrier. Hence, it is always recommended to maintain m between 0.9 and 0.95 to have maximum power in the sidebands.

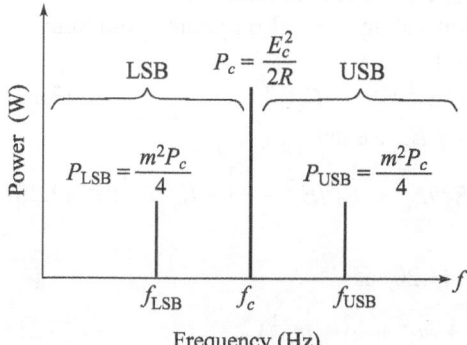

Fig. 3.7 Power spectrum: modulating signal with single frequency

Voltage and effective current

The effective or rms voltage E of the modulated wave is given by the equation

$$P_t = E^2/R \tag{3.14}$$

The effective or rms voltage E_c of the carrier wave is given by

$$P_c = E_c^2/R \tag{3.15}$$

Substitute $P_t = P_c (1 + m^2/2)$ in Eq. (3.14). Then

$$E^2/R = P_c (1 + m^2/2)$$

Now substituting the value of P_c from Eq. (3.15), we get

$$E^2/R = E_c^2/R(1 + m^2/2)$$

Therefore,
$$E = E_c (1 + m^2)^{1/2} \tag{3.16}$$

Similarly, we can find the effective current as follows:

$$\frac{P_t}{P_c} = I_t^2 R/ I_c^2 R = (I_t/I_c)^2 \tag{3.17}$$

Substituting the value of $P_t/P_c = 1 + m^2/2$ in Eq. (3.17), we get

$$I_t = I_c (1 + m^2)^{1/2} \tag{3.18}$$

Thus, by measuring the current or voltage with and without modulation, one can calculate the modulation index m.

Example 3.2 A 300 W carrier is modulated to a depth of 60%. Determine the total power in the modulated wave.

Solution

$$P_t = P_c \left(1 + \frac{m^2}{2} \right) = 300 \left(1 + \frac{0.6^2}{2} \right) = 354 \text{ W}$$

3.6 MODULATION BY SEVERAL SINE WAVES

The modulating wave will not have a single frequency, e.g., a speech signal contains a frequency in the range of 300 Hz to 3 kHz. When these signals modulate the carrier, the resulting power can be calculated as follows.

Let $E_1, E_2, E_3, \cdots, E_n$ be the modulation voltages of all frequency components, then the total modulating voltage E_t will be

$$E_t = (E_1^2 + E_2^2 + E_3^2 + \cdots + E_n^2)^{1/2} \tag{3.19}$$

Dividing both sides of the equation by E_c, we get

$$\frac{E_t}{E_c} = (E_1^2/E_c^2 + E_2^2/E_c^2 + E_3^2/E_c^2 + \cdots + E_n^2/E_c^2)^{1/2} \tag{3.20}$$

Since $\dfrac{E_m}{E_c} = m$, we can simplify Eq. (3.20) as

$$m_t = (m_1^2 + m_2^2 + m_3^2 + \cdots + m_n^2) \tag{3.21}$$

Then the total modulating power is

$$P_t = P_c \left[1 + 1/2(m_1^2 + m_2^2 + m_3^2 + \cdots + m_n^2)\right] \tag{3.22}$$

Hence, we can conclude that in the case of modulation by multiple frequencies, the carrier power is unaffected, but the total sideband power equals the sum of the individual sideband powers. However, we must ensure that the total modulation index should not exceed 1 to avoid distortion. The combined coefficient of modulation can be used to determine the total sideband and transmitted power as follows.

$$P_{\text{USB}t} = P_{\text{LSB}t} = P_t\, m_t^2/4 \tag{3.23}$$

and the total sideband power

$$P_{\text{SB}t} = P_t\, m_t^2/2 \tag{3.24}$$

where $P_{\text{USB}t}$ = total upper sideband power in watts
$P_{\text{LSB}t}$ = total lower sideband power in watts
$P_{\text{SB}t}$ = total sideband power in watts
P_t = total transmitted power in watts

Example 3.3 A transmitter radiates 10 kW with the carrier unmodulated and 11.5 kW when it is sinusoidally modulated. Calculate the modulation index. If another sine wave resulting in 50% modulation is transmitted simultaneously, find the total radiated power.

Solution

(a) Since $P_t = P_c \left(1 + \dfrac{m^2}{2}\right)$,

$$11.5\ \text{kW} = 10\ \text{kW}\left(1 + \frac{m^2}{2}\right)$$

or $1 + \dfrac{m^2}{2} = \dfrac{11.5}{10} = 1.15$

or $\dfrac{m^2}{2} = 0.15,$

or $m^2 = 0.30$

So, $m = \sqrt{0.30} = 0.54$

(b) $m_t = \sqrt{m_1^2 + m_2^2} = \sqrt{0.54^2 + 0.50^2} = 0.74$

With this, the new modulation index, $P_t = P_c\left(1 + \dfrac{m_t^2}{2}\right) = 10\left(1 + \dfrac{0.74^2}{2}\right)$

$$= 100\,(1 + 0.275)$$
$$= 12.75 \text{ kW}$$

Example 3.4 An AM transmitter with an unmodulated carrier power of 150 W is modulated simultaneously by four modulating signals with modulation indices 0.3, 0.4, 0.5, and 0.6, respectively. Find

(a) total modulation index,
(b) total transmitted power, and
(c) each sideband power.

Solution

(a) Total modulation index $= \sqrt{0.3^2 + 0.4^2 + 0.5^2 + 0.6^2} = \sqrt{0.86} = 0.92$

(b) Total transmitted power $= 150\left(1 + \dfrac{m^2}{2}\right) = 150(1 + 0.43) = 214.5\,\text{W}$

(c) Sideband power $= (0.92^2) \times \dfrac{150}{4} = 31.74 \text{ W}$

3.7 DOUBLE SIDEBAND SUPPRESSED CARRIER (DSBSC)

In this method, only the two sidebands, upper and lower, are transmitted. Thus, saving the power to the extent of the carrier power, DSBSC signal is obtained by multiplying the modulating signal with the carrier. Let the modulating signal be $E_m \sin \omega_m t$. The carrier signal be $E_c \sin \omega_c t$.

Multiplying both the signals, we get

$$y(t) = E_m \sin\omega_m t\, E_c \sin\omega_c t$$

$$= \frac{E_m E_c}{2}\, [(\cos(\omega_c + \omega_m)t - \cos(\omega_c + \omega_m)t]$$

$$= K[(\cos(\omega_c - \omega_m)t - \cos(\omega_c + \omega_m)t] \qquad (3.25)$$

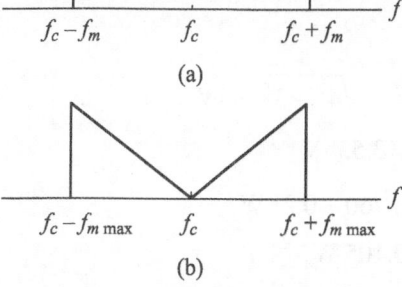

Fig. 3.8 Spectrum for DSBSC: (a) sinusoidal modulation and (b) general case

Equation (3.25) shows that there is no carrier component present, and it will be absent in the frequency spectrum. Thus, this type of amplitude modulation is called double sideband suppressed carrier (DSBSC). The absence of a carrier results in utilization of transmitter power efficiently than standard AM. But the bandwidth requirement remains the same. The suppressed carrier DSB frequency spectrum is shown in Fig. 3.8.

3.8 SINGLE SIDEBAND (SSB) SYSTEMS

The two AM techniques AMDSBFC and DSBSC have the following drawbacks. In the conventional AMDSBFC, the carrier power constitutes nearly 2/3 or more of the total transmitted power. Since the carrier contains no information, this amount of power is actually wasted. Moreover, since the information is contained only in the sidebands, it is enough to transmit only sidebands. Since both sidebands contain the same information, the question of redundancy arises in the case of DSBSC. Hence, just transmitting only a simple sideband saves power and bandwidth. There are several types of single sideband system. Each has its own merits and demerits.

3.8.1 Single Sideband with Carrier

In this system, the carrier is transmitted with full power along with only one of the sidebands. This transmission requires only half the bandwidth compared to a conventional AM system. The frequency spectrum and relative power distribution are shown in Fig. 3.9(b). It can be noted that for 100% modulation, the carrier power constitutes 80% of the total transmitted power P_t and only 20% of the total power is in the sideband. This, when compared to the conventional AMDSBFC brings out the fact that a single sideband with a carrier utilizes lesser power for information carrying part of the signal. Figure 3.10 shows the modulating signal, AM wave, suppressed carrier wave (DSB), and suppressed carrier SSB waveforms.

Example 3.5 If a carrier of 400 kHz is modulated with two frequencies 2.5 kHz and 5 kHz, find for a suppressed carrier SSB transmission the output frequency spectrum if only the lower sideband is transmitted. If E_1 and E_2 are the peak voltages of the modulating signal and are 4 V and 3 V, respectively, find the peak envelope power and the average output power across a load resistor of 60 Ω.

Solution

Two LSB frequencies are $f_{LSB1} = 395$ kHz
$$f_{LSB2} = 397.5 \text{ kHz}$$

$$E_t = \sqrt{E_1^2 + E_2^2} = \sqrt{4^2 + 3^2} = 5\,\text{V}$$

$$E_{rms} = 5 \times 0.707 = 3.535\,\text{V}$$

Peak envelope power, PEP $= (5 \times 0.707)^2/60 = 0.21$ W

Average power $=$ PEP $/2 = 0.105$ W

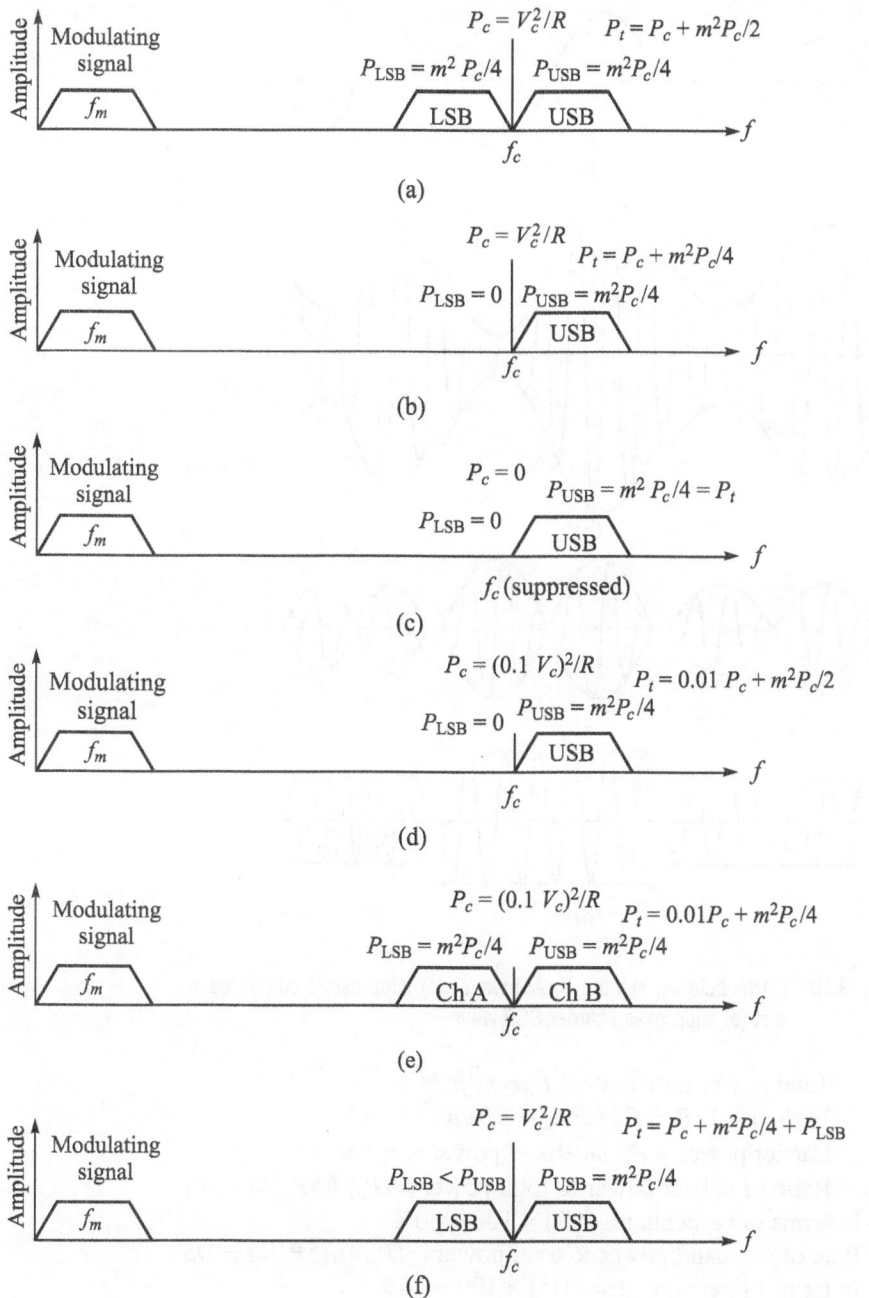

Fig. 3.9 Power spectrum for various systems: (a) DSBFC, (b) SSB with carrier, (c) SSB suppressed carrier, (d) SSB with reduced carrier, (e) independent sideband, and (f) VSB

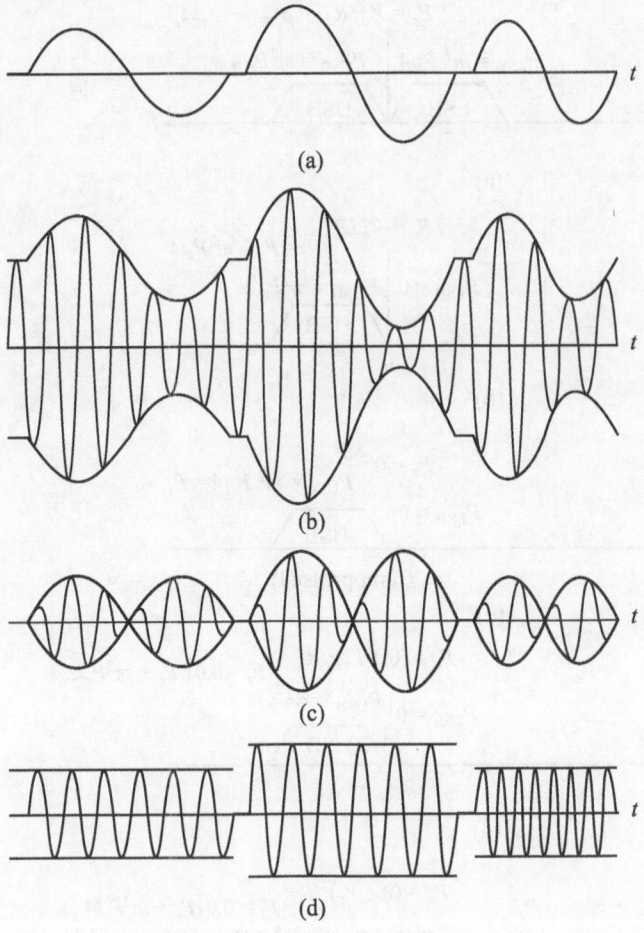

Fig. 3.10 (a) Modulating signal, (b) AM wave, (c) suppressed carrier wave, and (d) suppressed carrier SSB wave

Total power transmitted: $P_c = m^2 p_c/4$
With $m = 1$, $P_t = P_c + P_c/4 = 5P_c/4$
Carrier power $= P_c$, sideband power $= P_c / 4$
Ratio of carrier power to total power $= (P_c) /(5P_c/4) = 4/5$
In terms of percentage, $(4/5) \times 100 = 80\%$
Rate of sideband power to total power $= (P_c/4)/(5P_c/4) = 1/5$
In terms of percentage $= (1/5) \times 100 = 20\%$
Figure 3.11 shows the waveform for 100% modulated SSBFC with a single-frequency modulated signal. The envelope looks like a 50% double modulated sideband full carrier envelope. The maximum positive and negative peaks of an AMDSBFC waveform occur when the carrier and both sidebands reach their respective peaks at the same time and the peak change in the envelope is equal to the sum of the amplitudes of the upper and lower side frequencies.

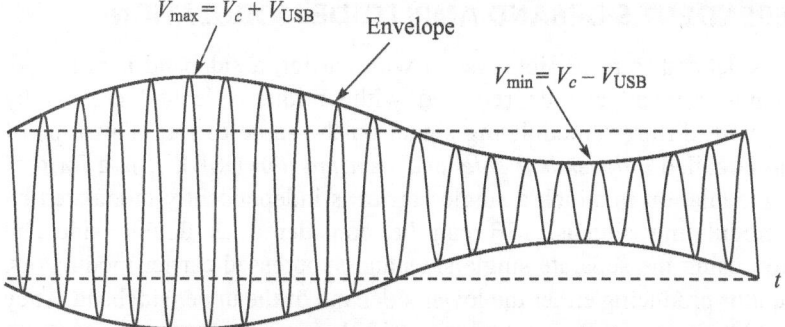

Fig. 3.11 100% modulation: SSBFC waveform

For a single sideband transmission, there is only one sideband, to add to the carrier; hence, the peak change in the envelope is only half of what it is for double side transmission.

The repetition rate of the envelope is equal to the frequency of the modulating signal and the depth of modulation is dependent on the amplitude of the modulating signal.

3.8.2 Single Sideband with Suppressed Carrier

Here the carrier is totally suppressed and only one of the sidebands is transmitted. This method also requires less bandwidth and less transmitted power. Also the sideband power is 100% of the total transmitted power. From Fig. 3.12, it can be seen that the waveform is not an envelope but a sine wave with a single frequency equal to the carrier frequency plus the modulating signal frequency or the carrier frequency minus the modulating signal frequency depending on which side the band is transmitted.

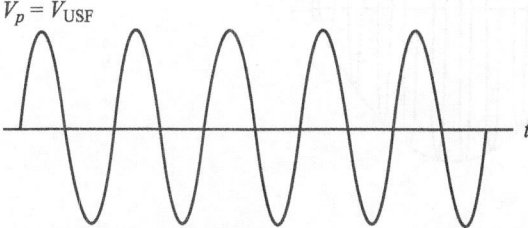

Fig. 3.12 SSBSC waveform

3.8.3 Single Sideband with Reduced Carrier

In this form of modulation, along with one of the sidebands, a reduced carrier of about 10% of its unmodulated amplitude is sent. Hence, the sideband has nearly 96% of the total power transmitted. While modulating, the carrier is totally suppressed and then reinserted. This carrier is called *pilot carrier*, which will be used at the receiver end for demodulation.

Figure 3.9(d) shows that the power in the sideband is nearly 100%. At the receiver end the reduced carrier transmitted is separated, amplified, and reinserted at a higher level. This reduced carrier transmission is called *escalated carrier* since the carrier is elevated in the receiver before demodulation. The carrier should be elevated so that its level is greater than that of the sideband signal. This type of modulation requires half as much bandwidth compared to the conventional AM. The power conservation results in this type of modulation due to the reduced carrier level in transmission.

3.9 INDEPENDENT SIDEBAND AMPLITUDE MODULATION

In a single sideband transmission system with carrier, a sideband is removed. The sideband removed can be replaced with another sideband created by modulating with different modulating signal on the same carrier. This type of modulation is called *independent sideband transmission* (ISB). It is a form of amplitude modulation in which a single carrier is independently modulated by different modulating signals. ISB can be considered as double sideband transmission using the separate single sideband suppressed carrier modulators, each modulator producing either the lower sideband or the upper sideband. They are combined to form a DSB signal. Figure 3.13 shows the transmitted waveform for two independent single frequency modulating signals (f_{m1} and f_{m2}) and Fig. 3.9(e) shows the frequency spectrum and power distribution for ISB. It can be seen that the waveform is similar to that of a double sideband suppressed carrier waveform except for a repetition rate equal to twice the modulating signal frequency. Since two information are transmitted with the same carrier, ISB conserves both transmit power, and bandwidth ISB is generally used for stereo AM transmission in which one channel is transmitted in the lower sideband, and the other channel in the upper sideband.

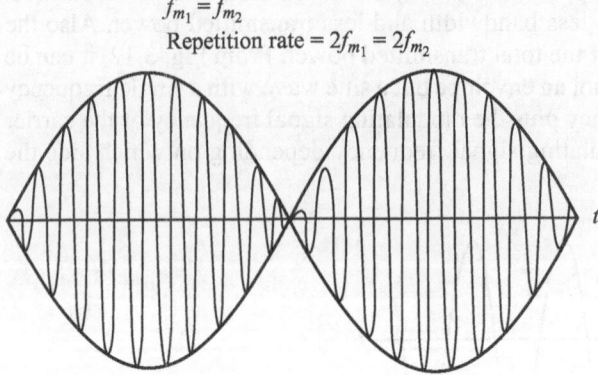

Fig. 3.13   ISB waveform

3.10 COMPARISON OF SSB AND AM

It is observed that lesser bandwidth and power efficiency of SSB suppressed carrier and reduced carrier transmission score over conventional DSBFC transmission. SSB transmission requires only half as much bandwidth as a double sideband. Suppressed and reduced carrier transmissions require less total transmitted power than full carrier AM. Figure 3.14 shows the waveform produced for a given modulating signal for DSBFC, DSBSC, and SSBSC. From the figure it is clear that the repetition rate of the DSBFC envelope is equal to the modulating signal frequency, the repetition of DSBSC envelope is equal to twice the modulating frequency, and the SSBSC waveform is a single frequency, equal to either the USB or LSB frequency.

As far as the power is concerned, the SSBSC uses only one-third of the power compared to 1.5 units of power in the case of DSBFC. From Fig. 3.15, we observe that it requires 0.5 units of voltage per sideband and 1 unit for the carrier with conventional AM for a total of 2 PEV (peak envelope volt) and only 0.707 PEV for a single sideband. The peak envelope power (PEP) is also seen.

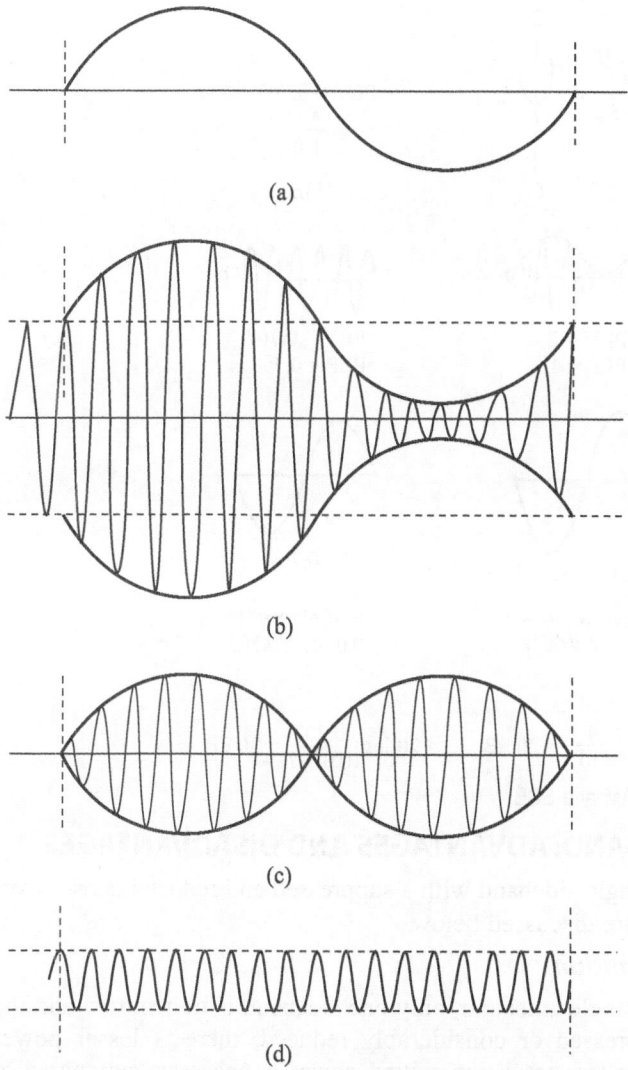

Fig. 3.14 AM transmission: (a) modulating signal, (b) DSBFC wave, (c) DSBSC wave, and (d) SSBSC wave

The demodulated signal at the output for DSBSC is proportional to the quadratic sum of voltages from the upper and lower sidebands, which is 1 PEV unit. For a single sideband reception, the demodulated signal is $0.707 \times 1 = 0.707$ PEV and the PEP is 0.5.

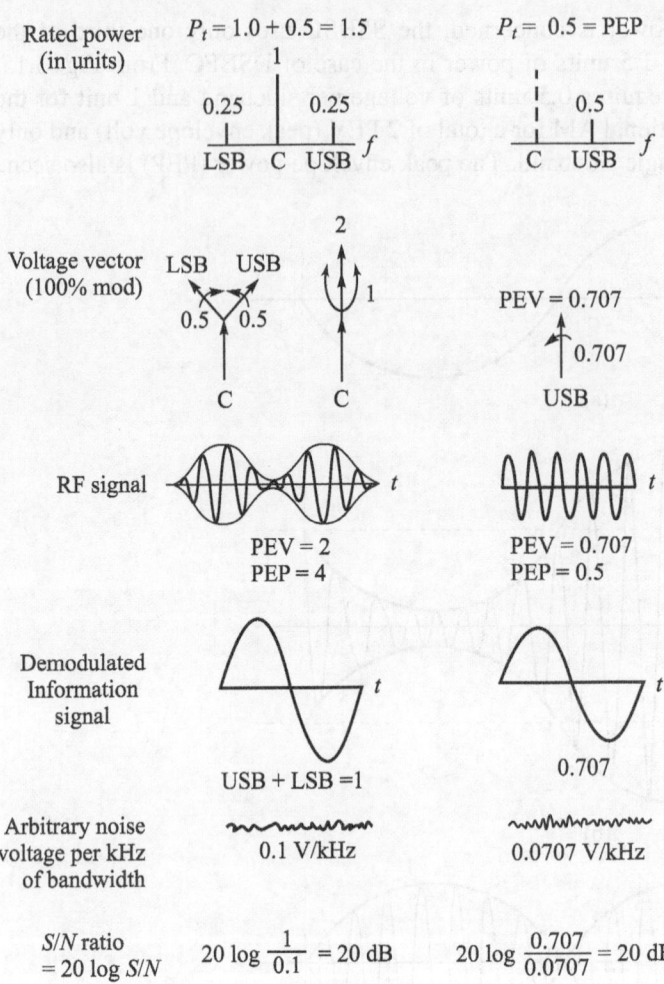

Rated power (in units)	$P_t = 1.0 + 0.5 = 1.5$	$P_t = 0.5 = PEP$

Fig. 3.15 Comparison of AM and SSB

3.11 SINGLE SIDEBAND: ADVANTAGES AND DISADVANTAGES

The advantages of a single sideband with a suppressed and reduced carrier over conventional DSBSC are discussed below.

Lesser power consumption

Since in the case of a single sideband, only one sideband is transmitted and the carrier is totally suppressed or considerably reduced, there is lesser power consumption and much less total transmitted power is necessary compared to DSBFC. Eliminating the carrier results in the power available in the sideband to increase by at least a factor of 3, thus providing 10 log 3 or 4.8 dB improvement in the signal to noise ratio.

Conservation of bandwidth

Single sideband transmission requires half as much bandwidth compared to DSBFC. Eliminating one sideband reduces the bandwidth by more than a factor of 2. For example, most of the modulating signal, including audio signals, rarely goes down up to 0 Hz. The range is generally from 300 to 3000 Hz, the bandwidth of 2700 Hz. Hence, an audio channel transmitted over a DSBFC system requires a bandwidth of 5.4 kHz compared to 2.7 kHz bandwidth required for an SSBSC system.

Fading

With DSB transmission, the two sidebands and a carrier may propagate through the transmission media by different paths and result in selective fading. This results in sideband fading wherein one sideband gets more attenuated than the other. This causes the reduction of signal amplitude at the output of the receiver modulator reducing the signal to noise ratio.

The second common and serious form of selective fading is called 'carrier amplitude fading'. Reduction of the carrier level of a 100% modulated wave will make the carrier voltage less than the modulating voltage and causes over-modulation which results in severe distortion of the demodulated wave.

The third cause of selective fading is carrier or sideband phase shift. This occurs when the relative position of the carrier and sideband signals change. This causes severe distortion in the demodulated signal.

In the case of single sideband transmission, since only one sideband is transmitted, there is no question of a carrier phase shift or carrier fading except for the occurance of sideband fading. This changes only the amplitude and frequency responses of the demodulated signal. Theses changes do not produce enough distortion to cause a loss a of intelligibility in the received signal. Hence, for SSBSC transmission there is no need to maintain a specific amplitude or phase relationship between the carrier and sideband signals.

Noise reduction

Because of a lower bandwidth in the SSBSC system, thermal noise power is reduced to half that of a DSBFC system. Both the bandwidth reduction and immunity to selective fading, an SSBSC system has the advantage of 12 db S/N ratio over the conventional AM.

Single sideband transmission also has some disadvantages. It requires complex receivers. At the receiver end, the carrier has to be generated for detection. If the reduced carrier is sent along with a sideband, this carrier has to be boosted at the synchronized receiver end. This increases the hardware cost.

3.12 SINGLE SIDEBAND GENERATION

Since in most single sideband systems, the carrier is either suppressed completely or reduces to a smaller percentage of original carrier signals, the method that immediately comes to the mind is to use a notch filter to remove the carrier. The characteristic of the notch filter should be critical as it has to only remove the carrier alone. But in practice it is very difficult to design a high Q notch filter with a very good attenuation characteristic outside the carrier frequency. Most of the practical filter removes some portion of the sideband along with the carrier. To circumvent this problem, the carrier is suppressed in the modulator itself and various modulator circuits that remove the carrier during the modulation process have been developed. These are called balanced modulators. Balanced modulators are one of the most popular methods of generating DSB with suppressed carrier. This method is also used in frequency and phase modulation system.

3.13 VESTIGIAL SIDEBAND (VSB) TRANSMISSION AND QUADRATURE AMPLITUDE MODULATION (QAM)

First we will discuss VSB transmission.

3.13.1 Vestigial Sideband Transmission

The stringent frequency response requirement on the sideband filter in an SSBSC system is the selective filtering method. This demands a DC null in the modulating spectrum. The attenuation characteristic of the filter should be large enough to suppress the other sideband. The generation of a DSB signal is much simpler, but requires twice the signal bandwidth. Hence, by allowing a portion of the other unwanted sideband, we get the advantage of DSB and SSB and avoid their disadvantage with a little extra cost. This resulting signal is called 'vestigial sideband' AM. Since *vestige* means a trace or small remaining bit, it pertains to a trace or small portion of unwanted sideband in SSB.

This type of modulation is most suitable for signals having a strong low-frequency component like video signals. This is the reason for using VSBAM for TV transmission. Hence, in this type of modulation carrier, one complete sideband and part of the other sideband are transmitted. The carrier is transmitted at full power. The frequency spectrum and relative power distribution of VSB are shown in Fig. 3.9(f). Comparison of VSB with DSB and SSB is shown in Fig. 3.16.

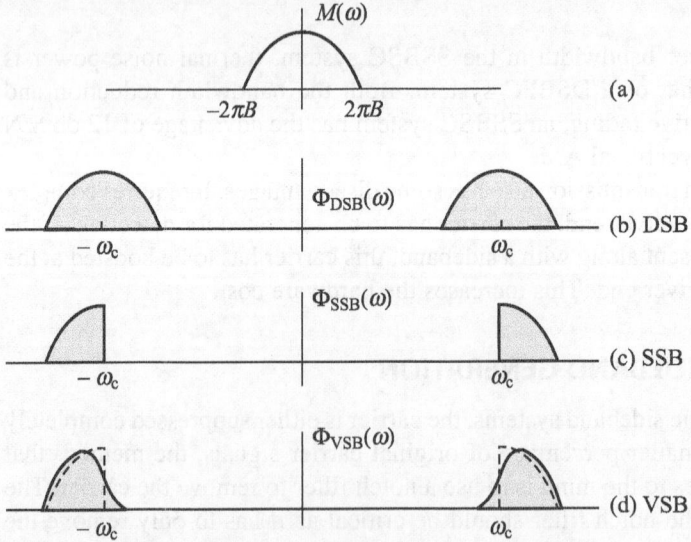

Fig. 3.16 Modulating signal spectrum, DSB, SSB, and VSB signals

Vestigial sideband AM–an analysis

The stringent frequency response requirements on the sideband filter in an SSB AM system can be relaxed by allowing vestige, which is a portion of the unwanted sideband, to appear at the output of the modulator. Thus, we simplify the design of the sideband filter at the cost of a modest increase in the channel bandwidth required to transmit the signal. The resulting signal is called vestigial sideband (VSB) AM. This type of modulation is appropriate for signals that have a strong low-frequency component, such as video signals. This is one of the prime reasons for this type of modulation being used in standard TV broadcasting.

To generate a VSB AM signal, we begin by generating a DSB SC AM signal and passing it through a sideband filter with a frequency response $H(f)$. In the time domain, the VSB signal may be expressed as

$$f(t) = [A_c m(t) \cos 2\pi f_c t] * h(t) \tag{3.26}$$

where $h(t)$ is the impulse response of the VSB filter. In the frequency domain, the corresponding expression is

$$F(f) = \frac{A_c}{2} [M(f - f_c) + M(f + f_c)]H(f) \tag{3.27}$$

To determine the frequency response characteristics of the filter, we will consider the demodulation of the VSB signal $u(t)$. We multiply $u(t)$ by the carrier component $\cos 2\pi f_c t$ and pass the result through an ideal low-pass filter. Thus, the product signal is

$$v(t) = u(t) \cos 2\pi f_c t \tag{3.28}$$

or equivalently

$$V(f) = \frac{1}{2} [U(f - f_c) + U(f + f_c)] \tag{3.29}$$

If we substitute $F(f)$, we obtain

$$V(f) = \frac{A_c}{4} [M(f - 2f_c) + M(f)]H(f - f_c)$$

$$+ \frac{A_c}{4} [M(f) + M(f + 2f_c)]H(f + f_c) \tag{3.30}$$

The low-pass filter rejects the double-frequency terms and passes only the components in the frequency range $|f| \leq f_m$. Hence, the signal spectrum at the output of the ideal low-pass filter is

$$V_l(f) = \frac{A_c}{4} M(f)[H(f - f_c) + H(f + f_c)] \tag{3.31}$$

The message signal at the output of the low-pass filter must be undistorted. Hence, the VSB filter characteristic must satisfy the condition

$$H(f - f_c) + H(f + f_c) = \text{constant} \quad \text{for } |f| \leq f_m \tag{3.32}$$

This condition is satisfied by a filter that has the frequency response characteristic shown in Fig. 3.16.

We note that $H(f)$ selects the upper sideband and a vestige of the lower sideband. It has an odd symmetry about the carrier frequency f_c in the frequency range $f_c - f_a < f < f_c + f_a$, where f_a is a conveniently selected frequency that is some small fraction of f_m, i.e., $f_a \ll f_m$. Thus, we obtain an undistorted version of the transmitted signal.

In practice, the VSB filter is designed to have some specified phase characteristics. To avoid distortion of the message signal, the VSB filter should have a linear phase over its passband

$$f_c - f_a \le |f| \le f_c + f_m \tag{3.33}$$

Demodulation of VSB signals

In VSB, a carrier component is generally transmitted along with the message sidebands. The existence of the carrier component makes it possible to extract a phase coherent reference for demodulation in a balanced modulator.

In applications such as a TV broadcast, a large carrier component is transmitted along with the message in the VSB signal. In such a case, it is possible to recover the message by passing the received VSB signal through an envelope detector.

3.13.2 Quadrature Amplitude Modulation (QAM)

This is a form of modulation in which two modulating signals are used to modulate a carrier of the same frequency but which are in phase quadrature. In this type of transmission, care has to be taken in the phase and frequency of the carrier at the receiving end so as not to lose the signal and avoid distortion. This method is also sometimes called *quadrature carrier multiplexing*. Thus, two baseband signals each of a *bandwidth B* Hz can be transmitted simultaneously over a bandwidth 2 *B* Hz by using DSB transmission and quadtrature multiplexing. The upper channel is known as in-phase signal and the lower channel is called quadrature channel.

The advantage of quadrature AM over conventional AM is conservation of bandwidth. This requires only half as much bandwidth compared to conventional AM. But the disadvantages are many. It is much more complex than conventional AM. The cost of the receiver is high since the demodulation circuit requires a carrier recovery circuit. This has to produce the same carrier frequency and the phase shift between the two carriers should be exactly 90° since the carriers are quadrature in phase.

Quadrature amplitude modulation is also sometimes called phase division multiplexing. It finds applications in modulation of analog carriers in data modem, stereo broadcasting, encoding colour signals in analog TV broadcasting systems, multiplexing the chrominance signals which carry information about colours, and also in digital satellite communications.

3.14 AM MODULATORS

There are many methods for generating AM modulated signals. We will be covering most of the methods. The process of modulation involves the generation of new frequency components which means that they are characterized by a non-linear system.

3.14.1 Square Law Modulation (Power Law Modulation)

In this method, the non-linear device that will be used is a normal P–N junction diode.

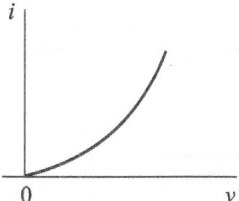

Fig. 3.17 *VI* characteristics of a PN diode

From the *VI* characteristics of the diode, shown in Fig. 3.17, it can be seen that the relation between the input and output follows the square law. If the input to such a device is a sum of the modulating signal $m(t)$ and the carrier $E_c \sin \omega_c t$. The non-linearity in the device characteristics will produce the product of message $m(t)$ with the carrier plus additional components. The desired modulated signal can be filtered out by passing the output through a bandpass filter. Block diagram of a power law AM modulator is shown in Fig. 3.18.

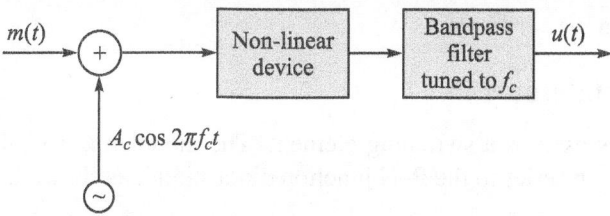

Fig. 3.18 Power law AM modulator

The functioning of a power law modulator can be analysed mathematically as shown below.

Let the input to the non-linear device be of the form

$$V_i(t) = m(t) + E_c \sin \omega_c t \qquad (3.34)$$

The output of the non-linear device is given by

$$V_o(t) = a_1 V_i(t) + a_2 V_i^2(t) \qquad (3.35)$$

where $V_i(t)$ is the input signal and $V_o(t)$ the output signal. Also a_1 and a_2 are constants. The non-linear device output is

$$V_o(t) = a_1[m(t) + E_c \sin\omega_c t] + a_2 [m(t) + E_c \sin\omega_c t]^2$$

$$= a_1 m(t) + a_2 m^2(t) + a_2 E_c^2 \sin^2 \omega_c t$$

$$+ E_c a_1[1 + (2a_2/a_1)m(t)]\sin \omega_c t \qquad (3.36)$$

This output of the non-linear device is passed through a bandpass filter with a bandwidth of $2B$ Hz centred at $\omega = \omega_c$ to generate the conventional AM output

$$V(t) = E_c a_1[1 + (2a_2/a_1) m(t) \sin \omega_c t] \qquad (3.37)$$

The operation of a square law demodulator is shown in Fig. 3.19.

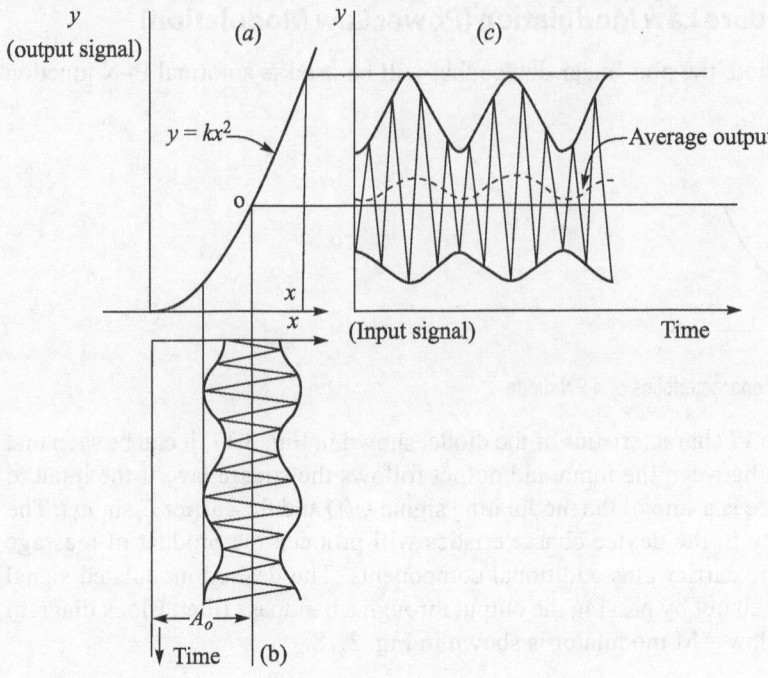

Fig. 3.19 Operation of square law demodulator

3.14.2 Switching Modulator

In this method, a diode is used as a switching element. The modulating signal and the carrier are applied in series to the P–N junction diode circuit as shown in Fig. 3.20(a).

The input–output of the characteristic is shown in Fig. 3.20(b). The output across the load resistor is

$$V_o(t) = V_i(t), \quad c(t) > 0$$
$$= 0, \qquad c(t) < 0 \tag{3.38}$$

The switching operation can be viewed mathematically as a multiplication of the input $V_i(t)$ and the switching function $S(t)$, i.e.,

$$V_o(t) = [m(t) + E_c \sin \omega_c t] \, S(t) \tag{3.39}$$

where $S(t)$ is shown in Fig. 3.20(c).

Since $S(t)$ is a periodic function, it can be represented in the Fourier series as

$$S(t) = \frac{1}{2} + \frac{2}{\pi} \sum_{n=1}^{\infty} \left[\{(-1)^{n-1}/(2n-1)\} \sin \omega_c t (2n-1) \right] \tag{3.40}$$

Substituting the value of $S(t)$ from Eq. (3.40) in (3.39), we get

$$V_o(t) = E_c / 2 [1 + (4/\pi E_c) m(t)] \sin \omega_c t + \text{other higher order terms} \tag{3.41}$$

The required AM modulated signal is obtained by passing $V_o(t)$ through a bandpass filter with a centre frequency $\omega = \omega_c$ and bandwidth $2B$. The output of bandpass filter is the required conventional AM signal.

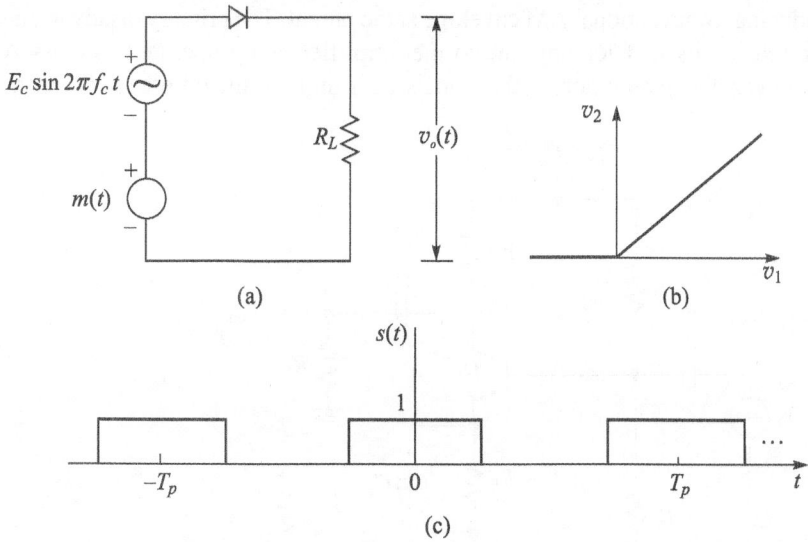

Fig. 3.20 Switching modulator: (a) circuit, (b) diode VI characteristic, and (c) switching function

3.14.3 Transistor Modulators

Transistors can be used as AM modulator circuits. The advantage is that amplification of the signal is also possible. The transistor configuration depends on the level of modulation. In the case of low level modulation, the modulation takes place at the initial stage itself prior to the output stage. In the case of high level and medium level modulation, the modulation takes place at the final element of the final stage. Here the carrier amplitude is at its maximum and thus requires a much higher amplitude modulated signal to achieve a reasonable modulation index. Low level modulation has the disadvantage of having all the stages following the modulator stage to be a highly linear amplifier.

Low level AM modulator

Figure 3.21(a) shows a class-A small signal amplifier, which can be used to perform amplitude modulation. This amplifier will have two inputs, one for the carrier and the other for the modulating signal. When no modulating signal is present, the circuit operates as a class-A amplifier, the output being the amplified carrier. But when the modulating signal is applied, the amplifier operates non-linearly and the signal multiplication occurs to generate an AM signal. In the above-mentioned circuit, the carrier is applied to the base and the modulating signal to the emitter. Hence, this circuit configuration is called *emitter modulation*.

The modulating signal is applied through an isolation transformer to the emitter of the transistor and the carrier is directly applied to the base. The modulating signal drives the amplifier into cut-off and saturation, thus producing non-linear amplification necessary for modulation. The collector output consists of the carrier and the upper and lower sidebands apart from the modulating signal. The coupling capacitor C_c in the collector circuit blocks the modulating signal, thus producing a conventional AM envelope at the output. The primary disadvantage of this circuit is its inefficiency due to the amplifier being operated as class-A and the power dissipation across the transistor is more at the quiescent current.

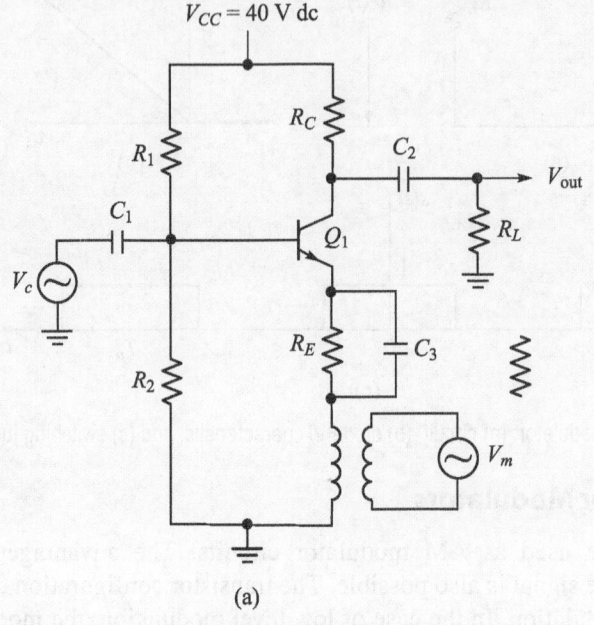

(a)

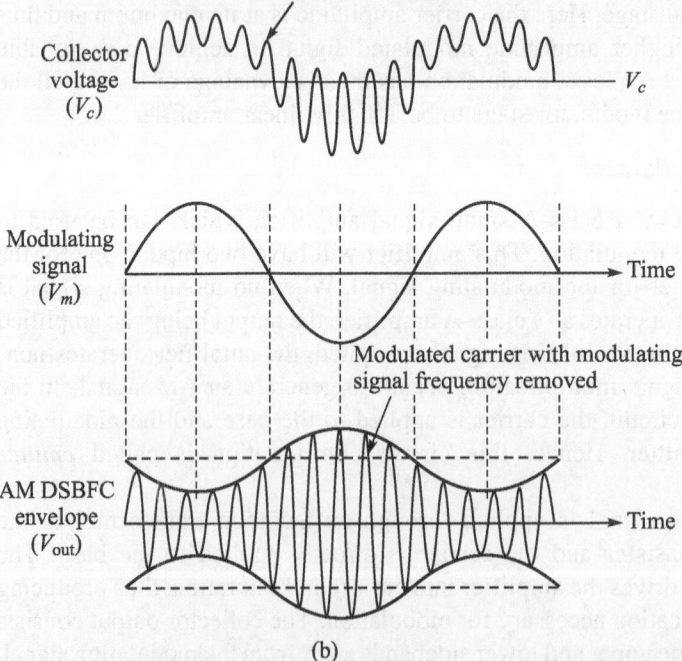

(b)

Fig. 3.21   Low level transistor AM modulator

The modulating signal that is fed to the emitter varies with the gain of the amplifier at a sinusoidal rate equal to the frequency of the modulating signal. The depth of modulation is proportional to the amplitude of the modulating signal. The voltage gain of this circuit can be expressed as

$$A_m = a \, (1 + m \sin \omega_m t) \tag{3.42}$$

where A_m is the voltage gain without modulation. Since $\sin \omega_m t$ goes from a maximum value of +1 to −1, Eq. (3.42) reduces to

$$A_m = A(1 + m) \tag{3.43}$$

where m, the modulation index, is equal to 1 (100% modulation). Then

$$A_{m\,max} = 2A, \; A_{m\,min} = 0$$

Figure 3.21 shows the waveform.

Medium power AM modulator

In this type of modulator, modulation takes place in the collector, the output element at the transistor. To achieve a high power efficiency normally, these modulators operate as class-C. Hence, their efficiency is as high as 80%. The circuit shown in Fig. 3.22(a) is a class-C amplifier with two inputs, a carrier V_c, and a single-frequency modulating signal V_m. Due to class-C operation, there will be a non-linear mixing. This circuit is called collector modulator because a modulating signal is applied directly to the collector.

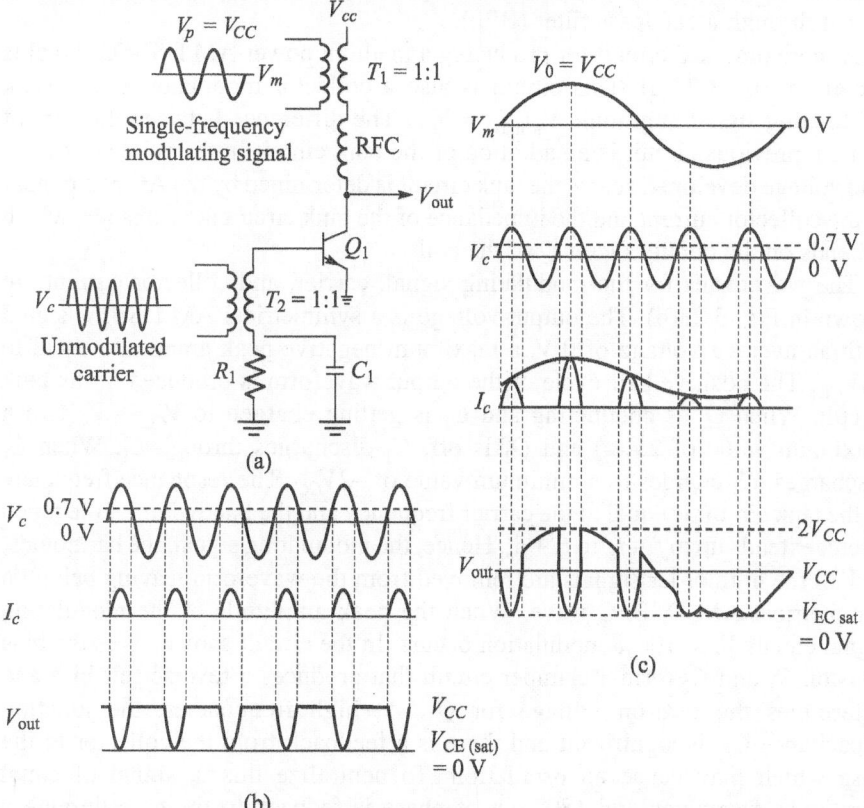

Fig. 3.22 Medium power Transistor AM modulator: (a) circuit, (b) collector waveform with no modulating signal, and (c) collector waveform with a modulating signal

The circuit shown in Fig. 3.22(a) operates as explained below.

When the amplitude of the carrier exceeds the cut-in voltage 0.7 V of the base-emitter junction, transistor Q_1 turns on and a collector current flows. When the carrier voltage falls below the base-emitter junction cut-in voltage, Q_1 gets switched off and the collector current stops flowing. Thus, Q_1 switches between the saturation and cut-off controlled by the carrier signal, and the collector current flows for less than 180° of each carrier cycle, thus achieving class-C operation.

Each successive cycle of the carrier turns Q_1 on for an instant and allows the current to flow for a short time, producing a negative-going waveform at the collector. The collector current and voltage waveform are shown in Fig. 3.22(b). The collector voltage waveform looks like a repetitive half-wave rectified signal with a fundamental frequency equal to f_c. The RFC is a radio frequency choke, which acts as a short circuit to direct current and open to other frequency. The modulating signal applied to the collector is in series with the collector DC supply and adds to and subtracts from V_{CC}. From Fig. 3.22(c), it is seen that the output voltage waveform swings from a maximum value of $2V_{CC}$ to approximately 0 V. The peak change in the collector voltage is equal to V_{CC}. The waveform here also resembles a half-wave rectifier superimposed onto a low frequency modulating signal. Since Q_1 is operating in a non-linear region, the collector waveform contains two original input frequencies $f_c \pm f_m$ and also the higher order harmonics and intermodulation components. The output is obtained by passing the collector output through a bandpass filter (BPF).

A more practical circuit for producing a medium power AM DSBFC signal is shown in Fig. 3.23(a). This circuit is also a collector modulator with a peak modulating signal amplitude $V_{m(max)} = V_{CC}$. The difference between this circuit and the previous circuit is an addition of the tank circuit in the collector of Q_1. The voltage developed across the tank circuit is determined by the AC component of the collector current and the impedance of the tank circuit at resonance, which depends on the quality factor Q of the coil.

The waveforms for the modulating signal, carrier, and collector current are shown in Fig. 3.23(b). The output voltage is a symmetrical AM DSBFC signal with an average voltage of 0 V, a maximum negative peak amplitude equal to $-2V_{CC}$. The positive half cycle of the output waveform is produced in the tank circuit. When Q_1 is conducting and C_1 is getting charged to $V_{CC} + V_m$ (to a maximum value of $2V_{CC}$) and Q_1 is off, C_1 discharges through L_1. When L_1 discharges, C_1 charges to a minimum value of $-2V_{CC}$. The resonance frequency of the tank circuit is equal to the carrier frequency and the bandwidth of the tuned circuit extends up to $f_c - f_m$ to $f_c + f_m$. Hence, the modulating signal, the harmonics, and all the high order signals are removed from the waveform leaving behind a symmetrical AM DSBFC wave. When the peak amplitude of the modulating signal equals V_{CC}, 100% modulation occurs. In the circuit shown, R_1 is the bias resistor. R_1 and C_2 form a clamper circuit that produces a reverse self-bias and determines the turn-on voltage for Q_1. At high frequencies, the junction capacitance C_{bc} is significant and there is a feedback from the collector to the base which may cause an oscillation. To neutralize this, a signal of equal amplitude, frequency and 180° out of phase is fedback to the base through a capacitor C_n, which is called a neutralizing capacitor. C_4 is an RF bypass capacitor. It isolates the DC supply from radio frequencies.

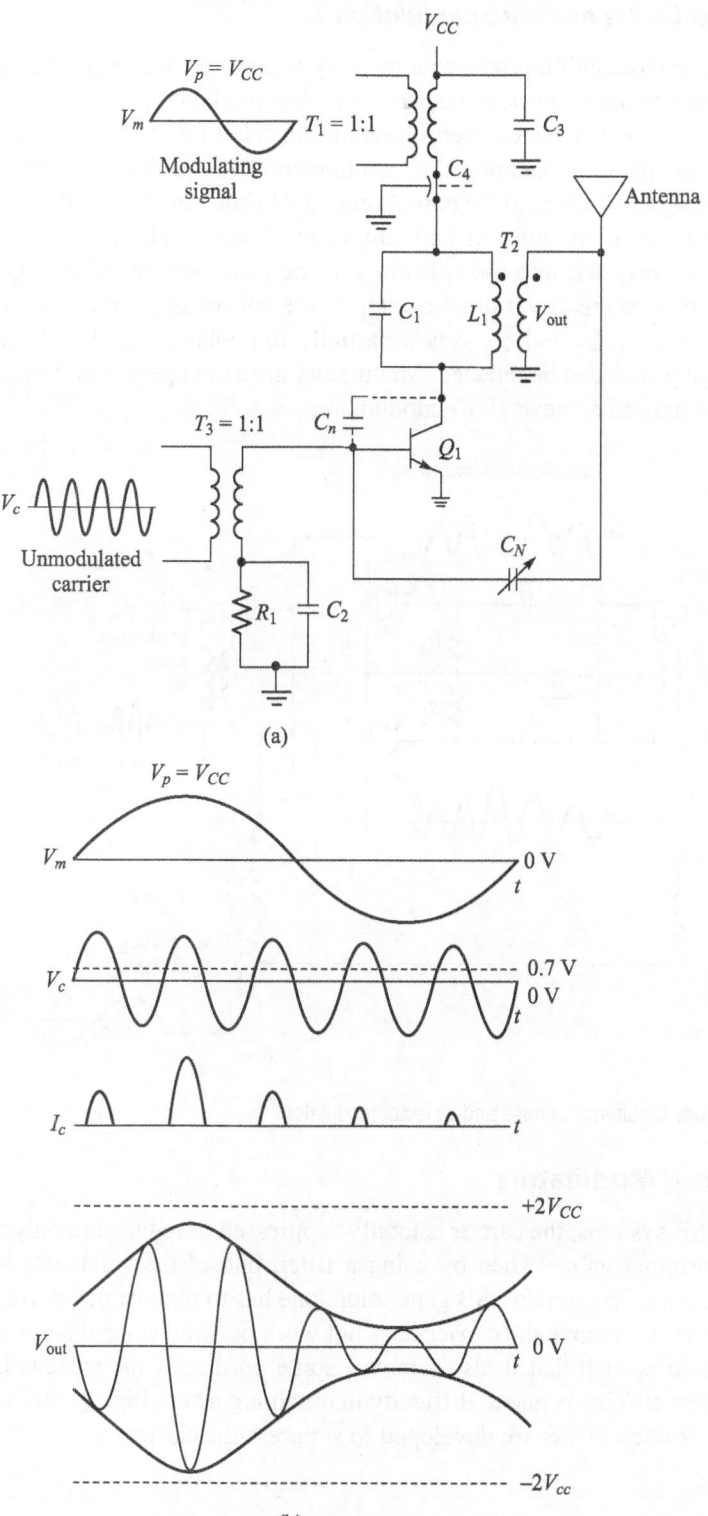

Fig. 3.23 Medium power AM DSBFC modulator: (a) circuit and (b) collector and output waveforms

Simultaneous collector and base modulation

Although collector modulation produces a more symmetric envelope compared to low power emitter modulators, it requires a higher modulating signal. This type of modulation cannot achieve either full saturation or cut-off, thus preventing 100% modulation from happening. To circumvent this problem, a little modulating drive signal is given to the base. Figure 3.24 shows an AM modulator with a combination of both collector and emitter modulation. The modulating signal is simultaneously fed into the collectors of the push-pull modulators Q_2 and Q_3 and collector of the driver amplifier Q_1. Collector modulation occurs in Q_1. Hence, the base of Q_2 and Q_3 gets a partially modulated signal and the modulating signal power can be reduced. Modulators need not operate over their entire operating curve to achieve 100% modulation.

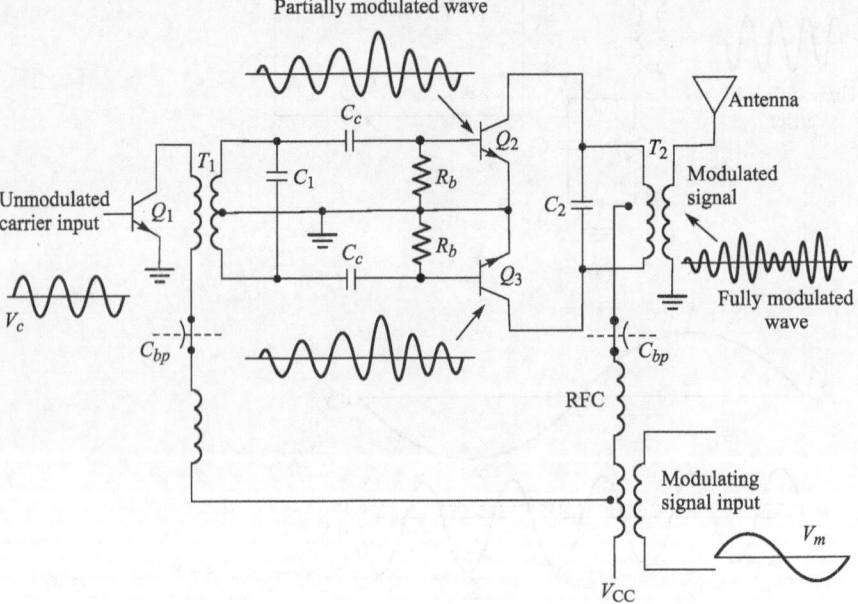

Fig. 3.24 High power simultaneous base and collector modulator

3.14.4 Balanced Modulators

In most of the SSB systems, the carrier is totally suppressed or reduced to only a fraction of its original value. Then by using a filter, one of the sidebands is removed. Hence, as a first step in SSB generation, one has to remove the carrier. Using a notch filter to remove the carrier does not work out well since the notch filter characteristic is such that it also removes some portion of the sidebands along with the carrier. This is due to difficulty in obtaining a very high Q for the filter. Hence, alternate circuits are developed to suppress the carrier.

Such circuits are called double sideband balanced modulators, which suppress the carrier and generates two sidebands. This is a relatively simple method for generating a DSBSC AM signal. Block diagram of a balanced modulator is shown in Fig. 3.25.

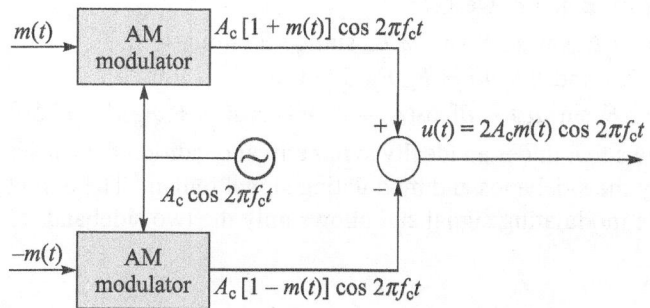

Fig. 3.25 Balanced modulator

Each AM modulator is a diode square law modulator of identical characteristics so that the carrier components cancel out at the summing junction. The circuit diagram of such a modulator is shown in Fig. 3.26.

The modulation voltage $E_m \sin \omega_m t$ is fed in push pull and the carrier voltage $E_c \sin \omega_c t$ is parallel to a pair of identical diodes. The carrier signal is thus applied in phase to both diodes and the modulating voltage with 180° out of phase to the diodes. The modulated output currents i_1 and i_2 of the two diodes get subtracted as shown in Fig. 3.26. If the circuit is exactly symmetrical, the carrier frequency component gets cancelled completely. Since it is very difficult to get a perfectly symmetric circuit, a 40 to 55 dB suppression of carrier is accepted. The output of the balanced modulator contains the two sidebands and other unwanted frequency components. The unwanted components are tuned out by the output transformer secondary winding. Let us see mathematically how the carrier gets suppressed.

The input voltage to D_1 is

$$E_m \sin \omega_m t + E_c \sin \omega_c t \qquad (3.44)$$

and the input voltage to D_2 is

$$E_c \sin \omega_c t - E_m \sin \omega_m t \qquad (3.45)$$

where $E_c \sin \omega_c t$ is the carrier signal and $\sin \omega_m t$ is the modulating signal. The output at each diode is

$$V_1(t) = a_1(E_m \sin \omega_m t + E_c \sin \omega_c t)$$
$$+ a_2(E_m \sin \omega_m t + E_c \sin \omega_c t)^2 \qquad (3.46)$$

$$V_2(t) = a_1(E_c \sin \omega_c t - E_m \sin \omega_m t)$$
$$+ a_2(E_c \sin \omega_c t - E_m \sin \omega_m t)^2 \qquad (3.47)$$

The output $V_o(t)$ is the difference between these two voltages.

Let $m = E_m \sin \omega_m t$ and $c = E_c \sin \omega_c t$. Then Equations (3.46) and (3.47) can be written as

$$V_1(t) = a_1(c + m) + a_2(c + m)^2 \qquad (3.48)$$

$$V_2(t) = a_1(c - m) + a_2(c - m)^2 \tag{3.49}$$

Since the output $V_o(t)$ is the difference between $V_1(t)$ and $V_2(t)$ and assuming the transformer turns ratio to be 1, we get

$$V_o(t) = V_1(t) - V_2(t) = 2a_1 m + 4a_2 mc$$

Substituting the values of m and c, we get

$$V_o(t) = 2a_1 E_m \sin \omega_m t + 4a_2 m\, E_m \sin\omega_m t\, E_c \sin\omega_c t$$

Let $\qquad A = 2a_1$ and $B = 4a_2 m\, E_m\, E_c$. Then

$$V_o(t) = AE_m \sin \omega_m t + B[\cos(\omega_c - \omega_m)t - \cos(\omega_c + \omega_m)t] \tag{3.50}$$

Equation (3.50) shows that under an ideally symmetrical condition the carrier gets cancelled and only the sidebands and modulating signal remain. The output tuned circuit rejects the modulating signal and allows only the two sidebands to appear.

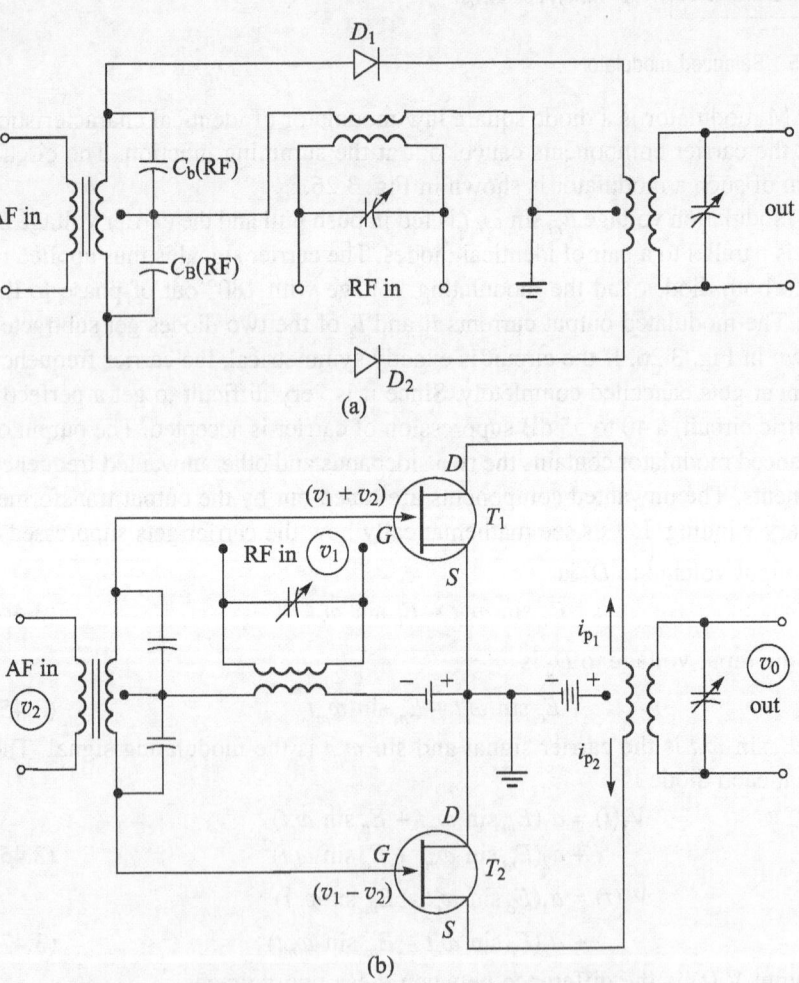

Fig. 3.26 Balanced modulators: (a) diode and (b) FET

Balanced ring modulator

The balanced modulator discussed above requires the output to be tuned. Thus, the rejection of a modulating signal depends on the Q of the tuned circuit. Since a high Q is difficult to get, inadvertently some part of modulating signal may appear at the output. To avoid the tuned circuit at the output, another type of balanced modulator called the ring modulator is used. The balanced ring modulator is shown in Fig. 3.27. It has got two inputs—a single frequency carrier and a modulated signal. For the circuit to operate efficiently, the amplitude of the carrier must be quite high (at least six times) compared to that of the modulating signal. This ensures that the carrier controls the on and off conditions of the diode.

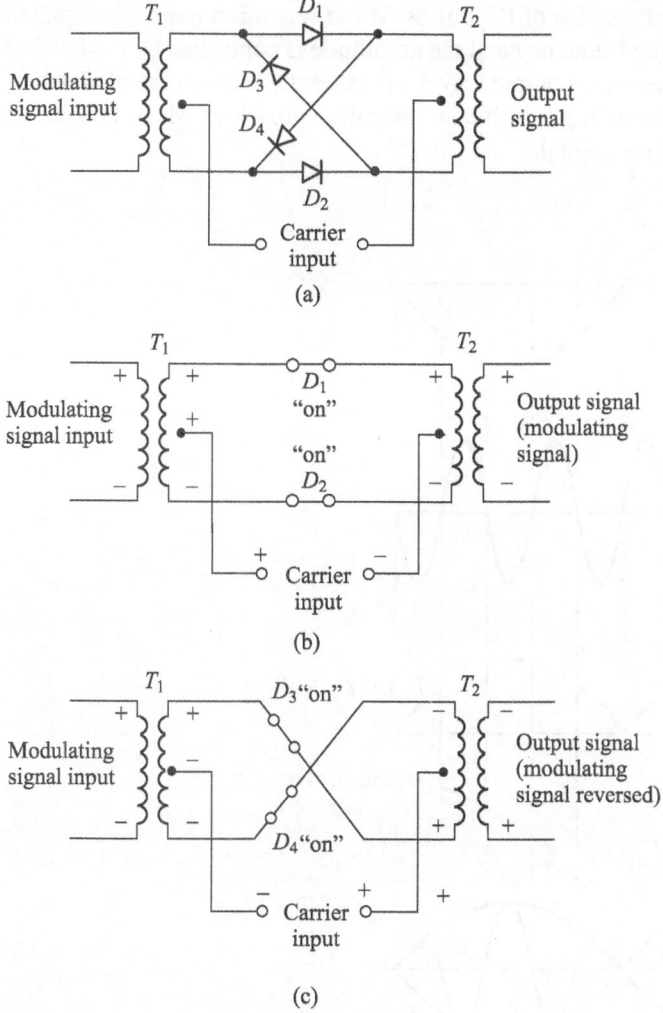

Fig. 3.27   Balanced ring modulator

The diodes in the circuit control whether the modulating signal is passed from the input transformer to the output transformer with or without 180° phase shift. When the left side of the carrier terminal is positive, diodes D_1 and D_2 conduct and D_3 and D_4 are off. This results in modulating signal passing on to the output without any phase reversal. When the polarity of the carrier gets reversed, diodes D_1 and D_2 are off, diodes D_3 and D_4 conduct, and the modulating signal phase undergoes 180° phase shift at the output. The carrier current flows in the opposite direction in the primary winding of the output transform for each positive and negative cycle, and thus the magnetic field gets cancelled in the secondary winding of the transformer suppressing the carrier.

If the diodes are not matched perfectly or if the transformer centre tap is not proper, then the carrier will not be suppressed totally. In practical circuits, there will always be a small component of the carrier at the output which is called 'carrier leak'. This suppression will be of the order of 45 to 60 DB. Fig. 3.28 shows the waveform of a ring modulator for a single frequency modulating signal. The output consists of a series of RF pulses whose repetition rate is determined by the carrier switching frequency and the amplitude is controlled by the level of modulating signal. Hence, the output waveform takes the shape of the modulating signal except with alternating positive and negative polarities which depend on the polarity of the carrier signal.

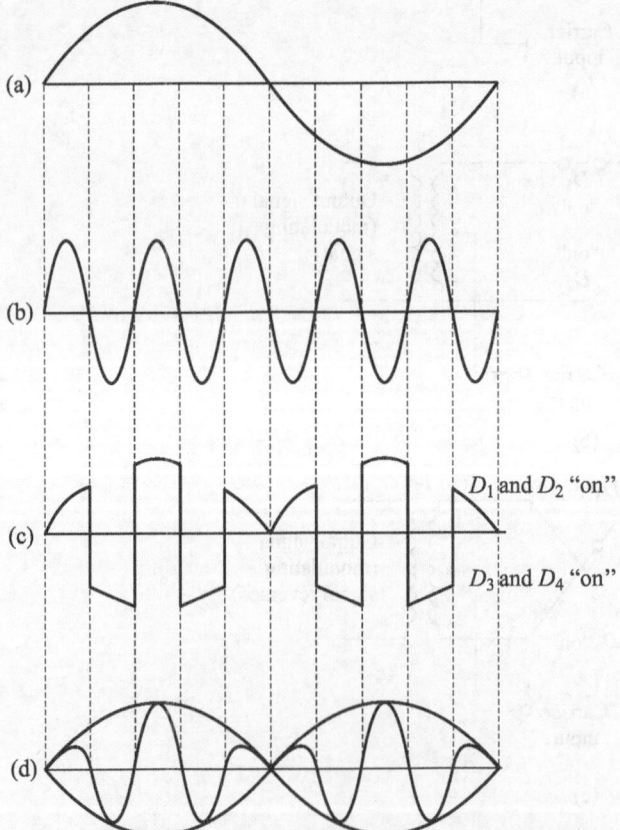

Fig. 3.28 Balanced modulator waveforms: (a) modulating signal,
(b) carrier, (c) output without filtering, and (d) output after filtering

Balanced bridge modulator

Figure 3.29 shows a balanced bridge modulator. When the carrier polarity is positive to the left of the terminal, all the diodes get reverse biased and the modulating signal appears at the output.

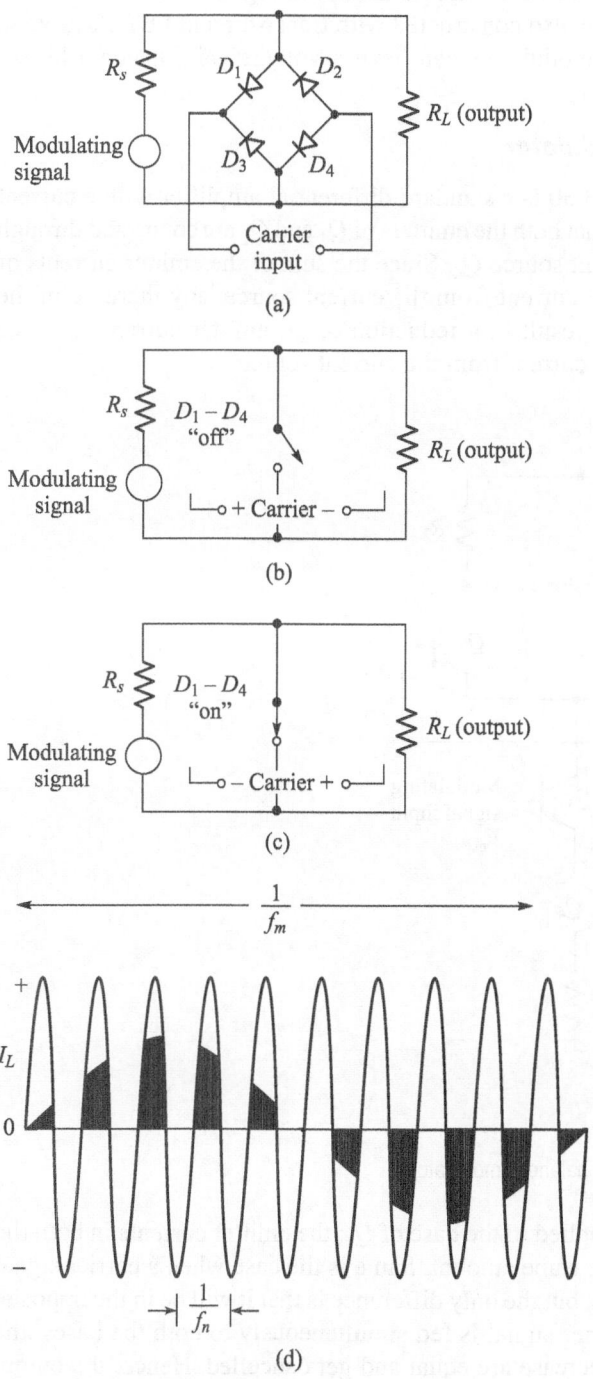

Fig. 3.29 Balanced bridge modulator

When the carrier polarity is positive to the right of the terminal, all the diode conducts and the load resistor is bypassed and there is no output across the load. As the carrier voltage changes from positive to negative, the output waveform contains a series of pulses which mainly comprise upper and lower sideband frequencies. The series of pulses are shown shaded in Fig. 3.29(d).

Balanced modulators are also constructed with transistor and FETs. Unlike in the case of diodes, these modulators can have a voltage gain, hence a higher power level.

Transistor balanced modulator

The circuit shown in Fig. 3.30 is a standard differential amplifier with a current source. It can be observed that both the emitters of Q_1 and Q_2 are connected through a transistor Q_3 to the current source Q_4. Since the sum of the emitter currents of Q_1 and Q_2 are equal to the current from the current source, any increase in the emitter current of Q_1 will result in a reduction of Q_2 emitter current and vice versa to maintain constant current from the current source.

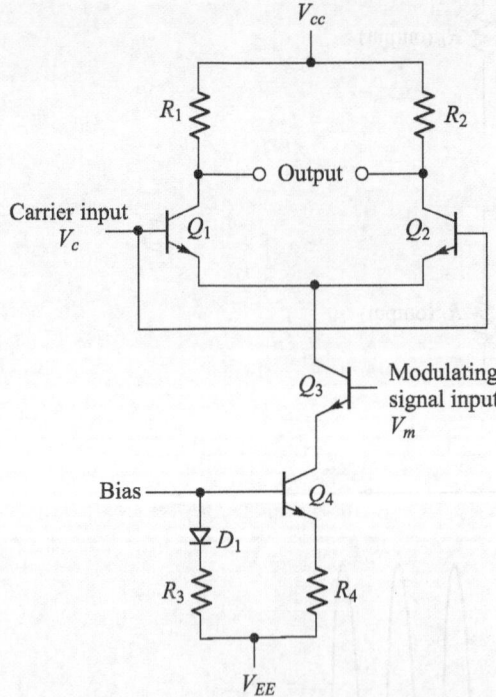

Fig. 3.30　Differential transistor balanced modulator

When a carrier signal is applied to the base of Q_1, the emitter currents in both the transistors will vary by the same amount. Same is the case when a carrier signal is applied to the base of Q_2, but the only difference is that it will be in the opposite direction. If the same carrier signal is fed simultaneously to both the bases, the respective increase and decrease are equal and get cancelled. Hence, the output voltage and current remain unchanged. If a modulating signal is applied to the base of Q_3, which is a part of the current source, there will be a corresponding increase or decrease of the collector currents of Q_1 and Q_2. The carrier signal and modulating signal get multiplied. This results in the appearance of sum and difference frequencies in the output. But the carrier and modulating frequency component does not appear at the output. Because of excellent common mode rejection ratio (CMRR) typically of the order of 80 DB and above, the carrier and modulating frequency component strength will be negligibly small.

FET balanced modulator

Field effect transistors (FETs) like diode exhibit square law properties and produce only second-order cross-product frequencies. They also produce only sidebands and suppress the carrier. Figure 3.31 shows a FET balanced modulator. From the figure, it is evident that it is a push-pull amplifier with only difference that a circuit has two sets of inputs, the carrier and modulating signal.

The carrier signal is fed through transformer T_2. The carrier signals to the gates are in phase and they produce current in both the top and bottom halves of the output transformer T_3, which are equal but 180° out of phase. Hence, they cancel each other and no carrier component is present in the output. On the other hand, the modulating frequency is applied in such a way that they are 180° out of phase to each gate. This causes an increase in the drain current in one FET and decrease in the drain current in the other FET.

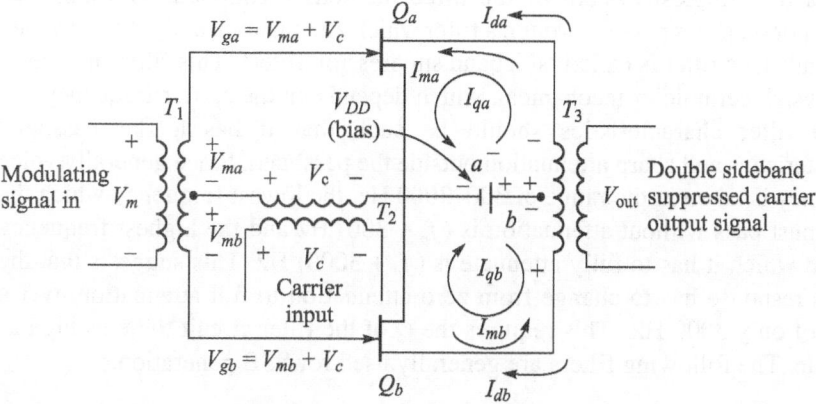

Fig. 3.31 FET balanced modulator

The carrier signal E_c produces currents I_{ca} and I_{cb}. These are in the same direction as the quiescent currents I_{qa} and I_{qb}. The modulating signals E_{ma} and E_{mb} produce a current I_{ma} in Q_a. That is in the same direction as I_{ca} and I_{qa} and a current I_{mb} in Q_b, is in the opposite direction to I_{cb} and I_{qb}.

Hence, the total current through the upper side primary winding of T_3 is $I_{ta} = I_{da} + I_{qa} + I_{ma}$ and the total current through the bottom half of the primary winding is $I_{tb} = -I_{ab} - I_{db} + I_{mb}$. This results in the net current flowing through the primary winding T_3 to be $I_{ta} + I_{tb} = I_{ma} + I_{mb}$. For a modulating signal with the opposite polarity, the drain current in Q_b will increase and that in Q_a will decrease. Ignoring the quiescent currents, the drain current in one FET is the sum of the carrier and modulating signal currents, and the drain current in the other FET is the difference of the carrier and modulating signal currents.

While T_1 is an audio transformer, T_2 and T_3 are radio frequency transformers. For a carrier signal to be suppressed completely, Q_a and Q_b must be perfectly matched and the centre tap of T_1 and T_3 must be exact. The carrier rejection in a FET balanced modulator is about 45 dB to 60 dB.

3.15 SSB GENERATION

The balanced modulator discussed so far suppresses the carrier and generates two sidebands. If we are able to suppress one of the sidebands from the output of a balanced modulator, the resultant output is a single sideband signal. There are three different methods to achieve this. They are

(a) the filter method,
(b) the phase shift method, and
(c) the 'third method'.

All of them will remove either the upper or lower sideband. Each method has its own advantage and disadvantage. Each of these systems is explained below.

3.15.1 The Filter Method

This is the simplest system of the three methods mentioned. The balanced modulator output is passed through a filter which attenuates heavily the unwanted sideband. This filter is called 'sideband suppression filter'. This filter may be of LC, crystal, ceramic or mechanical, which depends on the carrier frequency.

The filter characteristics should be such that it has a flat passband characteristics and sharp attenuation outside the passband. Since generally voice frequency is taken in the range of 300–3000 Hz, the lowest frequency which the filter must pass without attenuation is $(f_c + 300)$ Hz and the highest frequency beyond which it has to fully attenuate is $(f_c + 3000)$ Hz. This suggests that the filter's response has to change from zero attenuation to full attenuation over a range of only 3000 HZ. This requires the Q of the filter circuit to be as high as possible. The following filters are generally used for SSB generation.

LC filters

These are the simplest of them all. LC filters cannot be used beyond 100 kHz. Above this frequency, attenuation outside the passband is insufficient. They are currently used in HF equipment.

Crystal filters

Crystal lattice filters are commonly used in an SSB system. Figure 3.32(a) shows a typical crystal lattice filter. X_1 and X_2, X_3 and X_4 are two sets of matched crystal pairs. They are

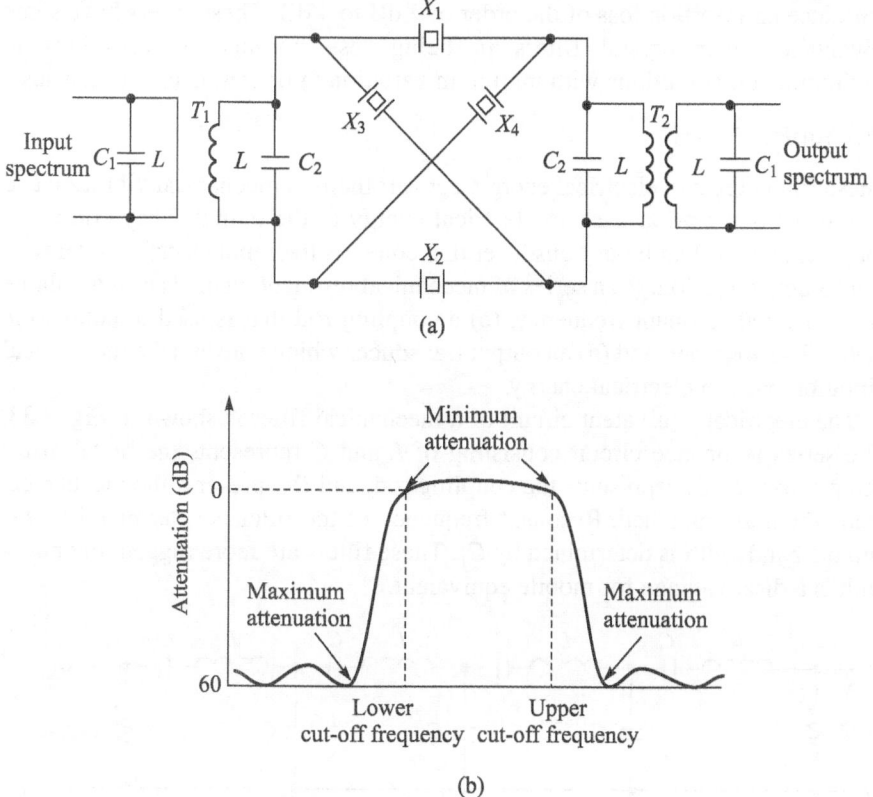

Fig. 3.32 Crystal lattice filter: (a) circuit and (b) frequency response

connected between tuned input and output transformers T_1 and T_2, respectively. Crystals X_1 and X_2 are connected in series and X_3 and X_4 are connected in parallel. Each of the crystals is matched in frequency within 10 Hz to 20 Hz.

X_1 and X_2 operate at the filter lower cut-off frequency and X_3 and X_4 operate at the filter upper cut-off frequency. Transformers T_1 and T_2 are tuned to the centre of the desired band. C_1 and C_2 are used to correct difference between the series and parallel resonant frequencies, and thus correct the overspreading of the frequency difference under the matched crystal condition.

This circuit works similar to a bridge circuit. When the bridge is balanced, there is no output or there will be maximum attenuation. At frequencies where the reactances of all arms are equal in magnitude but opposite in sign, the output will be maximum. Figure 3.32(b) shows the frequency response characteristic of the lattice filter. From the characteristics, it is clear that they have very good steep attenuation characteristics. One can obtain with crystal filters Q as high as 100,000. The insertion loss introduced by these filters are of the order of 1.8 dB to 3.5 dB.

Ceramic filters

They use piezoelectric effect. They are similar to crystal filters, except that they have a Q factor up to 2000, which is quite small compared to a crystal filter. They are less expensive, more rugged, and smaller than their crystal counterpart, but introduce an insertion loss of the order of 2 dB to 4 dB. These filters have some advantages over crystal filters in being less immune to variations in environmental conditions with minimum variation in operating characteristics.

Mechanical filters

These filters receive electrical energy, convert them to mechanical vibration and then convert vibration back to electrical energy at the output. They consist of four elements: (a) an input transducer that converts the input electrical energy to mechanical vibration, (b) a series of mechanical resonant metal discs that vibrate at the desired resonant frequency, (c) a coupling rod that is used to couple the metal disks together, and (d) an output transducer which converts the mechanical vibration back to electrical energy.

The electrical equivalent circuit of a mechanical filter is shown in Fig. 3.33. The series resonance circuit consisting of L and C represents the metal discs, shunt capacitor C_1 represents the coupling rod, and R represents the mechanical load which are matched. Resonant frequency of the filter is determined by LC and the bandwidth is determined by C_1. These filters are more rugged, but bulky with is a disadvantage for mobile equivalent.

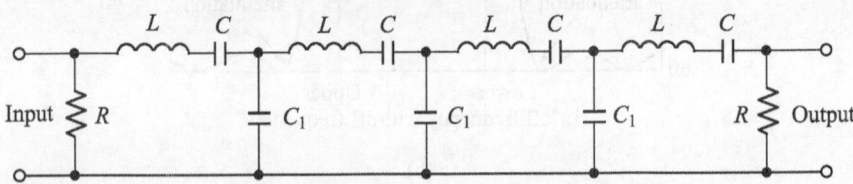

Fig. 3.33 Equivalent circuit: mechanical filter

Surface acoustic wave (SAW) filters

These filters use acoustic energy rather than electro-mechanical energy to provide excellent performance for precise passband filtering. They trap or guide acoustical waves along the surface. They can operate at centre frequencies up to several gigahertz and bandwidth up to 50 MHz. SAW filters have an extremely sharp roll off characteristic, with stop band attenuation up to 50 dB. They are used in single band multiple conversion superheterodyne receivers for both RF and IF filters and in SSB systems for filtering unwanted sideband.

The filter consists of transducers formed from a thin aluminium film deposited on the surface of a semiconductor crystal material. Due to the piezoelectric effect, there will be a physical deformation on the surface of the substrate, called ripples. These ripples vary at the frequency of the applied signal but travel along the surface of the material at the speed of sound. At the receiving end, a second crystal converts back this acoustic energy to electrical energy.

To provide filter action, metallic finger rows are deposited on the flat surface of the substrate as shown in Fig. 3.34. Finger centres are spaced at either a half or quarter wavelength of the desired centre frequency. As the acoustic waves travel across the surface of the substrate, they reflect back and forth, some of them getting cancelled and some of them getting aided. The bandwidth of the filter is determined by the thickness and the number of fingers.

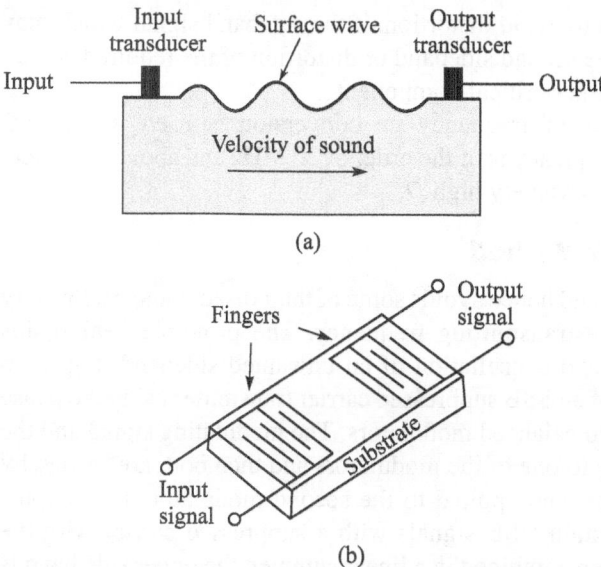

Fig. 3.34 SAW filter: (a) principle and (b) practical filter

A basic SAW filter is bidirectional, i.e., half the power is transmitted towards the output transformer and the other half radiated is towards the end of crystal substrate and has a high insertion loss of 25 to 35 dB, but it is highly reliable and very rugged. A more complex structure can be used to reduce the insertion loss by making the propagation unidirectional. Saw filters are not generally used for low level signals. They exhibit a much longer delay. Hence, sometimes they are used as delay lines.

Single sideband suppressed carrier transmitter

Figure 3.35 shows the block diagram of an SSB suppressed carrier transmitter using a BPF to eliminate the unwanted sideband. Initial modulation takes place in the balance modulator at a low frequency, e.g., at 120 kHz since it is difficult to get the required filter characteristics at higher frequencies. The filter shown in the block diagram is a BPF with a sharp cut-off at beyond the passband to obtain good adjacent sideband rejection. The filtered signal is up converted in a mixer which is nothing but another balanced modulator and then to a linear power amplifier and fed to the antenna for transmission through a matching network.

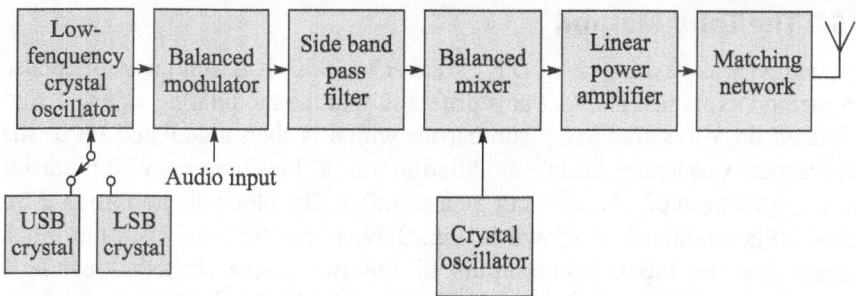

Fig. 3.35 SSB suppressed carrier transmitter (BPF is used to remove the other sideband.)

Linear amplifiers are used to avoid distortion of the sideband signal which may result in regeneration of the second sideband or distortion of the required signal. Sideband pass filter is another critical component.

Sometimes three stages of frequency up conversion is used in an SSB transmitter if the carrier frequency is of the order of 20 MHz and above. The final BPF used should have an extremely high Q.

3.15.2 The Phase Shift Method

This method avoids filters and hence avoids some of their disadvantages. Primary modulation is done at the transmitting frequency. The principle behind this method is phase shifting and cancellation of an unwanted sideband. Fig. 3.36 shows the block diagram of an SSB suppressed carrier transmitter using the phase shift method. There are two balanced modulators. The modulating signal and the carrier are applied directly to one of the modulators and then both are shifted by 90° through a phase shifter and applied to the second modulator. The outputs from both modulators contain DSB signals with a suppressed carrier with the proper phase such that when combined in a linear summer, the upper side band is cancelled and if combined in a subtractor, the lower sideband is cancelled. This result can be proved mathematically as follows:

Since the input to the first balanced modulator is $\sin \omega_m t$ and $\sin \omega_c t$, the output from this modulator is

$$\sin \omega_m t \sin \omega_c t = \frac{1}{2} \cos(\omega_c - \omega_m)t - \frac{1}{2} \cos (\omega_c + \omega_m)t \qquad (3.51)$$

Similarly the output from the second modulator is

$$\cos \omega_m t \cos \omega_c t = \frac{1}{2} \cos(\omega_c - \omega_m)t + \frac{1}{2} \cos (\omega_c + \omega_m)t \qquad (3.52)$$

The output of the linear summer is

$$\frac{1}{2} \cos(\omega_c - \omega_m)t - \frac{1}{2} \cos(\omega_c + \omega_m)t + \frac{1}{2} \cos(\omega_c - \omega_m)t + \frac{1}{2} \cos(\omega_c + \omega_m)t$$

$$(3.53)$$

which results in $\cos(\omega_c - \omega_m)t$, the lower sideband.

The SSB generating system thus explained is less preferred to that of the filter method. The reason is that the phase shift network must introduce precisely a $90°$ phase shift. If a 90° phase shift is not possible, then there will be some amount of unwanted sideband.

3.15.3 The Third Method

This method was developed by D.K. Weaver in 1950. It is similar to the phase shift method explained above, but it differs in that the modulating signal is first modulated on a low-frequency sub-carrier which is then modulated on to the high-frequency carrier. Initial modulation on a low-frequency sub-carrier eliminates the need of a 'wide band' phase shifter. The block diagram for a third method SSB modulator is shown in Fig. 3.37. It can be seen from the block diagram that the inputs and outputs of the two phase shifters are single frequencies $f_o, f_o + 90°$ and $f_c, f_c + 90°$. The input to the first balanced modulator is the modulating signal f_m and the first sub-carrier f_o phase shifted by 90°. The input to the second balanced modulator is the modulating signal f_m and the first sub-carrier f_o without any phase shift. The output of the first balanced modulator contains the upper and lower sidebands, each shifted in phase by 90°, i.e., $f_o \pm f_m$ + 90°. The output of the second balanced modulator contains both the sidebands. The upper sidebands are removed by their respective LPF which has the upper cut-off frequency equal to that of the suppressed first sub-carrier.

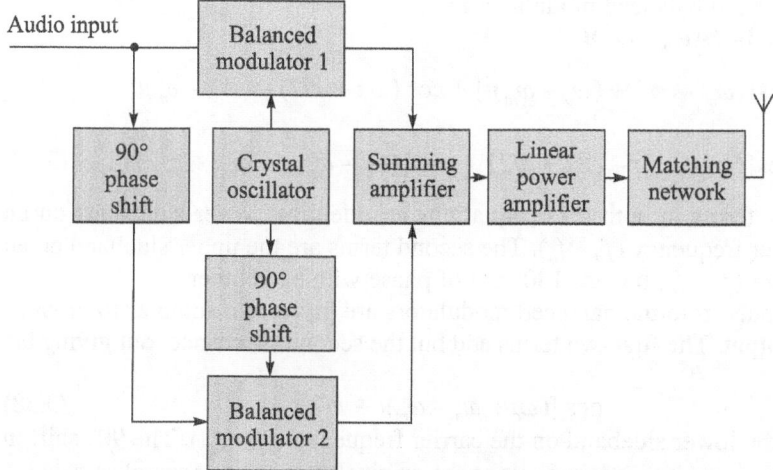

Fig. 3.36 Phase shift method of SSB suppressed carrier signal

The output from LPF1 $[(f_o - f_m + 90°)]$ is mixed with the RF carrier f_c in the balanced modulator 3 and the output of LPF2 $[(f_o - f_m)]$ is mixed with a 90° phase shifted RF carrier f_c + 90° in balanced modulator 4. The outputs from balanced modulators 3 and 4 are combined in the linear summer.

The output of the linear summer is the lower sideband on an offset carrier frequency $f_o + f_c$.

This can be shown mathematically as follows:

Let $\cos \omega_m t$ be the modulating signal and $\cos \omega_o t$ is the sub-carrier, then the output from balanced modulator 1 is

$$(\cos \omega_o t + \pi/2)(\cos \omega_m t) = \frac{1}{2}[\cos(\omega_o t + \omega_m t + \pi/2)] + [\cos(\omega_o t - \omega_m t + \pi/2)]$$

(3.54)

The output from balanced modulator 2 is

$$\cos \omega_o t \cos \omega_m t = \frac{1}{2}[\cos(\omega_o t + \omega_m t) + \cos(\omega_o t - \omega_m t)] \qquad (3.55)$$

Low-pass filter with a cut-off frequency set at sub-carrier frequency f_o removes the sum, the first term from each of the above signals leaving only the difference term as inputs to balanced modulators 3 and 4. These are lower sideband signals on f_o and are identical except that the signal applied to balanced modulator 3 is shifted by $\pi/2$. This process eliminates the need to provide a wide band $\pi/2$ phase shifting network for the baseband signals as in the case of the phase shifting method.

The RF carrier f_c is applied directly to balanced modulator 3 but shifted by $\pi/2$ and applied to balanced modulator 4. Output from balanced modulator 3 will be

$$\cos\omega_c t \cos\left[(\omega_o - \omega_m)t + \pi/2\,\right] = \frac{1}{2}\ [\cos\{\omega_c t + (\omega_o - \omega_m)t + \pi/2\}$$

$$+ \cos[\omega_c t - \{(\omega_o - \omega_m)t + \pi/2\}]$$

$$= \frac{1}{2}\ [\cos(\omega_c + \omega_o)t - \omega_m t + \pi/2]$$

$$+ \cos[(\omega_c - \omega_o)t + \omega_m t - \pi/2] \qquad (3.56)$$

The output from balanced modulator 4 is
$\cos(\omega_c t + \pi/2)\cos(\omega_o - \omega_m)t$

$$= \frac{1}{2}\ [\cos\{(\omega_c t + \pi/2) + (\omega_o - \omega_m)t\} + \cos\ (\omega_c t + \pi/2) - (\omega_o - \omega_m)t]$$

$$= \frac{1}{2}\ [\cos\{(\omega_c + \omega_o)t - \omega_m t + \pi/2\} + \cos\ \{(\omega_c - \omega_o)\ t + \omega_m t + \pi/2\}] \qquad (3.57)$$

The first terms in both these equations are identical lower sidebands on an offset carrier frequency $(f_c + f_o)$. The second terms are the upper sideband on an offset carrier $(f_c - f_o)$ but are $180°$ out of phase with each other.

The outputs from the balanced modulators are given to a summer to produce the final output. The first two terms add but the second two cancel out giving the output as

$$\cos\left[(\omega_c + \omega_o - \omega_m)t + \pi/2\,\right] \qquad (3.58)$$

This is the lower sideband on the carrier frequency $(f_c + f_o)$. The $90°$ shift in the output is of no consequence since the original carrier has been eliminated.

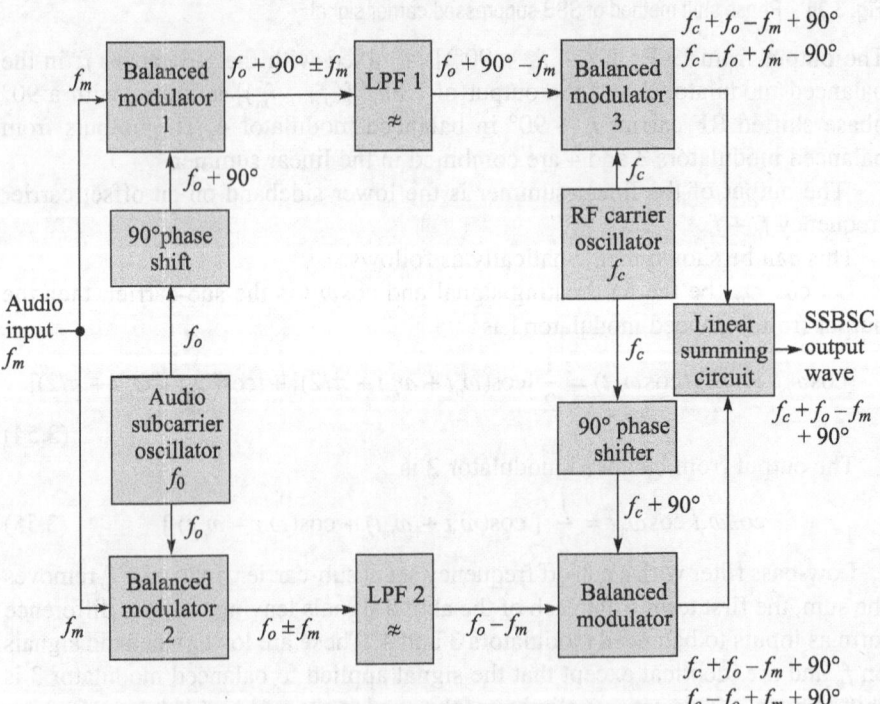

Fig. 3.37 Third method for SSB suppressed carrier generation

This signal is applied to a linear power amplifier and to the antenna for radiation. To get the upper sideband, before feeding to the summer, the outputs from balanced modulators 3 and 4 should be inverted.

3.16 INDEPENDENT SIDEBAND TRANSMITTER

Figure 3.38 shows a block diagram for an independent sideband (ISB) transmitter. It uses three stages of modulation. The transmitter uses the filter method to produce two independent SSB channels. The two channels are combined and a pilot carrier is reinserted. The composite ISB reduced carrier is up converted to RF with two additional stages of frequency translation. Two independent sources generate 5 kHz wide information signals. Channel 1 information signal modulates a 100 kHz LF carrier in balanced modulator 1. The output from this modulator passes through BPF 1. This filter is tuned to LSB, i.e., 95 kHz to 100 kHz.

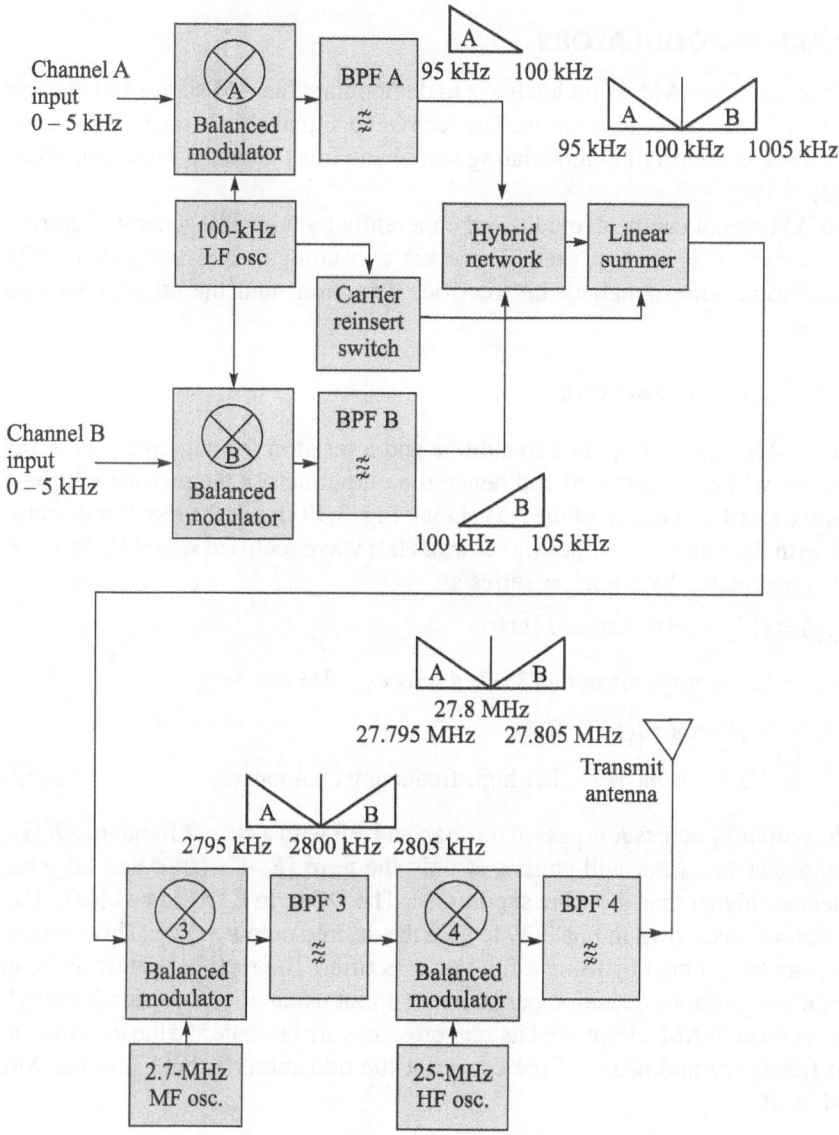

Fig. 3.38 ISB transmitter

The channel 2 information signal modulates the same LF carrier in balanced modulator 2. The output from this modulator passes through BPF 2, which is tuned to USB, i.e., 100 kHz to 105 kHz. These two sideband spectra are combined in a hybrid network to form a composite ISB suppressed carrier signal. The 100 kHz carrier is reinserted in the linear summer to form an ISB reduced carrier signal. The ISB spectrum is mixed with a 2.7 MHz MF carrier in balanced modulator 3. This output passes through BPF 3 to produce an ISB reduced carrier signal that extends from 2.795 to 2.805 MHz with a reduced 2.7 MHz pilot carrier. Balanced modulator 4, BPF 4, and HF carrier translate the MF spectrum to an RF band that extends from 27.795 to 27.8 MHz (channel 1) and 27.8 to 27.805 MHz (channel 2) with a 27.8 MHz reduced amplitude carrier.

3.17 AM DEMODULATORS

The function of an AM demodulator is to demodulate the AM signal and recover the original source of information. The recovered signal should contain the same frequencies as the original modulating signal and must have the same amplitude characteristic.

An AM signal can be demodulated coherently by a locally generated carrier. But this method is seldom used. There are two non-coherent methods of AM demodulation. One is called the 'rectifier detection' and the other 'envelope detection'.

3.17.1 Rectifier Detector

When an AM signal is applied to a diode and a resistor, the negative part of the AM wave will be suppressed and hence the output across the resistor will be a half wave rectified version of the AM signal. Fig. 3.39 shows the rectifier detector along with the input and output waveform. Half wave rectified signal V_R across R can be represented by a Fourier series as

$$V_R(t) = [\{E_c + m(t)\} \cos \omega_c t\,] \omega(t)$$

$$= [E_c + m(t)] \cos \omega_c t\, [1/2 + 2/\pi (\cos \omega_c t - 1/3 \cos 3\omega_c t$$

$$+ 1/5 \cos 5\omega_c t + \cdots)]$$

$$= 1/\pi [E_c + m(t)] + \text{other high-frequency components} \quad (3.59)$$

The output V_R across R is passed through an LPF with a cut-off frequency B Hz, the output of this filter will consist of only the term $[E_c + m(t)]/\pi$ and all other frequencies higher than B Hz are suppressed. The DC term E_c/π can be blocked by a capacitor C, as shown in Fig. 3.39 to give the desired output $m(t)/\pi$. This average output can be doubled by using a full wave rectifier. The rectified detection is, in effect, a synchronous detection performed without using a local carrier. The high carrier content in AM ensures that its zero crossings are periodic and the information about frequency and phase of the carrier at the transmitter is built into the AM signal itself.

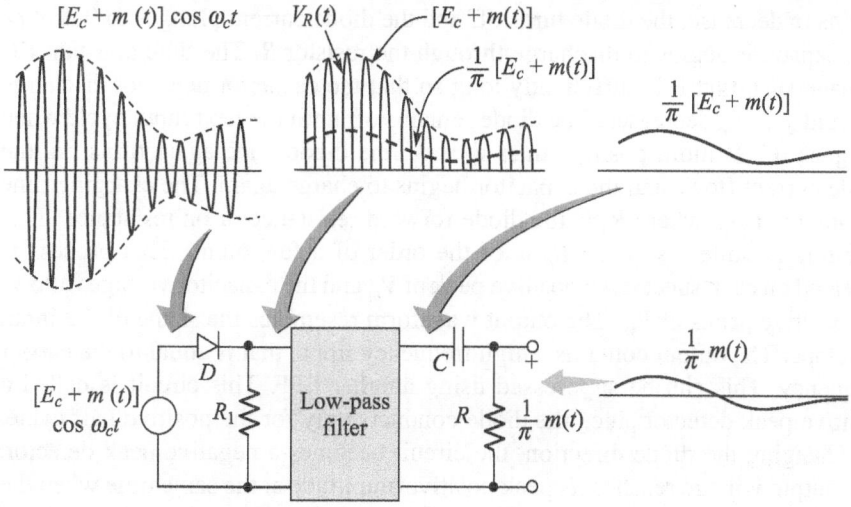

Fig. 3.39 Rectifier AM detector

3.17.2 Envelope Detector

The most common AM detector circuit is the diode envelope detector. A circuit diagram for an envelope detector is shown in Fig. 3.40. During the positive half cycle of the input signal, the diode conducts and the capacitor charges up to the peak value of the input signal. When the input voltage falls below that of the capacitor, the diode becomes reverse biased and the input disconnects from the output. When the diode is off, the capacitor discharges through R.

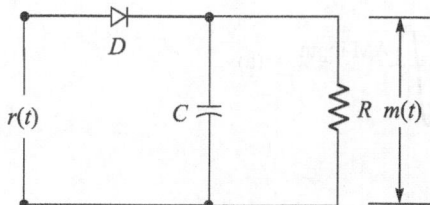

Fig. 3.40 Envelope detector

The amount of discharge depends on time constant RC of the circuit. On the next cycle of the carrier, the diode again conducts when the input signal exceeds the voltage across the capacitor. The capacitor again charges up to the peak value of the input signal and this process is repeated, since the output follows the envelope of the input. This circuit is called envelope detector. It is also called a peak detector since it detects the peak of the input envelope.

Figures 3.41(a) – (c) show the detctor input voltage waveform, the corresponding diode current waveform, and the detector voltage waveform.

At time t_0, the diode is reverse biased, the diode current is zero, the capacitor is fully discharged and the output is 0 V. As the input voltage exceeds 0.7 V, the cut-in voltage of the diode at t_1, the diode turns on and the diode current begins to flow. This current charges the capacitor. The capacitor voltage remains 0.7 V below the input voltage until V_m reaches its peak value. When the input voltage begins to decrease, the diode turns off and the diode current plunges to 0 A at t_2. The capacitor begins to discharge through the resistor R. The time constant RC is made such that it is sufficiently long so that the capacitor does not discharge as rapidly as V_m decreases. The diode remains off until the next input cycle when V_m goes 0.7 V more positive than V_c at t_3, the diode once again turns on, the diode current flows and the capacitor begins to charge again. The charging time constant is CR_d, where R_d is the diode forward resistance or on resistance. This charging is quite fast since R_d is of the order of a few ohms. This process is repeated on each successive positive peak of V_m and the capacitor voltage follows the positive peaks of V_m. The output waveform resembles the shape of the input envelope. The output contains a high frequency ripple that is equal to the carrier frequency. This can be suppressed using another LPF. This circuit is called a positive peak detector since the diode conducts only for the positive half cycle. By changing the diode direction, the circuit becomes a negative peak detector. The output voltage reaches its peak positive amplitude at the same time when the input envelope reaches its maximum positive value V_{max} and the output voltage goes to its minimum peak amplitude at the same time that the input voltage goes to its minimum V_{min}. For 100% modulation V_{out} swings from 0 V to $V_{max} - 0.7$ V.

The time constant RC plays a crucial role in the envelope detector. RC should be selected to follow the variations in the envelope of the modulated signal. If RC is too small, then the output of the filter falls very rapidly after each peak and will not follow the envelope of the modulated signal closely. This corresponds to the case where the bandwidth of the LPF is too large.

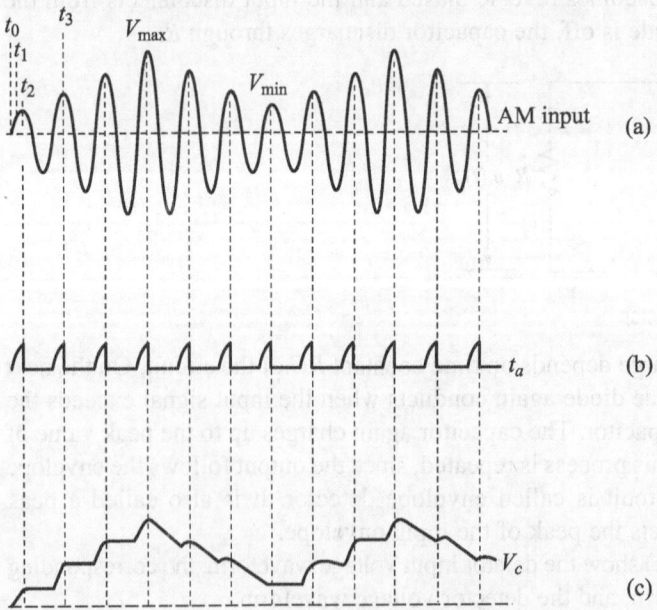

Fig. 3.41 Peak detector: input waveform, diode current, and output voltage

If RC is too large then the discharge of the capacitor is too low and again the output will not follow the envelope of modulating signal. This corresponds to the case where the bandwidth is too small. The effects of large and small RC values are shown in Fig. 3.42. For the best performance of envelope detector, the time constant RC should be in the range $1/f_c \ll RC \ll 1/B$, where f_c = carrier frequency and B = bandwidth of the signal.

3.17.3 Detector Distortion

When successive positive peaks of the detector input waveform are increasing, the capacitor should hold its charge between peaks. This is possible with a relatively long RC constant. However, when the positive peaks are decreasing in amplitude, the capacitor should discharge between successive peaks to a value less than the next peak. This requires RC to be quite small. Hence, a trade-off between a long and short time constant is required. If the RC time constant is too short, the output waveform resembles a half wave rectified signal. This is called *rectifier distortion*. If the RC time constant is too long, the slope of the output waveform cannot follow the trailing slope of the envelope. This type of distortion is called *diagonal clipping*. These are shown in Fig. 3.43. The RC network used in an envelope detector is an LPF. The slope of the envelope depends on both the modulating signal frequency and the modulation coefficient m. Hence, the maximum slope, i.e., the fastest rate of change occurs when the envelope is crossing the zero axis in the negative direction. The higher modulating frequency that can be demodulated by an envelope detector without attenuation is

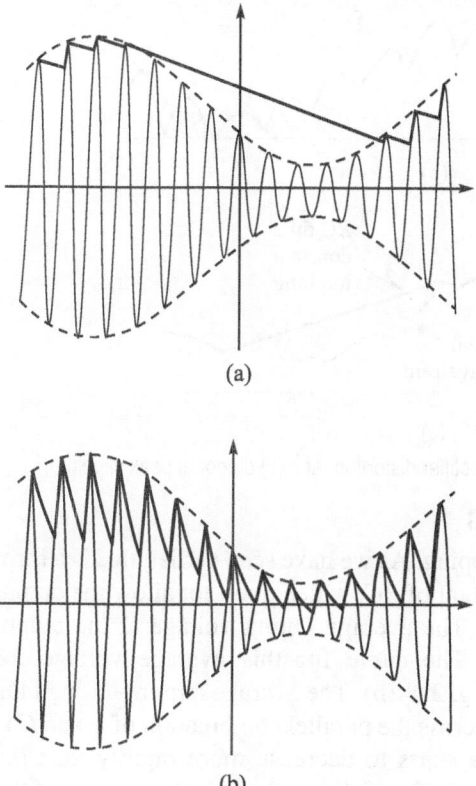

(a)

(b)

Fig. 3.42 Envelope detector waveforms: (a) RC too large and (b) RC too small

$$f_{m(\max)} = \sqrt{\frac{(1/m \times m) - 1}{2\pi RC}} \qquad (3.60)$$

where

$$f_{m(\max)} = \text{maximum modulating frequency in Hz}$$
$$m = \text{modulating index}$$
$$RC = \text{time constant in seconds}$$

For 100% modulation, $f_{m(\max)} = 0$, i.e., all the modulating signal frequencies are attenuated as they are demodulated.

For 70.7% modulation, $f_{m(\max)} = 1/2\pi RC$

This equation is used when designing envelope detectors to determine an approximate maximum modulating signal.

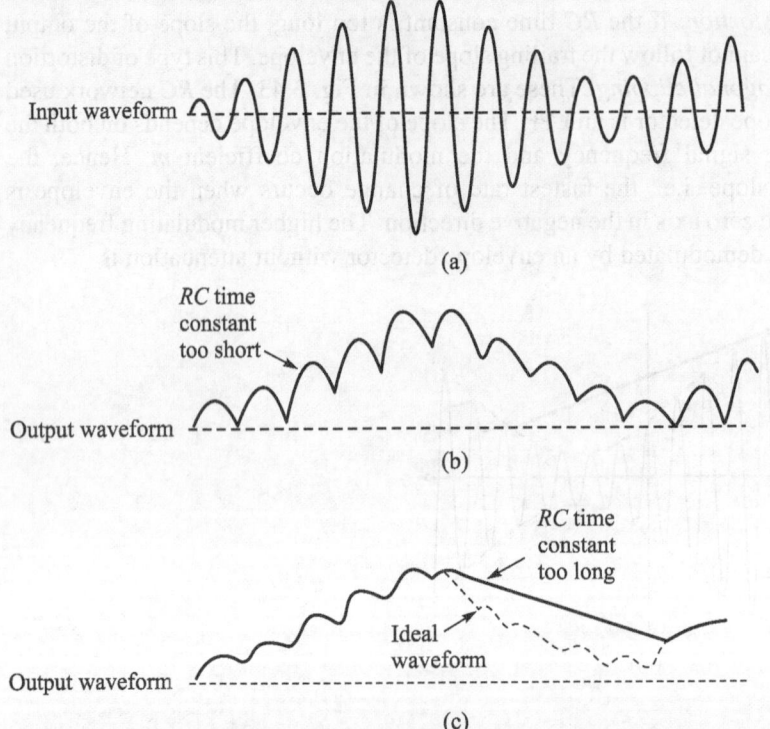

Input waveform

(a)

RC time
constant
too short

Output waveform

(b)

RC time
constant
too long

Ideal
waveform

Output waveform

(c)

Fig. 3.43 Detector distortion: (a) input, (b) rectifier distortion, and (c) diagonal peak clipping

3.17.4 Diagonal Peak Clipping

Figure 3.44 shows diagonal peak clipping. As we have seen earlier, this is a form of distortion due to a long time constant of RC, which does not allow the output to follow the modulation envelope. The average output voltage at the output follows the modulation envelope. The curve for this average voltage for sinusoidal modulation is shown in Fig. 3.44(b). The average current through the diode is I_{av} and the average voltage across the parallel combination of R and C is V_{av}. At t_A, the modulation envelope starts to decrease more rapidly than the capacitor discharge. The output voltage then follows the discharge curve of the RC network until time t_B when it meets the modulation envelope and the curve once again increases.

To avoid this diagonal peak clipping, the modulation index m should be less than or equal to $|Z_m|/R$, where Z_m is the impedance of parallel combination of RC at the modulation frequency and R is the value of the resistor. This relation can be proved as follows.

Because of RC network, the current leads the voltage as shown in Fig. 3.44(c). The average current consists of two components, a DC component I_{DC} and an AC component that has a peak value of I_P. The DC component of voltage is approximately equal to the maximum unmodulated carrier voltage or $V_{DC} = E_{C\max}$ and the direct current is $I_{DC} = V_{DC}/R$.

The peak value of the average output voltage is $V_P = mE_{C\max}$ and the peak current is $I_P = V_P/Z_m$.

If the envelope falls faster than the capacitor discharge, the diode stops to conduct and the current supplied by the diode falls to zero, i.e., $I_{av} = 0$. During this period, the output voltage follows the discharge curve of the capacitor. This results in a diagonally clipped peak as shown in Fig. 3.44(b). Hence, to avoid this, the direct current has to be greater than the peak current, i.e., $I_{DC} = |I_P|$.

Hence, $$\frac{V_{DC}}{R} \geq \frac{mV_{DC}}{|Z_m|} \quad \text{or} \quad m \leq \frac{|Z_m|}{R} \tag{3.61}$$

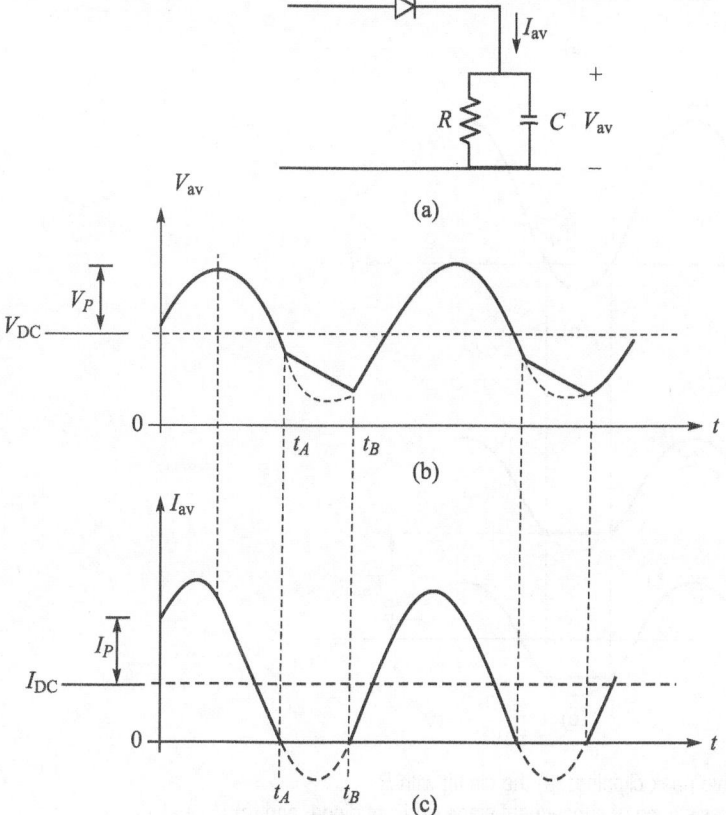

Fig. 3.44 Diagonal peak clipping: (a) circuit, (b) voltage waveform, and (c) current waveform

3.17.5 Negative Peak Clipping

This is due to the loading of the next stage. The circuit and the V_{DC}/R waveform are shown in Fig. 3.45. Capacitor C_1 is the DC blocking capacitor and resistor R_1 is due to the input resistance of the next stage. At modulating frequency, the reactance of C_1 is very small compared to that of C. Hence, the AC impedance Z_m is a parallel combination of R and R_1, i.e., $R_P = RR_1/R + R_1$. The modulation index must satisfy the condition: $m = R_P/R$.

Under this condition, current I_{av} is in phase with V_{av} and over the period when $I_{av} = 0$, the voltage across capacitor C_1 approximately remains constant at V_{DC}. The voltage across R is then $V_{Rmin} = V_{DC} R/R_1$. It is this voltage that keeps the diode off. The voltage V_{av} and V_1 are shown in Fig. 3.45. It can be seen from the figure that both V_{av} and V_1 get clipped at the negative portion and hence the name 'negative peak clipping'.

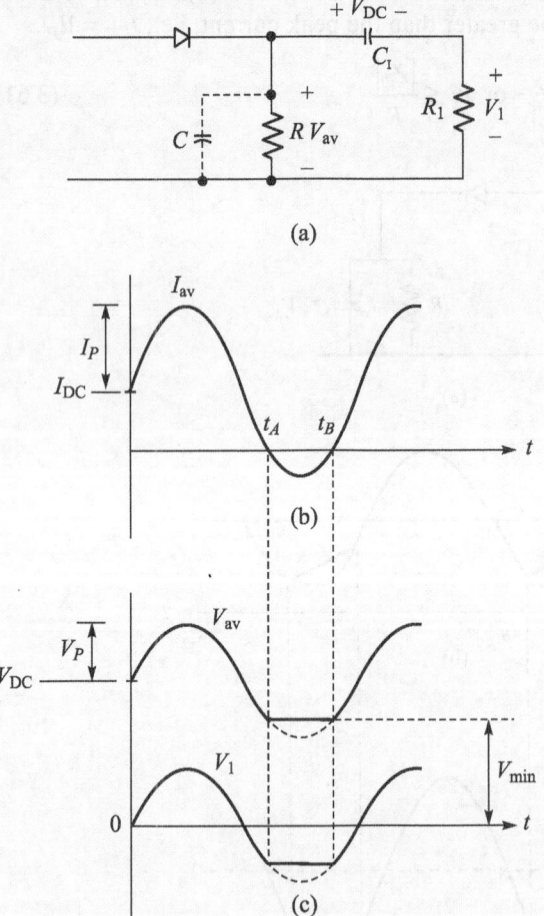

(a)

(b)

(c)

Fig. 3.45   Negative peak clipping: (a) the circuit with R_1
input resistance of subsequent stage,(b) I_{av} of diode, and (c) V_{av} and V_1

3.18 SSB RECEPTION

There are two methods to extract the modulating signal from the transmitted SSB signal. One method employs the use of a phase coherent carrier signal. In the other method, a rectifier is used to detect the SSB signal with a carrier. Both methods are explained in detail below.

3.18.1 Coherent Detection

Figure 3.46 shows the block diagram of this method. In this method, the carrier signal for coherent detection is locally generated if the SSB signals are transmitted with carrier completely suppressed. This requires extreme stability for the local oscillator signals used for demodulation. A crystal oscillator or phase locked loop (PLL) can be used to generate this carrier. For higher stability, PLL is preferred. The coherent carrier is given as one of the inputs to the balance modulator, the other input being the SSB signal. The output of the balanced modulator contains the baseband signal and the other unwanted SSB signal with a carrier $2\omega_c$. A low-pass filter at the output suppresses the unwanted SSB signals giving the desired baseband signal, i.e., modulating signal. This can be mathematically proved as follows:

$$\text{SSB signal} = \cos(\omega_c - \omega_m)t$$
$$= \cos \omega_c t \cos \omega_m t + \sin \omega_c t \sin \omega_m t \qquad (3.62)$$

Let the carrier signal be $\sin \omega_c t$. The output of the balance modulation is

$$\sin \omega_c t\{\cos \omega_c t \cos \omega_m t + \sin \omega_c t \sin \omega_m t\}$$
$$= \cos \omega_m t \sin \omega_c t \cos \omega_c t + \sin \omega_m t \sin^2 \omega_m t$$
$$= (1/2) \cos \omega_m t \sin 2 \omega_c t + (1/2) (1- \cos 2\omega_c t) \sin \omega_m t$$
$$= (1/2) \sin \omega_m t + (1/2) (\cos \omega_m t \sin 2\omega_c t - \sin \omega_m t \cos 2\omega_c t) \qquad (3.63)$$

When this signal is passed through an LPF, it allows the modulating signal $(1/2)$ $\sin \omega_m t$ and rejects other signals.

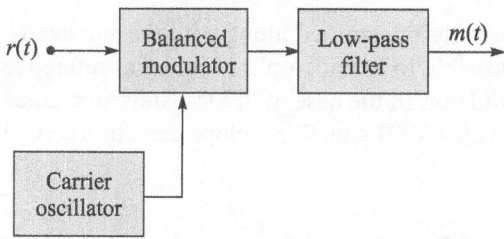

Fig. 3.46   Coherent detection

3.18.2 SSB Reception with Pilot Carrier

In the case of a pilot carrier system, a low level carrier signal is transmitted along with the single sideband. This pilot carrier is used to synchronize, the carrier oscillator or the PLL in the receiver, thus eliminating any modulation distortion due to an incorrect carrier frequency. Rest of the analysis for this method is the same as the coherent method of detection.

Envelope detection of SSB signal with a carrier

SSB signal can be expressed as

$$\cos \omega_m t \cos \omega_c t + \sin \omega_m t \sin \omega_c t = \cos \omega_m t \cos \omega_c t + \cos (\omega_m t - \pi/2) \sin\omega_c t \qquad (3.64)$$

Let $m(f) = \cos \omega_m t$. Then $\sin \omega_m t = \cos(\omega_m t - \pi/2)$ can be represented as $m_h(t)$, where $m_h(t)$ is the Hilbert transform of $m(t)$. Hence, the SSB signal in terms of the Hilbert transform is

$$m(t) \cos \omega_c t + m_h(t) \sin \omega_c t \qquad (3.65)$$

To this signal, we add a carrier signal $A\cos \omega_c t$. Then the resultant signal is

$$A\cos \omega_c t + [m(t)\cos \omega_c t + m_h(t) \sin \omega_c t] \qquad (3.66)$$

If the carrier amplitude A is large enough, $m(t)$ can be recovered by envelope detection using a non-linear device like a diode. This can be shown as follows:

The carrier plus SSB signal $= A\cos \omega_c t + [m(t)\cos \omega_c t + m_h(t)\sin \omega_c t]$ (3.67)

$$= [A + m(t)] \cos \omega_c t + m_h(t) \sin \omega_c t$$

$$= E(t) \cos(\omega_c t + \theta)$$

where $E(t)$, the envelope of the combined signal, is given by

$$E(t) = [\{A + m(t)\}^2 + m^2_h(t)]^{1/2}$$

$$= A \left[1 + \frac{2m(t)}{A} + m^2(t)/A^2 + m^2_h(t)/A^2 \right]^{1/2} \qquad (3.68)$$

If $A >> m(t)$, then A is also $>> |m_h(t)|$ and the terms $m^2(t)/A^2$ and $m^2_h(t)/A^2$ can be ignored. Thus, $E(t)$ can be approximated as $E(t) = A[1 + 2m(t)/A]^{1/2}$. Using binomial expansion and ignoring higher order terms, we get

$$E(t) = A[1 + m(t)/A] = A + m(t) \qquad (3.69)$$

Hence, for a larger carrier, SSB can be demodulated by an envelope detector. Since A, the carrier amplitude, is quite larger than $m(t)$ the efficiency of this method is low compared to the coherent phase method.

3.19 DEMODULATION OF VSB SIGNALS

In VSB, a carrier component is generally transmitted along with the sidebands. The existence of carrier makes it possible to extract a phase coherent reference for demodulation in a balanced modulator. In the case of TV transmission, since a large carrier is transmitted along with a VSB signal, envelope detection can be used.

3.20 DETECTION OF ISB SIGNALS

ISB signals are similar to DSB signals. The only difference is that each sideband consists of a different modulating signal. Hence, to recover the ISB signal, two separate BPFs are used.

3.21 TRANSMITTERS

We have discussed so far how AM, DSB, SSB, VSB, and ISB signals are generated. Now we will discuss transmitters. Transmitters transmit these modulated signals with a required power over a distance. Most of the transmitters use class C amplifier to improve the efficiency of transmission. The audio signal is amplified by a series of low level audio amplifiers and a power amplifier. This power amplifier controls the power being delivered to the final amplifier. Hence, it should have the power driving capacity, i.e., the collector supply must deliver one half the maximum power to the RF amplifier under 100% modulation conditions.

A low power transmitter with the output power up to 1 kW uses low power modulation and audio amplifiers. But class C amplifier cannot be used to amplify an already modulated carrier due to its non-linearity. This non-linearity introduces distortion in the modulation envelope. Normally, a linear power amplifier such as a class B push-pull amplifier is used. But this is at the cost of efficiency.

The output of the final amplifier is passed through an impedance matching network. The Q of the matching network should be low enough to allow all the sidebands of the signals without amplitude/frequency distortion, but at the same time must have considerable attenuation for the unwanted signal. The bandwidth required is 3 dB at ±5 kHz around the carrier. For AM broadcast transmitters, the response is broadened so that the sideband will be down as less as 1dB at 5 kHz. This is because in music programmes one expects a low level of distortion. To reduce distortion in class C amplifiers, a negative feedback is quite often used. A sample of RF signal given to the antenna is demodulated and fedback to the modulator power amplifier along with the modulating input.

3.21.1 AM Transmitters

AM transmitters are classified as low level and high level transmitters. In the low level transmitters, the modulation process is done at a lower power level and then the modulating signal is passed through a high level power amplifier.

In a high level transmitter, modulation and power amplification are done at a higher level. This requires the modulating signal and carrier signal to be brought to a certain power level before modulation is effected.

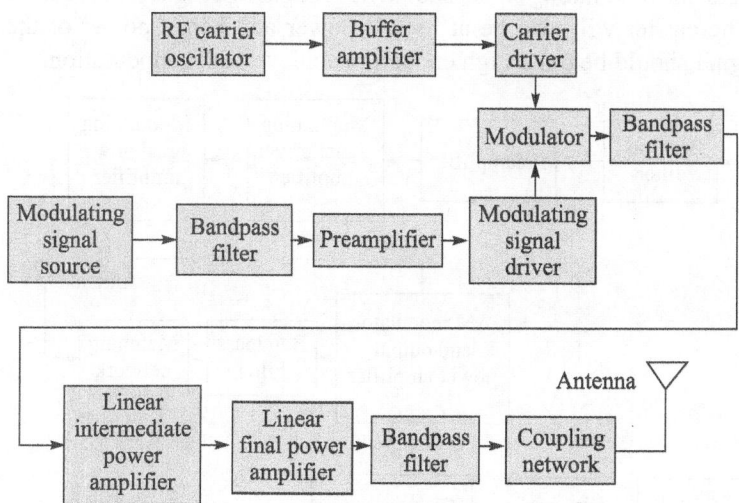

Fig. 3.47 Low level AM DSBFC transmitter

Low level transmitters

Figure 3.47 shows a block diagram for a low level AM double sideband with a full carrier transmitter. These types of transmitters are predominantly used for a low-power, low-capacity system such as wireless intercoms, remote control units, radio pagers, and short range walkie-talkies. The modulating signal is obtained from a microphone, or an auxiliary output of a magnetic tape recorder or a phonographic record. The preamplifier is a typically sensitive class A linear voltage amplifier, the input stage being a differential amplifier. This amplifier must have a high input impedance. The purpose of the preamplifier is to bring the source signal to such a level so that the input to the driver amplifier is noise and distortion free. The driver of the modulating signal is also a linear amplifier which amplifies the modulating signal to an adequate level to sufficiently drive the modulator.

The RF carrier signal is from an oscillator, which can be of any one of conventional type crystal oscillators, or a phase locked loop. This can be used to generate the carrier whose frequency stability is quite high. The buffer amplifier is a low-gain, high-input impedance linear amplifier that isolates the oscillator from the high-power amplifiers. The emitter followers or of late operational amplifiers are used as buffers. Modulators can be either emitter or collector modulation type. The intermediate and final power amplifiers are generally class A or class B push-pull type. This helps to maintain symmetry in the AM envelope. The output impedance of the final power amplifier is matched to the antenna by using an antenna matching network.

High level transmitters

Figure 3.48 shows the block diagram of a high level transmitter. The modulating signal goes through the same stages as in the case of low-power transmitters except for the addition of a power amplifier. This is due to the fact that for high-level transmitters, the modulating signal should be brought to a higher level before modulation. The carrier will also be at its full power and hence power of the modulating signal should be quite high enough to achieve 100% modulation.

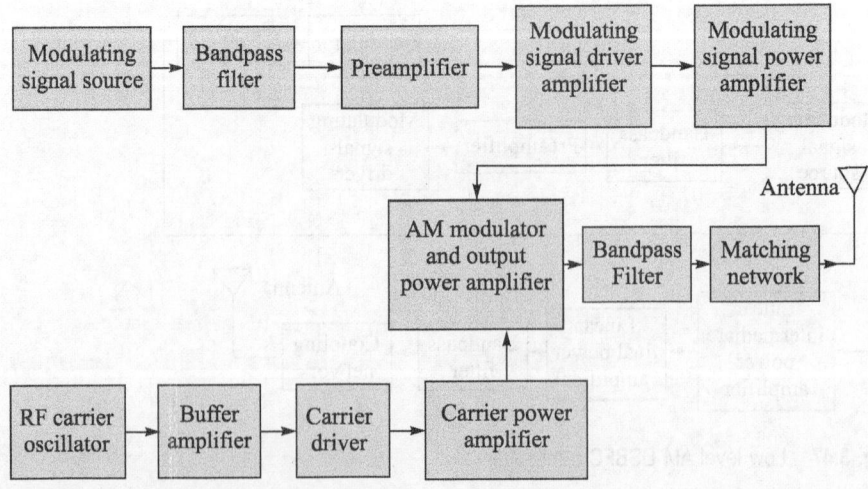

Fig. 3.48 High level AM DSBFC transmitter

An RF oscillator and its associated circuit are similar to that of a low-level transmitter. The carrier signal also requires an additional power amplifier before it is given to the modulator. The final power amplifier is the actual modulator. Collector modulator class C type has a very good efficiency. The matching network is used to couple power amplifier output through a BPF with the antenna.

3.21.2 SSB Transmitters

A typical pilot carrier SSB transmitter is shown in Fig. 3.49. The balanced modulator output is given to the USB filter. The linear summer inserts the attenuated carrier (pilot carrier) to the USB signal. The USB signal plus the attenuated carrier is translated to a higher RF frequency by another balanced modulator. The output of this balanced modulator is fed to the linear power amplifier through a BPF and coupled with the antenna for transmission.

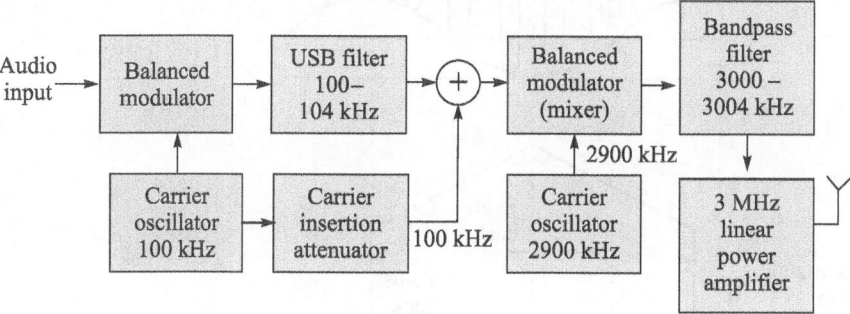

Fig. 3.49   SSB transmitter with a pilot carrier

3.22 TRAPEZOIDAL PATTERNS

These are used for observing the modulation characteristics of AM transmitters. The characteristics are modulation symmetry and coefficient of modulation. We can observe the waveforms on an oscilloscope. The trapezoidal pattern is the one that is more easily and accurately interpreted. In this method, the time base of the oscilloscope is not used. Instead an external modulating signal is applied to the external horizontal input disabling the internal horizontal sweep circuit. The AM wave is applied to the vertical input of the oscilloscope. Thus, the horizontal sweep rate is determined by the modulating signal frequency and the magnitude of the horizontal deflection is proportional to the amplitude of the modulating signal. The amplitude and rate of change of modulated signal determine the vertical deflection. The modulated signal and the modulating signal produce a trapezoidal pattern on the cathode ray tube (CRT) as shown in Fig. 3.50. Initially, when there is no modulated and modulating signal, the electron beam is located at the centre of the CRT. As the modulating signal goes positive, the beam deflects to the right. At the same time, the modulated signal is going positive, which deflects the beam upward. The beam continues to deflect to the right until the modulating signal reaches its maximum peak value at the instant t_1. While the beam moves towards the right, it is also deflected up and down as the modulated signal alternates between positive and negative. It can be seen that the modulated signal reaches a higher magnitude on each successive alternation than the previous one.

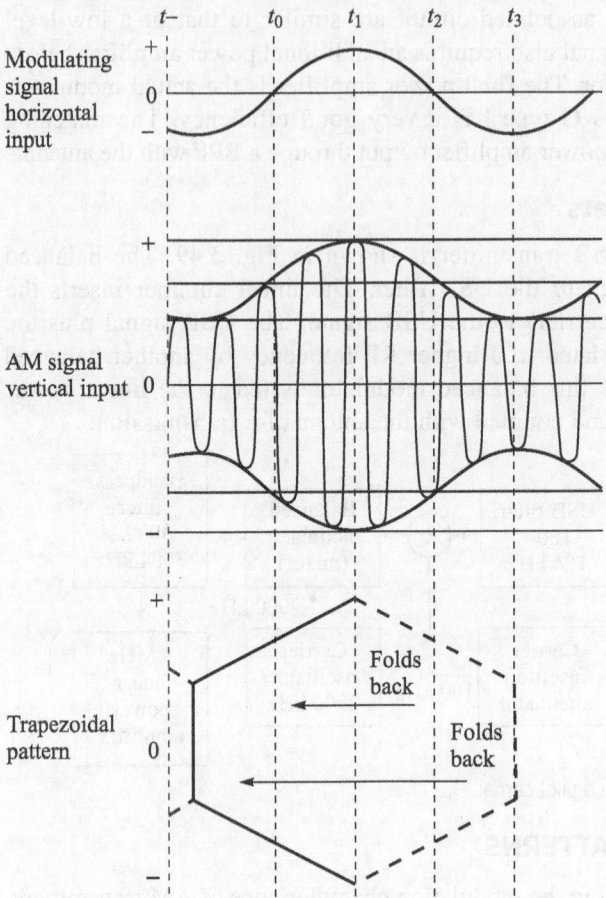

Fig. 3.50 Trapezoidal pattern on a CRT

Hence, as the CRT beam is deflected to the right, its peak to peak vertical deflection increases with each successive cycle of the modulated signal. As the modulating signal becomes less positive, the beam is deflected towards the left. Simultaneously, the modulated signal alternately swings between positive and negative, deflecting the beam up and down except that each successive alternation is lower in amplitude than the previous one. Now as the beam moves horizontally towards the centre of CRT, the vertical deflection decreases. The modulating and modulated signals pass through 0 V at the same time and the beam is again at the centre of the CRT. This is at the instant t_2. The beam is deflected to the left side of the CRT as the modulating signal goes negative and on each successive alternation the modulated signal is decreasing in amplitude at the same time. The modulated signal reaches its minimum amplitude at the instant t_3 and at the same time the modulating signal reaches its maximum negative value. The trapezoidal pattern shown between t_1 and t_3 folds back on top of the pattern displayed during the time interval $t-$ and t_1. Thus, a complete trapezoidal pattern is displayed on the screen after both the left to right and right to left horizontal sweeps are complete.

Trapezoidal patterns are shown in Fig. 3.51 for different modulation indices. Figure 3.51(a) shows that if the modulation is symmetrical , the top half of the modulated signal is a mirror image of the bottom half. Figure 3.51(b) shows the trapezoidal pattern for 100% modulation. Figure 3.51(c) shows the trapezoidal pattern for overmodulation. Figure 3.51(d) shows the pattern for both the modulating and modulated signal when they are out of phase. Figure 3.51(e) shows the pattern when the magnitudes of the positive and negative alternations of the modulated signal are not equal. We can see that the percentage modulation and modulation symmetry are seen clearly with a trapezoidal pattern compared to a standard oscilloscope display of the modulated signal.

$$\% \text{ Modulation} = \frac{V_{max} - V_{min}}{V_{max} + V_{min}} \times 100 \qquad \% \text{ Modulation} = \frac{V_{max} - 0}{V_{max} + 0} \times 100$$

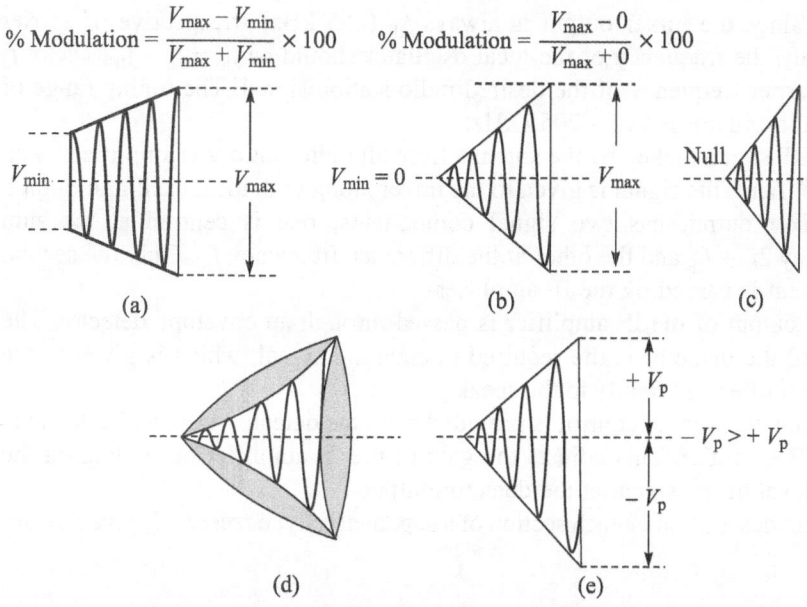

Fig. 3.51 Trapezoidal patterns: (a) 50% linear AM, (b) 100% AM, (c) greater than 100% AM, (d) phase relationship improper, and (e) AM envelope non-symmetric

$$G_{dB} = \text{gains}_{dB} - \text{losses}_{dB} \qquad (3.69)$$

3.23 RECEIVERS

Most of the AM and SSB receivers are superheterodyne. They are generally commercial receivers.

3.23.1 AM Receivers

Most of the AM receivers cater to the reception of AM radio broadcasting. They operate in the frequency band 535–1605 kHz for transmission of voice and music with the carrier frequency in the range 540–1600 kHz with10 kHz spacing.

The baseband signal is limited to a bandwidth of 5 kHz and hence the carrier frequencies are spaced at 10 kHz from each other. The block diagram of a superheterodyne receiver is shown in Fig. 3.52. It consists of a radio frequency tuned amplifier, a mixer, a local oscillator, an intermediate frequency (IF) amplifier, an envelope detector, an audio frequency amplifier, and a speaker. Heterodyne means to mix two frequencies together in a non-linear device or to translate one frequency to another using non-linear mixing.

Tuning of a desired radio frequency is provided by a ganged condenser which simultaneously tunes the RF amplifier and the local oscillator.

The advantage of a superheterodyne receiver is that all radio frequencies are converted to a single frequency called intermediate frequency of 455 kHz. This conversion helps the use of a single tuned IF amplifier for signals from any radio station in the frequency band. The mixer output is always 455 kHz. The IF amplifier has the bandwidth of 10 kHz matching with that of the transmitted signal. Since the mixer output is always f_{IF} (455 kHz) irrespective of carrier frequency, the frequency of the local oscillator should be $f_{lo} = f_c + f_{IF}$, where f_c is the carrier frequency of the desired radio station signal. The tuning range of the local oscillator is 995 – 2055 kHz.

The RF signal picked by the antenna from all radio stations is amplified by an RF amplifier. This signal is given to the mixer along with local oscillator output. The mixer output has two signal components; one is centred at the sum frequency $2f_c + f_{IF}$ and the other at the difference frequency f_{IF}. Only the second component is passed by the IF amplifier.

The output of the IF amplifier is passed through an envelope detector. The output of the detector is the required modulating signal, which is given to the audio amplifier and finally to the speaker.

Automatic volume control is provided from the detector as a feedback signal to the IF amplifier. This adjusts the gain of the IF amplifier depending on the power level of the signal at the detector output.

A brief description of each section of a superheterodyne receiver is given below.

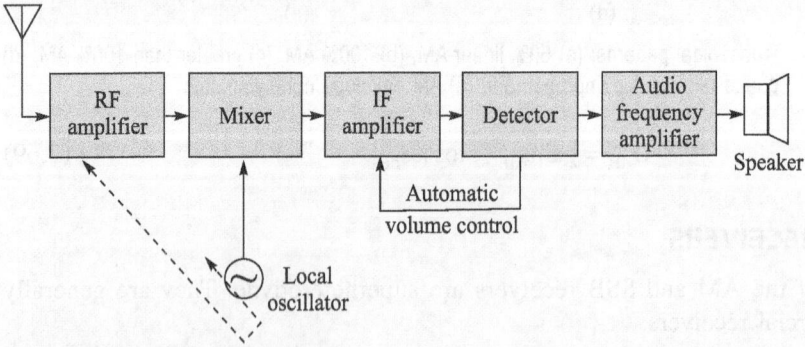

Fig. 3.52 Superheterodyne receiver

RF section It generally consists of a pre-selector and an amplifier stage. They can be a separate circuit or can be a combined circuit. The pre-selector is a bandpass filter with an adjustable centre frequency that is tuned to the desired carrier frequency. The primary purpose of the pre-selector is to provide enough initial bandlimiting to prevent a specific unwanted carrier frequency called the image frequency from entering the receiver. It also reduces the noise bandwidth of the receiver and provides the initial step towards reducing the overall receiver bandwidth to the minimum bandwidth required to pass the information signal. The RF amplifier determines the sensitivity of the receiver. Since it is a first stage, which is the primary contributor of noise, care has to be taken so that the noise figure of the RF stage is taken care of in the designing process. A receiver can have more than one amplifier. The advantages of including an RF amplifier in a receiver are as follows:

(a) Greater gain and hence better sensitivity
(b) Improved image frequency rejection
(c) Better signal to noise ratio
(d) Better selectivity

Mixer/converter   This section includes a local oscillator stage and a mixer/converter stage. The mixer is a non-linear device and converts the radio frequency to the intermediate frequency. Heterodyning takes place here and radio frequencies are downconverted to intermediate frequencies. The intermediate frequency used in AM radio receiver is 455 kHz.

IF section   The IF section consists of a series of amplifiers and bandpass filters. Most of the receiver gain and selectivity are achieved in the IF section. The IF centre frequency and bandwidth are constant for all stations and are chosen so that their frequency is less than any of the RF signal to be received. It is always important to have IF frequencies to be lower than the RF because it is easier and less expensive to construct high-gain, stable amplifiers for a low frequency signal. Moreover low frequency IF amplifiers are less prone to oscillation.

Detector section   This section converts the IF signals to audio signals, i.e., into the original signal. The detector can be a simple envelope detector or a more advanced detector such as balanced modulator and PLL.

Audio amplifier section   This section consists of several cascaded audio amplifiers. The power output depends on the requirement. Present-day audio amplifiers use integrated circuits which occupy less space, more rugged, and reliable.

AM receivers using ICs

With the advent of integrated circuits, the design of AM receiver became much simpler. The advantages of using integrated circuits are miniaturization, reliability, and less power consumption. Except the audio amplifier stage and the tuned circuit, all the blocks shown in the previous AM receivers are integrated in a single IC. The output of the IC is given to an audio amplifier which also can be one of the many ICs available. One such IC is Signetics^{TM} TDA 1072A. It is an AM receiver circuit, which performs the active function and part of filtering function of an AM receiver. It is 16 PIN plastic DIP. and supports oscillator frequency up to 50 MHz and RF signal up to 500 mV. Symmetric design minimizes the sensitivity to interference and RF radiation. The VCO (voltage control oscillator) provides signals with extremely low distortion over the entire frequency range with the help of Varicap diodes. Selectivity is obtained by using a block filter before the IF amplifier. It consumes only 875 mW of power and works with a supply voltage in the range of 7.5 to 18 V, typical being 8.5 V DC. Other features are: gain controlled RF stage, double balanced mixer, separately buffered VCO with temperature compensation, gain control IF stage with wide AGC range, and full wave balanced envelope detector.

The output from TDA 1072A can be connected to TDA 1520B audio amplifier. For more details, the readers are requested to refer to Signetics™ Linear Data Manual, Volume 1, Communications.

AM receivers using phase locked loop

A phase locked loop (PLL) can also be used to receive amplitude modulated signal. It basically consists of a VCO, a phase detector, a low-pass filter, and an amplifier.

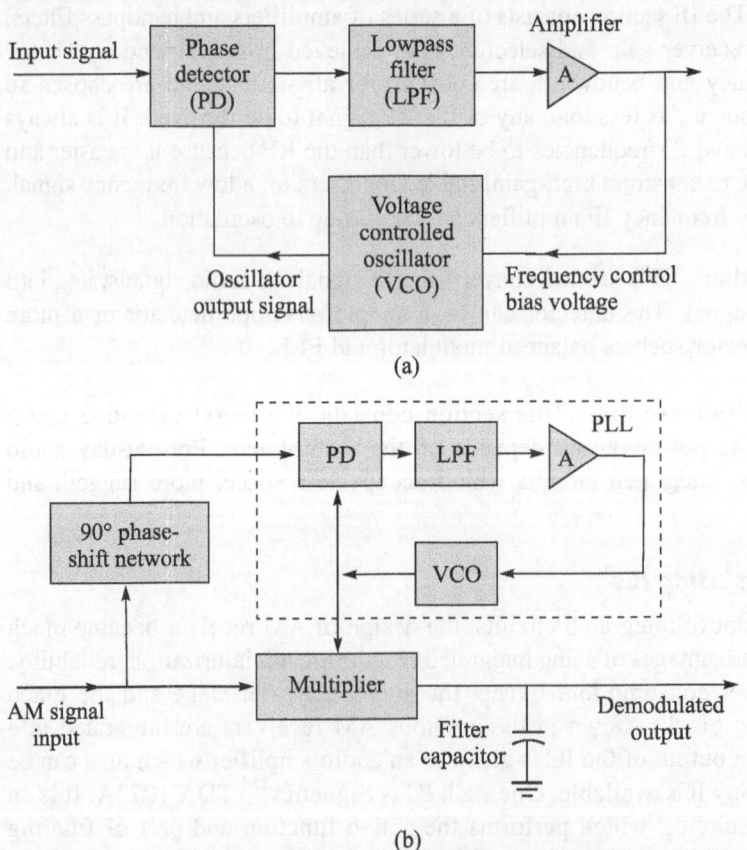

Fig. 3.53 PLL used as AM receiver: (a) block diagram of PLL and (b) AM receiver

The phase detector, which compares the input signal and the VCO output, generates an error signal whenever the input signal frequency and the VCO frequency do not match. This error signal is filtered, amplified, and fed to the VCO as DC bias to correct the VCO frequency to be equal to the input signal frequency. Figure 3.53 shows an AM receiver using PLL.

The AM signal is given as the input signal. When this signal is locked by a PLL, the VCO frequency adjusts automatically to that of the carrier frequency. The VCO output voltage is approximately 90° out of phase with the AM carrier voltage. To compensate for this phase shift introduced by the PLL, the AM signal input is passed through a 90° phase-shift network before giving to the PLL. The VCO output and AM signal are given to a multiplier circuit which is nothing but a balanced modulator. The output of the multiplier passes through an LPF to give the required modulating signal. The receiver using a PLL can also be regarded as synchronous or coherent detection. Most popular PLL is Signetics^{TM} NE 567. It is a 14 PIN DIP IC. For more information on this PLL, the readers are requested to refer to the Signetics Linear Data Manual, Volume 1, Communications.

Image rejection

The mixer of the superheterodyne receiver produces a signal component at the IF frequency, which is the difference between the oscillator frequency and signal frequency. The signal frequency may be either below or above the oscillator frequency, which can still produce an IF signal. If the desired signal is located at $(f_0 - IF)$, a strong signal at $(f_0 + IF)$ will interfere with it. This second signal is called the *image*.

The bandpass characteristic of RF circuits will be narrow enough to reject this image and prevent it from reaching the mixer. However, if the Q of the RF circuit is low, its bandpass characteristic will be wide and if the IF is small, then the image will fall within the bandpass of the tuned circuit and will not be rejected. Once this signal reaches the mixer and is converted it cannot be separated from the desired signal. This problem can be avoided if the IF frequency is selected high just below the signal band. But as the IF is made higher, Q becomes smaller and the bandwidth increases. Beyond an HF band at about 30 MHz, it becomes impossible to acquire the required image rejection using an ordinary tuned circuit. Hence, single conversion superheterodyne is rarely used above 20 MHz.

Net receiver gain

Net receiver gain is the ratio of the demodulated signal level at the audio output of the receiver to the RF signal level at the input to the receiver. It can be also expressed in dBm as the difference between the audio signal level in dBm and the RF signal level in dBm. Hence, the net receiver gain is the sum of all gains in the receiver minus the sum of all losses expressed in dB. Receiver losses typically include the pre-selector loss, mixer loss due to the conversion gain and detector loss. The gains include RF gain, IF gain, and the audio amplifier gain.

We can mathematically express the net receiver gain as

$$G_{dB} = gains_{dB} - losses_{dB} \tag{3.70}$$

3.23.2 SSB Receiver with Pilot Carrier

At the receiver, the received signal from the antenna is down converted to an IF signal using double conversion so that the sideband remains an upper sideband. A USB filter passes this sideband to the balanced modulator. The other input to the modulator is the carrier signal generated through a PLL. The pilot carrier signal is used as a reference to lock the local oscillator PLL. The result is that the local oscillator signal may not be exactly 100 kHz, but will be locked to the received signal, thus ensuring proper demodulation of the signal. The receiver block diagram is shown in Fig. 3.54.

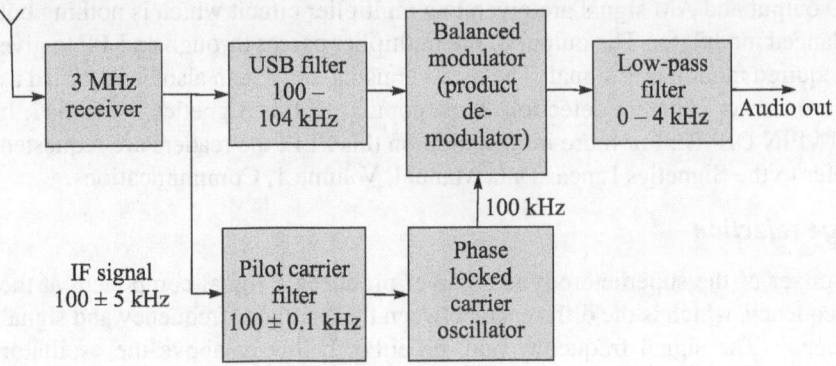

Fig. 3.54 SSB pilot carrier receiver

3.23.3 Communication Receivers

These are receivers of high quality, more sensitivity, and good image rejection ability. Double conversion receivers are used to have good image rejection and good adjacent channel selectivity. A typical double conversion superheterodyne receiver is shown in Fig. 3.55.

The image rejection is obtained in the first conversion by ensuring that the image frequency is well outside the fixed tuned RF amplifier bandpass characteristic. Adjacent channel rejection is obtained from the second IF narrow band characteristic. These receivers have additional circuits for delayed AGC and squelch circuits to take care of the noise output when there is no signal.

3.23.4 Receiver Parameters

Although there are several parameters generally used to evaluate the ability of the receiver to produce truly the received signal in the presence of noise, selectivity and sensitivity are the most important parameters.

Selectivity

The ability of a receiver to accept a given band of frequencies and reject all others is called its selectivity. The selectivity of a receiver is measured by mentioning the bandwidth of the receiver at its –3 dB point. But this bandwidth may not necessarily indicate how well the receiver will reject the unwanted frequencies. To circumvent this, a factor, called shape factor, is defined.

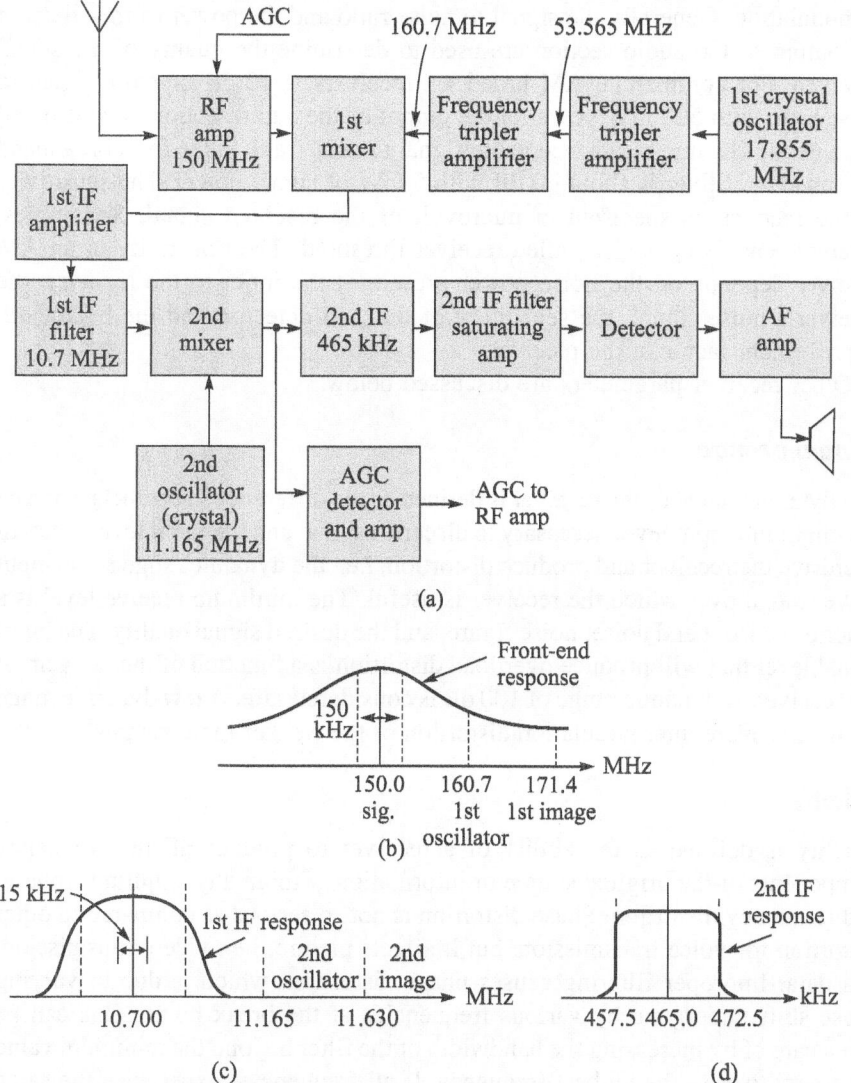

Fig. 3.55 Communication receiver: (a) block diagram frequency response, (b) RF stage, (c) first IF stage, and (d) second IF stage

The ratio between the bandwidth of the receiver at −3 dB and −60 dB points is the *shape factor*. Ideally, this factor should be 1. But this value is difficult to achieve in practice and a factor of 2 is generally achieved. But sophisticated receivers like satellite, two way, and microwave achieve factors close to this ideal value. The bandwidth for commercial AM broadcast receivers is 10 kHz. The receiver must restrict its bandwidth to this value. For commercial FM broadcast, it is 200 kHz.

Sensitivity

The sensitivity of a receiver is the minimum RF signal level that can be detected at the input to the receiver and still produce a valid information signal after demodulation. Generally the signal to noise ratio and the power of the signal at the output of the audio section are used to determine the quality of the signal received. For commercial AM broadcast receivers, a 10 dB or more signal to noise ratio with 500 mW power at the output of the audio section is considered to be good. The minimum acceptable signal to noise ratio value for a broadband microwave receiver is about 40 dB with 5 mW of signal power. The sensitivity of the receiver is specified in microvolt of the received signal. Sometimes, receiver sensitivity is also called receiver threshold. The sensitivity of an AM receiver depends on the noise power present at the input to the receiver, the receiver's noise figure, the sensitivity of the AM detector, and the bandwidth improvement factor of the receiver.

Other receiver parameters are discussed below.

Dynamic range

The dynamic range of the receiver is defined as the difference in decibels between the minimum input level necessary to discern a signal and the input level that will overdrive the receiver and produce distortion, i.e., the dynamic range is the input power range over which the receiver is useful. The minimum receive level is a function of front end noise, noise figure, and the desired signal quality. The input signal level that will produce over load distortion is a function of the net gain of the receiver. A dynamic range of 100 dB is considered safe. A low dynamic range results in severe intermodulation distortion of the weaker input signals.

Fidelity

Fidelity is defined as the ability of a receiver to produce all the frequency components of the original source of information without any amplitude, phase, and frequency distortion. Phase distortion is not so serious as compared to other distortion for voice transmission, but it affects picture, i.e., video transmission, and data. Improper filtering causes phase distortion, which is due to varying phase shift undergone by various frequencies at the break point. This can be taken care of by increasing the bandwidth of the filter beyond the minimum value necessary to pass the highest frequency. If all frequencies experience the same phase delay, it causes only a delay. But if the frequencies experience different phase delays, the received signal will have what is known as phase distortion.

Amplitude distortion is caused by an unequal gain of the receiver for various frequency components. This distortion is avoided by not overdriving the receiver.

Frequency distortion occurs when frequencies other than the one present in the original source of information are present at the received signal. This distortion is due to harmonic and intermodulation distortion and caused by non-linear amplification. Second-order products $2f_1, 2f_2, f_1 \pm f_2$, and so on are only a problem in a broadband network since they fall outside the narrowband system. However, the third-order products fall within the system bandwidth and produce a third-order intermodulation distortion. These are produced when the second harmonic of one signal is added to the fundamental frequency of the another signal. By using a square law modulator, frequency distortion can be minimized as they produce only a second harmonic distortion.

Bandwidth

Since thermal noise is the most prevalent noise in the receivers, there is a tendency to reduce the bandwidth, which in turn reduces this noise. This increases the signal to noise power ratio, thus improving the system performance. However, there is a limitation as to how much the bandwidth can be reduced. Hence, the circuit bandwidth must exceed the bandwidth of the information signal. Otherwise the frequency content and hence the power content of the information signal will be reduced. This in turn degrades the system performance.

Generally the signal to noise ratio is calculated at a receiver input using the RF stage bandwidth which is generally wider than the bandwidth of the rest of the receiver. The noise reduction ratio achieved by reducing the bandwidth is effectively equivalent to improving the noise figure of the receiver. This noise reduction ratio is called bandwidth improvement (BI) and is given by

$$\text{BI} = \frac{B_{RF}}{B_{IF}} \tag{3.71}$$

where

B_{RF} = RF bandwidth
B_{IF} = IF bandwidth

and 10logBI gives the noise figure improvement in dB.
As BI is a ratio, it is unitless.

Noise temperature and equivalent noise temperature

Since thermal noise is predominant among the various kinds of noise encountered in the receiver, the noise temperature can be taken as

$$T = \frac{N}{KB} \tag{3.72}$$

where

T = ambient temperature in degree kelvin
N = the noise power in watts
K = Boltzmann's constant (1.38×10^{-23} joules/kelvin)
B = bandwidth in Hz

The equivalent noise temperature T_e is a hypothetical value that cannot be measured directly but used in low noise radio receivers in place of noise figure. The lower the equivalent noise temperature, the better the quality of the receiver. T_e can be expressed as

$$T_e = T\,(F - 1) \text{ kelvin} \tag{3.73}$$

where

T = ambient temperature in kelvin
F = noise figure

3.24 AUTOMATIC GAIN AND VOLUME CONTROL CIRCUITS

First we will deal with automatic gain control (AGC) circuits.

3.24.1 Automatic Gain Control (AGC)

Minor variations in the RF signal level can be compensated by an automatic gain control (AGC) circuit. For a weak signal, the AGC circuit automatically increases the receiver's gain and for strong RF signals, it automatically reduces the receiver's gain. An excessive signal can cause overdriving resulting in non-linear distortion and saturation. Weak signals can become immersed in noise and make signal detection impossible. There are several types of AGCs like direct or simple AGC, delayed AGC, and forward AGC.

Simple AGC

A block diagram of an AM superheterodyne receiver with a simple AGC is shown in Fig. 3.56. The AGC circuit monitors the received signal and a feedback signal is sent to RF and IF amplifiers to adjust their gain automatically. AGC is a form of degenerative circuit that introduces a negative feedback. AGC takes care of different signals in a typical radio transmission, which are of different signal strength. The AGC circuit produces a voltage that adjusts the receiver gain and keeps the IF carrier power at the input to the AM detector at a relatively constant level. AGC is independent of modulation and totally unaffected by normal changes in the modulating signal amplitude.

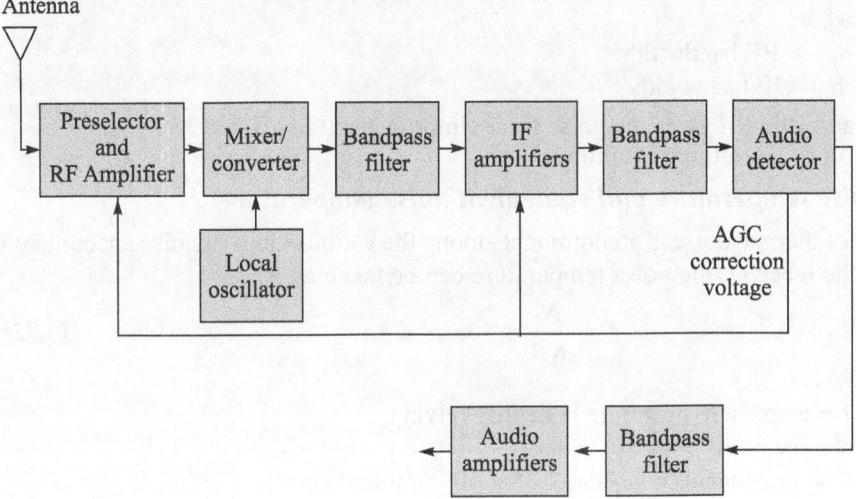

Fig. 3.56 AM receiver with a simple AGC

Figure 3.57 shows a schematic diagram for a simple AGC circuit. It can be observed that the AGC circuit is a peak detector. The AGC correction voltage is taken from the output of the audio detector. The DC voltage at the output of a peak detector is equal to the peak unmodulated carrier amplitude less the drop across the diode and totally independent of the depth of modulation. The circuit shown in Fig. 3.57 is a negative peak detector and produces a negative voltage at the output. This voltage is fedback to the IF stage where it controls the bias voltage on the base of Q_1. When the carrier amplitude increases, the voltage on the base of Q_1 becomes less positive, causing the emitter current to decrease. This result in increasing r_e and the amplifier gain r_c/r_e decreases resulting in a reduction in the carrier amplitude.

When the carrier amplitude decreases, the AGC voltage becomes less negative, the emitter current decreases, and the amplifier gain increases. The audio bypass capacitor C_1 prevents changes in the AGC voltage due to modulation from affecting the gain of Q_1. It can be seen from the figure that the AGC correction voltage is given to both the IF and RF amplifiers.

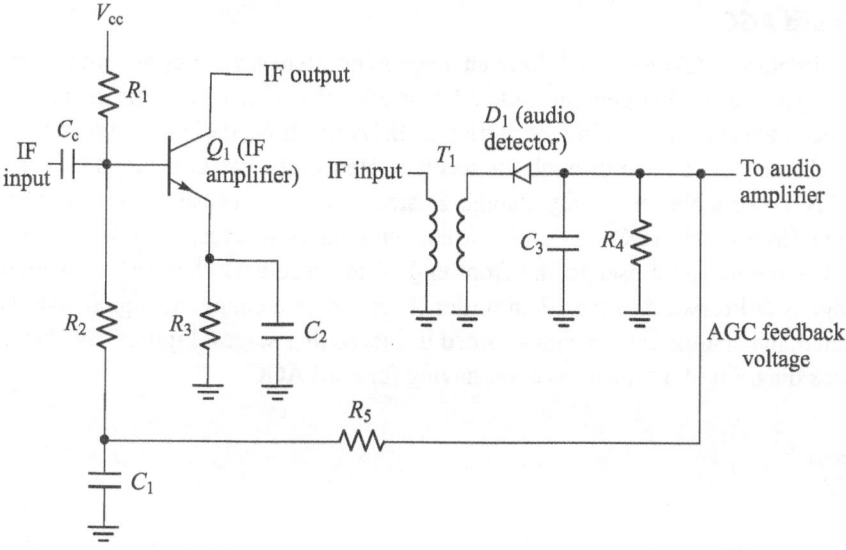

Fig. 3.57 Simple AGC circuit

Delayed AGC

Inexpensive radio receivers use simple AGC. In delayed AGC, the AGC bias begins to increase as soon as the received signal level exceeds the thermal noise of the receiver. This results in the receiver becoming less sensitive. Delayed AGC prevents the AGC feedback voltage reaching the RF or IF amplifier until the RF level exceeds the predetermined value. Once this limit is exceeded by the carrier signal, the delayed AGC voltage is proportional to the carrier signal strength. Figure 3.58 shows the response characteristics for both simple and delayed AGC. It can be seen that with delayed AGC, the receiver gain is unaffected until the threshold level is exceeded. But in the case of simple AGC, the received gain is immediately affected.

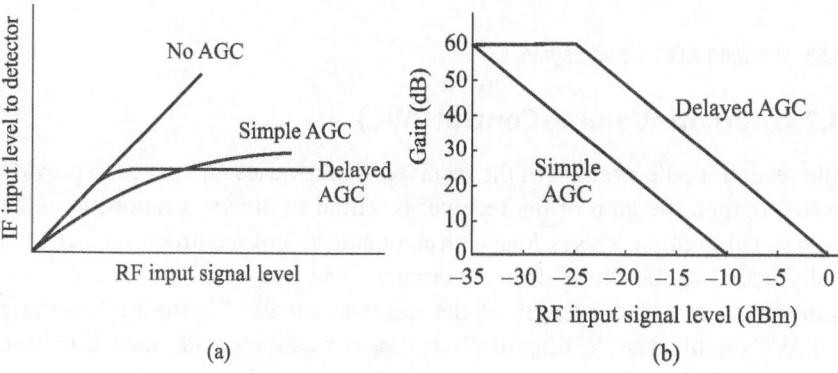

Fig. 3.58   (a) AGC response characteristic and (b) RF input signal versus IF gain

Forward AGC

Both simple and delayed AGCs have an inherent problem since they are both 'post-AGC'. The circuit that generates the AGC is after the IF stage and hence there is an inherent delay in correcting the carrier level. By this time, the carrier would have been already propagated through the receiver. Hence, we can conclude that post-AGC is not suitable for rapidly changing carriers. In such a case, we go for what is known as forward AGC. It is similar to conventional AGC except that the received signal is monitored closer to the front end of the receiver. Then the correction voltage is fed forward to the IF amplifier. Thus, when a change in signal level is detected, the change can be compensated in succeeding stages. Figure 3.59 shows a block diagram of a typical receiver having forward AGC.

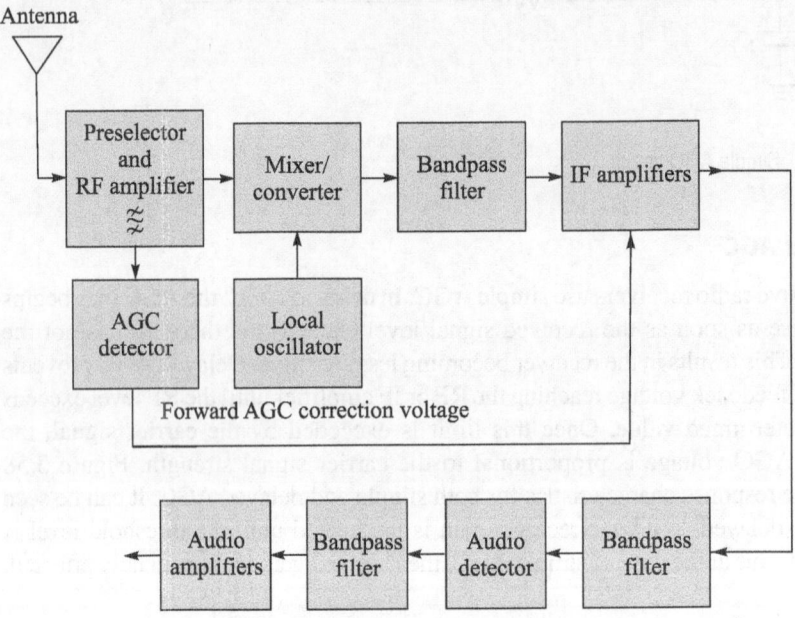

Fig. 3.59 Forward AGC block diagram

3.24.2 Automatic Volume Control (AVC)

In radio reception, the strength of the received signal varies and hence to provide a constant output, the gain of the receiver is varied in inverse proportion to the received signal strength. This is done with automatic volume control (AVC) circuits normally added to the diode detector circuits. The output of the detector is a measure of the received strength of the received carrier. Figure 3.60 shows a typical AVC circuit. The *RC* filter of a long time constant separates the DC voltage from the modulation. The AVC circuit samples the '*a*' fraction of the detector output and generates the AVC bias voltage. The negative DC component is applied to the RF and IF stage. As the input of the received signal increases, the AVC bias voltage also increases which in turn increases the negative bias to the RF and IF stages, thus reducing the gain. The disadvantage of a simple AVC circuit is that it operates even for a low input signal at the input of the receiver. This reduces the sensitivity of the receiver. To overcome this problem and ensuring that the AVC circuit operates only when the signal is strong, a forward biased diode D_2 is added to the AVC filter circuit as shown in the figure. The AVC operates only when the signal is more than the diode D_2 bias voltage.

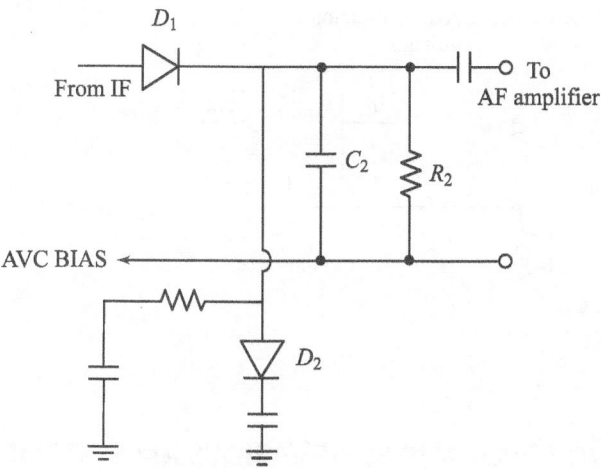

Fig. 3.60   Amplified and delayed AVC

3.24.3 Squelch Circuit

When we employ an AGC circuit in a receiver in the absence of an RF signal, the AGC circuit adjusts the receiver for maximum gain. This amounts to the receiver amplifying its own noise and then demodulating it. A squelch circuit keeps the audio section of the receiver turned off in the absence of a received signal. This is also called muting the receiver. The drawback of a squelch circuit is that a weak RF signal does not produce any audio output. Figure 3.61 shows a schematic diagram of a typical squelch circuit.

This circuit monitors the AGC voltage to know about the received RF signal level. The strength of AGC voltage determines how strong the RF signal is. When the AGC voltage drops below a preset level, the squelch circuit is activated and disables the audio portion of the receiver. From Fig. 3.61, it can be seen that the squelch detector uses a resistive voltage divider to monitor the AGC voltage. When the RF signal drops below the squelch threshold, Q_2 turns on, thus shutting off the audio amplifier. When the RF signal level increases above the squelch threshold, the AGC voltage becomes more negative turning off Q_2, and enabling the audio amplifier. The squelch threshold level can be adjusted with R_3

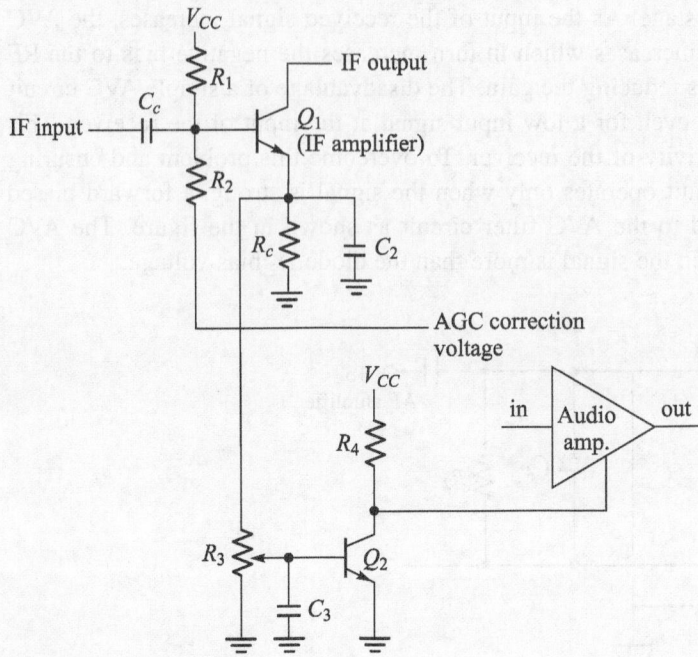

Fig. 3.61 Squelch circuit

3.25 COMPARISON AND APPLICATIONS OF VARIOUS AM SYSTEMS

Various AM systems are compared below.

1. The AM DSBFC (double sideband with full carrier) is the simplest system although the carrier is transmitted along with the sidebands, thus increasing the transmitted power and the bandwidth. The modulator and demodulator for this system are simple and straightforward.
2. In the case of DSB with a suppressed carrier, the carrier is suppressed. Hence, compared to AM DSBFC, the carrier power is saved but the bandwidth remains the same. The modulator and demodulator are slightly more complex compared to its conventional AM counterpart.
3. In the case of an SSB system, we save both power and bandwidth compared to AM DSBFC and DSB with a suppressed carrier. Bandwidth conservation and power efficiency are the obvious advantages of an SSB suppressed carrier and reduced carrier transmission over conventional AM.
4. A VSB system is in between SSB and DSB systems and is used where the bandwidth requirement is more compared to an SSB system.

AM systems have the following applications.

(a) Conventional AM systems are generally used in radio broadcasting.
(b) SSB systems are used in high quality communication equipment like police wireless systems.
(c) VSB systems are used in commercial television transmissions.

3.26 FREQUENCY TRANSLATION

We have seen that the modulation process shifts the modulating frequency to a higher frequency, which in turn depends on the carrier frequency, thus producing upper and lower sidebands. Hence, in the transmitter, information signals are up converted from low frequencies to high frequencies and down converted from high frequencies to low frequencies in the receiver. The process of converting a frequency or a band of frequencies to another location in the frequency spectrum is called *frequency translation.*

3.27 COSTAS LOOP

In telecommunication, a Costas loop is a phase locked loop used for carrier phase recovery from suppressed carrier modulation signals such as double sideband suppressed carrier signal. It was invented by John P. Costas at General Electric in 1950s. It consists of a local voltage-controlled oscillator providing quadrature outputs, one to each of two phase detectors (product detectors). The same phase of the input signal is also applied to both phase detectors and the output of each phase detectors is passed through a low-pass filter. The outputs of these low-pass filters are inputs to another phase detector, the output of which passes through a loop filter before being used to control the voltage-controlled oscillator. Binary phase shift keying (BPSK) in terms of noise immunity per unit bandwidth is one of the most efficient binary modulation techniques. However, because of its complexity as compared to a simple FSK modulator, BPSK is not generally preferred. But using the costas loop technique simplifies the design of BPSK demodulator. Carrier's phase shifts by $180°$ for one data symbol, while no shifting for the other. Shifting the phase of the carrier by $180°$ is the same mathematical process as reversing the magnitude of a carrier for one symbol and not for the other. With identical results, the following amplitude modulation can be substituted interchangeably.

3.27.1 Carrier Recovery

The two common methods for BPSK carrier recovery are: (i) squaring the BPSK signal and then dividing by two and (ii) the $180°$ Costas loop. The first technique as shown in Fig. 3.62 relies on the fact that because the BPSK modulation causes $±180°$ phase transitions, its second harmonic will be phase modulated by an ambiguous $±360°$. The second harmonic is an unmodulated carrier at twice the frequency. Dividing this second harmonic of the carrier by 2 will result in a theoretically phase-coherent carrier. The advantage of the squaring and then dividing the circuit is that it is mathematically simple to analyse. However, in practice, controlling the phase offset will be somewhat complicated and layout dependent. The recovered carrier takes a different path from the demodulator path and this creates a time differential that will result in a phase error. Also, several filters are required making it difficult to maintain a proper phase over the range of operating frequencies. While the first method is a feed-forward technique, the costas loop relies on feedback concepts related to the PLL. The costas loop shown in Fig. 3.63 offers an inherent ability to self-correct the phase (and frequency) of the recovered carrier and, in the end, its implementation is no more complicated than the first technique. Its main disadvantage is involvement of a loop settling time.

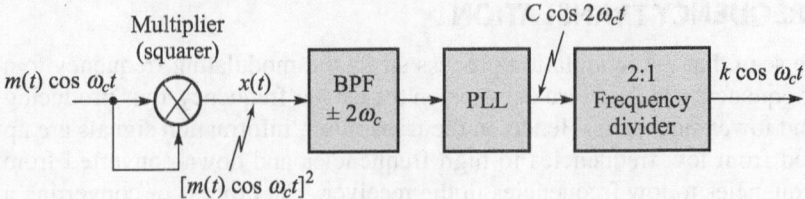

Fig. 3.62 Squaring method

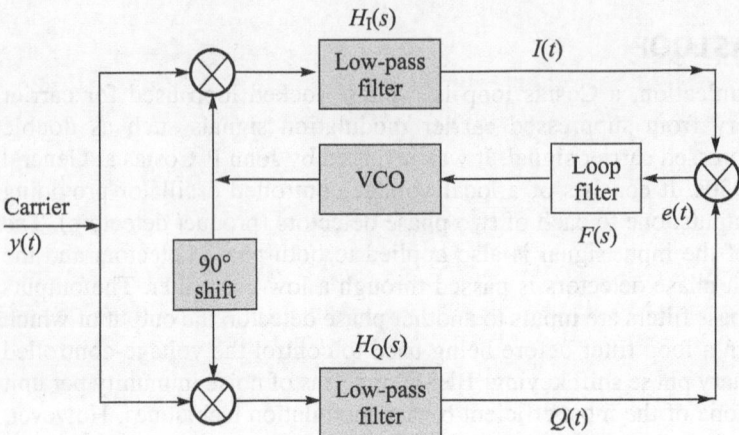

Fig. 3.63 Costas loop basic block diagram

The Costas loop is analysed by assuming that the VCO is locked to the input suppressed carrier frequency ω_c with a constant phase error of ϕ_e. The bandwidth of the two low-pass filters is predetermined by the data rate. The two quadrature output signals are multiplied together and filtered with a low pass filter that has a cut-off frequency near DC so that the filter acts as an integrator to produce the necessary DC control voltage, $k \sin(2\phi_e)$. The costas loop has a 180° phase ambiguity. Whenever the loop is energized, it is just as likely to phase lock so that the binary 1s come out as binary 0s, and vice versa. One of the two methods can be used to resolve this 180° phase ambiguity. A known test signal could be sent over the system after the loop is turned on so that the sense of the polarity can be determined, or differential coding and decoding may be used.

3.27.2 Digital Implementation

Owing to continuing advances in high-speed digital technology, digital implementations of the Costas loop are becoming increasingly attractive. Advantages of digital implementations include their relative insensitivity to temperature variations and ageing. More important, however, is the unique advantage that the loop design parameters, such as loop gain and loop filter time constant, are programmable. The low-pass filters, $H_I(s)$ and $H_Q(s)$, each have the same format consisting of a single pole, and this can be expressed as

$$H(s) = \frac{H(0)}{1 + \dfrac{s}{\omega_a}} \tag{3.74}$$

where ω_a is the cut-off frequency and

$$H(0) = \{\tan(\omega_a T/2) + 1\}/\tan(\omega_a T/2) \tag{3.75}$$

The signal flow graph for this first-order system is shown in Fig. 3.64.

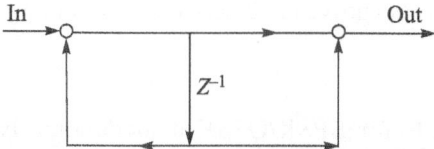

Fig. 3.64   First-order low-pass filter model

In Fig. 3.64, b is calculated as follows:

$$b = \frac{\left(1 - \tan\dfrac{\omega_a T}{2}\right)}{\left(1 + \tan\dfrac{\omega_a T}{2}\right)} \tag{3.76}$$

The signal flow graph of the first-order loop filter is shown in Fig. 3.65.

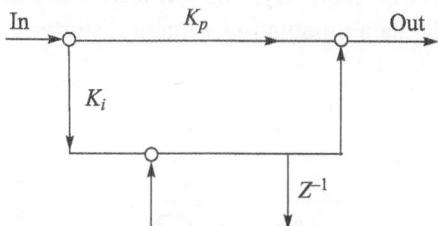

Fig. 3.65   First-order loop filter model

In the above figure, K_p is the proportional gain, which equals g_c, and K_i is the integral gain, which can be expressed as

$$K_i = g_c T \omega_z \tag{3.77}$$

where ω_z is the zero frequency of the loop filter.

3.27.3   Traditional Design Method

Design of the Costas algorithm includes compromises between algorithm complexity and performance objectives. Typically, the designer sketches out a signal flow graph of the algorithm using 'black boxes' to represent signal processing operations. The computational requirements can be estimated from the signal flow graph by counting the number of multiplies, multiply-accumulate, and additions. A block diagram of the system can then be drawn out. Once the algorithm has been worked out on paper, a simulation program may be written to verify whether the concept is correct. The simulation has quite often been written in the past using a high level language such as C and Fortran. Unless the designer is highly skilled in programming, the simulation software can require a long time to write and debug. This is because errors in the algorithm can be mistaken for programming errors or vice versa. Analysing algorithmic trade-offs is more difficult because the software must be modified and debugged while changing or modifying subsystems. Viewing and analysing the results of a simulation typically require the use of a different software package or writing and debugging special display programs. And finally, testing can also prove to be very time consuming.

3.27.4 Detailed Description

The classical Costas loop that is suitable for BPSK/QPSK demodulation is shown in Fig. 3.66. The system involves two parallel tracking loops operating simultaneously from the same VCO (voltage-controlled oscillator) or NCO (numerically-controlled oscillator). The first loop, called the in-phase loop (or I arm), uses the VCO as in a PLL (phase locked loop), and the second, called the quadrature loop (or Q arm) uses a 90° shifted VCO. The I and Q mixer outputs are filtered by single pole Butterworth low-pass filters. The I and Q arm filter outputs are multiplied together and the product is scaled and filtered to produce the loop error used to control the VCO. The loop error should settle to a value when the loop is locked. A negative loop error decreases the VCO increment resulting in a lower VCO frequency, and similarly, a positive loop error increases the VCO increment resulting in a higher VCO frequency. The low pass filters in each arm must be wide enough to pass the data modulation without distortion.

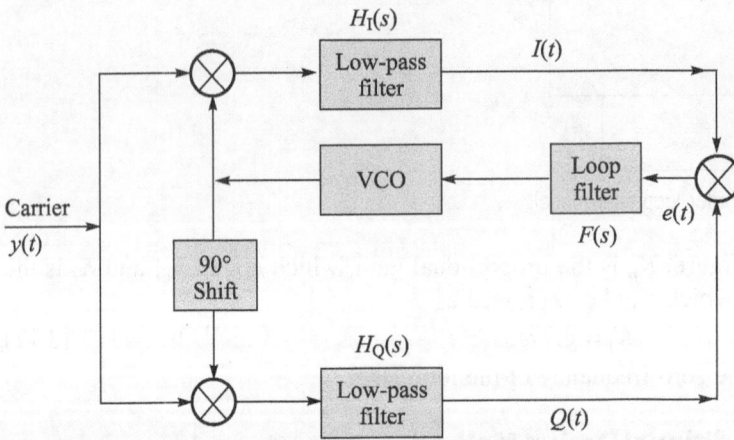

Fig. 3.66 Costas loop block diagram

In the circuit, both the low-pass filters act similarly to a second-order loop (the combined effect of both adds a second pole to the loop response). The filtered Q signal moves just slightly above or below zero and is multiplied by the filtered I product. Its doubled sine phase detection response allows two stable locking points—180° phase error and zero degrees—both produce a redundant output that drives the VCO to the correct phase/frequency. Both the low-pass filters must pass the modulation (the direct result of filters that are too narrow is ISI as $\omega_{LP12} \geq 2\pi B_M$, where B_M, the modulation bandwidth, is half the data rate before the loop has settled. Whether a PLL or a Costas loop, the phase detection response must be one that, based on the phase relationship between the VCO and the input signal, guides the VCO to a stable locking phase and frequency. If we were to apply a signal whose phase is reversing by 180° to an ordinary PLL, the phase detector result would constantly reverse polarity and the phase error magnitude is unlikely to converge on any stable value (i.e., the PLL will 'track' in opposite directions for opposite phases). We might refer to a conventional phase detector as 360° periodic. This means that the phase of the incoming carrier would have to be modulated with 360° phase transitions (which is no phase transition at all because a sinusoidal carrier has a period of 360°) not to upset the tracking so that the loop error may converge.

The input to the costas loop in waveform written as

$$y(t) = m(t) \sin(\omega_c(t) + \Psi(t)) + n(t) \tag{3.78}$$

where $m(t)$ is the BPSK modulation and $n(t)$ is a white bandpass noise. The in-phase mixer generates

$$I(t) = m(t) \cos \Psi_e + n_{mc}(t) \tag{3.79}$$

while the quadrature mixer generates

$$Q(t) = m(t) \sin \Psi_e + n_{ms}(t) \tag{3.80}$$

where the mixer noise $n_{mc}(t)$ and $n_{ms}(t)$ are low pass demodulated noise processes in the carrier noise $n(t)$. The output of the multiplier is then

$$I(t)Q(t) = m^2(t)\sin\left(\frac{2\Psi_e}{2}\right) + n_{sq}(t) \tag{3.81}$$

where $n_{sq}(t)$ represents all the signal and noise cross-products. The multiplier of the costas loop can be thought of as allowing the bit polarity of the in-phase loop to correct the phase error orientation of the tracking loop, thereby removing the modulation. When the phase error $\psi_e(t)$ is small, the Costas loop has the equivalent linear model as shown in Fig. 3.67.

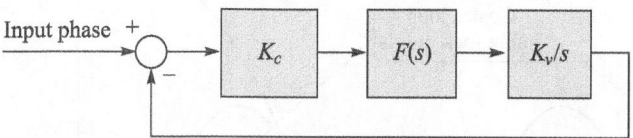

Fig. 3.67 BPSK Costas equivalent loop model

In Fig. 3.67, K_c is the closed loop gain, which can be expressed as

$$K_c = \left(\frac{m(t)}{2}\right)^2 H_I(o)H_Q(o)g_c \tag{3.82}$$

where g_c is defined as follows:

$$g_c = \left(\frac{4\omega_c}{K_v m^2(t)H_I(o)H_Q(o)}\right) \tag{3.83}$$

and ω_c is the cross-over frequency, K_v is the gain of VCO, and $F(s)$ is the transfer function of the loop filter, which is expressed in the following equation:

$$F(s) = \left(\frac{sT+1}{sT}\right) \tag{3.84}$$

where T is the sampling interval. The transfer function of the VCO is K_v/s.

3.27.5 Costas Versus Conventional Loop

Conversely, the Costas loop phase detection response is 180° periodic—there are two stable tracking points. BPSK modulation shifts the costas loop input by 180°, which is the next period of the phase detection function, where the loop tracking response is identical. Therefore, the costas loop is able to track a BPSK modulated carrier (loops can also be derived that track higher order phase modulation schemes such as QPSK). The only catch is that the loop-phase doubled response means that it has a 50% chance of generating an upside-down carrier. Figure 3.68 displays simulation results of how a costas loop and an ordinary PLL with similar loop parameters would behave with a BPSK signal as an input. Because a loop filter is not part of the control loop (and not the PLL loop filter), it must not have a frequency response that falls within the loop bandwidth. Its purpose is only to remove the excess noise products produced by the three previous multipliers and two imperfect filters. This filter constitutes an undesired S-plane pole that would cause the loop to oscillate, but if its response is far outside of the loop response, then it will not cause problems. A rule-of-thumb recommendation for a safe, out of the loop, response would be to set the pole of loop filter to a minimum of four times that of what the closed loop response would be without this filter. Exactly how the VCO will settle depends on the initial phase and frequency of the VCO as it relates to the incoming BPSK signal, as well as to the noise characteristics. Although not apparent, the behaviour of any practical implementation of this circuit will also be affected by the actual data

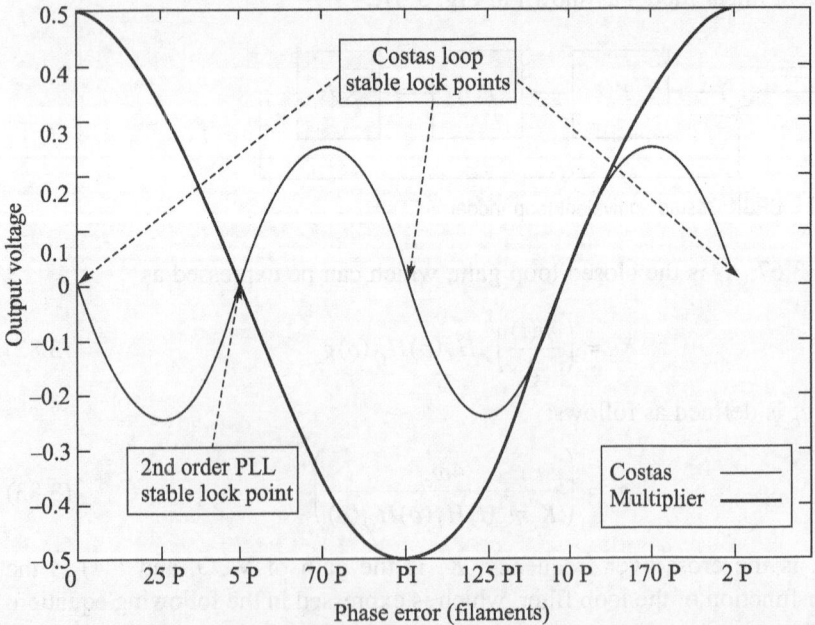

Fig. 3.68 Simulation results: Costas loop versus ordinary PLL (with similar loop parameters)

that has been modulated. Real-world communications are usually bandlimited, and the abrupt 180° phase shifts of BPSK, which the Costas loop is immune to, would require an infinite bandwidth. A more realistic version of a BPSK signal is one in which a bit transition will cause the carrier amplitude to sweep slowly from its current phase to the opposite phase through the zero-amplitude point.

The phase-detector contribution to the loop gain (although realistic phase detectors are not perfect multipliers, they still have minimum input level requirements) is diminished as the input signal level shrinks. Every BPSK phase transition will cause a costas loop 'dropout' at and near the zero-crossing instant during the interval between the two discrete phase levels. If the loop is still in the locking phase at this point (i.e., when the VCO phase does not match that of the carrier), such a 'glitch' could allow a phase slippage and may temporarily allow the loop to track in the wrong direction. Other design issues include the effect of realistic (non-ideal) filters. Some 'high side' product will always 'leak' through and affect the circuit's performance; their respective responses will not be identical, and there will be ISI (a 1-0-1-0 pattern will not quite produce 180° phase transitions). Further, it is not realistic to assume that the quadrature components of the VCO will have a perfect 90° offset or that the phase detector is an ideal multiplier free from DC offset. A second-order PLL analysis (where the loop filter is the same as the low-pass filters) of a carrier will approximate the settling characteristics of a costas loop, but a computer simulation is recommended if the designer needs accurate information. This is because 'mathematical' building blocks may need to be substituted with commonly available and inexpensive components. Figures 3.69 and 3.70 show the simulated timing waveforms of a Costas loop operating under noise-free and noisy conditions, respectively.

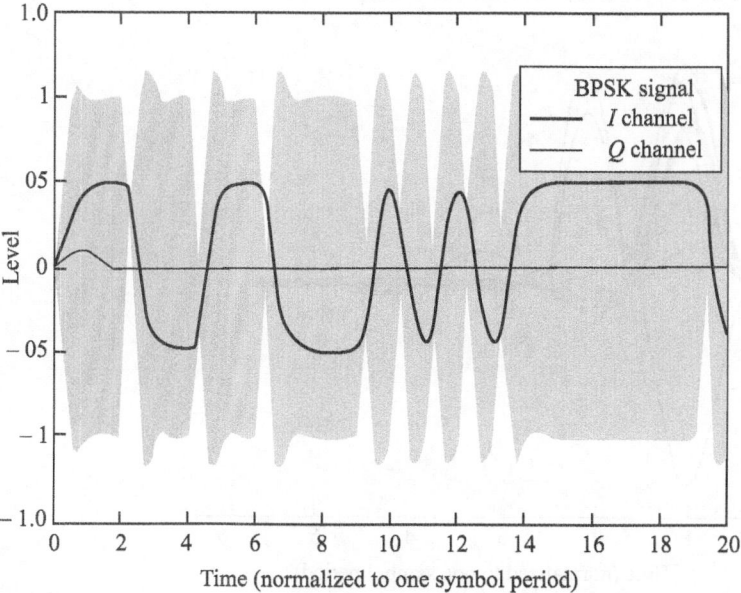

Fig. 3.69 Simulated timing waveforms of a Costas loop operating under noise-free conditions

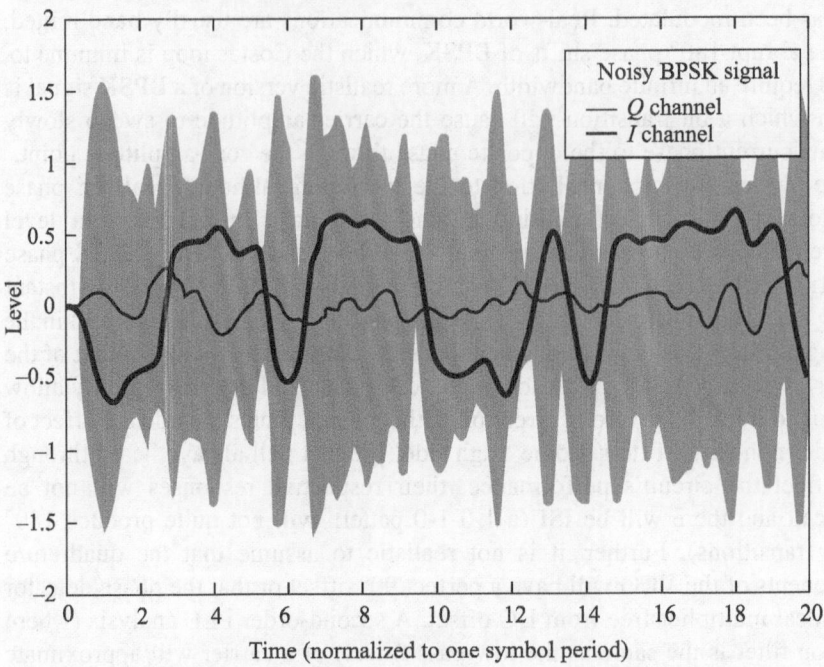

Fig. 3.70 Simulated timing waveforms of a Costas loop operating under noisy conditions

Figure 3.71 is a plot of the VCO settling function where the loop parameters are identical for each run, but the VCO starting phase/frequency and the modulation data are randomized over 10 trials. Realistic costas loop behaviour is somewhat chaotic for the reasons mentioned previously, depending on when BPSK phase transitions occur during the lock phase.

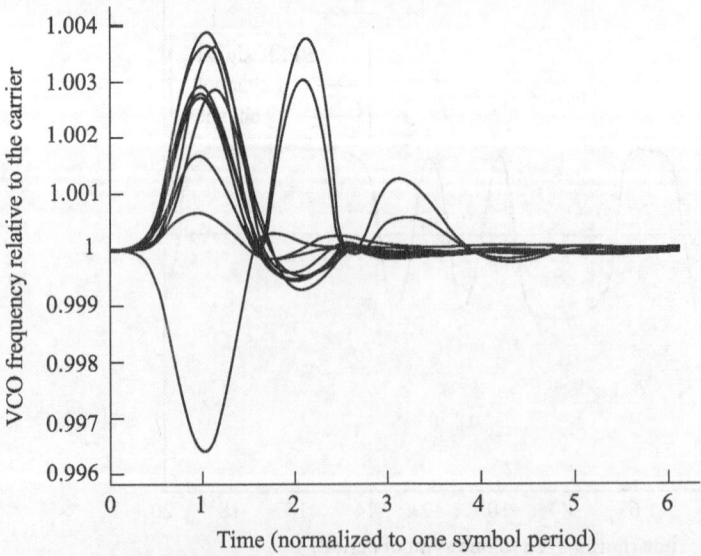

Fig. 3.71 Ten costas loop settling patterns under identical parameters, but randomized initial conditions and BPSK modulation data

3.27.6 Design Considerations for Costas Loop

Similar to the PLL design, the Costas loop design considerations are noise performance, settling time, and a reliable lock range. As a demodulator, the noise performance is maximized when the least amount of noise is allowed in the loop. This is accomplished by setting the low-pass filters response to their maximum SNR. This corresponds to –3 dB cut-off equaling half the data rate for a single-pole *RC*. For loop settling purposes, this cut-off is also the minimum allowable for the loop filter. Additionally, this is an attractive choice because this filter also serves the purpose of a data filter. The loop gain must now be set. Because a low-pass filter connected to the output of the balanced modulator having a carrier without phase shifting as the input serves the dual purpose of also being the data filter and is required to pass BPSK modulation, a compromise has been made.

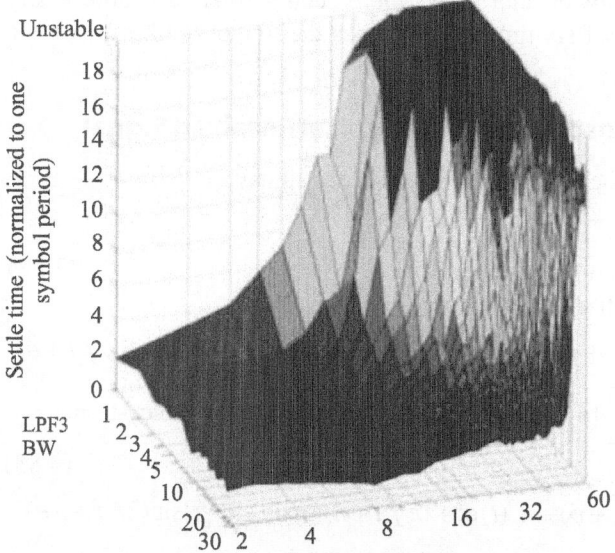

Fig. 3.72   Averaged results of simulation comparing Costas loop settle time to the bandwidth of LPF3 and VCO gain

The critical damping point (the point where minimum settling time occurs) for a PLL using the same lag type filter as the low-pass filters is when the pole of this filter equals the closed-loop bandwidth. Based on the settling time to a particular threshold, simulation shows that setting the pole of lag filter to half the DC forward gain results in a quicker lock (this is a point where the loop is slightly underdamped). Therefore, these same parameters were used in the costas loop design as a starting point for simulation. The Costas loop phase detector gain, under unity input conditions is $(1/4)V/r$, as stated. So, the VCO gain is the variable that needs to be determined. Solving for this parameter the unity output VCO should have a gain of eight times the filter's pole frequency in terms of radians per second for a unity BPSK input for this theoretical circuit containing perfect multipliers as the phase detector. Simulation confirms this result for the Costas loop (see Fig. 3.72). The fastest achievable settle time is one in which the VCO has a gain of eight times that of the LPF1/LPF2 pole frequency with the above the phase detector gain parameters and a random BPSK input. Using the costas loop parameters presented here, where the filter poles and loop gain are all in a fixed relationship to the data rate, the regenerated carrier will settle in less than three bit times. Of course if the phase detector has some gain or gain function other than that of the standard one, it should be appropriately modified. The loop filter must then be specified. This filter should have its pole at a low enough frequency that the costas loop will not be too noisy nor be subject to carrier phase reversals in the presence of noise (the costas loop is equally stable in both phases) while high enough that it does not cause the loop to oscillate. Setting this pole to four times K (or eight times the low-pass filter pole) is the point at which this filter will negligibly affect the loop, simulation shows this (Fig. 3.72). To be cautious, particularly at lower data rates where such a filter can more easily create problems, a factor of six times K (12 times the low-pass filter pole) is a better choice.

3.27.7 Analysis of a Costas Loop for a Typical Received Signal

Figure 3.73 shows the block diagram. The received signal

$$f(t) = A_c m(t) \cos(2\pi f_c t + \phi) + n(t)$$

is multiplied by $\cos(2\pi f_c t + \hat{\phi})$ and $\sin(2\pi f_c t + \hat{\phi})$ (3.82)
which are outputs from the VCO. The two products are

$$y_c(t) = [A_c m(t)\cos(2\pi f_c t + \phi) + n_c(t)\cos 2\pi f_c t - n_s(t)\sin 2\pi f_c t]\cos(2\pi f_c t + \hat{\phi})$$

$$= \frac{A_c}{2} m(t)\cos \Delta\phi + \frac{1}{2}[n_c(t)\cos \hat{\phi} + n_s(t)\sin \hat{\phi}] + \text{double frequency terms}$$

(3.83)

$$y_s(t) = [A_c m(t)\cos(2\pi f_c t + \phi) + n_c(t)\cos 2\pi f_c t - n_s(t)\sin 2\pi f_c t]\sin(2\pi f_c t + \hat{\phi})$$

$$= \frac{A_c}{2} m(t)\sin \Delta\phi + \frac{1}{2}\left[n_c(t)\sin \hat{\phi} - n_s(t)\cos \hat{\phi}\right] + \text{double frequency terms} \quad (3.84)$$

where $\Delta\phi = \hat{\phi} - \phi$. The double frequency terms are eliminated by the low-pass filters following the multiplications.

An error signal is generated by multiplying the two outputs $y_c'(t)$ and $y_s'(t)$ of the low-pass filters. Thus,

$$e(t) = y\mathcal{C}_c(t)\, y\mathcal{C}_s(t)$$

$$= \frac{A_c^2}{4} m^2(t)\sin 2\Delta\phi + \frac{A_c}{4} m(t)\left[n_c(t)\cos \hat{\phi} + n_s(t)\sin \hat{\phi}\right]\sin \Delta\phi$$

$$+ \frac{A_c}{4} m(t)\left[n_c(t)\sin \hat{\phi} - n_s(t)\cos \hat{\phi}\right]\cos \Delta\phi$$

$$+ \frac{1}{4}\left[n_c(t)\cos \hat{\phi} + n_s(t)\sin \hat{\phi}\right]\left[n_c(t)\sin \hat{\phi} - n_s(t)\cos \hat{\phi}\right]$$

(3.85)

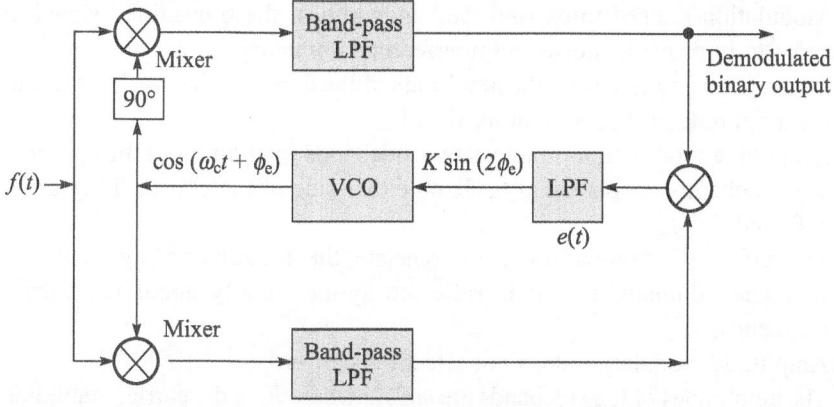

Fig. 3.73   Costas loop for a typical received signal

This error signal is filtered by the loop filter whose output is the control voltage that drives the VCO.

We note that the error signal into the loop filter consists of the desired term $(A_c^2 m^2(t)/4) \sin 2\Delta\phi$ and the terms that involve signal × noise and noise × noise. These terms are similar to the two noise terms at the input of the PLL for the squaring method. In fact, if the loop filter in the costas loop is identical to that used in the squaring loop, the two loops are equivalent. Under this condition, the probability density function of the phase error and the performance of the two loops are identical.

In conclusion, the squaring PLL and the Costas PLL are two practical methods for deriving a carrier phase estimate for the synchronous demodulation of a DSB-SC AM signal.

SUMMARY

- Baseband communication does not involve any carrier.
- Baseband communication is used only when the distance is quite small.
- For a large distance and to improve the efficiency of transmission, modulation is required.
- To simplify the structure of the transmitter, modulation is needed.
- Modulation is needed to
    (i) gain an ability to transmit huge amount of information using a single carrier frequency.
    (ii) increase the transmission length
    (iii) gain an ability to multiplex signals
    (iv) use shorter aerial length
    (v) provide security in transmission
    (vi) improve quality of service
- Modulation enables simultaneous transmission of signals from several message sources by means of FDM (frequency division multiplexing).

- Modulation is used to expand the bandwidth of the transmitted signal in order to improve its noise and interference immunity.
- In amplitude modulation, the amplitude of the carrier varies with respect to the amplitude of the modulating signal.
- Amplitude modulation produces two sidebands in addition to the carrier.
- The sidebands are placed at f_m Hz on either side of the carrier. They are $f_c + f_m$ and $f_c - f_m$.
- The effect of modulation is to translate the modulating signal in the frequency domain, i.e., it is reflected symmetrically about the carrier frequency.
- Amplitude modulation does not affect the carrier.
- The amplitudes of the sidebands are $mE_c/2$, where E_c is the carrier amplitude and m is the modulation index, which is equal to E_m/E_c, where E_m is the modulating signal amplitude and E_c is the carrier amplitude.
- The value of m varies from 0 to 1. With $m = 0$, there is no modulation. When $m = 1$, there is 100% modulation and when $m > 1$, it is called overmodulation.
- The carrier power after modulation is $P_c = E_c^2/2R$ and the sideband power is $P_{LSB} = P_{VSB} = (m^2/4)P_c$.
- It is evident that the power of the carrier is unaffected by modulating process. Total power in an AM envelope increases with modulation index.
- The power in a sideband is equal to 1/4th of the power in the carrier for 100% modulation. Thus, for 100% modulation, 50% of the transmitted power does not carry any information. This is the major disadvantage of AM.
- When several frequencies simultaneously modulate, the combined coefficient of modulation is the square root of the quadratic sum of individual modulation index, i.e., $m_t = \sqrt{m_1^2 + m_2^2 + \cdots + m_n^2}$.

  Total sideband power in this case is $P_t = P_c \left\{ 1 + \dfrac{m_t^2}{2} \right\}$.

- Use of a simple semiconductor diode results in power law modulator. If the switching property of the diode is used, it is called switching modulator.
- Transistor modulator types are low level, medium level, and high level.
- Out of the total transmitted power with $m = 1$, the carrier contains two-thirds of the transmitted power and sidebands contain one-third of the transmitted power. Since carrier contains no information, this is a wasteful power. Similarly, there is wasteful bandwidth. Moreover, as the sidebands are symmetrically placed on either side of the carrier and the bandwidth required is twice that of the message signal, it is enough if a single band is transmitted from which we can extract the message signal. Hence, as far as the transmission of information is concerned, only one side band is necessary. This means that the channel needs to provide only the same bandwidth as the baseband signal. Hence, carrier and a sideband have to be suppressed. This results in increased system complexity.

- We have to trade complexity with bandwidth and power. This resulted in DSBSC, SSB, and VSB. In DSB, carrier is suppressed and only upper and lower sidebands are transmitted. In SSB, carrier is suppressed and one of the sidebands is transmitted. In VSB, one sideband and part of the other sideband are transmitted.

- AM signals are detected using diode envelope detector. It is a half-wave rectifier with a capacitive filter. The carrier signal is filtered out and the modulating signal appears as an envelope.

- In a rectifier detector circuit, the AM signal is applied to a diode and a resistor circuit, then to a low pass filter which suppresses all harmonics except the modulating signal. However, diode detector has distortion called diagonal peak clipping and negative peak clipping. To avoid diagonal peak clipping the modulating index m should be $\leq /Z_p // R$, where Z_p is the impedance of the RC circuit at the modulating frequency. To avoid negative peak clipping, the modulating index m should be $\leq R_p/R$, where R_p is the parallel combination of R and R_1.

- SSB signals can also be detected with an envelope detector with carrier. This method is used generally for a larger carrier.

- In the coherent detection method, SSB signal is detected by multiplying a synchronous carrier with the SSB signal. This carrier signal is generated locally at the receiver.

- In the pilot carrier method, a low level carrier signal is transmitted with a single side band. This pilot carrier is used to synchronize the local carrier oscillator used for demodulation.

- VSB signals are demodulated using the carrier component that is generally transmitted. In the case of TV transmission, since a large carrier is transmitted along with the VSB signal, envelope detection can be used.

- AM transmitters are generally of Class C type to have efficient transmission. They are classified as low level and high level.

- In low level modulation, modulation is done at lower power level. Then this modulated signal is passed through a high level power amplifier.

- In high level transmitters, both the modulating and carrier signals are brought to higher power levels and then modulated.

- An SSB transmitter has a pilot carrier and the demodulation is done by coherent detection.

- AM receivers are generally superheterodyne type. In this type of receivers, all RF frequencies are converted to a single frequency called intermediate frequency (IF) 455 kHz. This conversion helps to have a single tuned IF amplifier for signals from any RF frequency band and has a bandwidth of 10 kHz.

- With the advent of integrated circuits, the design of AM receivers becomes a simple one. Other advantages include miniaturization, reliability, and low power consumption.

- The mixer of a superheterodyne receiver produces a signal component at the IF frequency which is the difference between the oscillator frequency and signal frequency. The signal frequency may be either below or above oscillator frequency, i.e., it can be $(f_o - IF)$ and $(f_o + IF)$. This signal at f_o + IF can interfere with the desired signal at f_o − IF. This signal is called the image. One method to eliminate is to have the band pass characteristic of the RF circuit to be as narrow as possible.

- For high quality, more sensitivity, and good image rejection capability double conversion receivers called communication receivers are used.

- Another form of amplitude modulation is ISB (independent sideband). Here the same carrier is modulated with two different modulating signals. This is similar to a DSB except that each side band is independent of each other. Generally they are used for stereo AM transmission.

- In quadrature amplitude moullation (QAM), signals from two different sources modulate the same carrier frequencies which are 90° out of phase. These types of modulations are exclusively used for digital modulation of analog carriers in data modem.

- Average power is the power of the modulated output averaged over a long period of time (at least over one cycle of a periodic modulating signal).

- Maximum instantaneous power is the maximum peak power of the modulated waveform.

- Peak envelope power is the maximum short term average power where the average is taken only over approximately one to several modulated carrier signals, it is always one half of the maximum instantaneous power

- Total RMS power, $P_T = P_c + P_{VSB} + P_{ISB}$

- Carrier power, $P_c = \dfrac{\left(\dfrac{E_c}{\sqrt{2}}\right)^2}{R}$

- Total power = carrier power + sideband power, i.e.,

$$P_T = \frac{E_c^2}{2R}\left\{1 + \frac{m^2}{2}\right\} = P_c\left\{1 + \frac{m^2}{2}\right\}$$

IMPORTANT FORMULAE

- Modulating coefficient or index, $m = \dfrac{E_m}{E_c}$

- AM wave $= E_c\sin\omega_c t + \dfrac{mE_c}{2}\cos(\omega_c - \omega_m)t - \dfrac{mE_m}{2}\cos(\omega_c + \omega_m)t$

- Average carrier power $P_c = E_c^2/2R$

- Average sideband power $= \dfrac{m^2}{4}P_c$

- Total power, $P_t = P_c\left(1+\dfrac{m^2}{2}\right)$

- For AM modulated by several sine waves:

$$m_t = \sqrt{m_1^2 + m_2^2 + m_3^2 + \cdots + m_{n-1}^2 + m_n^2}$$

$$P_t = P_c\left(1+\dfrac{m_t^2}{2}\right)$$

DSB signal $= \dfrac{mE_c}{2}\cos(\omega_c - \omega_m)t - \dfrac{mE_m}{2}\cos(\omega_c + \omega_m)t$

- Effective voltage, $E = E_c\sqrt{1+\dfrac{m^2}{2}}$

- Effective current, $I = I_c\sqrt{1+\dfrac{m^2}{2}}$

ADDITIONAL EXAMPLES

1. The carrier frequency for an AM DSBFC modulator is 500 kHz. The maximum modulating signal is 1 kHz. Find (a) the upper and lower sideband frequencies and (b) the bandwidth. (c) If the modulating signal is a single frequency 1500 Hz, find the lower and upper sideband frequencies and (d) draw the output frequency spectrum for $f_m = 1$ kHz.
   Solution
   (a) USB = 500 + 1 = 501 kHz
       LSB = 500 − 1 = 499 kHz
   (b) The bandwidth is 2 kHz.
   (c) USB = 500 + 1.5 = 501.5 kHz
       LSB = 500 − 1 = 498.5 kHz
   (d) The output frequency spectrum is as shown in Fig. 3.74.

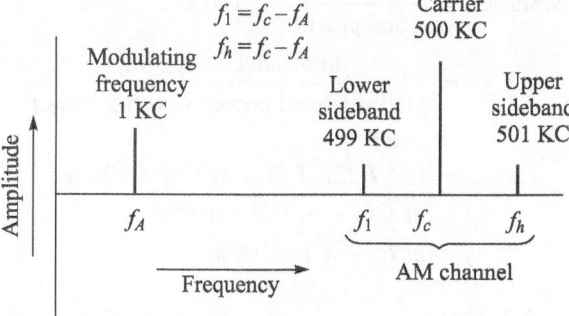

Fig. 3.74 The amplitude spectrum

2. A 3 kHz sine wave of peak amplitude $E_A = 5$ V modulates a carrier wave of frequency 8 MHz and a peak amplitude of $E_R = 10$ V. What is the modulation index? Draw the amplitude spectrum.

Solution

(a) $m = \dfrac{E_A}{E_R} = \dfrac{5}{10} = 0.5$

(b) Sideband frequencies = $8 \pm 0.003 = 8.003$ and 7.997 MHz

(c) The amplitude of each sideband = $0.5 \times 10/2 = 2.5$ V

(d) The output spectrum is as shown in Fig. 3.75.

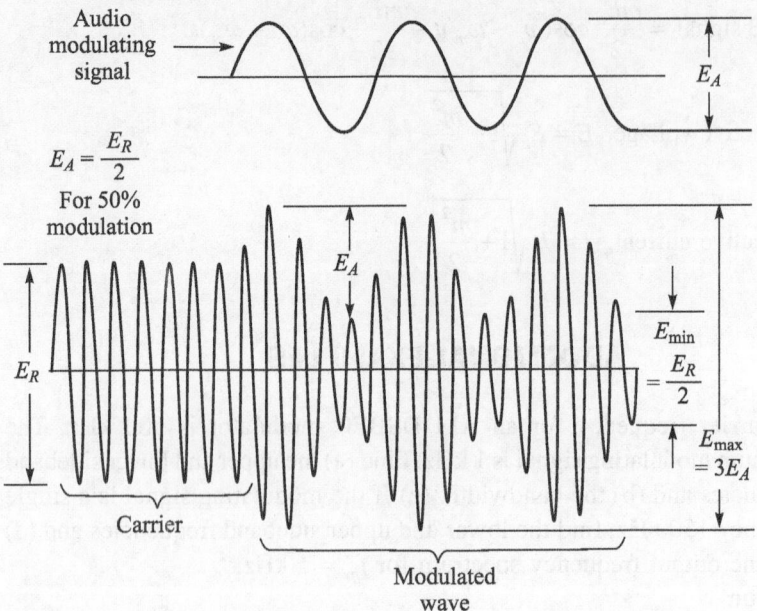

Fig. 3.75 AM wave with 50% modulation

3. Derive an expression for efficiency of transmission η in DSBFC in terms of modulation index m.

Determine η and the percentage of total power carried by the sidebands of the DSBFC wave for the modulation index $m = 0.7$.

Solution

Since efficiency of transmission $= \dfrac{\text{useful power}}{\text{total power}} \times 100\%$

$$= \dfrac{\text{total sideband power}}{\text{total sideband power} + \text{carrier power}} \times 100\%$$

$$= [(m^2 P_c/2)/ P_c(1 + m^2/2)] \times 100\%$$

$$= [m^2/2(1 + m^2/2)] \times 100\%$$

i.e., $$\eta = (m^2/2 + m^2) \times 100\%$$

For $m = 0.7$,

$$\eta = \dfrac{0.7 \times 0.7}{2 + (0.7 \times 0.7)} \times 100\% = 19.6\%$$

Hence, only 19.6% of the total power is in sidebands.

4. For an DSBFC wave with a peak unmodulated carrier voltage 8 V peak a load resistance of $R_L = 8$ Ω and modulation index $m = 1$, determine

(a) Power of the carrier and the sidebands
(b) Total sideband power
(c) Total power of the modulated wave
(d) Efficiency percentage

Solution

(a) The carrier power $= 8^2/ 2(8) = 4$ W
   Each sideband power $= (1)^2(4)/4 = 1$ W
(b) Total sideband power $= 2$ W
(c) Total power of the modulated wave $= 6$ W
(d) Efficiency percentage $= (2/6) \times 100\% = 33.33\%$

5. A carrier with an unmodulated power 80 W is modulated simultaneously by four modulating signals with coefficient of modulation $m_1 = 0.3$, $m_2 = 0.4$, $m_3 = 0.5$, $m_4 = 0.6$. Find

(a) Total coefficient of modulation
(b) Sideband powers
(c) Total transmitted power
(d) Efficiency of transmission

Solution

(a) Total coefficient of modulation $= (m_1^2 + m_2^2 + m_3^2 + m_4^2)^{1/2}$
$$= (0.09 + 0.16 + 0.25 + 0.36)^{1/2}$$

i.e., $\qquad m_t = 0.92$

(b) Sideband powers $= (0.92)^2 \times 80/2 = 37$ W
(c) Total transmitted power $= 117$ W

(d) Efficiency percentage $= \left(\dfrac{37}{117}\right) \times 100\% = 31.6\%$

6. The signal $m(t) = 2 \cos (400\pi t) + \sin (800\pi t)$ modulate a carrier $2\cos(2 \times 10^6 t)$. The modulation index is $m = 0.6$. Find out the power in the carrier and in the sideband components of the modulated signal.

Solution

First we have to determine $m_n(t)$, the normalized signal. To find this, we have to find the maximum of $|m (t)|$. For this, we have to differentiate $m(t)$ and make it equal to zero, i.e.,

$$\frac{dm}{dt} = - 800 \pi \sin (400\pi t) + 800 \pi \cos (800\pi t) = 0$$

This gives $\cos (800 \pi t) = \sin(400 \pi t)$

$$= \cos\left(\frac{\pi}{2} - 400\pi t\right)$$

Comparing LHS and RHS of the equation, we can infer that $1200 \pi t = \dfrac{\pi}{2}$

Hence, $t = \dfrac{1}{2400}$

Substituting this value into $m(t)$, we get

$$m\left(\dfrac{1}{2400}\right) = 2\cos\left(\dfrac{\pi}{6}\right) + \sin\left(\dfrac{\pi}{3}\right) = 2.598$$

which is the maximum value of the signal $m(t)$. Therefore,

$$m_n(t) = \dfrac{2\cos 400\pi t + \sin 800\pi t}{2.598}$$

$$= 0.7698 \cos 400\pi t + 0.3849 \sin 800\pi t$$

The power in the sum of the two sinusoids with different frequencies is the sum of powers in them. Hence,

$$P_{mn} = 0.5(0.7698^2 + 0.3849^2) = 0.3703$$

The power in the carrier component of the modulated signal is

$$E_c \times E_c = \dfrac{2 \times 2}{2} = 2$$

and the power in the sideband is

$$(E_c \times E_c)m^2 P_{mn} = 2 \times 2 \times 0.6 \times 0.6 \times 0.3703 = 0.53323$$

7. A Diode detector has got a parallel combination of 660 pF capacitor and a 15 K resistor. If the modulating frequency is 10 kHz, calculate the maximum depth of modulation to avoid diagonal clipping.

Solution

The admittance of RC combination is

$$Y = \dfrac{1}{R} + j2\pi f_m C$$

$$= (0.066 \times 10^{-3}) + j\, 4.146 \times 10^{-5} \text{ mhos}$$

$$|Z| = \dfrac{1}{Y} = 12.83\ K\Omega$$

Hence, the maximum modulation index without diagonal peak clipping is

$$m = \dfrac{|Z|}{R} = \dfrac{12.83}{15} = 0.855$$

8. The modulating signal for an SSB system is $m(t) = \cos 2\pi f_m t$. Find out the two possible SSB AM signals. Assume the carrier frequency to be $E_c \cos 2\pi f_c t$.

Solution

The modulated signal will be

$$f(t) = E_c \cos 2\pi f_m t \cos 2\pi f_c \pm E_c \sin 2\pi f_m t \sin 2\pi f_c t$$

By taking the upper sign, we get the upper sideband

$$f_u(t) = E_c \cos 2\pi (f_c + f_m)t$$

The lower sideband is obtained by taking the + sign

$$f_l(t) = E_c \cos 2\pi (f_c - f_m)t$$

REVIEW QUESTIONS

1. What is the need for modulation?
2. Define modulation index for AM.
3. Draw the frequency spectrum of an AM wave.
4. Draw the phasor representation of an AM wave.
5. What are the different degrees of modulation?
6. What is the difference between high and low level modulation?
7. What is the percentage of the saving power in SSB modulation?
8. List the advantages of SSB transmission and a conventional DSB system.
9. How much power is saved in an SSB system compared to an AM system?
10. Draw the circuit diagram of a balanced modulator and give the output expression.
11. Draw the circuit diagram of an envelope detector.
12. What is the difference between DSB and VSB?
13. What is the primary requirement for a modulating signal?
14. What is the range of modulation index and useful power?
15. Mention the relative merits of high level modulation.
16. What is VSB transmission? How is it used in TV broadcast?
17. Define selectivity and sensitivity of a receiver.
18. What are the features of a communication receiver?
19. List the advantages of superheterodyne reception over TRF reception.
20. What do you mean by image frequency rejection?
21. What are negative peak clipping and diagonal peak clipping in a diode detector?
22. What do you mean by linear detector?
23. What is square law detector?
24. What is an AGC? Explain it briefly.
25. What do you mean by tracking in a radio receiver?
26. What is meant by modulation depth?
27. What is overmodulation? What does it result?
28. With the help of a simple block diagram, explain the production of SSB SC.
29. Explain the merits of VSB system over normal AM in the case of video transmission.
30. What is the IF for the standard broadcast AM receiver?
31. What is the maximum efficiency of transmission in a DSBFC type of modulation?
32. To avoid distortion, what should be the maximum modulating index?
33. Does the modulation have effect on the amplitude of the carrier?
34. What is the advantage of base modulation over collector modulation?
35. What is the principle of balanced modulator?
36. What do you mean by baseband transmission?
37. Why is modulation necessary in a communication system? List the different types of modulation schemes.

38. Explain with a neat waveform, the principle of amplitude modulation. Write the expression for an instantaneous voltage of an AM wave. Write a note on the frequency spectrum of AM wave.

39. What is amplitude modulation? Derive the expression for (a) modulation index and (b) transmitted power in terms of carrier power and modulation index.

40. With necessary equations, explain the principle of DSB and SSB generation. Draw their frequency spectra.

41. How is SSB signal generated by the phase shift method? Explain in detail with a block diagram and necessary equation. Give the advantages and disadvantages of this method.

42. Discuss the different methods of SSBSC generation with a relevant block diagram.

43. Explain a vestigial sideband system. How is it generated? What is its advantage over DSB and SSB? What are its applications?

44. Draw a circuit diagram for collector modulation and explain its working in detail.

45. Write a detailed note on collector, base, and emitter modulation. Mention the advantages and disadvantages of each method.

46. Explain with a neat circuit diagram, the working of a push-pull balanced modulator. Obtain the expression for the current at the secondary of the output transformer.

47. Explain a square law modulator with a neat block diagram and derive an expression for the output.

48. Explain with a neat circuit diagram the working of switching modulator. Obtain an expression for its output.

49. Explain in detail the working of a ring modulator with a neat circuit diagram.

50. With a block diagram, explain the functioning of a synchronous detector. Derive an expression for the output voltage. Hence show that any shift in phase or frequency of the locally generator carrier from that of the transmitter carrier results in phase distortion or delay.

51. Explain the principle and operation of envelope detector used for AM detection. Give its advantages and disadvantages.

52. With a block diagram, explain the working of high power AM transmitter.

53. Draw the block diagram of an SSB transmitter and explain its working in detail.

54. What are three methods of generating SSB? Compare their merits and demerits.

55. Describe with the help of a block diagram, the working of a superheterodyne receiver. Explain the function and purpose of each block. Describe how the image signal is formed and how it can be minimized.

56. With the help of a neat block diagram, describe a basic communication system.

57. The phase shift SSB modulator shown in Fig. 3.36 has a carrier frequency f_c and is modulated by a single frequency f_m. Find out the expression for

the output voltage of the two balanced modulators and show that the output summation produces only the lower sideband.

58. Explain how the information signal is recovered from an SSB carrier by a demodulating circuit.
59. Explain how a suitable train of current pulses fed to a tuned circuit will result in an amplitude modulated output wave. Draw the waveforms.
60. Describe in detail the quadrature carrier multiplexing of transmitting message signals.

PROBLEMS

1. An amplitude modulated wave has a peak to peak voltage of 600 V and valley to valley voltage of 100 V. Find the percentage depth of modulation.
2. How many stations can be accommodated in a 100 kHz bandwidth with the highest modulating frequency being 5 kHz?
3. If a 100 kW amplitude modulated transmitter is modulated sinusoidally by 50%, what is the total RF power delivered?
4. A sinusoidal carrier of frequency 1 MHz and amplitude 100 V is modulated by a sinusoidal voltage of frequency 5 kHz producing 50% modulation. Calculate the amplitude and frequency of USB and LSB.
5. For 80% modulation, what is the ratio of carrier voltage to sideband voltage?
6. Find the modulating index, carrier, and sideband frequency for the amplitude modulated wave given by the equation

$$f(t) = 8 \sin (2\pi 400 \times 10^3 t) - 2\cos(2\pi 390 \times 10^3 t) + 2\cos (2\pi 410 \times 10^3 t)$$

7. If the maximum bandwidth of an AM system is 30 kHz, what must be the maximum frequency of the modulating signal?
8. What is the power saved if the carrier of a 80% modulated AM wave is suppressed?
9. What is the total modulating index if a carrier is simultaneously modulated by two sine waves with modulating indices of 0.3 and 0.4?
10. If audio sine waves with 400 Hz, 600 Hz, and 800 Hz simultaneously modulate a 120 kHz carrier, what frequencies will you find at the output?
11. What will be the radiated power of an AM transmitter at 70% modulation if it radiates 70 kW of power.
12. An AM transmitter produces 15 kW with 60% modulation. How much of it is carrier power? How much power is saved if SSB transmission takes place?
13. A diode detector has a load 0.01 μF in parallel with a 10 kΩ resistor. What is the maximum depth of sinusoidal modulation that the detector can handle without diagonal peak clipping when the modulating frequency is at 800 Hz and 8 kHz?
14. Draw the waveform of an AM signal with V_{max} = 80 V, and V_{min} = 20V. Assume $f_c = 10 f_m$. Obtain the modulation index, magnitude of the carrier wave, and magnitude of sideband components.

15. A 500 W, 100 kHz carrier is modulated to a depth of 60% by modulating signal of frequency 1 kHz. Calculate the total power transmitted. What are the sideband components of the AM wave?

16. A 220 W carrier is modulated to a depth of 65%. Calculate the total power in the modulated wave.

17. A carrier of 1 MHz with 400 W of its power is amplitude modulated with a sinusoidal signal of 2500 Hz. The depth of modulation is 75%. Calculate the sideband frequencies, the bandwidth, the power in the sidebands, and the total power in the modulated wave.

18. An AM radio transmitter gives a power output of 8 kW when modulated to a depth of 80%. If after modulation by a speech signal, which produces an average depth of 30%, the carrier and one sidebands are suppressed, determine the average power in the remaining output.

19. A transistor Class C amplifier has maximum permissible collector dissipation of 30 W and collector efficiency of 75%. It is to be collector modulated to a depth of 85%. Calculate the maximum unmodulated carrier power and the sideband power generated.

20. The output voltage of a transmitter is given by $500(1+ 0.4 \sin 3140t)$ $\sin 6.28 \times 10^7 t$. Determine (a) the carrier frequency, (b) modulating frequency, (c) carrier power, (d) mean power output, and (e) peak power output.

21. For the ISB transmitter shown in Fig. 3.38, channel A input frequency f_a = 0 to 4 kHz, channel B input frequency f_b = 0 to 4 kHz, LF carrier frequency f_{LF} = 200 kHz, MF carrier frequency f_{MF} = 4 MHz, and HF carrier frequency f_{HF} = 30 MHz. (a) Sketch the frequency spectrum for the following points: BPFA output, BPFB output, BPF 3 output, and BPF 4 output. (b) For an A channel input frequency f_a = 1.5 kHz and B channel frequency f_b = 2 kHz, determine the frequency components at the following points: BPF A output, BPFB output, BPF 3 output, and BPF 4 output.

22. The per cent modulation of an AM wave changes from 35% to 70%. The original power content of the carrier frequency was 800 W. Find out the new power content of the carrier frequency and each sidebands after the percent modulation has risen to 70%.

23. The normalized signal $f_m(t)$ has a bandwidth of 15 kHz and its power content is 0.4 W. The carrier $E_c \cos 2\pi f_c(t)$ has a power content of 15 W. Find the bandwidth and the power content of the modulating signal (a) for SSB amplitude modulation and (b) for DSBSC amplitude modulation.

24. For the balanced ring modulator shown in Fig. 3.27(a), a carrier input frequency f_c = 400 kHz and modulating input signal frequency f_m = 300 to 3300 Hz. Find (a) output frequency range and (b) output frequency for a single input frequency f_m = 2200 Hz.

25. In an SSB transmitter using the third method shown in Fig. 3.37, the audio carrier frequency and the radio carrier frequency are 3 kHz and 300 kHz, respectively. Audio signals of 400 and 4000 Hz are modulated simultaneously. Find (a) the frequency components of all components present in the output and (b) the carrier position frequency and the two band edge frequencies.

Answers to problems

1. 71.4%
2. 10 channels ideal (8 channels with 2.5 kHz guard band)
3. 112.5 kW
4. 995 kHz, 1005 kHz amplitude 25V
5. 2.5
6. Carrier = 400 kHz, upper sideband = 410 kHz, lower sideband = 390 kHz, $m = 0.5$
7. 15 kHz.
8. 75%
9. 0.5
10. carrier = 120 kHz, sidebands 120.4 kHz, 119.6 kHz, 120.6 kHz, 119.4 kHz, 120.8 kHz, and 119.2 kHz
11. 56 kW
12. 12.71 kW, power saved = 13.86 kW
13. $m = 0.847$ and 0.195
14. $m = 0.6$, carrier amplitude = 50 V, and sideband amplitude = 15V
15. Total power = 590 W, sideband power = 45 W
16. Total power = 266.2 W
17. sideband frequencies = 1.0025 MHz, 997.5 kHz, bandwidth = 5 kHz, and total power = 505 W
18. 88.2 W, 31.76 W
20. 125 kW, 135 kW, 270 kW
21. 196 kHz – 200 kHz, 200 kHz – 204 kHz, 196 kHz – 200 kHz – 204 kHz, 4.196 MHz – 4.2 MHz – 4.204 MHz, and 34.196 MHz – 34.2 MHz – 34.204 MHz
22. 800 W, 98 W
23. 0.16, 96 mW and 192 mW
24. 396.7 kHz to 403.3 kHz

    397.8 kHz to 402.2 kHz
25. Taking carrier frequency to be 303 kHz and modulating frequencies to be 400 Hz, 1 kHz, 2 kHz, 3 kHz, and 4 kHz, the range of frequencies are

    302.6 kHz and 303.4 kHz

    302 kHz and 304 kHz

    301 kHz and 305 kHz

    300 kHz and 306 kHz

    299 kHz and 307 kHz

MATLAB EXAMPLES

```
1.%Amplitude modulation:DSB-C with tone input%
  fc=100000;
  fs=1000000;
  f=1000;
  m=0.5;
  A=1/m;
  opt=-A;
  t=0:1/fs:((2/f)-(1/fs));
  x=cos(2*pi*f*t);
  y=modulate(x,fc,fs,'amdsb-tc',opt);
  subplot(2,2,1); plot(x); title('modulating signal-
timedomain');
  subplot(2,2,2); plot(y); title('DSB-C signal-
timedomain,m=0.5');
  m=1.0; opt=-1/m;
  y=modulate(x,fc,fs,'amdsb-tc',opt);
  subplot(2,2,3); plot(y); title('DSB-C signal-
timedomain,m=1.0');

  m=1.2; opt=-1/m;
  y=modulate(x,fc,fs,'amdsb-tc',opt);
  subplot(2,2,4); plot(y); title('DSB-C signal-
timedomain,m=1.2');
```

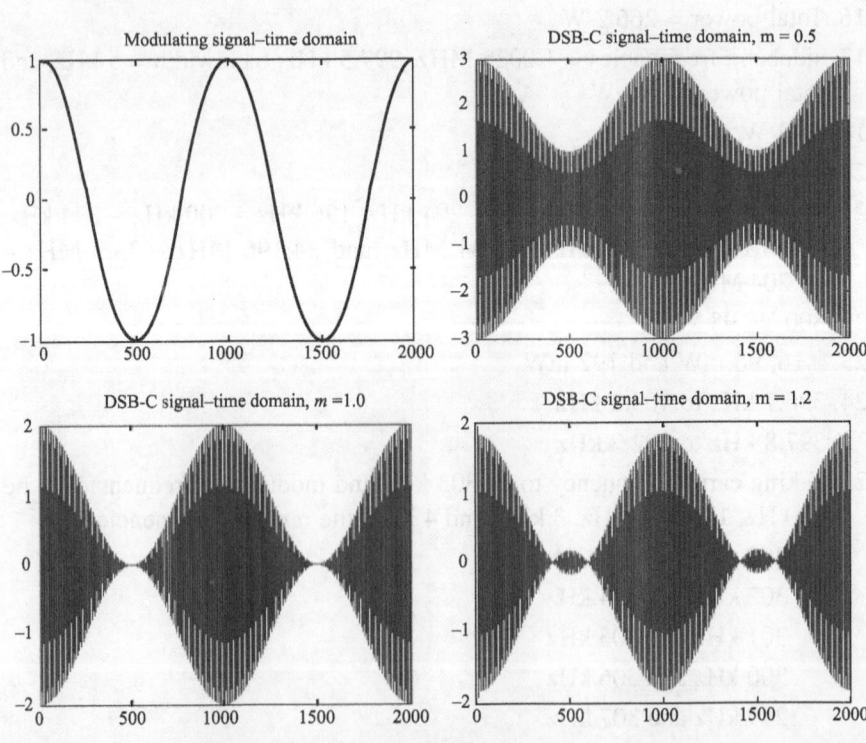

2. `%program for QAM%`
```
fc=500;
fs=5000;
f1=100; f2=200;
t=0:1/fs:((4/f1)-(1/fs));
x1=cos(2*pi*f1*t);
```

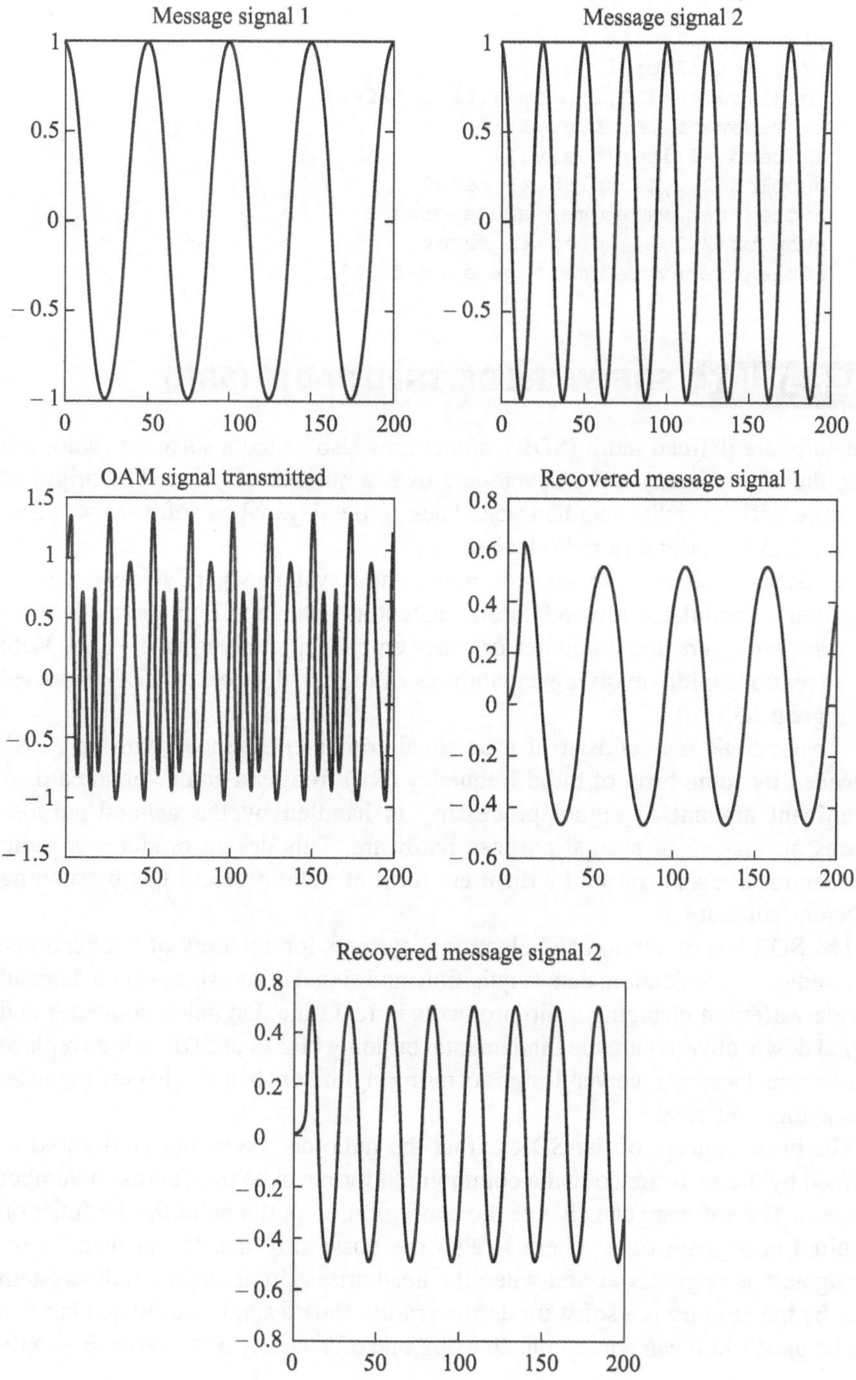

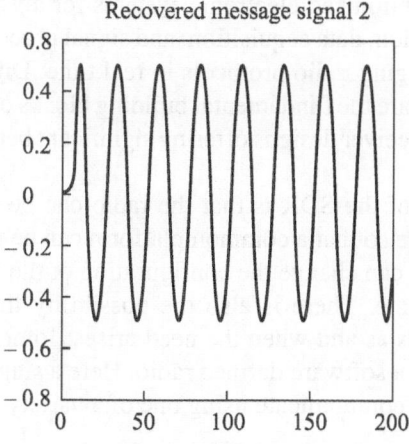

```
x2=cos(2*pi*f2*t);
%modulation%
y=x1.*cos(2*pi*fc*t)+x2.*sin(2*pi*fc*t);
subplot(2,3,1); plot(x1); title('message signal 1');
subplot(2,3,2); plot(x2); title('message signal 2');
subplot(2,3,3); plot(y); title('QAM signal TRANSMITTED');
%DEMODULATION%

y1=y.*cos(2*pi*fc*t);
y2=y.*sin(2*pi*fc*t);
[b,a]=butter(10,2*pi*max(f1,f2)/fs);
x1_recov=filter(b,a,y1);
x2_recov=filter(b,a,y2);
subplot(2,3,4); plot(x1_recov);
title('recovered message signal 1');
subplot(2,3,5); plot(x2_recov);
title('recovered message signal 2');
```

CASE STUDY SOFTWARE DEFINED RADIO (SDR)

The software defined radio (SDR), sometimes also called a software radio, has been the aim of many radio developers over a number of years. The origin of software defined radios can be traced back to the days when software was first used within radios and in radio technology.

An SDR system is a radio communication system where mixers, filters, amplifiers, modulators/demodulators, detectors. etc. are implemented using software on a personal computer or other embedded computing devices. With the advent of rapidly evolving capabilities of digital electronics, SDR is poised for a great take-off.

A basic SDR may consist of a personal computer equipped with an ADC, preceded by some form of radio frequency (RF) front end and a sound card. A significant amount of signal processing is handled by the general-purpose processor, instead of special-purpose hardware. This design produces a radio that can receive and transmit a different form of radio protocol just by running different software.

The SDR has revolutionized electronic systems for a variety of applications, including communication, data acquisition, and signal processing, which demand a wide variety of changing radio protocols in real time. Digital upconverter and digital downconverter are the fundamental building blocks of SDR, which replace conventional analog receiver designs offering significant benefits in performance, packaging, and cost.

The basic concept of the SDR is that the radio can be totally configured or defined by the software so that a common platform can be used across a number of areas. The software can change the configuration of the radio for the function required at a given time. There is also the possibility that it can then be re-configured as upgrades as and when the need arises. Joint tactical radio system used by the military is a software defined radio. Here a single hardware platform can be used and it can communicate using one of a variety of waveforms simply by reloading or reconfiguring the software for a particular application required. This is particularly useful for coalition style operations where forces from different countries may operate together on a joint mission and radios could be re-configured to enable communications to occur between troops from different countries.

The SDR concept is equally applicable for commercial applications. One application may be for cellular base stations where standard upgrades frequently occur. By having a generic hardware platform, upgrades of standards can easily be incorporated. Migrations, for example from universal mobile telecommunications system (UMTS) to high speed packet access (HSPA) and onto long term evolution (LTE), could be accommodated simply by uploading new software and reconfiguring it without any hardware changes, despite the fact that different modulation schemes and frequencies can be used.

There can be many areas where the SDR concept can be used. As time progresses and the technology moves forward, it will be possible to use the concept in new areas.

Software Defined Radio: Definition

It is necessary to produce a robust definition for many reasons, including regulatory applications and standards issues, and for enabling the SDR technology to move forward more quickly. Many definitions have appeared that might cover a definition for a software defined radio. The SDR Forum defines the two main types of radios containing software as below.

- **Software Controlled Radio**   It is the radio in which some or all of the physical layer functions are software controlled. In other words, this type of radio uses software only to provide control of the various functions that are fixed within the radio.
- **Software Defined Radio**   It is the radio in which some or all of the physical layer functions are software defined. In other words, the software is used to determine the specification of the radio and what it does. If the software within the radio is changed, its performance and function may change.

Another definition that seems to encompass the essence of the SDR is that it has a generic hardware platform on which software runs to provide functions (including modulation and demodulation), filtering (including bandwidth changes), and other functions such as frequency selection and even frequency hopping, if required. By reconfiguring, i.e. changing the software, the performance of the radio gets changed.

To achieve this, the software defined radio technology uses software modules that run on a generic hardware platform consisting of digital signal processing (DSP) processors as well as general-purpose processors to implement the radio functions of transmitting and receiving signals.

In an ideal situation in a transmitter, a digital signal processor would generate a stream of numbers. These would be sent to a DAC connected to a radio antenna, and similarly for reception, the signal from the antenna would be directly converted to digital form and all the processing is undertaken under software control. In this way, there are no limitations introduced by the hardware. To achieve this, the digital-to-analog conversion for transmission needs to have a relatively high power, depending on the application and it would also demand very low noise for receivers. As a result, full software definition is not normally possible. Figure 3.76 shows the block diagram of an ideal SDR.

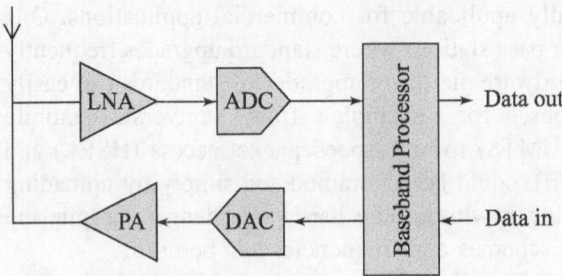

Fig. 3.76 Block diagram of an 'ideal' software defined radio

Levels of SDR

It is not always feasible or practicable to develop a radio that incorporates all the features of a fully software defined radio. Some radios may only support a number of features associated with SDRs, whereas others may be fully software defined. In order to give a broad idea of the level at which a radio is classified, the SDR Forum has defined a number of tiers. These tiers can be explained in terms of what is configurable.

- **Tier 0:** This tier has a non-configurable hardware radio, i.e. one that cannot be changed by software.
- **Tier 1:** It has a software controlled radio where limited functions are controllable. These may be power levels, interconnections, etc. but not mode or frequency.
- **Tier 2:** In this tier of software defined radio, a significant proportion of the radio is software configurable. Often, the term *software controlled radio* (SCR) may be used. There is software control of parameters, including frequency, modulation, and waveform generation/detection, wideband/narrowband operation, security, etc. The RF front end still remains hardware based and non-reconfigurable.
- **Tier 3:** It has ideal software radio (ISR), where the boundary between the configurable and non-configurable elements exists very close to the antenna and the 'front end' is configurable. It could be said to have full programmability.
- **Tier 4:** It has the ultimate software radio (USR), which is a stage further on from the ISR. Not only does this form of software defined radio have full programmability but it is also able to support a broad range of functions and frequencies at the same time. With many electronic items such as cell phones having many different radios and standards, a software definable multifunctional phone would fall into this category.

Although these SDR tiers are not binding in any way, they give a way of broadly classifying the different levels of software defined radios that may exist.

SDR Architecture

Although there are many different levels of SDR and many ways in which a software defined radio may be designed, it is possible to give some generalized comments about the basic structures that are used.

Apart from the control and system software and its associated hardware, a software defined radio (SDR) can be considered to contain a number of basic functional blocks as detailed below.

- **RF Amplification** These elements are the RF amplification of the signals travelling to and from the antenna. On the transmit side, the amplifier is used to increase the level of the RF signal to the required power level for transmission. It is unlikely that direct conversion by the DAC will give the required output level. On the receive side, signals from the antenna need to be amplified before passing further into the receiver. If antenna signals are directly converted into digital signals, quantization noise becomes an issue even if the frequency limits are not exceeded.

- **Frequency Conversion** In many designs, some analog processing may be required. Typically, this involves converting the signal to and from the final radio frequency (RF). In some designs, this analog section may not be present and the signal will be converted directly to and from the final frequency from and to the digital format. Some intermediate frequency processing may also be present.

- **Digital Conversion** It is at this stage that the signal is converted between the digital and analog formats. This conversion is in many ways at the heart of the equipment. When undertaking these conversions, there are issues that need to be considered. On the receive side, the maximum frequency and the number of bits to give the required quantization noise are of great importance. On the transmit side, the maximum frequency and the required power level are some of the major issues.

- **Baseband Processing** The baseband processor is at the very centre of the software defined radio. It performs many functions such as digitally converting the incoming or outgoing signal in frequency. The elements, known as digital upconverter (DUC), are used for converting the outgoing signal from the base frequency up to the required output frequency and then converting it from digital to analog. On the receive side, a digital downconverter (DDC) is used to bring the signal down in frequency. The signal also needs to be filtered and demodulated and the required data is extracted for further processing. One of the key issues of the baseband processor is the amount of processing power required. The greater the level of processing, the higher the current consumption, which in turn requires additional cooling, etc. This may have an impact on what can be achieved, if power consumption and size are the limitations. Also the hardware is to be considered, such as general processors, DSPs, ASICs, and in particular FPGAs. The FPGAs are of particular interest because they may be reconfigured to change the definition of the radio.

- **SDR SCA–Software Communications Architecture** The software communications architecture (SCA) developed for SDRs came out of the joint tactical radio system (JTRS). It basically describes the software components within a software defined radio and how they interface with each other. It is effectively an architecture within which the SDR software can be written and to which it complies. This provides two main advantages:

(a) The ability to reuse some elements of the code in other products and on other platforms

(b) The ability to obtain software elements from different sources and know that they will interface with each other

The software that is within a software defined radio and hence falls within the SCA can be categorized into one of the three categories.

(a) **Management** The software which falls into this SCA category is used for managing the radio system. Various applications may include plug-and-play deployment and configuration software.

(b) **Node** This software may comprise such applications as bootstrapping and access to hardware.

(c) The third category includes the software used particularly for the signal processing. Examples of this are waveform generation, demodulation, frequency translation, etc.

Of the three different categories of SCA software, only the application software is fully transportable between different platforms as it is platform independent. The other two are platform dependent and cannot be transported. However, all of them conform to the same interface standards and can be used with each other.

Middleware such as CORBA and common object request broker architecture is used to facilitate inter-module communications.

In order that any software can be declared as SCA software, it needs to be tested for SCA compliance. In this way, the API (applications programming interface) can be determined as compliant and it will operate with other SCA complaint software. Also its performance is tested for correct operation.

SDR Security

Another area of growing importance is that of SDR security. Many military radios and often many commercial radio systems will need to ensure that the transmissions remain secure, and this is an issue that is important for all types of radios. However, when using an SDR, there is another element of security, namely that of ensuring that the software within the radio is securely upgraded. With the growing use of the Internet, many SDRs will use this medium to deliver their updates. This presents an opportunity for malicious software to be delivered, which could modify the operation of the radio or prevent its operation altogether. Accordingly, the SDR software security needs to be considered if the Internet is used for software delivery and whenever there is going to be security weaknesses.

Hardware

The ideal receiver scheme would be to connect an ADC to an antenna. A digital signal processor would read the converter. Then its software would transform the stream of data from the converter to any other form as the application demands.

An ideal transmitter would be similar. A digital signal processor would generate a stream of numbers. These would be sent to a DAC connected to a radio antenna. An ideal scheme as usual is practically impossible to realize.

Most of the SDR receivers utilize a variable frequency oscillator, mixer, and filter to tune the desired signal to a common baseband, i.e. an intermediate frequency. It is then sampled by the ADC. In some applications, it is not necessary to tune the signal to an intermediate frequency and the radio frequency signal after amplification is directly sampled by the ADC.

Since the ADC available cannot pick up radio signals of the order of few microvolts and, in some cases, even a few nanovolts, a low-noise amplifier preceeds the ADC. Introduction of low-noise amplifier has its own problem such as spurious signal and introduction of distortion. Hence, to avoid this, a bandpass filter is introduced between the anenna and the low-noise amplifier. But this reduces the radio's flexibility thus defeating the very purpose of going in for SDR. Real software radios mostly have two or three analog channels that are switched in and out. They contain matched fiters, amplifiers, and mixers.

A typical amateur software radio uses a direct conversion receiver. Unlike the usual direct conversion receivers, the mixer technologies used are based on the quadrature sampling detector and the quadrature sampling exciter. The receiver performance of SDR is directly related to the dynamic range of the analog to digital converters (ADCs) used. Radio frequency signals are downconverted to the audio frequency band, which is sampled by a high-performance audio frequency ADC. The newer software defined radios use embedded ultra-high performance ADCs, which provide higher dynamic range and are more immune to noise and RF interference. A fast PC performs the digital signal processing (DSP) operations using software specific for the radio hardware.

The SDR software performs all of the demodulation, filtering (both radio frequency and audio frequency), and signal enhancement like equalization and binaural presentation. The uses include every common amateur modulation—morse code, AM, SSB, and FM—and a variety of digital modes such as radioteletype, slow-scan television, and packetradio. DDCs (digital downconverters) and DUCs (digital upconverters) are the fundamental building blocks of SDR, which can replace conventional analog receiver designs offering significant benefits in performance, density, and cost. In order to fully appreciate the benefits of SDR, a conventional analog receiver system will be compared to its digital receiver counterpart, highlighting similarities and differences.

Conventional Analog Receiver

The conventional heterodyne radio receiver is shown in Fig. 3.77. First the RF signal from the antenna is amplified, typically with a tuned RF stage that amplifies a range of the frequency band of interest. This amplified RF signal is then fed into a mixer stage. The other input to the mixer comes from the local oscillator whose frequency is determined by the tuning control of the radio. The mixer translates the desired input signal to the intermediate frequency (IF). The IF stage is a bandpass amplifier that lets only one signal or radio station through. The centre frequencies for IF stages are 455 kHz and 10.7 MHz for commercial AM and FM broadcasts. The demodulator recovers the original modulating signal from the IF output using one of several different schemes. For example, AM uses an envelope detector and FM uses a frequency discriminator. In a typical home radio, the demodulated output is fed to an audio power amplifier that drives the speaker.

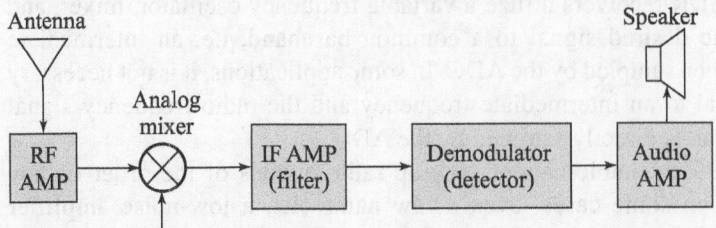

Fig. 3.77 Analog radio receiver

The mixer performs an analog multiplication of the two inputs and generates a difference frequency signal. The frequency of the local oscillator is set so that the difference between the local oscillator frequency and the desired input signal (the radio station you want to receive) equals the IF. For example, if one wants to receive an FM station at 100.7 MHz and the IF is 10.7 MHz, one would tune the local oscillator to: 100.7 - 10.7 = 90 MHz. This is called *downconversion* or *translation* because a signal at a high frequency is shifted down to a lower frequency by the mixer. The IF stage acts as a narrowband filter, which only passes a 'slice' of the translated RF input. The bandwidth of the IF stage is equal to the bandwidth of the signal (or the 'radio station') that you are trying to receive. For commercial FM, the bandwidth is about 100 kHz and for AM, it is about 5 kHz. This is consistent with channel spacing of 200 kHz and 10 kHz, respectively.

SDR Receiver

Figure 3.78 shows a block diagram of a software defined radio receiver. The RF tuner converts analog RF signals to analog IF frequencies, the same as the first three stages of the analog receiver. The A/D converter that follows digitizes the IF signal, thereby converting it into digital samples. These samples are fed to the next stage, which is the digital downconverter (DDC) shown within the dotted lines. The digital downconverter is typically a single monolithic chip or FPGA IP and is a key part of the SDR system.

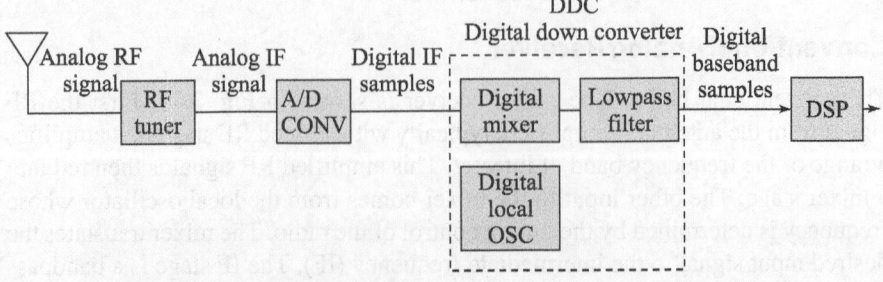

Fig. 3.78 Block diagram of an SDR receiver

A conventional DDC has three major sections:
- A digital mixer
- A digital local oscillator
- An FIR low-pass filter

The digital mixer and local oscillator translate the digital IF samples down to baseband. The FIR low-pass filter limits the signal bandwidth and acts as a decimating low-pass filter. The digital downconverter includes hardware multipliers, adders, and shift register memories. The digital baseband samples are then fed to DSP, which performs tasks such as demodulation, decoding, and other processing tasks. Previously, these needs have been handled with dedicated application specific ICs (ASICs), and programmable DSPs. At the output of the mixer, the high-frequency wideband signals from the A/D input have been translated down to DC as complex I and Q components with a frequency shift equal to the local oscillator frequency. This is similar to the analog receiver mixer except that the mixing was done down to an IF frequency. Here, the complex representation of the signal allows us to go right down to DC. By tuning the local oscillator over its range, any portion of the RF input signal can be mixed down to DC. In effect, the wideband RF signal spectrum can be 'slid' around 0 Hz, left and right, simply by tuning the local oscillator still preserving the upper and lower sidebands.

Since the local oscillator uses a digital phase accumulator, it has some very nice features. It switches between frequencies with phase continuity, so that one can generate FSK signals or sweeps very precisely with no transients as shown in Fig. 3.79. The frequency accuracy and stability are determined entirely by the A/D clock. So, it is inherently synchronous to the sampling frequency. There is no aging, drift, or calibration since it is implemented entirely with digital logic. Since the output of the FIR filter is bandlimited, the Nyquist theorem allows one to lower the sample rate. If only one out of every N samples is kept as shown in Fig. 3.80, one has dropped the sampling rate by a factor of N. This process is called *decimation* and it means that keeping one out of every N signal samples. If the decimated output sample rate is kept higher than twice the output bandwidth, no information will be lost. The advantage is that the decimated signals can be processed easily, can be transmitted at a lower rate, or requires less memory for storage. As a result, the decimation can dramatically reduce system costs.

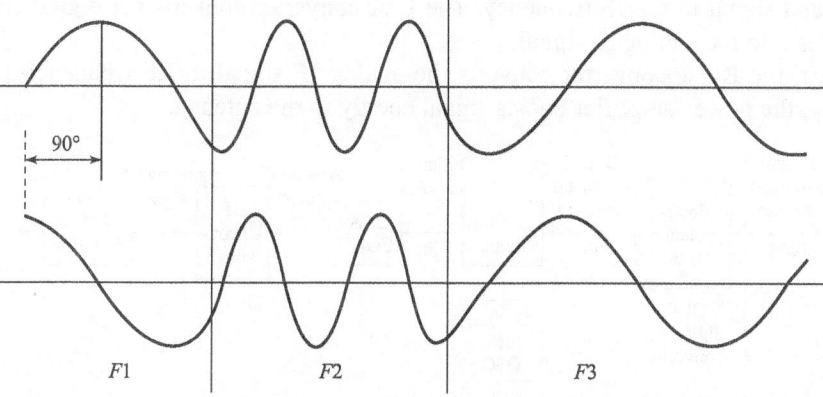

Fig. 3.79 Local oscillator frequency switching

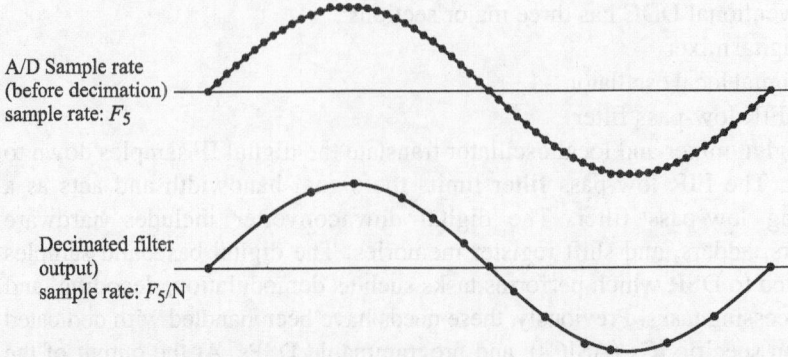

Fig. 3.80 FIR filter decimation

As shown in Fig. 3.81, the DDC performs two signal processing operations:
(a) Frequency translation with the tuning controlled by the local oscillator
(b) Low-pass filtering with the bandwidth controlled by the decimation setting

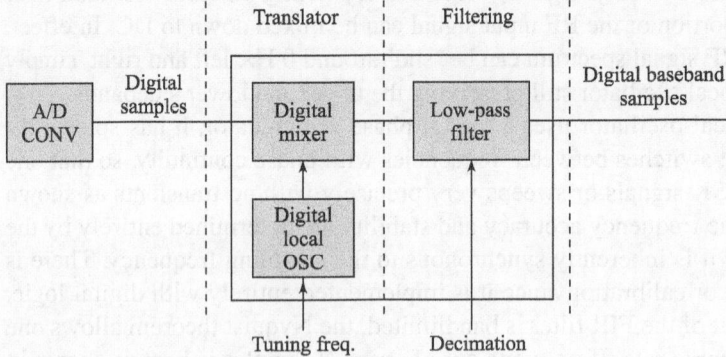

Fig. 381 Digital downconverter

Software Defined Radio Transmitter

Let us now see software defined radio transmitter.

The input to the transmit side of an SDR system is a digital baseband signal, typically generated by a DSP stage as shown in Fig. 3.82. The digital hardware block in the dotted lines is a DUC (digital upconverter), which translates the baseband signal to the IF frequency. The D/A converter converts the digital IF samples into the analog IF signal.

Next, the RF upconverter converts the analog IF signal to RF frequencies. Finally, the power amplifier boosts signal energy to the antenna.

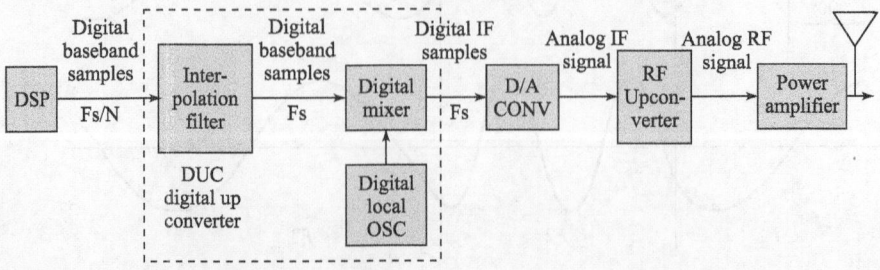

Fig. 3.82 SDR transmitter block diagram

In the DUC shown in Fig. 3.82, the digital mixer and the local oscillator at the right translate baseband samples up to the IF frequency. The IF translation frequency is determined by the local oscillator. The mixer generates one output sample for each of its two input samples and the sample frequency at the mixer output must be equal to the D/A sample frequency f_S. Therefore, the local oscillator sample rate and the baseband sample rate must be equal to the D/A sample frequency f_S. The local oscillator already operates at a sample rate of f_S, but the input baseband sample frequency at the left is usually much lower. This problem is solved with the *interpolation filter.*

Summary

The software defined radio (SDR) is a reality today, and it is being used in many areas. However, there are a number of limitations that prevent them being used in as many applications as one would like. One is the sheer processing power that is required and the resulting power consumption. It is necessary to undertake power consumption/processing power trade-off, and this is one of the core decisions that needs to be made at the outset. This is one of the reasons for not using SDR for cell phone designs, but cell phone base stations are using them since the power consumption and the space are normally not a constraint and the software can be upgraded to enable to keep in touch with the latest technology. Also SDRs are being used by the military, and there are already some handheld designs available. As technology progresses, SDRs will be used in more applications; yet there will always be a decision to be made as the SDR is not the right decision for all radios. For small cheap radios, where changes will be few, the SDR is definitely not the choice. But for more complicated systems, where length of the service is important and where changes are likely, the SDR is definitely a good option to be considered

4

Angle Modulation

LEARNING OBJECTIVES

This chapter will enable the students to
- Know angle modulation and its theory
- Describe the angle modulation process
- Understand the difference between frequency and phase modulation
- Observe and draw the FM and PM waveforms
- Know the similarity between AM and narrowband FM
- Observe that the carrier power is distributed over sidebands and that for certain value of modulation index, the carrier component is zero and the power is carried by sidebands only
- Know that phase modulation can be obtained from frequency modulation and vice versa
- Become familiar with different methods of FM and PM generation and detection
- Observe that the conventional method of generation can be replaced with modern methods using integrated circuits
- Determine and calculate modulation index
- Analyse the frequency spectrum for wideband FM using Bessel function
- Know about various types of FM transmitters and receivers
- Determine the maximum bandwidth from the spectrum
- Know about Carson's formula
- Understand the difference between AM, PM, and FM

4.1 INTRODUCTION

In Chapter 3, we considered amplitude modulation of the carrier as a means of transmitting the message signal. Any analog signal has three properties—amplitude, phase, and frequency. Any of these quantities can be varied.

If the amplitude of the signal is varied, we call it an amplitude modulation. This modulation was discussed in detail in the previous chapter. Another class of modulation method includes frequency and phase modulation. In frequency modulation (FM), the frequency of the carrier f_c is changed by the message signal. In phase modulation (PM), the phase of the carrier Φ is changed according to the variation in the message signal. Unlike amplitude modulation, which is linear, both FM and PM are non-linear. They are often jointly called angle modulation. Because of non-linearity, these modulation methods are more difficult to implement and much more difficult to analyse. Analysis is approximate. Bandwidth requirements are large. In most of the cases, the effective bandwidth of the modulated channel is usually many times the bandwidth of the modulating signal. Angle modulation, although it is more complex and occupies larger bandwidth than AM, is preferred for its high degree of noise immunity, improved system fidelity, and more efficient use of power. The system sacrifices the bandwidth for all the above-mentioned advantages. Also it has the advantage of constant envelope, which is beneficial when amplified by non-linear amplifiers. This is one of the reasons why angle modulation is preferred widely for high-fidelity music broadcasting, point-to-point communication systems, TV sound transmission, cellular radio, micro wave, and satellite communication systems. Angle modulation was first introduced in 1931. E.H. Armstrong developed the first successful FM radio system in 1936.

4.2 INSTANTANEOUS FREQUENCY

The sinusoidal signal has a constant frequency and hence the variation of frequency with time appears to be contradictory to the conventional definition of a sinusoidal signal frequency. The concept of a sinusoidal signal can be extended to a generalized function whose frequency may vary with time.

In FM, the carrier frequency is varied in proportion to the modulating signal $m(t)$. This means that the carrier frequency is changing continuously at every instant. This is against the definition of frequency, since to define a frequency, we must have a sinusoidal signal at least over one cycle with the same frequency. This is similar to instantaneous velocity, where we consider the velocity as a variable (and not a constant) over an interval, changing at every instant. Let us consider a generalized sinusoidal signal

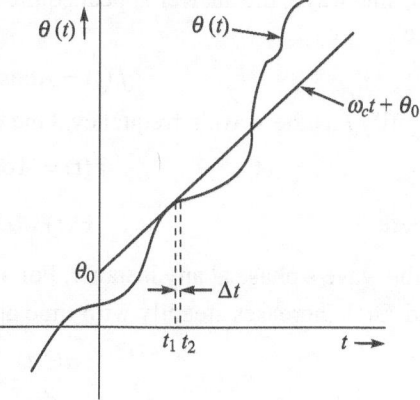

Fig. 4.1 Instantaneous frequency

$$f(t) = A \cos \theta(t) \tag{4.1}$$

where A is the amplitude of the signal and $\theta(t)$ is the generalized angle, a function of t. Figure 4.1 shows $\theta(t)$ versus t.

A sinusoidal signal is also represented with its initial phase angle as

$$F(t) = A \cos(\omega_c t + \theta_0) \qquad (4.2)$$

where θ_0 is called the phase shift and ω_c is the frequency.

From the figure, it is seen that a straight line with slope ω_c intercepts $\theta(t)$ axis at θ_0. The plot of $\theta(t)$ happens to be the tangential to the angle $(\omega_c t + \theta)$ at some instant t. Over a small interval $\Delta t \to 0$, the signal $f(t) = A \cos \theta(t)$ and the sinusoid $A \cos(\omega_c t + \theta_0)$ are identical, i.e.,

$$f(t) = A \cos(\omega_c t + \theta_0), \quad t_1 < t < t_2 \qquad (4.3)$$

It can be justified that over this small interval Δt, the frequency of $f(t)$ is ω_c. Since $(\omega_c t + \theta_0)$ is tangential to $\theta(t)$, the frequency of $f(t)$ is the slope of its angle $\theta(t)$ over this small interval. This concept can be generalized at every instant, and the instantaneous frequency ω_i at any instant t is the slope $\theta(t)$ at t. Thus, for $f(t)$ in Eq. (4.1),

$$\omega_i(t) = \frac{d\theta}{dt} \qquad (4.4)$$

$$\theta(t) = \int_{-\infty}^{t} \omega_i(\alpha) d\alpha \qquad (4.5)$$

There is the possibility of transmitting the information $m(t)$ by varying the angle θ of a carrier. This type of modulation when the angle of the carrier is varied in some manner with modulating signal $m(t)$ is called angle modulation or exponential modulation. Two possibilities are phase modulation and frequency modulation.

4.3 FM AND PM SIGNALS

First of all, let us define what is the frequency of a waveform. For a simple wave like sine wave, the answer appears quite obvious. For a wave using an expression like

$$f(t) = A \cos(2\pi f t + \phi) \qquad (4.6)$$

identify f as the wave's frequency. One can also represent such a wave as

$$f(t) = A \cos\{\Theta(t)\} \qquad (4.7a)$$

where

$$\Theta(t) = 2\pi f t + \phi \qquad (4.7b)$$

is the wave's phase at any instant t. For a simple sine wave, f has a constant value and $\Theta(t)$ increases steadily with time at the rate

$$\frac{d\Theta(t)}{dt} = 2\pi f \qquad (4.8)$$

We can define the FM wave produced when we modulate a carrier frequency f_c with a modulating signal $m(t)$ to be

$$f(t) = A \cos\{2\pi f_i(t) t + \phi\} \qquad (4.9a)$$

where

$$f_i = f_c + k_f m(t) \qquad (4.9b)$$

It is the instantaneous frequency of the wave at the instant t. The term k_f is a constant, whose value depends upon the modulating system and which determines how many Hz of frequency change one gets for each volt of modulating signal.

The FM wave can now be used to convey information about the modulating pattern in a manner similar to the AM variations. It can be seen that unlike an AM wave, an FM wave does not have a single frequency value. This makes an FM wave obviously different to an AM one. In case of an FM wave, we can define two distinct quantities—its unmodulated (i.e., carrier) frequency and its modulated frequency $f_i(t)$—which can change from instant to instant. The instantaneous phase of the modulated wave at any instant can be obtained by substituting Eq. (4.9b) into Eq. (4.8) and integrating to get

$$\Theta(t) = 2\pi f_c t + 2\pi k_f \int_0^t m(t)dt \tag{4.10}$$

with the assumption that the phase is zero at $t = 0$. In a similar way, we can define a phase modulated (PM) wave as having the form

$$f(t) = A\cos\{\Theta(t)\} \tag{4.11a}$$

where

$$\Theta(t) = 2\pi f_c t + k_p m(t) \tag{4.11b}$$

and k_p determines the phase change per volt of modulation that the PM modulator being used produces. The instantaneous frequency of the PM wave will, therefore, be

$$f_i(t) = f_c + k_p \frac{dm(t)}{dt} \tag{4.12}$$

Comparing Equations (4.9b) and (4.10) with (4.11b) and (4.12), we find that unless we know something about the modulation in advance, we may not be able to make out whether the signal is FM or PM. In both cases, the wave's frequency and phase vary from moment to moment. Mathematically speaking, FM and PM are almost like identical twins. The only difference is that one corresponds to a modulation pattern that is differential of the pattern produced by the other. However, FM and PM arguments can be mixed and most of the conclusions about one apply to the other. But we have to know in advance which type of modulation is being used if we want to recover the modulated information correctly. In general, however, both FM and PM waves are obviously different to an AM wave.

4.3.1 Spectrum of an FM Signal

From Eq. (4.10), we can write an FM wave in the form

$$f_{FM}(t) = A\cos[2\pi f_c t + 2\pi k_f \int_0^t m(t)dt] \tag{4.12a}$$

Let us take the example of a simple sine wave modulation at a modulation frequency f_m

$$m(t) = E_m \cos\{2\pi f_m(t)\} \qquad (4.12b)$$

which produces an instantaneous FM signal frequency of

$$f_i(t) = f_c + k_f E_m \cos\{(2\pi f_m(t)\} \qquad (4.12c)$$

We can conclude from the above-mentioned expression that f_i swings up and down either side of f_c over a range of $\pm k_f E_m$. This range is usually described in terms of the modulated signal's *peak frequency deviation* value defined as

$$\Delta f = k_f E_m \qquad (4.12d)$$

since it indicates the largest swing or deviation in frequency either side of f_c. Note that its value depends upon the magnitude of the modulation E_m but not upon the modulating frequency f_m.

By combining Equations (4.10), (4.12c), and (4.12d), we get the instantaneous phase of the sine wave modulated FM wave as

$$\Theta(t) = 2\pi f_c t + \frac{\Delta f}{f_m} \sin\{2\pi f_m(t)\} \qquad (4.12e)$$

The term $\dfrac{\Delta f}{f_m}$ is defined as modulation index β, i.e.,

$$\beta = \frac{\Delta f}{f_m} \qquad (4.12f)$$

We can then write the FM wave in the form

$$f_{\text{FM}}(t) = A \cos[2\pi f_c t + \beta \sin\{2\pi f_m(t)\}] \qquad (4.12g)$$

This provides us with information on how the modulated signal varies with time. However, we often need to know the frequency spectrum of the modulated wave in order to be able to determine the bandwidths of any filters, amplifiers, etc. that we require. Similarly, we can get the expression for phase modulated sine wave as

$$f_{\text{PM}}(t) = A \cos[2\pi f_c t + m \cos\{2\pi f_m(t)\}] \qquad (4.12h)$$

where m is the phase modulation index and defined as

$$m = k_p E_m \text{ rad}$$

where

$$k_p = \frac{\Delta\theta}{E_m}$$

Substituting this value of k_p into Eq. (4.12h), we get

$$f_{\text{PM}}(t) = A \cos[2\pi f_c t + \Delta\theta \cos\{2\pi f_m(t)\}] \qquad (4.12i)$$

Both β and m are explained in detail in Section 4.4.

4.3.2 Concept of Angle Modulation

From equations of FM and PM, it is clear that PM and FM are not only similar but are also inseparable. Replacing $m(t)$ with $\int m(t)$ changes PM to FM. Thus, a signal that is an FM wave corresponding to $m(t)$ is also the PM wave corresponding to $\int m(\alpha)d\alpha$. Similarly, a PM wave corresponding to $m(t)$ is the

FM wave corresponding to $dm(t)/dt$. Hence, by looking at an angle modulated carrier, there is no way of telling whether it is FM or PM. In both FM and PM, the angle of the carrier is varied in proportion to some measure of $m(t)$. In PM, it is directly proportional to $m(t)$, whereas in FM, it is proportional to the integral of $m(t)$. The relation between FM and PM is shown by a block diagram in Fig. 4.2.

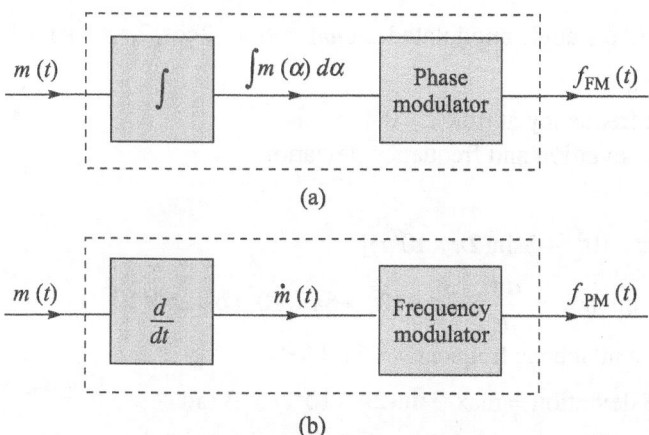

(a)

(b)

Fig. 4.2 Angle modulation: (a) frequency modulator and (b) phase modulator

Figure 4.3 shows the phase and frequency modulation of a sine wave carrier by a single frequency modulating signal. It is evident from Fig. 4.3 that we cannot conclude merely from the waveform whether it is FM or PM wave without knowing the characteristic of the modulating signal. In FM, the maximum frequency deviation, i.e., change in carrier frequency occurs during the positive and negative peaks of the modulating signal, the frequency deviation being proportional to the amplitude of the modulating signal.

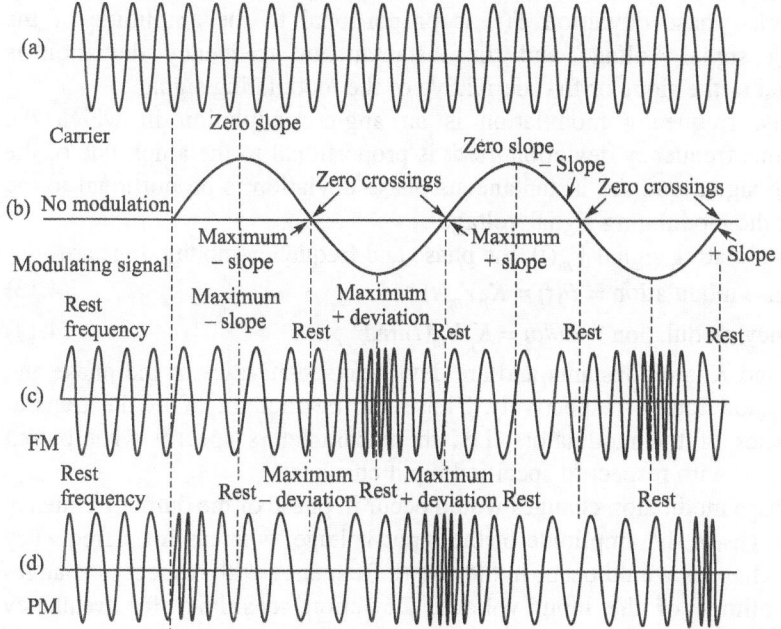

Fig. 4.3 Angle modulation of a sine wave carrier by a sine wave modulating signal

With PM, the maximum frequency deviation occurs during the zero crossing of the modulating signal, i.e., the frequency deviation is proportional to the slope or first derivative of the modulating signal. For both frequency and phase modulation, the rate at which the frequency change occurs is equal to the modulating signal frequency.

Example 4.1 Consider an angle modulated signal $x(t) = 2\cos[2\pi \times 10^5 t + 3 \sin(2\pi \times 10^2 t)]$ and find
(a) its instantaneous frequency at time $t = 0.4$ ms and
(b) maximum phase deviation and frequency deviation.

Solution

(a) Phase $\Phi(t) = [2\pi \times 10^5 t + 3\sin(2\pi \times 10^2 t)]$

Instantaneous frequency $= \dfrac{d\phi}{dt} = 2\pi \times 10^5 + 6\pi \times 10^2 \cos(2\pi \times 10^2 t)$

At $t = 0.4$ ms, instantaneous frequency $= 11.3$ kHz

(b) Maximum phase deviation $= \max 3\sin(2\pi \times 10^2 t) = 3$ rad

Maximum frequency deviation $= \max 6\pi \times 10^2 \cos(2\pi \times 10^2 t)$

$$= 6\pi \times 10^2 \text{ rad} = 300 \text{ Hz}$$

4.4 MODULATION INDEX

In this section, we will discuss deviation sensitivity, frequency deviation, and percentage modulation.

4.4.1 Deviation Sensitivity

By earlier definition, the phase modulation is an angle modulation in which the instantaneous phase deviation $\theta(t)$ is proportional to the amplitude of the modulating signal voltage and the instantaneous frequency deviation is proportional to the slope or first derivative of the modulating signal.

Similarly, frequency modulation is an angle modulation in which the instantaneous frequency deviation $d\theta/dt$ is proportional to the amplitude of the modulating signal and the instantaneous phase deviation is proportional to the integral of the modulating signal voltage.

For a modulating signal $V_m(t)$, the phase and frequency modulations are

Phase modulation $= \theta(t) = K_p V_m(t)$ rad $\hspace{2cm}$ (4.13)

Frequency modulation $= d\theta/dt = K_f V_m(t)$ rad $\hspace{2cm}$ (4.14)

where K_p and K_f are constants and are deviation sensitivities of the phase and frequency modulators, respectively. These can be thought of as the transfer characteristics of the modulators, i.e., these parameters specify what output changes occur with respect to specified input changes.

For a phase modulator, changes would occur in phase of the output frequency due to changes in the amplitude of the input voltage, whereas for a frequency modulator changes would occur in the output frequency with respect to changes in the amplitude of the input voltage. Deviation sensitivity for frequency modulation is

$$K_f = (\Delta\omega/\Delta V) \tag{4.15}$$

Deviation sensitivity for phase modulation is

$$K_p = (\Delta\theta/\Delta V) \tag{4.16}$$

Since for frequency modulation,

$$f(t) = V_c \cos [\omega_c t + \int d\theta(t)/dt] \tag{4.17}$$

substituting the value of $d\theta/dt$ from Eq. (4.14) into Eq. (4.17), we get

$$f(t) = V_c \cos[\omega_c t + \int K_f V_m(t)\ dt] \tag{4.18}$$

Let the modulating voltage be $V_m (t) = V_m \cos \omega_m t$. Then

$$f(t) = V_c \cos[\omega_c t + \int K_f V_m \cos \omega_m t] = V_c \cos [\omega_c t + K_f V_m/\omega_m \sin \omega_m t]$$
$$\tag{4.19}$$

For phase modulation,

$$f(t) = V_c \cos[\omega_c t + \theta(t)] \tag{4.20}$$

or $\qquad\qquad f(t) = V_c \cos[\omega_c t + K_p V_m \cos \omega_m t] \tag{4.21}$

The general expression for a carrier that is being phase or frequency modulated by a single modulating signal can be written as

$$f(t) = V_c \cos (\omega_c t + m \cos \omega_m t) \tag{4.22}$$

where $m \cos \omega_m t$ is the instantaneous phase deviation $\theta(t)$. If the modulating signal is a single frequency sinusoid, then the phase angle of the carrier varies from its demodulated value in a simple sinusoid fashion. In Eq. (4.22), m represents peak phase deviation in radians for a phase modulated carrier. Peak phase deviation is called *modulation index* or *index of modulation*. Frequency and phase modulations differ in the way their modulation index is defined.

For PM, the modulation index is proportional to the amplitude of the modulating signal and independent of its frequency. It is expressed mathematically as

$$m = K_p V_m \text{ rad} \tag{4.23}$$

where m = modulation index and peak phase deviation $\Delta\theta$ (in rad)

K_p = deviation sensitivity (in rad/volt)

V_m = peak modulating signal amplitude (in volt)

Thus, $m = K_p \left(\dfrac{\text{radian}}{\text{volt}}\right) V_m = \text{radian}$

Hence, the equation for PM can be written as

$$f(t) = V_c \cos\{\omega_c t + K_p V_m \cos \omega_m(t)\} \tag{4.24}$$
$$= V_c \cos\{\omega_c t + \Delta\theta \cos \omega_m(t)\} \tag{4.25}$$
$$= V_c \cos\{\omega_c t + m \cos \omega_m(t) \} \tag{4.26}$$

For a frequency modulated carrier, the modulation index is directly proportional to the amplitude of the modulating signal and is inversely proportional to the frequency of the modulating signal. Hence, for FM, the modulation index is expressed as

$$m = K_f V_m / \omega_m \tag{4.27}$$

where m = modulation index

K_f = deviation sensitivity (radian /second per volt)

V_m = peak amplitude of modulating signal

ω_m = radian frequency

Hence, $m = K_f V_m / \omega_m = \dfrac{\text{radian}}{\text{volt}} \times \dfrac{\text{volt}}{\text{radian}} = \text{unitless}$ (4.28)

Therefore, the FM modulation index is unitless and used only to describe the depth of modulation due to the modulating signal with a given peak amplitude and radian frequency. The deviation sensitivity can be expressed in hertz per volt in more practical form as

$$m = K_f V_m / f_m \qquad (4.29)$$

Instead of ω_m in Eq. (4.28), we have replaced f_m, frequency of the modulating signal. Since $\omega_m = 2\pi f_m$, only the constant value K_f gets changed. Again, we can observe that for FM, the modulation index m is unitless.

4.4.2 Frequency Deviation

It is the change in frequency that occurs in a carrier due to the modulating signal frequency. The frequency deviation is described as a peak frequency shift in HZ (Δf). The peak-to-peak frequency deviation $2\Delta f$ is called *carrier swing*. For an FM, the deviation sensitivity is often given in hertz per volt. Hence, the peak frequency deviation is simply the product of the deviation sensitivity and the peak modulating signal voltage. This is expressed mathematically as

$$\Delta f = K_f V_m \text{ (Hz)} \qquad (4.30)$$

Hence, $K_f = \Delta f / V_m$ (4.31)

Substituting this value of K_f in Eq. (4.29), the expression for modulating index m can be written as

$$m = \Delta f / f_m \qquad (4.32)$$

which is a unitless quantity.

Therefore, for FM, the equation can be written as

$$f(t) = V_c \cos \{\omega_c t + K_f V_m / f_m \sin \omega_m t\} \qquad (4.33)$$

$$= V_c \cos \{\omega_c t + \Delta f / f_m \sin \omega_m t\} \qquad (4.34)$$

Therefore, $f(t) = V_c \cos \{\omega_c t + m \sin \omega_m t\}$ (4.35)

From the expression for modulating index m for both PM and FM, it can be inferred that the modulation indices for PM and FM relate to the modulating signal differently. With PM, both the modulating index and peak phase deviations are directly proportional to the amplitude of the modulating signal and the modulating frequency does not have any effect. However, with FM, both the modulation index and frequency deviation are directly proportional to the amplitude of the modulating signal and inversely proportional to its frequency. Figure 4.4 shows the relationship among modulation index and peak phase deviation for both PM and FM.

We can conclude from the above-mentioned mathematical relation for both PM and FM that if the amplitude of the modulating signal changes, the modulation index for both frequency and phase modulated waves will change proportionally. However, if the frequency of the modulating signal changes, the modulating index for the frequency modulated wave is inversely proportional to the modulating frequency, while the modulation index of the phase modulated wave is unaffected. Hence, under identical condition both FM and PM are indistinguishable for a single frequency of modulating signal changes. This is due to the fact that the PM modulation index remains constant but the modulating index of FM wave increases as the modulating signal frequency decreases and vice versa. To distinguish between the modulating index between AM and FM/PM, the modulating index for AM is defined as m and that for FM as β. We will use β whenever we refer to angle modulation.

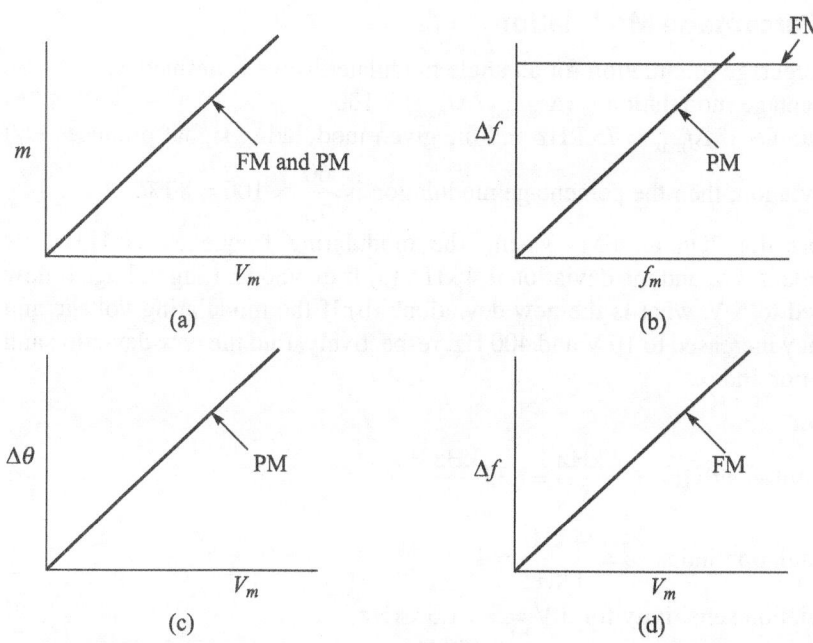

Fig. 4.4 Relationship between m, Δf, and $\Delta \theta$ w.r.t. the modulating frequency and amplitude: (a) modulation index vs amplitude, (b) frequency deviation vs modulating frequency, (c) phase deviation vs amplitude, and (d) frequency deviation vs amplitude

Example 4.2 Find out the peak frequency deviation Δf and modulation index β for an FM modulator with a deviation sensitivity of 4 kHz/volt. The modulating signal is $f(t) = 3 \cos(2\pi\,1500t)$.

Solution
Since the peak frequency deviation is equal to the product of the deviation sensitivity and the peak amplitude of the modulating signal,

$$\Delta f = \frac{4\,\text{kHz}}{V} \times 3\,\text{V} = 12\,\text{kHz}$$

and

$$\beta = \frac{\Delta f}{f_m} = \frac{12\,\text{kHz}}{1.5\,\text{kHz}} = 8$$

Example 4.3 Determine the peak phase deviation β_p for a PM modulator with a deviation sensitivity of 3.5 rad/volt and the modulating signal $f(t) = 2.5 \cos(2\pi 1800\ t)$.

Solution
We have

$$\beta_p = 3.5\ \frac{\text{rad}}{\text{volt}} \times 2.5 = 8.75$$

4.4.3 Percentage Modulation

The percentage modulation for an angle modulated wave is defined as

Percentage modulation = $(\Delta f_{\text{actual}} / \Delta f_{\text{max}}) \times 100$ (4.36)

For example, if $\Delta f_{\text{max}} = 75$ kHz and the given modulating signal produces ± 60

kHz deviation, then the percentage modulation is $\dfrac{60}{75} \times 100 = 80\%$.

Example 4.4 On an FM system, the modulating frequency is 1kHz, its amplitude is 3 V, and the deviation is 4 kHz. (a) If the modulating voltage is now increased to 5 V, what is the new deviation? (b) If the modulating voltage and frequency increased to 10 V and 400 Hz, respectively, find the new deviation and modulation index.

Solution

Deviation sensitivity = $\dfrac{4\ \text{kHz}}{3\ \text{V}} = 1.33\ \dfrac{\text{kHz}}{\text{V}}$

Modulation index, $\beta = \dfrac{4\ \text{kHz}}{1\ \text{kHz}} = 4$

(a) Deviation sensitivity for 5 V = 5×1.33 kHz
$$= 6.65\ \text{kHz}$$

Modulation index, $\beta = \dfrac{6.65\ \text{kHz}}{1\text{kHz}} = 6.65$

(b) When $f_m = 400$ Hz and the modulating voltage increased to 10 V,
Deviation sensitivity = $10 \times 1.33 = 13.3$ kHz

$$\beta = \frac{13.33\ \text{kHz}}{400\ \text{Hz}} = 33.25$$

It is clear that the change in modulating frequency affects only the modulation index and not the frequency deviation since it is independent of the modulating frequency.

4.5 BANDWIDTH REQUIREMENTS FOR ANGLE MODULATED WAVES

From the previous discussion, it can be seen that the bandwidth of an angle modulated wave is a function of the modulating signal frequency and the modulating index. In an angle modulation, multiple sets of sidebands are produced and hence the bandwidth can be significantly wider than that of an AM wave with the same modulating signal.

Angle modulated waveforms are generally classified low, medium, or high, depending on their modulating index. If the modulation index is <1, it is called low index and if the modulation index >10 is called high index. The modulation index lying between 1 and 10 is called medium index.

Low index FM systems are called narrowband FM and high index FM systems are called wideband FM. The bandwidth for wideband depends on the number of significant sidebands. For narrowband FM, the bandwidth required will be approximately $2f_m$ Hz (which will be proved in due course), which is same as for double sideband (DSB) in AM, i.e.,

$$\text{Bandwidth} = 2f_m \text{ Hz} \tag{4.37}$$

For wideband FM,

$$\text{Bandwidth} = 2\Delta f \text{ Hz} \quad \text{(approximately)} \tag{4.38}$$

where $2\Delta f$ is the peak-to-peak frequency deviation. It will be proved later that the actual bandwidth for a wideband FM will be

$$\text{Bandwidth} = 2(n \times f_m) \text{ Hz} \tag{4.39}$$

where n = the number of significant sidebands

f_m = modulating signal frequency

Hence, the actual bandwidth required to pass all the significant sidebands for an angle modulated wave is equal to two times the product of the highest modulating signal frequency and the number of significant sidebands n determined from the Bessel function table.

A general rule to estimate the bandwidth for angle modulated wave was given by Carson in 1939. This rule approximates the bandwidth required to transmit an angle modulated wave as twice the sum of the peak frequency deviation and the highest modulating frequency, i.e.,

$$\text{Bandwidth} = 2(\Delta f + f_m) \text{ Hz} \tag{4.40}$$

For low modulation index, in case of narrowband FM since $2\Delta f \ll f_m$, Eq. (4.40) reduces to, Bandwidth = $2f_m$ and for wideband FM, where $\Delta f \gg f_m$, Eq. (4.40) reduces to, Bandwidth = $2\Delta f$. This rule gives the transmission bandwidth that is slightly less than the bandwidth determined using the Bessel table.

4.6 SINUSOIDAL FM: NARROWBAND AND WIDEBAND

Many characteristics of FM can be found by using sinusoidal modulating signal.

4.6.1 Narrowband FM

Consider an FM modulation system in which the deviation constant K_f and the message signal $m(t)$ are such that for all t, we have the modulation index $\beta \ll 1$. Then we can use simple approximation to expand $f(t)$ for FM wave.

Since for an FM wave as explained earlier $f(t) = V_c \cos \{\omega_c t + \beta \sin \omega_m t\}$, where β is the modulation index for FM = $\Delta f / f_m$. Using trigonometric identities, we can expand $f(t)$ as $f(t) = V_c \{\cos \omega_c t \cos (\beta \sin \omega_m t) - \sin \omega_c t \sin (\beta \sin \omega_m t)\}$.

Since $\beta \ll 1$, then the expression can be simplified as

$$f(t) = V_c (\cos \omega_c t - \beta \sin \omega_c t \sin \omega_m t) \tag{4.41}$$

Since for $\theta \ll 1$, $\cos \theta = 1$, and $\sin \theta = \theta$, the equation reduces to

$$f(t) = V_c \left[\cos \omega_c t - \frac{\beta}{2} \cos (\omega_c - \omega_m) t + \frac{\beta}{2} \cos (\omega_c + \omega_m) t \right] \quad (4.42)$$

From Eq. (4.41), it is obvious that the narrowband FM signal resembles AM wave signal except for the fact that the message signal modulates the sine of the carrier rather than the cosine carrier, i.e., the modulating signal modulates the original carrier after a $\pi/2$ shift. The bandwidth of this signal is similar to the bandwidth of the conventional AM signal, which is twice the bandwidth of the modulating signal. This bandwidth is only an approximation of real bandwidth of FM signal.

Compared to the conventional AM, the narrowband FM has far less amplitude variation and this is also due to the first-order approximation we have assumed. It will be shown later that narrowband FM system does not provide better noise immunity than a conventional AM system. Hence, narrowband system is seldom used. It is used as intermediate stage for the generation of wideband FM system. Figure 4.5 shows phasor diagram for narrowband FM and conventional AM.

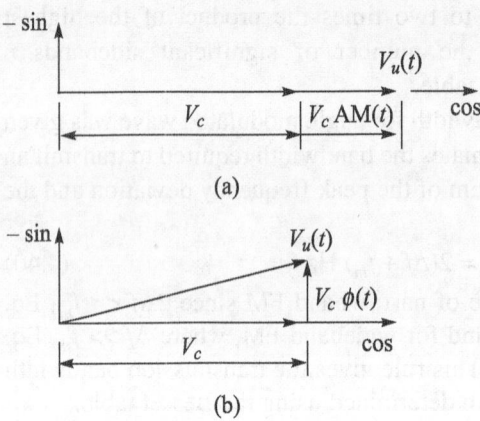

(a)

(b)

Fig. 4.5 Phasor diagram: (a) conventional AM and (b) narrowband FM

From the figure, it is clear that in both the cases, we have a carrier term and sideband centred round ω_c. The sideband spectrum for FM has a phase shift of $\pi/2$ with respect to the carrier, but for AM, it is in phase with the carrier. However, it must be noted that they both have different waveforms. In AM signal, the frequency is constant and the amplitude varies with time, whereas in FM signal, the amplitude remains constant and the frequency varies with time.

Figure 4.6 shows the phasor diagram for a narrowband FM signal showing individual sidebands and the carrier. Since $\beta \ll 1$,

$$f(t) = \cos (\omega_c t + \beta_m \sin \omega_m t) \quad (4.43)$$

or $\quad f(t) = \cos \omega_c t - (\beta/2) \cos (\omega_c - \omega_m) t + (\beta/2) \cos (\omega_c + \omega_m)t \quad (4.44)$

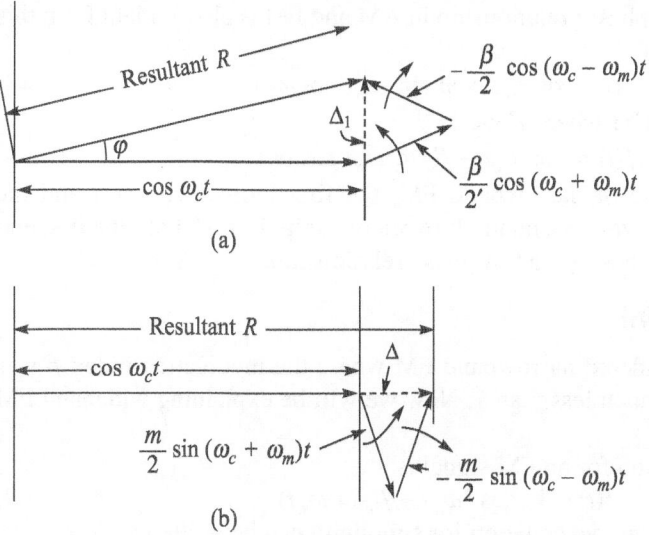

Fig. 4.6 Phasor diagram: (a) narrowband FM and (b) conventional AM

Assuming a coordinate system that rotates counterclockwise at an angular velocity ω_c, the phasor for the carrier frequency term in Eq. (4.44) is fixed and oriented along the horizontal axis. The phasor for the term $(\beta/2) \cos(\omega_c + \omega_m)t$ rotates in a counterclockwise direction at an angular velocity ω_m, whereas the phasor term for $(\beta/2) \cos(\omega_c - \omega_m)t$ rotates clockwise at an angular velocity ω_m.

At time $t = 0$, both phasors, which represent the sideband components, have maximum projections in the horizontal directions. At this time, one is parallel to and the other is not parallel to the phasor representing the carrier, so that the two cancel each other out. Figure 4.6(a) shows the phasors shortly after $t = 0$. At this time, the relation of the sideband phasors that are in opposite directions give rise to a sum phasor Δ_1. If the carrier phasor is stationary as shown, the phasor Δ_1 is always perpendicular as shown in Fig. 4.6(a) and has the magnitude

$$\Delta_1 = \beta \sin \omega_m t \qquad (4.45)$$

The carrier and Δ_1 combined to give rise to a resultant R. The angle between the carrier and the resultant R is φ. From Fig. 4.6(a), it can be observed that since $\beta \ll 1$, the maximum value of φ is approximately equal to tan φ, which is equal to β. The small variation in the amplitude of the resultant is due to the fact that we have neglected higher order sidebands.

The phasor diagram of AM shows that there is a 90° phase shift in the phases of sidebands between FM and AM. This is because an AM signal is represented as

$$f(t) = \cos \omega_c t + (m/2) \sin(\omega_c + \omega_m)t - (m/2) \sin(\omega_c - \omega_m)t \qquad (4.46)$$

The sum Δ of the sideband phasor is given by

$$\Delta = m \sin \omega_m t \qquad (4.47)$$

Thus, the important difference between AM and FM is that in FM, the sum Δ_1 is always perpendicular to the carrier phasor, whereas in the case of AM, it is in line with the carrier phasor and the resultant R does not rotate with respect to the carrier phasor but instead varies in amplitude between $(1 + m)$ and $(1 - m)$.

This difference of phasor relationship in AM and FM is also evident from this equation, i.e., for AM

$$f(t) = \cos \omega_c t + m \sin \omega_m t \cos \omega_c t \qquad (4.48)$$

and for narrowband FM where $\beta \ll 1$,

$$f(t) = \cos \omega_c t - \beta \sin \omega_m t \sin \omega_c t$$

That is to say in case of narrowband FM, the first term is $\cos \omega_c t$ and the second term $\sin \omega_c t$ involves a quadrature relationship. In AM, both the first and second terms involve $\cos \omega_c t$, an in-phase relationship.

4.6.2 Wideband FM

So far we have considered narrowband FM where the modulation index β was assumed to be very much less than 1. Now we will be explaining wideband FM for which $\beta > 10$.

The general equation for an FM signal is

$$f(t) = V_c \cos(\omega_c t + \beta \sin \omega_m t) \qquad (4.49)$$

Assuming $V_c = 1$, the above equation for simplicity can be written as

$$f(t) = \cos(\omega_c t + \beta \sin \omega_m t) \qquad (4.50)$$

Expanding this equation, we get

$$f(t) = \cos \omega_c t \cos(\beta \sin \omega_m t) - \sin \omega_c t \sin(\beta \sin \omega_m t) \qquad (4.51)$$

First let us consider the expression $\cos(\beta \sin \omega_m t)$, which appears as a factor in the first term on the RHS of Eq. (4.51). It is an even periodic function having an angular frequency ω_m. It is possible to expand this as a Fourier series in which $\omega_m / 2\pi$ is the fundamental frequency. Since the evaluation of Fourier coefficients for this function is quite complex, we will not go into this aspect and instead take them for granted. These coefficients are, of course, functions of β and since the function is even, the coefficients of odd harmonics are zero. Hence, we can expand $\cos(\beta \sin \omega_m t)$ as

$$\cos(\beta \sin \omega_m t) = J_0(\beta) + 2J_2(\beta) \cos 2\omega_m t + 2J_4(\beta) \cos 4\omega_m t + \cdots$$
$$+ 2J_{2n}(\beta) \cos 2n\omega_m t + \cdots \qquad (4.52)$$

Similarly, since $\sin(\beta \sin \omega_m t)$ is an odd function, the expansion contains only odd harmonics and is given by

$$\sin(\beta \sin \omega_m t) = 2J_1(\beta) \sin \omega_m t + 2J_3(\beta) \sin 3\omega_m t + 2J_5(\beta) \sin 5\omega_m t$$
$$+ 2J_{2n-1}(\beta) \sin(2n-1)\omega_m t + \cdots \qquad (4.53)$$

The function of $J_n(\beta)$ occurs often in solution to engineering problems. It is known as Bessel function of the first kind of order n. Substituting the values of $\cos(\beta \sin \omega_m t)$ and $\sin(\beta \sin \omega_m t)$ in Eq. (4.51) and using the trigonometric identities,

$$\cos A \cos B = (1/2) \cos(A - B) + (1/2) \cos(A + B)$$
$$\sin A \sin B = (1/2) \cos(A - B) - (1/2) \cos(A + B)$$

we find

$$f(t) = J_0(\beta) \cos \omega_c t - J_1(\beta)\{ \cos(\omega_c - \omega_m)t + \cos(\omega_c + \omega_m)t\} + J_2(\beta)$$
$$\{\cos(\omega_c - 2\omega_m)t + \cos(\omega_c + 2\omega_m)t\} - J_3(\beta) \{ \cos(\omega_c - 3\omega_m)t$$
$$- \cos(\omega_c + 3\omega_m)t \}+ \cdots \qquad (4.54)$$

The spectrum is composed of a carrier with an amplitude $J_o(\beta)$ and a set of sidebands spaced symmetrically on either side of the carrier at frequency separation of $\omega_m, 2\omega_m, 3\omega_m, \cdots$ and so on.

Figure 4.7 shows the Bessel function $J_n(\beta)$ plotted as a function of $\beta = 0, 1, 2, \cdots$ and Fig. 4.8 gives the table for the value of the Bessel function of the first kind.

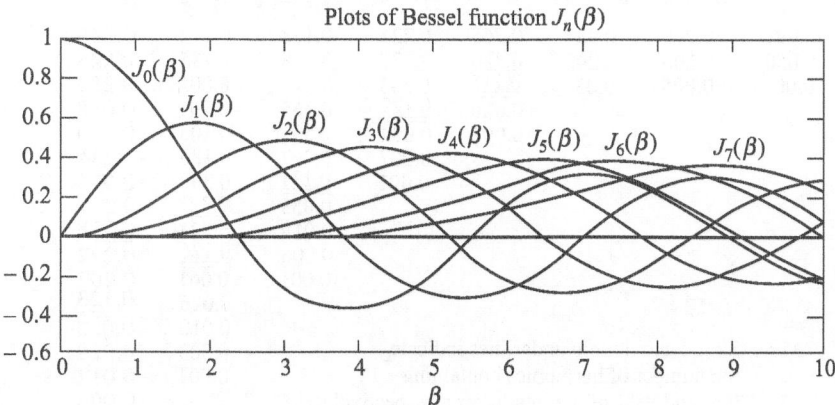

Plots of Bessel function $J_n(\beta)$

Fig. 4.7   Bessel function as a function of β

We have seen that wideband FM signal consists of the carrier and infinite number of pairs of sidebands each preceded by Bessel coefficients J. The Bessel functions happen to be of first kind and of the order denoted by their subscript with the argument β. $J_n(\beta)$ may be shown to be a solution of an equation of the form

$$(\beta^2)\, d^2y/d\beta^2 + \beta\, dy/dm + (\beta^2 - n^2)y \tag{4.55}$$

The solution for this equation, i.e., the formula for the Bessel function is

$$J_n(\beta) = \left(\frac{\beta}{2}\right)^n \left[\frac{1}{n} - \frac{\left(\frac{\beta}{2}\right)^2}{\underline{1}} \frac{1}{\underline{(n+1)}} + \frac{\left(\frac{\beta}{2}\right)^4}{\underline{2}} \frac{1}{\underline{(n+2)}} - \frac{\left(\frac{\beta}{2}\right)^6}{\underline{3}} \frac{1}{\underline{(n+1)}} + \cdots \right] \tag{4.56}$$

In order to evaluate the value of a given pair of sidebands, it is necessary to know the value of the corresponding Bessel function. This is obtained from Table 4.1. From the table, the following points are made for a wideband FM signal.

(a) We note that for $\beta = 0$, $J_c(0) = 1$, while all other J_n's are zero. This is evident since when there is no modulation, only the carrier of normalized amplitude unity is present, while all sidebands have unit amplitude.

(b) Unlike AM, where there are only three frequencies, the carrier and the two sidebands, FM has an infinite number of sidebands along with the carrier. These sidebands are separated from the carrier by $f_m, 2f_m, 3f_m, \cdots$.

(c) The J coefficients eventually decrease in value as n increases but not in a simple manner.

As seen in Fig. 4.7, the value fluctuates on either side of zero, gradually diminishing since each coefficient represents the amplitude of a particular pair of sidebands. They also eventually decrease, but only past a certain value of n. It is the modulation index β, which determines how many sideband components have significant amplitudes.

Table 4.1 Bessel function values

n	$\beta = 0.1$	$\beta = 0.2$	$\beta = 0.5$	$\beta = 1$	$\beta = 2$	$\beta = 5$	$\beta = 8$	$\beta = 10$
0	0.997	0.990	0.938	0.765	0.224	−0.178	0.172	−0.246
1	0.050	0.100	0.242	<u>0.440</u>	<u>0.577</u>	−0.328	0.235	0.043
2	0.001	0.005	0.031	<u>0.115</u>	0.353	0.047	−0.003	0.255
3				0.020	<u>0.129</u>	0.365	−0.291	0.058
4				0.002	0.034	<u>0.391</u>	−0.105	−0.220
5					0.007	0.261	0.186	−0.234
6					0.001	<u>0.131</u>	0.338	−0.014
7						0.053	<u>0.321</u>	0.217
8						0.018	0.223	<u>0.318</u>
9						0.006	<u>0.126</u>	0.292
10						0.001	0.061	0.207
11							0.026	<u>0.123</u>
12							0.010	0.063
13		*Note:* Single and double underlines indicate					0.003	0.029
14		the number of harmonics containing					0.001	0.012
15		70% and 98% of the total power, respectively.						0.004
16								0.001

(d) It can also be observed that sidebands at equal distance from f_c have equal amplitude so that the sideband distribution is symmetrical about the carrier frequency. The J coefficients sometimes have negative value. This signifies an 180° phase change for that particular pair of sidebands.

(e) From Table 4.1, it can be seen that as β increases so does the value of particular J coefficient such as J_{12}. Since β is inversely proportional to the modulating frequency, we see that the relative amplitude of distant sidebands increases when the modulation frequency is lowered assuming that the deviation (the modulating voltage) has remained constant.

(f) In the case of AM, the sideband power gets increased as the modulation depth increases, thus increasing the total transmitted power. In FM, the total transmitted power always remains constant, but with increased depth of modulation the required bandwidth is increased. This is obvious since as β increases, more sideband pairs are added, which acquire significant amplitudes.

(g) From the equation of FM wave, it can be concluded that the theoretical bandwidth required for an FM system is infinite, i.e., the bandwidth used is one that has been calculated to allow for all significant amplitudes of sideband components under the most exacting conditions. This ensures that with maximum deviation by the highest modulating frequency, no significant sideband components are chipped off.

(h) In FM, unlike AM the amplitude of the carrier component does not remain constant. The carrier coefficient is J_o, which is a function of β. This is due to the fact that the envelope of FM signal has constant amplitude. Therefore, the power of such a signal is constant independent of the modulation since the power of a periodic waveform depends only on the square of its amplitude and not on its frequency. The power of unit amplitude signal is $P_c = 1/2$ and is independent of β. When the carrier is modulated to generate FM signal, the power in the sidebands may appear only at the expense of the power originally in the carrier.

(i) For a certain value of β, the carrier component in FM waves disappears completely. This can be observed in Fig. 4.7. The values of β at which the carrier disappears are called *eigen values*. The eigen values can be obtained from Fig. 4.7. They are approximately at $\beta = 2.4$, 5.6, 8.6, 11.7, and so on. This disappearance of the carrier for a specific value of β forms a handy basis for measuring deviation.

4.7 SPECTRAL CHARACTERISTIC OF A SINUSOIDAL MODULATED FM SIGNAL

In principle, for an FM signal, the number of sidebands is infinite and the bandwidths required to accommodate these sidebands are infinite. But in practice the number of sidebands depends on the modulation index β. The higher sidebands, which have relatively insignificant amplitudes, can be ignored and the signal is band-limited. It is found experimentally that the distortion resulting from bandlimiting an FM signal can be tolerated as long as 98% or more of the power is passed by bandlimiting filter. Hence, for any value of β, only those J_n need to be considered that contribute significant power to sidebands. Using Table 4.1, it is possible to evaluate the size of the carrier and each sideband for each specific value of the modulation index β. Once this is done, the frequency spectrum of the FM wave for that particular value of β is plotted in Fig. 4.8. The figure shows the spectra for increasing deviation (Δf) for constant f_m and then for decreasing the modulating frequency with deviation (Δf) being constant.

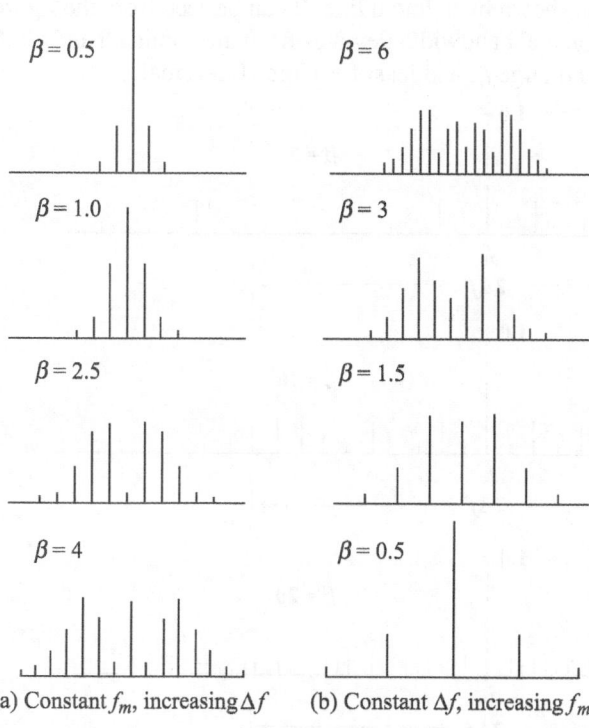

$\beta = 0.5$ $\beta = 6$

$\beta = 1.0$ $\beta = 3$

$\beta = 2.5$ $\beta = 1.5$

$\beta = 4$ $\beta = 0.5$

(a) Constant f_m, increasing Δf (b) Constant Δf, increasing f_m

Fig. 4.8   FM spectrum

We can observe that as the modulation depth increases so does the bandwidth and also the reduction in the modulating frequency increases the number of sidebands though not necessarily the bandwidth. Although the number of sideband components is theoretically infinite, in practice a lot of high sidebands have insignificant relative amplitudes and are not shown in the spectra. To calculate the required bandwidth accurately, we have to look at Table 4.1 to see J, the last coefficient shown for that value of the modulation index.

4.7.1 Spectrum of Constant Bandwidth FM

Let us consider $V_m \cos \omega_m t$ as the modulating signal with V_m, the peak voltage. In a phase modulating system, the phase angle $\phi(t)$ would be proportional to this modulating signal so that $\phi(t) = K_p V_m \cos \omega_m t$, where K_p is a constant. The phase deviation $\beta = K_p V_m$ and V_m is constant. So, the bandwidth occupied increases linearly with the modulating frequency, since bandwidth $= 2\beta f_m = 2K_p V_m f_m$. This variation of bandwidth with the modulating frequency can be avoided by making

$\phi(t) = (K/2\pi f_m) V_m \sin \omega_m t$, K being a constant.

Here $\qquad\qquad b = KV_m/2\pi f_m$ $\qquad\qquad\qquad\qquad\qquad$ (4.57)

and the bandwidth $= (2K/2\pi)V_m$, independent of f_m. The instantaneous frequency is $\omega = \omega_c + KV_m \cos \omega_m t$. Since the instantaneous signal is proportional to the modulated signal, the initial angle modulated signal has become an FM signal.

Thus, a signal intended to occupy a nominally constant bandwidth is a frequency modulated rather than a PM signal.

Figure 4.9 shows the spectrum for three values of β, keeping the product βf_m constant. The nominal bandwidth $2\Delta f \approx 2\beta f_m$ is hence constant. The amplitude of the demodulated carrier f_c is shown by a dotted line. It can be seen from the figure that the extent to which the actual bandwidth deviates from the nominal bandwidth is the greatest for small β and large f_m and least for large β and small f_m.

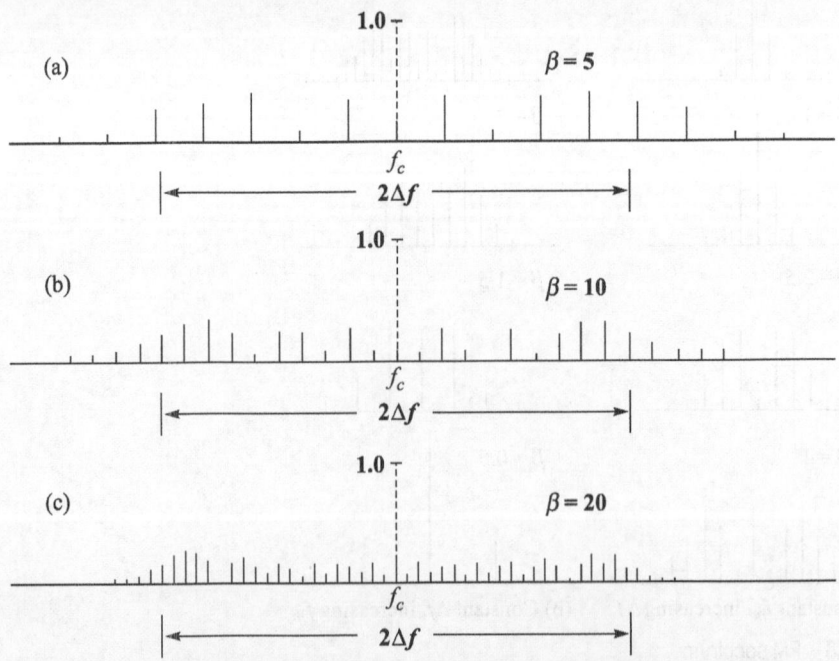

Fig. 4.9 Sinusoidally modulated FM signals – the spectrum keeping the bandwidth fixed

For commercial broadcasting, the maximum frequency deviation Δf allowed is 75 kHz. If the highest audio frequency to be transmitted happens to be 15 kHz, then at this frequency, $\beta = \Delta f / f_m = 75/15 = 5$. For all other modulating frequencies, β is larger than 5. When $\beta = 5$, there are $\beta + 1 = 6$ significant sideband pairs. Hence, at $f_m = 15$ kHz, the bandwidth required is $2 \times 6 \times 15 = 180$ kHz, which is more than the nominal bandwidth $2\Delta f = 150$ kHz.

When $\beta = 20$, there are 21 significant sideband pairs and the bandwidth required is $2 \times 21 \times (15/4) = 157.5$ kHz. In the limiting case of very large β and correspondingly very small f_m, the actual bandwidth becomes equal to the nominal bandwidth $2\Delta f$.

Example 4.5 What is the bandwidth required for an FM signal with a modulating frequency 3 kHz and with a maximum deviation of 15 kHz. Compare it with Carson's rule.

Solution

We have
$$\beta = \frac{\Delta f}{f_m} = \frac{15 \text{ kHz}}{3 \text{ kHz}} = 5$$

From the Bessel table, it can be seen that the highest J coefficient for $\beta = 5$ is J_8. Thus, bandwidth $= 8 \times 3$ kHz $= 24$ kHz

As per Carson's rule, bandwidth $= 2(\Delta f + f_m)$

$= 2(15+3) = 36$ kHz

4.8 AVERAGE POWER IN SINUSOIDAL FM

The peak voltages of the spectrum component for an FM wave is given by $V_{n\max} = J_n(\beta)V_{c\max}$. Since the RMS values V_n and E_c are proportional to the peak values, they are also related as

$$V_n = J_n(\beta)V_{ic} \tag{4.58}$$

For a fixed load resistor R, the average power of any one spectral component is $P_n = V_n^2/R$. The total average power is the sum of all such components. Since there is only one carrier component and a pair of sidebands, the total average power is

$$P_T = P_o + 2(P_1 + P_2 + \cdots) \tag{4.59}$$

In terms of the demodulated carrier and Bessel function coefficients, the total power becomes

$$P_T = V_c^2 J_0^2(\beta)/R + 2(V_c^2/R)\{J_1^2(\beta) + J_2^2(\beta) + J_3^2(\beta) + \cdots\}$$

$$= (V_c^2/R)[J_0^2(\beta) + 2\{J_1^2(\beta) + J_2^2(\beta) + J_3^2(\beta) + \cdots\}]$$

$$= P_c[J_0^2(\beta) + 2\{J_1^2(\beta) + J_2^2(\beta) + J_3^2(\beta) + \cdots\}] \tag{4.60}$$

where P_c is the unmodulated power V_c^2/R. As we know that the property of the Bessel function is that the sum $[J_0^2(\beta) + 2\{J_1^2(\beta) + J_2^2(\beta) + J_3^2(\beta) + \cdots\}] = 1$. Hence, the total average modulating power is equal to the carrier power. This is due to the fact that the amplitude of the wave remains constant irrespective of modulation or no modulation. In fact, when modulation is there, the total power that was originally in the carrier is redistributed between all the components of the spectrum. As we have seen from the graphs of the Bessel function at certain value of β, the carrier component becomes zero and that in these instances, the power is carried by the sidebands only.

Example 4.6 A 10 W unmodulated carrier is frequency modulated with a modulating voltage resulting in a peak frequency deviation of 5 kHz. The frequency of the modulating signal is 1 kHz. Calculate the average power of the modulated signal. Also prove that this power is equal to the unmodulated carrier power.

Solution
For $\beta = 5$ from the Bessel table, the Bessel function value is taken up to J_9.

$$P_T = 10\left[0.18^2 + 2(0.33^2 + 0.05^2 + 0.36^2 + 0.39^2 + 0.26^2 + 0.13^2\right.$$
$$\left. +0.05^2 + 0.02^2 + 0.01^2)\right]$$
$$= 10\left[0.0324 + 2(0.1089 + 0.0025 + 0.1296 + 0.1521 + 0.0676\right.$$
$$\left. +0.0169 + 0.0025 + 0.0004 + 0.0001)\right]$$
$$= 10\ (0.0324 + 0.9612) = 10\ (0.9936)$$
$$= 10\ \text{W}$$

The result shows that the average power of the modulated signal is equal to the unmodulated carrier power.

4.9 DEVIATION RATIO FOR NON-SINUSOIDAL FREQUENCY MODULATION

Inter-modulation products are formed in FM modulation process when the modulating signal is non-sinusoidal. This results in beat frequencies occurring between various sidebands. But the bandwidth requirements are determined by the maximum frequency deviation and modulation frequency present in the modulating signal. The ratio of the maximum deviation to the maximum frequency component is termed as the *deviation ratio* and is given by

$$D = \Delta F / f_m \tag{4.61}$$

where ΔF is the maximum frequency deviation and f_m is the highest frequency component in the modulating signal. The bandwidth is

$$\text{BW}_{max} = 2\ (D + 1)\ f_m \tag{4.62}$$

Substituting the value of D from Eq. (4.61) in Eq. (4.62), we get

$$\text{BW}_{max} = 2\ (\Delta F + f_m) \tag{4.63}$$

This is Carson's rule, which we have explained earlier.

4.10 PHASE MODULATION

The sinusoidal signal $A \cos \omega_c t$ can be considered as the real part of $Ae^{j\omega_c t}$ and represented as

$$A \cos \omega_c t = \text{Re}(Ae^{j\omega_c t}) \tag{4.64}$$

The function $(Ae^{j\omega_c t})$ is a vector represented on a complex plane as a phasor of length A and an angle $(\omega_c t)$ measured counterclockwise from the real axis. The phasor rotates in the counterclockwise direction with an angular velocity of ω_c. The phasor will look stationary if the coordinate system is also assumed to rotate counterclockwise with the same angular velocity ω_c. Since a sinusoidal wave is represented as $f(t) = A \cos (\omega_c t + \Phi)$, where Φ is the phase shift or the phase angle. If Φ is independent of the time and is constant, then $f(t)$ can be regarded as a rotating vector with an angular velocity ω_c. But suppose $\Phi = \Phi (t)$ and it changes with time, then $f(t)$ will be represented by a phasor of amplitude A, which runs ahead of and falls behind the phasor representing $A \cos \omega_c t$. Hence, we may consider that the angle $\omega_c t + \Phi(t)$ of $f(t)$ undergoes a modulation around the angle $\theta = \omega_c t$. The waveform of $f(t)$ is, therefore, a representation of a signal that is modulated in phase. If the phasor of angle $\omega_c t + \Phi(t)$ alternately runs ahead of and falls behind the phasor $\theta = \omega_c t$, then the first phasor must alternately be rotating more or less rapidly than the second phasor. This can be considered as if the angular velocity of the phasor $f(t)$ undergoes a modulation around the nominal angular velocity ω_c. Hence, the signal $f(t)$ is a angular velocity modulated waveform. Figure 4.10 shows the phasor representation of a carrier of $E_{c\,\max}$ with an initial phase shift of Φ_c. Effect of applying phase modulation is shown in Fig. 4.10(b). This figure can be explained as follows.

Let the unmodulated carrier be

$$f(t) = E_{c\max} \cos(\omega_c t + \Phi_c) \qquad (4.65)$$

The phase angle Φ_c is arbitrary and indicates only the reference line for the vector. When the phase modulation is applied, it has the effect of moving the reference line as shown in Fig. 4.10(b). The phase modulation mathematically is represented as

$$\Phi (t) = \Phi_c + K_p e_m(t) \qquad (4.66)$$

where K_p is the phase deviation constant. K_p should have units of rad per volt when Φ_c is measured in rad. The constant phase angle Φ_c has no effect on the modulation process and hence can be dropped. With this, the expression for a phase modulated wave becomes

$$f(t) = E_{c\,\max} \cos[\omega_c t + K_p e_m(t)] \qquad (4.67)$$

4.10.1 Sinusoidal Phase Modulation

For sinusoidal phase modulation, the carrier is $E_c \cos \omega_c t$ and the modulating signal is

$$E_m (t) = E_m \sin \omega_m t \qquad (4.68)$$
$$K_p E_m (t) = K_p E_m \sin \omega_m t = \Delta\Phi \sin \omega_m t \qquad (4.69)$$

where $$\Delta\Phi = K_p E_m \qquad (4.70)$$

Here $\Delta\Phi$ is defined as *peak phase deviation* and is directly proportional to the peak modulating signal. The sine signal is used for the modulation signal rather than the cosine expression since this brings out clearly the equivalence in the spectra for FM and PM.

The expression for sinusoidal PM is

$$f (t) = E_c \cos(\omega_c t + \Delta\Phi \sin \omega_m t) \qquad (4.71)$$

This is identical to the expression for FM wave with only difference being that we have $\Delta\Phi$ in place of β. Hence, the trigonometric expression will be similar to that for sinusoidal FM containing a carrier term and sideband frequencies $f_c \pm n f_m$. The amplitudes are also given in terms of the Bessel function of the first kind $J_n (\Delta\Phi)$. It can be seen that in the case of PM, the argument is the peak phase deviation $\Delta\Phi$ instead of the frequency modulation index β.

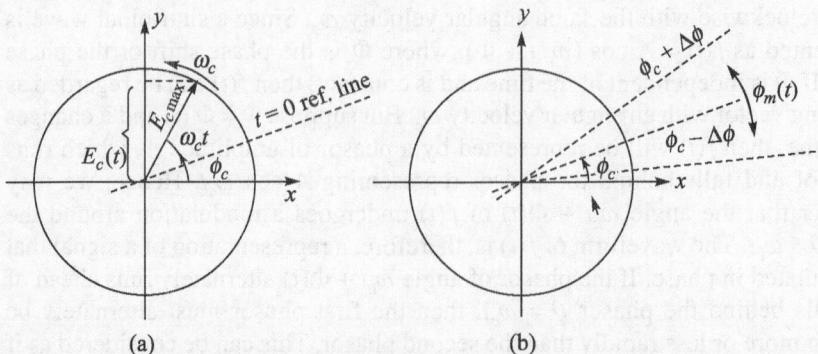

Fig. 4.10 (a) Rotating phasor representation of carrier E_{cmax} and phase lead ϕ_c and (b) phase modulation

Therefore, the magnitude and extent of the spectrum components for the PM wave will be the same as for the FM wave for which $\Delta\Phi$ is numerically equal to β. Similarly, the power relationship developed for sinusoidal FM also applies to PM.

The phase modulation is used chiefly as a stage in generation of frequency modulation. The other major area of application for phase modulation lies in the digital modulation of carriers.

Example 4.7   A carrier signal of 10 kHz with maximum amplitude of 8 V is phase modulated by a modulating signal $f(t) = 4 \cos (2\pi \times 1.5 \times 10^3 t - 90°)$. The phase deviation constant is 3 rad/V. Give the expression for the modulating waveform.

Solution

The phase modulation function is $\Phi_m(t) = 3 \times 4 \cos(2\pi \times 1.5 \times 10^3 t - 90)$

$$= 12 \sin(2\pi \times 1.5 \times 10^3 t)$$

Hence, the modulated wave function is $8 \cos(2\pi \times 10^4 t) + 12 \sin(3\pi \times 10^3 t)$

4.10.2 Digital Phase Modulation

Phase modulation is widely used in digital systems. Phase shift keying (PSK) is another form of phase modulation except that with PSK, the input phase for a single-frequency carrier depends on the number of bits transmitted. Single bit is used for binary phase shift keying and two bits are used for QPSK (quaternary phase shift keying) and four bits for sixteen-phase PSK. These topics come under digital communication.

4.11 COMPARISON OF FM AND PM

The difference between FM and PM is the way in which the modulation index is defined. In phase modulation, the phase deviation is proportional to the amplitude of the modulating signal and independent of its frequency. Since the phase modulated vector sometimes leads and sometimes lags behind the reference carrier vector, its instantaneous angular velocity must be continuously changing between the limits imposed by Φ_m. This may result in some form of frequency change.

In frequency modulation, the frequency deviation is proportional to the amplitude of the modulating voltage. The reference vector rotating with constant angular velocity, which corresponds to the carrier frequency, will have a phase lead or lag with respect to the reference since its frequency oscillates between $f_c - \delta$ and $f_c + \delta$. Hence, FM must be a form of PM. With this close similarity, let us see the difference.

If we consider FM as a form of PM, then we must determine the reasons for phase change in FM. The larger the frequency deviation, the larger the phase deviation so that the latter depends, at least to certain extent, on the amplitude of the modulating signal just as in PM. This difference can be explained by comparing the definition of PM, which states that the modulation index is proportional to the modulating voltage only, whereas in the case of FM, the modulating index is also inversely proportional to the modulating frequency. From this, we can infer that under ideal conditions, FM and PM are indistinguishable for a single modulating frequency. But when the modulating frequency is changed, the modulation index for PM will remain constant and the modulating index for FM will increase as the modulation frequency is reduced and vice versa.

The practical aspects of the above-mentioned consideration can be explained as follows.

Let us assume that we receive the FM wave on a PM receiver. Since the low frequencies of the modulating signal will have more deviation in phase in FM than a PM transmitter would have given them, these bass frequencies will get boosted in the PM receiver, since the output of the PM receiver would be directly proportional to the phase deviation.

When we receive PM wave on an FM receiver, the bass frequencies of the modulating signal will not be boosted since the phase deviation of a PM wave depends only on the amplitude of the modulating signal and not on its frequency. This deficiency can be corrected by bass boosting the modulating signal prior to the phase modulation.

4.12 FM GENERATION

Any modulation and demodulation process involves the generation of new frequencies that are not present in the input signal. This is true for amplitude and frequency/phase modulation system. If we consider a modulating system with a modulating signal $m(t)$ and carrier as the input and the modulated signal $f(t)$ as the output, this system has frequencies on its output that were not present in the output. Hence, in this case, modulator cannot be a linear time variant system because a linear time invariant system cannot produce any frequency components at the output that are not present in the input signal. FM modulators are generally time varying and non-linear systems. One method to generate an FM signal is to design an oscillator whose frequency changes with the input voltage. With the applied voltage zero, the oscillator generates a sinusoidal signal with frequency f_c. When the input voltage changes, this frequency changes accordingly. This method is called *direct method*.

Another method for generating an FM signal is to generate a narrowband FM signal and then change it to a wideband signal. This method is known as *indirect method*. Another indirect method for FM generation is to use phase modulation to indirectly generate FM. In this method, narrowband FM (NBFM) is generated by integrating $m(t)$, the modulating signal and using it to phase modulate a carrier. The NBFM is then converted to wideband FM (WBFM) by using frequency multipliers.

4.12.1 Direct Method

The general method to generate a sinusoidal carrier is a tuned circuit oscillator. In such oscillator circuits, the frequency of the carrier to a large extent is determined by the resonant frequency of the turned circuit. The frequency of oscillation is given by $f = 1/2\pi\sqrt{LC}$, where L is the inductance and C is the capacitance. Hence, by changing the value of either C or L by a modulating voltage, the oscillator frequency can be changed. The capacitor or inductors, which change due to the modulation voltage, are shunted across the LC turned circuit.

Varactor diodes are used as a variable capacitor, pin diodes and FET are used as a variable resistor, and saturable reactors are used as a variable inductor. This type of generation is also called *parameter variation method* because in this method any one of the parameters L, C, or R is changed.

Varactor diode implementation

A varactor diode is a capacitor whose capacitance changes with the applied voltage when reverse biased. Hence, if this varactor diode is used in a tuned circuit of the oscillator and the message signal is applied to it, the frequency of the tuned circuit and the oscillator will change in accordance with the modulating signal. A typical circuit is shown in Fig. 4.11. Let L be the inductance of the tuned circuit and C_a be the capacitance across L. The varactor diode is connected across L and C_a. The capacitance of the varactor diode is given by $C(f) = C_0 + K_0 m(t)$, where C_0 is the capacitor when no modulating signal is applied to the diode.

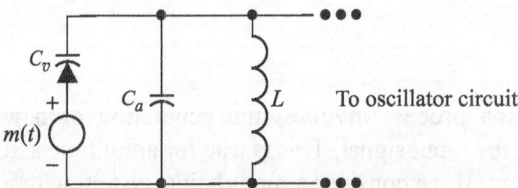

Fig. 4.11 Angle modulation using varactor diode

Hence, the frequency of the tuned circuit without any modulating signal is

$$f_c = \frac{1}{2\pi\sqrt{LC}} \tag{4.72}$$

where

$$C = C_a + C_0$$

This frequency is the carrier frequency. Now if we apply the modulating signal, then the frequency of the tuned circuit changes as

$$f(t) = \frac{1}{2\pi\sqrt{L[C_a + C_0 + K_0 m(t)]}} \tag{4.73}$$

$$= \frac{1}{2\pi\sqrt{L[C + K_0 m(t)]}}$$

$$= \frac{1}{2\pi\sqrt{LC}} \frac{1}{\sqrt{1 + \dfrac{K_0 m(t)}{C_0}}}$$

or
$$f(t) = f_c \frac{1}{\sqrt{1 + \dfrac{K_0 m(t)}{C_0}}} \qquad (4.74)$$

Assuming that $\Delta = \dfrac{K_0}{C_0} m(t) \ll 1$,

$$f(t) = f_c \frac{1}{\sqrt{1 + \Delta}} = \frac{f_c}{(1 + \Delta)^{1/2}} \qquad (4.75)$$

Using the approximation $(1 + \Delta)^{1/2} = 1 + \Delta/2$ and $\dfrac{1}{1 + \Delta} = 1 - \Delta$, we obtain

$$\frac{1}{(1 + \Delta)^{1/2}} = 1 - \frac{\Delta}{2}$$

Substituting this value of $\dfrac{1}{(1 + \Delta)^{1/2}}$ in Eq. (4.75), we get $f = f_c (1 - \Delta/2)$

Therefore, $f = f_c \left(1 - \dfrac{K_0}{2C_0} m(t)\right)$ \qquad (4.76)

which is nothing but the relationship for frequency modulated signal.

Varactor diode modulators

Figure 4.12 shows a schematic diagram of a direct FM generation that uses a varactor diode to deviate the frequency of a crystal oscillator. Reverse bias for the diode is developed by R_1 and R_2 combination. This reverse bias determines the carrier frequency of the oscillator.

The modulating signal is applied to the base as shown. This signal is super-imposed on the reverse bias of the diode, thus adding or subtracting to the DC bias. This changes the capacitance of the diode and thus the frequency of the oscillation. Positive change in the modulating signal increases the reverse bias on the diode, which results in reduction of the diode capacitance and hence the increase in the frequency of oscillation. Conversely, negative swing of the modulating voltage decreases the frequency of oscillation. These types of FM modulators are quite popular because they are reliable and simple to use and have stability of a crystal oscillator. The only drawback is the small value of the frequency deviation because of the crystal. Hence, they are generally used for low index application like in two-way mobile radio.

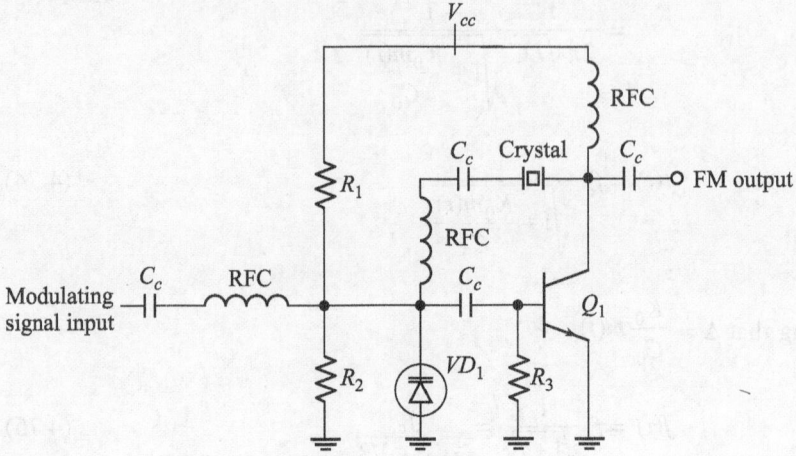

Fig. 4.12 FM modulator using varactor diode

Figure 4.13 shows a clap FM oscillator using a varactor diode. The fixed bias for varactor diode is adjusted by means of the variable resistor in the diode bias circuit. RFC choke prevents the radio frequency (RF) signal from entering the modulating circuit. C_3 is a decoupling capacitor, which blocks the DC and varactor and the varactor diode capacitor is in series with C_3. The diode is biased into its reverse bias region at some fixed bias. The modulating voltage is superimposed on this. The variation in the modulation voltage changes the reverse bias of the diode, which in turn changes its capacitance and thus causes a change in frequency.

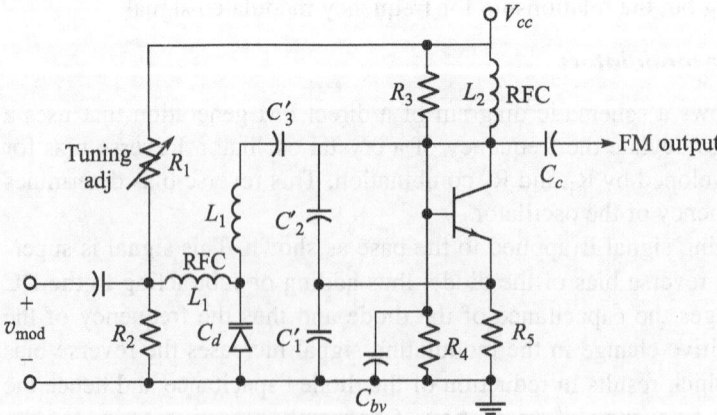

Fig. 4.13 Varactor diode clap FM oscillator

FM reactance modulator

Figure 4.15 shows a schematic diagram for a reactance modulator using JEET, since JEET looks like a variable reactance load to the tank circuit. This circuit is called a reactance modulator. The modulating signal varies the reactance of JEET which causes a corresponding change in resonant frequency of the resonant circuit.

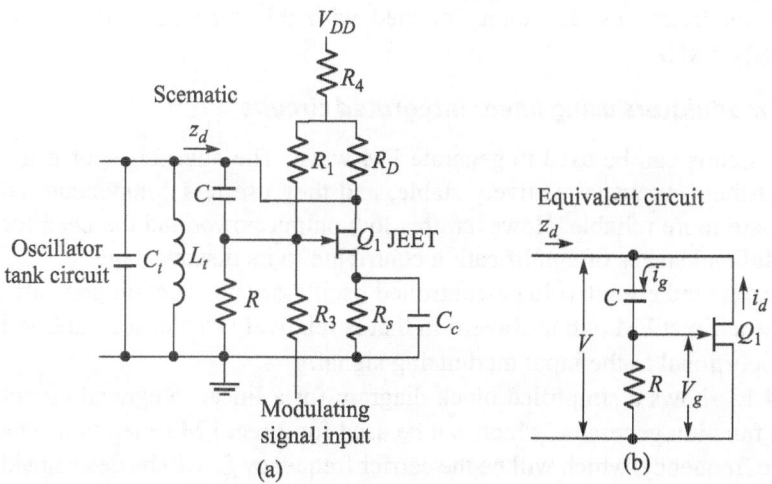

Fig. 4.15 FM reactance modulator

The resistances R_1, R_3, and R_4 provide the DC bias for the JEET. R_S can be omitted from AC equivalent circuit since it is bypassed by C_c. Assuming the gate current I_g to be zero,

$$V_g = I_g R \tag{4.77}$$

$$I_g = V/(R - jX_c) \tag{4.78}$$

and $\quad V_g = VR/(R - jX_c) \tag{4.79}$

The drain current i_d is given by

$$i_d = g_m V_g = g_m[V/(R - jX_c)]R \tag{4.80}$$

where g_m is the transconductance of the JEET. The impedance between the drain and ground is

$$Z_d = V/i_d \tag{4.81}$$

Substituting the value of i_d from Equations (4.80) and (4.81), we obtain

$$Z_d = (R - jX_c)/g_m R = (1/g_m)(1 - jX_c/R)$$

Assuming $R \ll X_c$,

$$Z_d = -jX_c/g_m R = -j/2\pi f_m g_m RC \tag{4.82}$$

As $g_m RC$ is equivalent to a variable capacitance, Z_d is inversely proportional to resistance R. The angular velocity ω_m of the modulating signal and the transconductance g_m of FET vary with the gate-to-source voltage. When the modulating signal is applied in series with R_3, the gate-to-source voltage is varied accordingly causing a proportional change in g_m. Hence, the equivalent circuit impedance Z_d is a function of the modulating signal. This makes the resonance frequency of the oscillator circuit to be a function of the amplitude of the modulating signal. The rate at which it changes is equal to f_m of the modulating signal. R and C can be interchanged to get variable reactance to be inductive rather than capacitive. However, this does not affect the output FM waveform. The maximum frequency deviation obtained with this type of modulator is approximately 5 kHz.

Direct FM modulators using linear integrated circuits

Integrated circuits can be used to generate FM wave. The advantages of using ICs are that their output is relatively stable, and they use less component and hence they are more reliable. However, this low output power and the need for several additional stages of amplification contribute to its disadvantage.

Linear integrated circuit voltage-controlled oscillator and function generator can generate a direct FM output waveform that is relatively stable, accurate, and directly proportional to the input modulating signal.

Figure 4.15 shows a simplified block diagram for a linear integrated circuit monolithic function generator, which can be used for direct FM generation. The VCO centre frequency, which will be the carrier frequency f_c, will be determined by external resistor and capacitor. The modulating signal is directly proportional to the input of the VCO, where it deviates the carrier frequency f_c and produces an

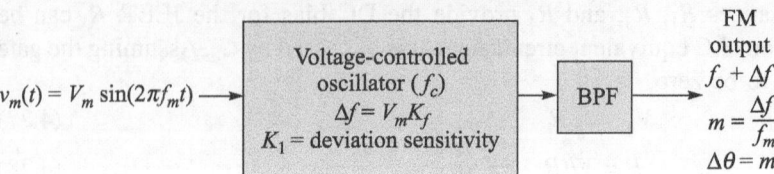

Fig. 4.15   Direct FM modulator using linear integrated circuit

FM signal at the output. The peak frequency deviation depends on the peak amplitude of the modulating signal V_m and the deviation sensitivity of VCO K_f. The modulator output is

$$\text{FM}_{\text{out}} = f_c + \Delta f \qquad (4.83)$$

where f_c is the carrier frequency, i.e., the VCO output without any modulating voltage and Δf is the peak frequency deviation ($V_m K_f$ Hz), K_f being the deviation sensitivity in Hz/volt.

Figure 4.16(a) shows a schematic diagram of a monolithic FM transmitter. This IC, made by Motorola, is a complete FM modulator on a single 8 PIN DIP IC. Its type number is MC1376. It can operate with carrier frequencies between 1.4 MHz and 14 MHz. It is meant for producing direct FM waves for low-power application. The output power can be increased as high as 600 mW if the auxiliary transistor is connected to a 12 V supply voltage. Figure 4.16(b) shows the output frequency versus the input voltage curve for the internal VCO. From the curve, it can be observed that it is fairly linear between 2 V and 4 V and can produce peak frequency deviation of nearly 150 kHz.

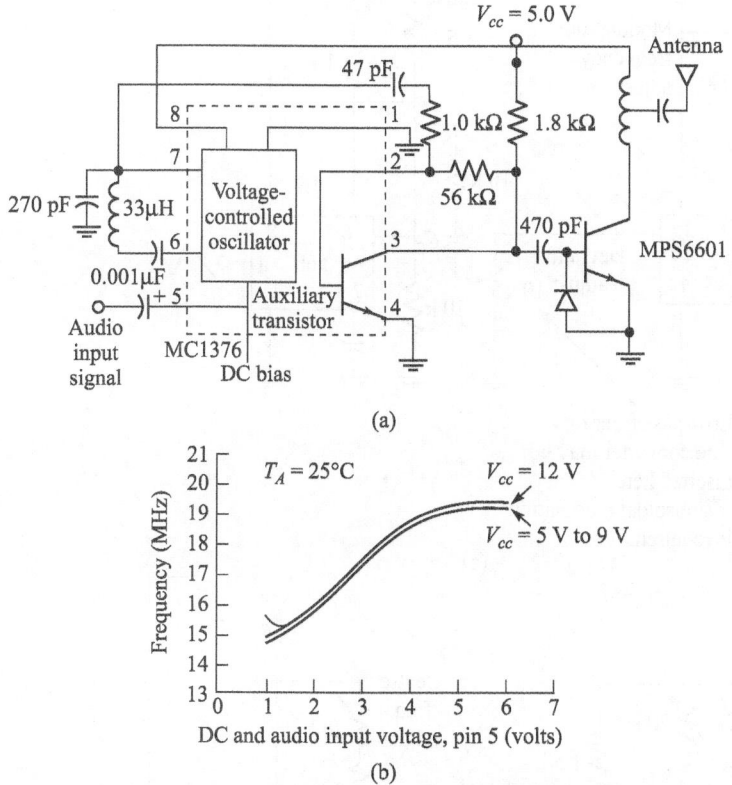

Fig. 4.16 FM transmitter using MC1376: (a) schematic diagram and (b) VCO response

Low frequency FM generators using NE 566 function generator

Figure 4.17 shows an FM generator for low frequency (less than 0.5 MHz centre frequency). This circuit uses a NE 566 was Signetics function generator IC. The first IC NE 566 was used to generate carrier signal. Capacitor C_1 selects the modulation range C_1' selects the carrier frequency. Capacitor C_2 is a coupling capacitor which has to be large enough to avoid distorting the modulating signal. High slew rate Op. Amp. 531 with associated circuit is an LPF. Some converter is inserted between the two NE 566 if sinusoidal modulation is required. More information on NE 566 and NE 531 can be obtained from Signetics Linear Data Manual.

FM modulator using NE 564 phase locked loop(PLL)

NE 564 is a versatile highly reliable PLL designed to operate up to 50 MHz. It consists of a VCO, a phase comparator, a limiter, and a post-detection processor as shown in Fig. 4.18.

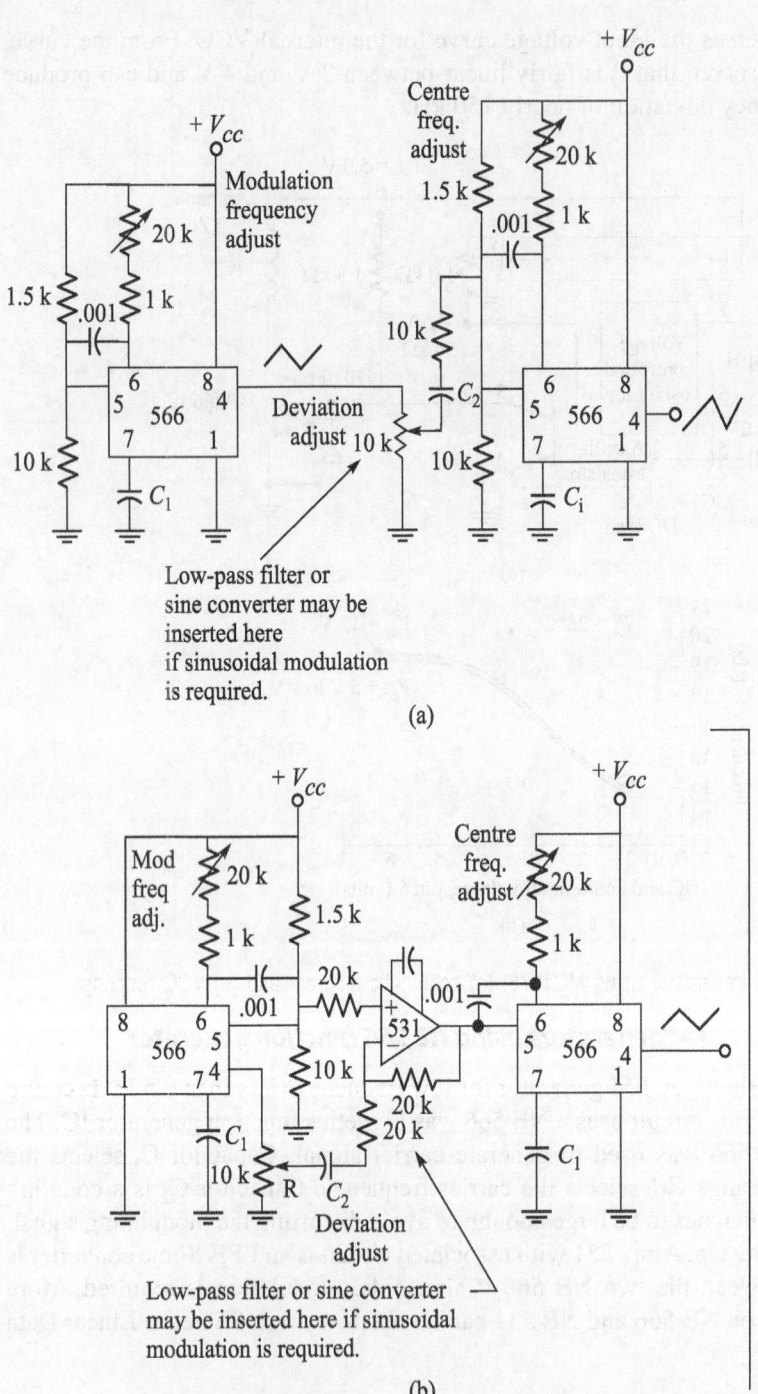

Fig. 4.17 Low-frequency FM generator: (a) small frequency deviations up to ±20% and (b) large frequency deviations up to ±100%

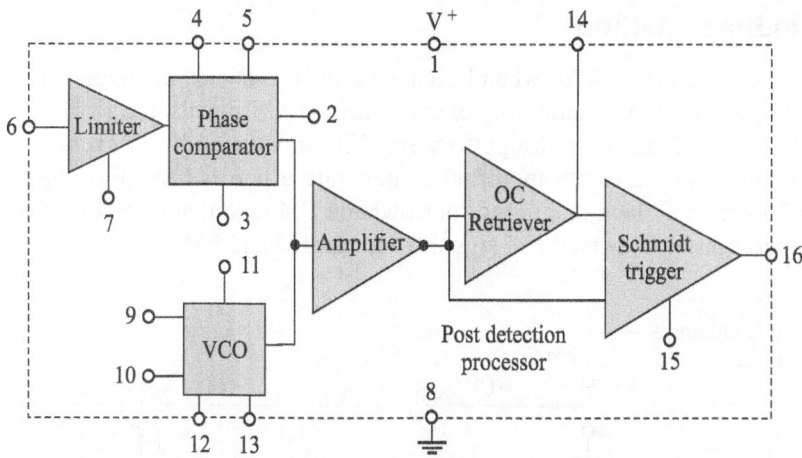

Fig. 4.18   NE 564 PLL block diagram

To use NE 564 as FM modulator, VCO frequency is set to the carrier frequency by frequency set capacitor connected between the terminals 12 and 13 of the PLL IC. Modulating signal is given to PIN 6 since the other output to the phase detector PIN 3 (which will be normally connected to VCO output) is grounded, depending on the modulating signal VCO output changes, which is nothing but the FM wave. This output is available from PIN 9, which is a transistor-transistor logic (TTL) output. Figure 4.19 shows the circuit arrangement for the modulator.

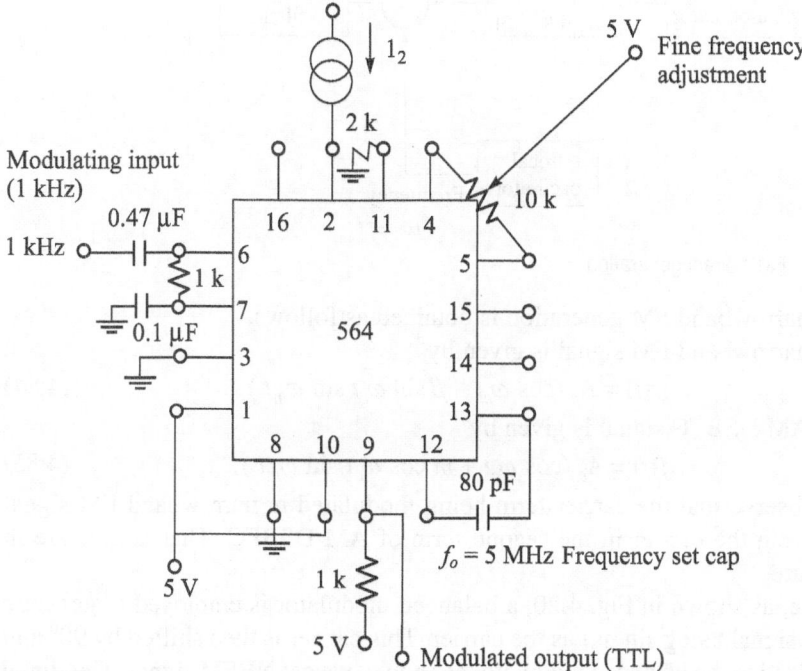

Fig. 4.19   PLL FM modulator

4.12.2 Indirect Method

This method generates a narrowband signal initially and then changes it to a wideband signal. Due to the similarity of conventional AM signals, the generation of narrowband FM signal is straightforward. The modulator for conventional AM generation can be easily modified to generate a narrowband FM signal. Figure 4.20 shows a block diagram of narrowband FM modulator and Fig. 4.21 shows generation of wideband FM signal from narrowband FM.

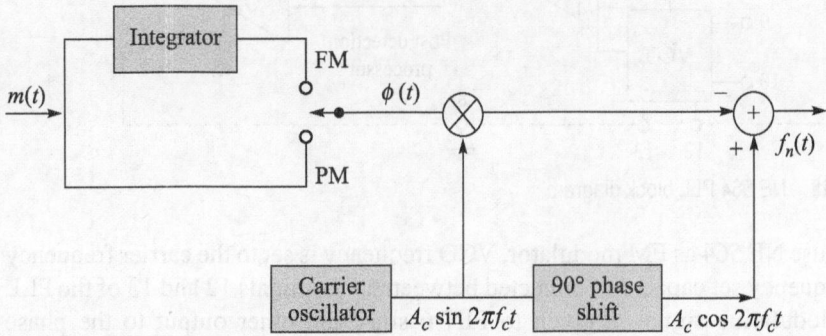

Fig. 4.20 Narrowband FM generation

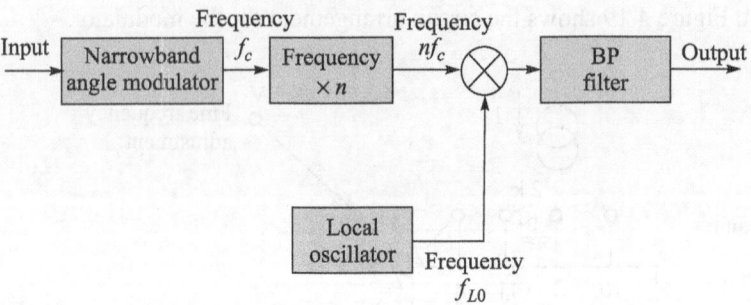

Fig. 4.21 FM indirect generation

The narrowband FM generation is obtained as follows.

The narrowband FM signal is given by

$$f(t) = A_c (\cos \omega_c t - \beta \sin \omega_c t \sin \omega_m t) \qquad (4.84)$$

The AM DSBFC signal is given by

$$f(t) = A_c (\cos \omega_c t + m \cos \omega_c t \sin \omega_m t) \qquad (4.85)$$

We observe that the carrier term being modulated in narrowband FM signal differs from the carrier in the second term of AM DSBFC. That is they are in quadrature.

Hence, as shown in Fig. 4.20, a balanced modulator is employed to generate DSBSC signal using $\sin \omega_c t$ as the carrier. This carrier is then shifted by 90° and when added to the balanced modulator as shown, we get NBFM signal. The signal thus generated is actually phase modulated. However, since we have integrated the modulating signal $m(t)$, i.e., $\sin \omega_m t$, before feeding to the balanced modulator, the resultant signal will be frequency modulator.

The narrowband FM signal is passed through a frequency multiplier circuit, which multiplies the instantaneous frequency of the input by some constant n. This multiplication is obtained by applying the input signal to a non-linear element, and then passing its output through a BPF tuned to the desired central frequency. Since the FM signal has a carrier frequency f_c, which ranges through a frequency deviation of $\pm \Delta f$. When the FM signal is multiplied by n, then the output will have a carrier frequency nf_c and will range through the deviator $\pm n\Delta f$, i.e., the modulator also multiplies both the carrier and deviation frequency. Since the modulation index is proportional to the frequency deviation, for a fixed modulation frequency, the multiplier increases the modulating index by the same factor n.

4.13 PHASE MODULATORS

It this section, we will discuss direct PM modulators.

4.13.1 Varactor Diode Direct PM Modulators

In direct PM, the instantaneous phase of the carrier is changed proportional to the modulating signal. Figure 4.22 shows the circuit for a direct PM modulator. A varactor diode (VD) is in series with the parallel combination of L_1 and R_1. This series parallel combination appears as a resonant circuit to the output frequency from the crystal oscillator.

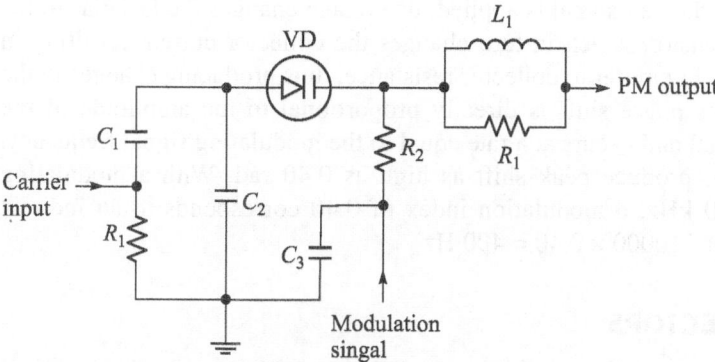

Fig. 4.22 PM modulator: direct method

The modulating signal is applied to VD, which changes its capacitance and hence the phase angle seen by the carrier varies. This results in corresponding phase shift in the carrier. The phase is directly proportional to the amplitude of the modulating signal. The disadvantage of this circuit is the non-linear relationship between the varactor voltage and the capacitance. They resemble closely to a square root function. Hence, to minimize the distortion, the amplitude of the modulating signal must be quite small. This limits the phase deviation to rather smaller values and hence this circuit is generally used for low-index narrowband application.

4.13.2 PM Modulator: Direct Method with Transistor

Figure 4.23 shows a simple transistor direct PM modulator. This is a common emitter amplifier with modulating signal given to the base and the carrier signal given to the collector. The phase shifter consists of C_1, R_t (the collector to emitter resistor) and R_E. The circuit is designed such that at the carrier input frequency f_c, the sum of R_T and R_E equals the capacitance reactance of C_1. This results in 45° phase shift to the carrier.

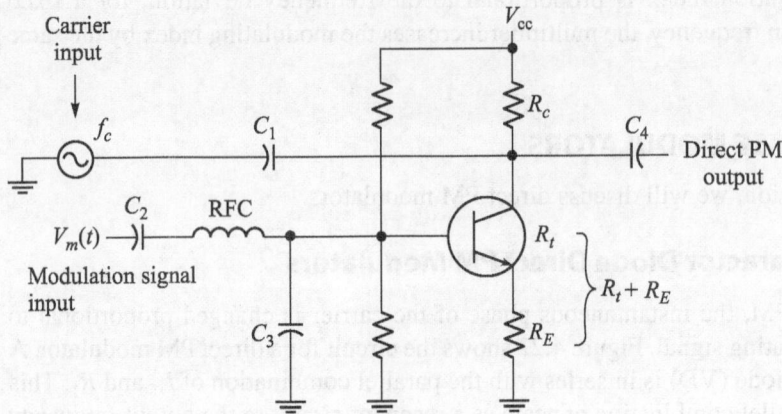

Fig. 4.23 PM modulator with transistor: direct type

When the modulated signal is applied, its voltage changes the DC bias to the base of the transistor, which in turn changes the collector current resulting in dynamic change in emitter to collector resistance, thus producing changes in the phase shift. This phase shift is directly proportional to the amplitude of the modulating signal and occurs at a rate equal to the modulating signal frequency. This circuit can produce peak shift as high as 0.40 rad. With a modulating frequency of 10 kHz, a modulation index of 0.40 corresponds to an indirect frequency shift of $10000 \times 0.40 = 400$ Hz.

4.14 FM DETECTORS

The information in an FM signal is contained in the instantaneous frequency $\omega_i = \omega_c + K_f m(t)$. Hence, a frequency selective network with a transfer function of the form $|H(\omega)| = a\omega + b$ over the FM band would yield an output proportional to the instantaneous frequency, i.e., FM demodulators are implemented by generating an AM signal whose amplitude is proportional to the instantaneous frequency of the FM signal and then using an AM demodulator to recover the message signal. There are many circuits to implement FM to AM conversion. One such circuit is a simple differentiator. The expression for an FM signal is given by

$$f(t) = A \cos \left[\omega_c t + K_f \int_{-\infty}^{t} m(\tau) \, d\tau \right] \tag{4.86}$$

This signal is applied to a differentiator and the output is

$$\frac{df(t)}{dt} = \frac{d}{dt} \left[A\cos[\omega_c t + K_f \int_{-\infty}^{t} m(\tau)d\tau] \right] \tag{4.87}$$

$$\frac{df(t)}{dt} = A \left[\omega_c + K_f m(t) \right] \sin[\omega_c t + K_f \int_{-\infty}^{t} m(\tau)d\tau] \qquad (4.88)$$

The signal $df(t)/dt$ is both amplitude and frequency modulated as shown in Fig. 4.24(a). The envelope of this wave is $A[\omega_c + K_f m(t)]$. Since $\Delta\omega = K_f m(t) > 0$ for all t, $m(t)$ can be obtained by envelope detection of $df(t)/dt$. This is shown in Fig. 4.24(b).

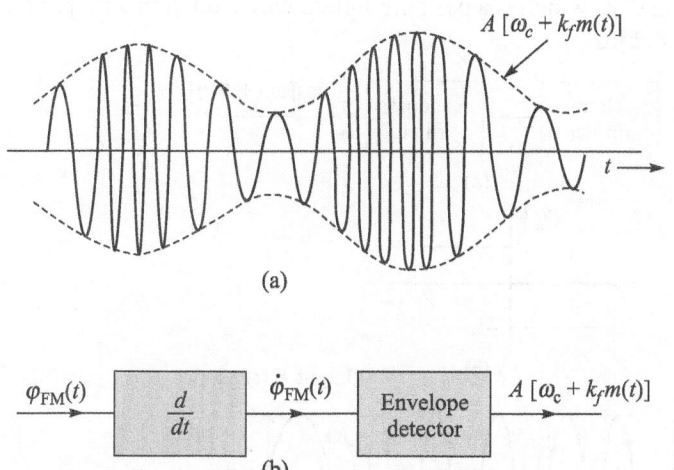

(a)

(b)

Fig. 4.24   FM demodulator principle: (a) differentiator output and (b) block diagram

The amplitude A of the incoming FM carrier is assumed to be constant. If it is constant and a function of time, there will be an additional term dA/dt in the RHS of Eq. (4.88). If this term is neglected, the envelope of $df(t)/dt$ will be $A(t) [\omega_c + K_f m(t)]$ and the envelope detector output would be proportional to $A(f)m(t)$. Hence, it is important to make A constant and any variation in the value of A, if any, should be removed before applying the signal to the FM detector.

The amplitude variations of an FM carrier can be eliminated by a circuit known as bandpass limiter.

4.14.1 Bandpass Limiter

This consists of a hard limiter followed by a BPF as shown in Fig. 4.25(a). The input/output characteristic of a hard limiter is shown in Fig. 4.25(b). The bandpass limiter output to a sinusoidal signal will be a square wave of unit amplitude regardless of the incoming signal amplitude. Also the zero crossings of the incoming sinusoids are preserved in the output, because when the input is zero, the output is also zero. Thus, a frequency modulated sinusoidal input $V_i(t) = A(t) \cos \theta(t)$ results in a constant amplitude FM square wave $V_o(t)$ as shown in Fig. 4.25(c). Such a non-linear operation preserves the frequency modulation information. When $V_o(t)$ is passed through a BPF centred as ω_c, the output is a constant amplitude FM wave. To prove this, let us consider the incoming FM wave

$$V_i(t) = A(t) \cos \theta(t) \qquad (4.89)$$

where $$\theta(t) = \omega_c t + K_f \int_{-\infty}^{t} m(\tau)\, d\tau \tag{4.90}$$

The output $V_o(t)$ of the hard limiter is $+1$ or -1 depending on whether $V_i(t)$ $= A(t) \cos \theta(t)$ is positive or negative. Since $A(t) \geq 0$, $V_o(t)$ can be expressed as a function of θ.

$$V_o(\theta) = 1, \quad \cos \theta > 0$$
$$= -1, \quad \cos \theta < 0 \tag{4.91}$$

So, V_o is a function of θ, which is a periodic square wave function with period 2π as shown in Fig. 4.25(d).

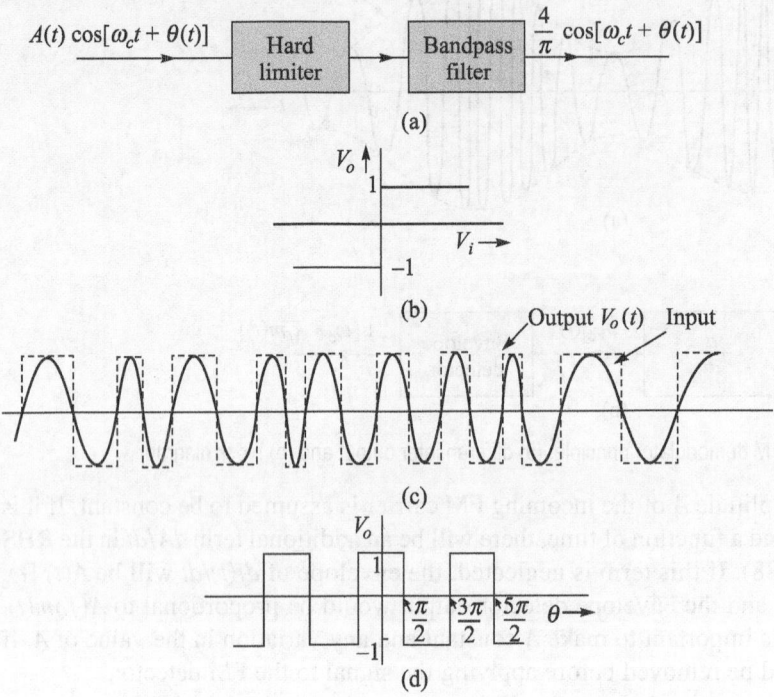

Fig. 4.25   Bandpass limiter: (a) hard limiter and BPF removing amplitude variation in FM wave, (b) limiter's transfer characteristic, (c) limiter's input with corresponding output, and (d) hard limiter output as a function of θ

This can be expanded by a Fourier series

$$V_o(\theta) = (4/\pi)\{\cos \theta - (1/3) \cos 3\theta + (1/5) \cos 5\theta + \cdots\} \tag{4.92}$$

This is valid for any real variable θ. At any instant t,

$$\theta = \omega_c t + K_f \int_{-\infty}^{t} m(\tau)\, d\tau$$

and the output is

$$V_o \left[\omega_c t + K_f \int m(\tau)\, d\tau \right]$$

Hence, the output function as a function of time is given by

$$V_o[\theta(t)] = V_o \left[\omega_c t + K_f \int m(\tau)\, d\tau \right]$$

$$= (4/\pi)[\cos\{\omega_c t + K_f \int m(\tau)\, d\tau\} - (1/3) \cos 3\{\omega_c t + K_f \int m(\tau)\, d\tau\}$$

$$+ (1/5) \cos 5\{\omega_c t + K_f \int m(\tau)\, d\tau\} + \cdots] \tag{4.93}$$

Hence, the output has the original FM wave plus a frequency multiplied FM wave with multiplication factor 3, 5, 7, and so on. The output of the hard limiter is passed through a BPF with a centre frequency ω_c and a bandwidth B_{FM}. The filter output is the desired FM carrier with constant amplitude, which is equal to

$$(4/\pi)\cos[\omega_c(t) + K_f \int \Delta\, m(t)\, dt] \tag{4.94}$$

The BPF not only maintains the constant amplitude of FM carrier but also partially suppresses the channel noise.

4.14.2 Practical Frequency Demodulators

A simple tuned circuit followed by an envelope detector can also serve as a frequency detector as shown in Fig. 4.26. The rising half of the frequency characteristic of the tuned circuit is used in this method. In this region, the frequency characteristic is approximately linear as shown in Fig. 4.27. Since this operation is on the slope of the characteristic. This method is called *slope detection*.

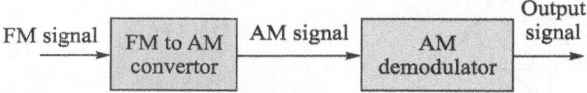

Fig. 4.26   FM demodulator: general circuit

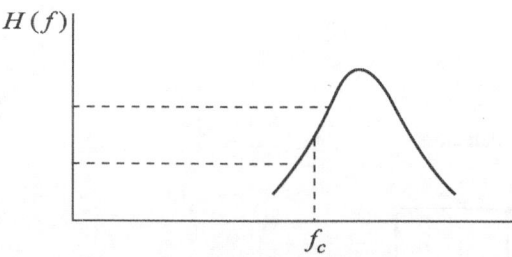

Fig. 4.27   Tuned circuit characteristic

To obtain linear characteristics over a wide range of frequencies, usually two circuits tuned at two frequencies f_1 and f_2 are connected in a configuration known as *balanced discriminator*. A balanced discriminator with the corresponding frequency characteristic is shown in Fig. 4.28.

Another balanced demodulator, known as *ratio detector*, gives better protection against carrier amplitude variation than does the discriminator. Zero crossing detectors are also used. There are frequency counters designed to measure the instantaneous frequency by the number of crossings. The rate of zero crossing is equal to the instantaneous frequency of the input signal.

Phased locked loop is also used as FM demodulator. Because of its low cost and superior performance, especially in the presence of noise, this method is widely used.

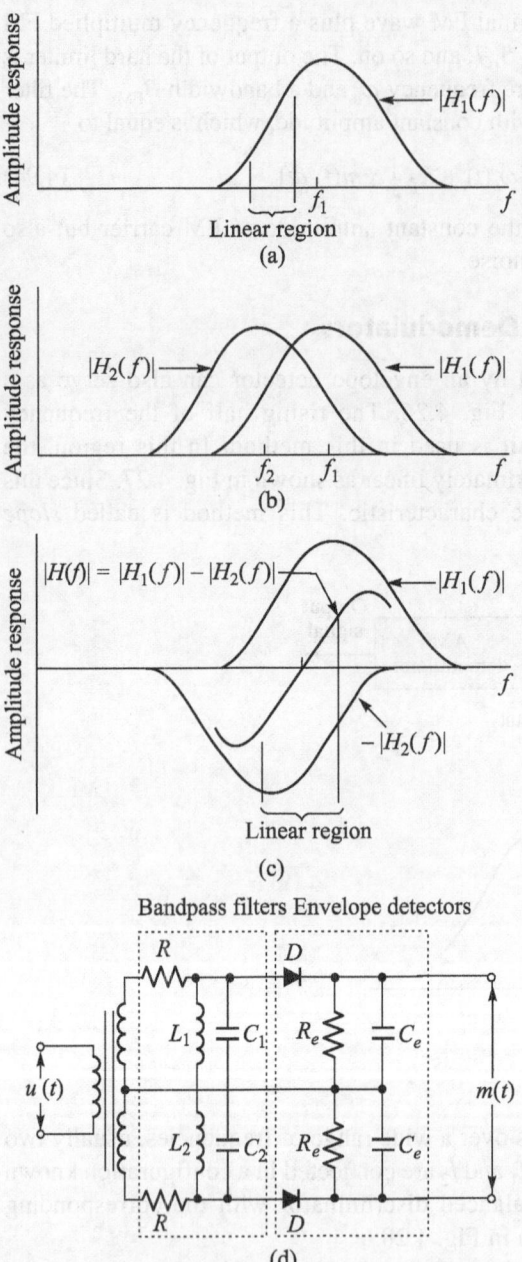

Fig. 4.28 Balanced discriminator

4.14.3 Slope Detector

The schematic diagram for a single ended slope detector is shown in Fig. 4.29. It is nothing but a tuned circuit followed by a peak detector.

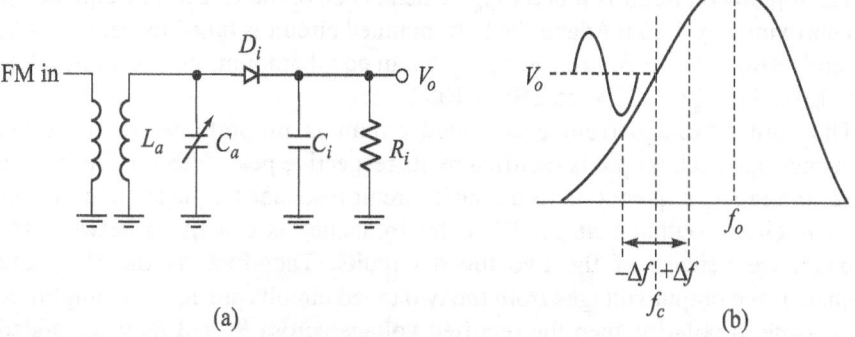

Fig. 4.29 Slope detector: (a) schematic diagram and (b) output characteristic

The tuned circuit L_a and C_a produces an output voltage that is proportional to the input frequency. At the resonant frequency f_c of the tank circuit, the maximum output voltage occurs. The output increases or decreases proportionately with respect to the input frequency deviation $\pm\Delta f$. The circuit is designed so that the intermediate centre frequency f_c falls in the centre of the most linear portion of the frequency characteristic. When the intermediate frequency deviates above f_c, the output voltage increases and when the intermediate frequency (IF) deviates below f_c, the output voltage decreases. Hence, tuned circuit converts frequency variation to amplitude variation. The amplitude variation thus obtained is converted to an output voltage that varies at a rate equal to that of the input frequency change and whose amplitude is proportional to the magnitude of the frequency change. Since this detector has the most non-linear voltage versus frequency characteristics, it is seldom used.

4.14.4 Balanced Slope Detector

Figure 4.30(a) shows the circuit diagram of a balanced slope detector. It is also known as *round travis detector*. This circuit combines two circuits of individual single slope detectors in parallel and the FM input is fed 180° out of phase. Centre tap winding on the tuned secondary side of the transformer T_1 introduces phase inversion. The tuned circuit consists of L_a, C_a and L_b, C_b and converts the FM input to AM signal. The balanced peak detectors D_1, C_1, R_1 and D_2, C_2, R_2 extract the information from AM envelope.

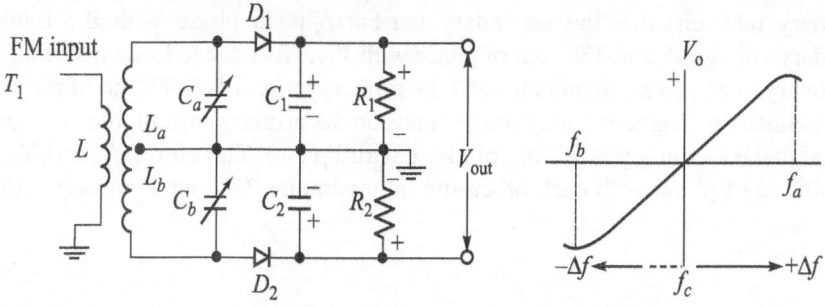

Fig. 4.30 Balanced slope detector: (a) circuit diagram and (b) output waveform

The top tuned circuit is tuned to f_a, which is above the IF centre frequency f_o by approximately $1.33 \times \Delta f$ and the bottom tuned circuit is tuned to frequency f_b, which is below the IF centre frequency by an equal amount. For example, if Δf = 75 kHz, then f_a = 1.33×75 kHz ~ 100 kHz.

The output voltage from each tuned circuit is proportional to the input frequency and each output is rectified by its respective peak detector. Hence, the closer the input frequency is to the tank circuit resonant frequency, the greater the tank circuit output voltage. IF centre frequency is exactly in between the resonant frequencies of the two tuned circuits. Therefore, at the IF centre frequency, the output voltages from the two tuned circuits are equal in amplitude or opposite in polarity, then the rectified voltage across R_1 and R_2 when added produces a higher output voltage than the bottom circuit and the output voltage goes positive. When the IF deviates below resonance, the output from the bottom resonance circuit is larger than the output voltage from the top tank circuit and the resultant output is negative. The output response curve against frequency is shown in Fig. 4.30(b).

The above-mentioned circuit has several disadvantages like poor linearity, difficulty in tuning, and poor limiting. Because of poor limiting this circuit produces an output that is proportional to the amplitude as well as the frequency variation in the output signal. Hence, a separate limiter stage has to be preceded. The balanced slope detector is aligned by tuning C_a and C_b for zero volts with IF frequency equal to f_0. Then f_a and f_b are alternately injected while C_a and C_b are tuned for maximum and equal output voltage with opposite polarity.

4.14.5 Foster–Seeley Discriminator

The phase shift between primary and secondary voltages of a tuned transformer is a function of frequency and the Foster–Seeley discriminator utilizes this frequency phase relation for the recovery of the modulating signal. This circuit is also called a *phase shift discriminator*. The operation of the discriminator is similar to that of the balanced slope detector. Figure 4.32(a) shows the schematic diagram for a Foster–Seeley discriminator. Values for C_c, C_1, and C_2 are chosen such that they are short circuits for the IF centre frequency. Hence, the junction of L_3, C_1, C_2, R_1, and R_2 is at AC ground potential. The IF signal V_i is fed directly in phase across L_3. Hence, $V_{L3} = V_i$.

The incoming IF is inverted $180°$ by transformer T_1 and divided equally between L_a and L_b. At the IF centre frequency (the resonant frequency of the secondary tank circuit), the secondary current I_S is in phase with the total secondary voltage V_S and $180°$ out of phase with V_{L3}. Also due to loose coupling, the primary of T_1 acts as an inductor and the primary current I_P is $90°$ out of phase with V_i. Since the magnetic induction depends on the primary current, the voltage induced in the secondary is $90°$ out of phase with $V_i(V_{L3})$. Therefore, V_{La} and V_{Lb} are $180°$ out of phase with each other and in quadrature ($90°$ out of phase) with V_{L3}.

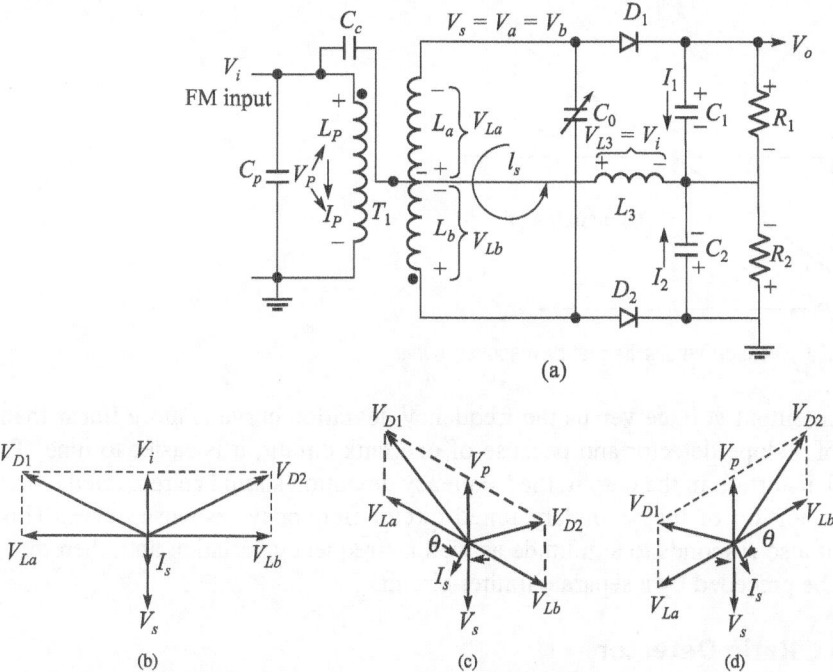

Fig. 4.31 Foster–Seeley discriminator: (a) schematic diagram, (b) $f_i = f_o$, (c) $f_i > f_o$, and (d) $f_i < f_o$

The voltage across diode D_1 is V_{D1}, the vector sum of V_{L3} and V_{La}, and the voltage across diode D_2 is V_{D2}, the vector sum of V_{L3} and V_{Lb}. The vector diagram is shown in Fig. 4.31(b).This figure shows that the voltages V_{D1} and V_{D2} are equal. Hence, at resonance, I_1 and I_2 are equal and C_1 and C_2 get charged to equal magnitude voltages except with opposite polarities. Therefore, the output voltage V_o, $= V_{c1} - V_{c2}$, which is equal to zero. When the IF goes above resonance, the secondary tank circuit impedance becomes inductive and the secondary current lags behind the secondary voltage by θ, which is proportional to the magnitude of the frequency deviation. The corresponding phasor diagram is shown in Fig. 4.31(c). The figure shows that the vector sum of the voltages across D_1 is greater than the vector sum of the voltages across D_2. Hence, C_1 charges while C_2 discharges and V_o goes positive. When the IF goes below resonance, the secondary tank circuit impedance becomes capacitive and the secondary current leads the secondary voltage by θ, which is again proportional to the magnitude of the change in frequency. The corresponding phasor is shown in Fig. 4.31(d). It can be seen that the vector sum of the voltages across D_1 is now less than the vector sum of the voltages across D_2. Consequently, C_1 discharges while C_2 gets charged and V_o goes negative. Foster–Seeley discriminator is tuned by injecting a frequency equal to IF centre frequency and tuning C_o for getting $V_o = 0$ volts.

Figure 4.32 shows a typical voltage versus frequency response curve for a Foster–Seeley discriminator. From the figure, it can be clearly seen that the output voltage from a Foster–Seeley discriminator is directly proportional to the magnitude and direction of the frequency deviation. This voltage versus frequency deviation response curve is called S curve for obvious reason.

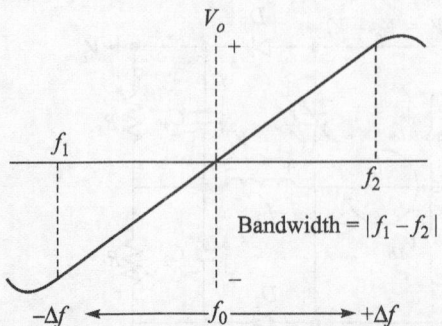

Fig. 4.32 Voltage versus frequency response curve

The output voltage versus the frequency deviation curve is more linear than that of a slope detector and because of one tank circuit, it is easier to tune. To avoid distortion in the output, the frequency deviation should be restricted to the linear portion of the secondary tuned circuit frequency response curve. This circuit also responds to amplitude as well as frequency variation and, therefore, must be preceded by a separate limiter circuit.

4.14.6 Ratio Detector

By reversing the direction of one diode in the Foster–Seeley circuit, we can improve the limiting action. But this will be at the expense of output level. The reversal of a diode causes both diodes to conduct in series and the detector circuit then provides a damping action, which in turn tends to maintain a constant secondary voltage. Figure 4.33(a) shows the schematic diagram of a ratio detector. This has also got a single tuned circuit in the transformer secondary. The voltage vectors for D_1 and D_2 are identical to those of Foster–Seeley discriminator, but because of a diode reversal, the current I_d can flow around through D_1, D_2, transformer secondary R_S and C_S. Hence, after several cycles of the input signal, shunt capacitor C_S charges to approximately peak secondary voltage of T_1. The time constant $R_S C_S$ is sufficiently long so that rapid changes in the amplitude of the input signal due to thermal noise or other interfering signals are shorted to the ground and have no effect on the average voltage across C_S. C_1 and C_2 charge and discharge proportional to frequency changes in the input signal and are relatively immune to amplitude variations. Also the output voltage is always positive, since it is taken with respect to ground. At resonance, the output voltage is divided equally between C_1 and C_2 and is redistributed as the input frequency is deviated above and below resonance. Therefore, changes in V_o are due to the changing ratio of the voltages across C_1 and C_2 while the total voltage is clamped by C_S.

The output frequency response curve for the ratio detector as shown in Fig. 4.33(b) clearly indicates that the output voltage is not equal to zero at resonance but equal to one half of the voltage across the secondary winding of T_1. Since a ratio detector is relatively immune to amplitude variations, it is preferred to a discriminator.

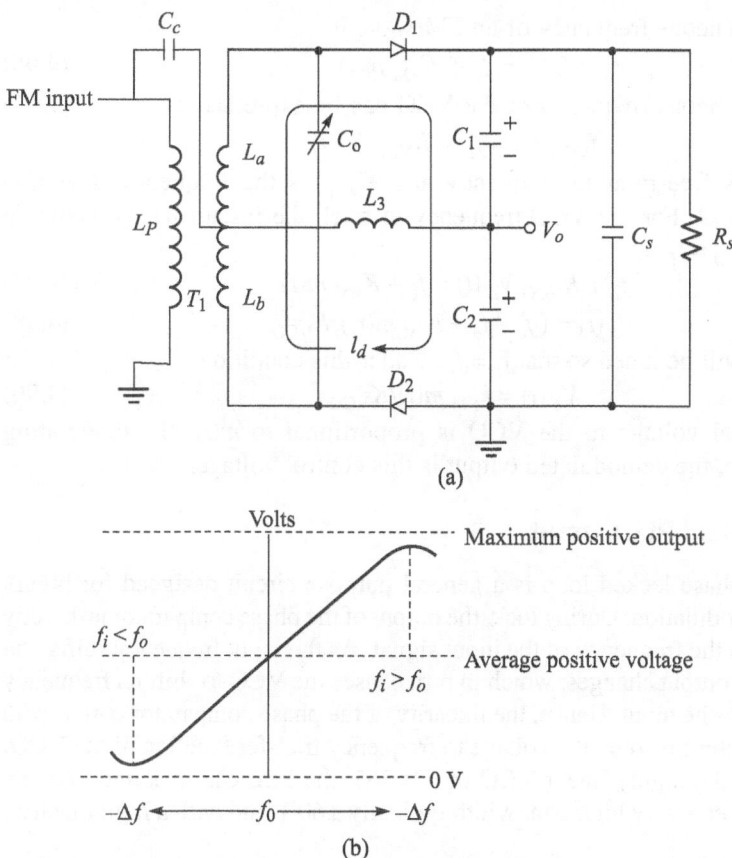

Fig. 4.33   Ratio detector: (a) schematic diagram and (b) output voltage response

4.14.7 FM Demodulator Using a PLL

A phase locked loop can be used to demodulate FM signal. A block diagram of a PLL is shown in Fig. 4.34. The phase detector which is basically a balanced modulator produces an average low frequency output voltage that is a linear function of the phase difference between the two input signals. The lowpass filter shown allows the low frequency components and also removes much of the noise. The filtered signal is amplified and the amplifier voltage is fed as the control voltage to the voltage controlled oscillator. When the loop is in the lock, the VCO frequency follows the incoming signal frequency. The DC correction voltage fed to the VCO is proportional to frequency deviation and thus the demodulated information signal.

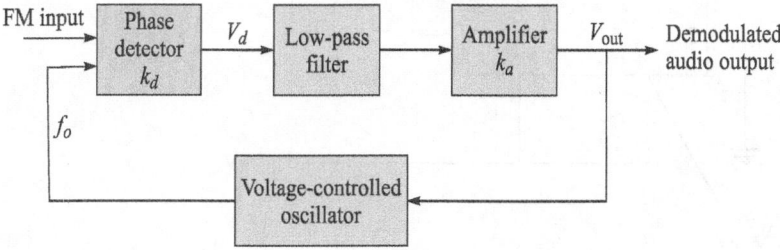

Fig. 4.34   PLL block diagram

The instantaneous frequency of an FM wave is

$$f_i(t) = f_c + K_{FM} m(t) \tag{4.95}$$

The instantaneous frequency of the VCO can be expressed as (4.96)

$$f_{VCO}(t) = f_o + K_{VCO} V_c(t)$$

where f_o is its free-running frequency and K_{VCO} is the frequency deviation constant of VCO. For the VCO frequency to track the instantaneous incoming frequencys $f_{VCO} = f_i$ or

$$f_o + K_{VCO} V_c(t) = f_c + K_{FM} m(t) \tag{4.97}$$

Hence, $$V_c(t) = (f_c - f_o + k_{FM} m(t))/K_{VCO} \tag{4.98}$$

The VCO will be tuned so that $f_c = f_o$. Under this condition,

$$V_c(t) = k_{FM} m(t)/K_{VCO} \tag{4.99}$$

i.e., the control voltage to the VCO is proportional to $m(t)$, the modulating voltage. Hence, the demodulated output is this control voltage.

4.14.8 Practical PLL Circuit

The NE 565 phase locked loop is a general purpose circuit designed for highly linear FM demodulation. During lock, the output of the phase comparator is directly proportional to the frequency of the input signal. As the input frequency shifts, the phase detector output changes, which in turn causes the VCO to shift its frequency to match that of the input. Hence, the linearity of the phase comparator output with frequency is determined by the voltage to frequency transfer function of the VCO.

Because of the highly linear VCO of NE 565, the PLL can lock and track an input signal over a very high bandwidth typically ±60% and with a high linearity of 0.5%.

Figure 4.35 shows a typical FM demodulator circuit using NE 565 PLL. The VCO frequency is given approximately by the relation

$$f_c = (1.2/4)R_1 C_1 \tag{4.100}$$

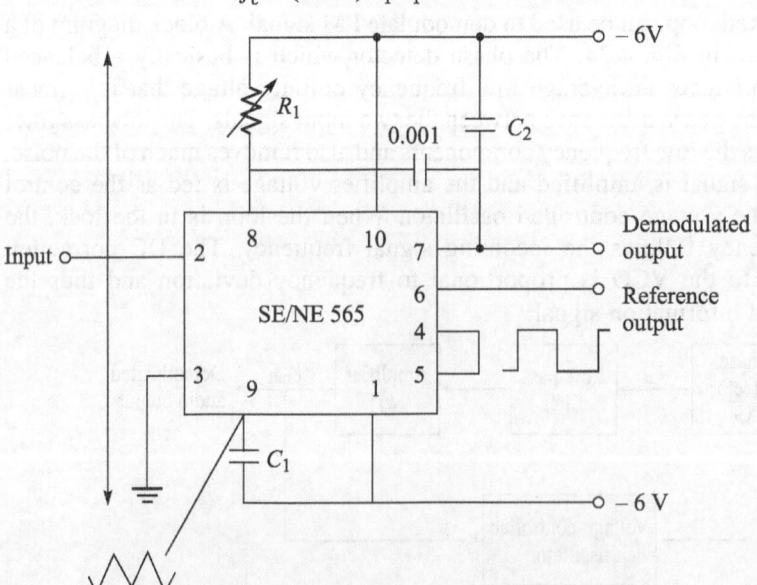

Fig. 4.35 FM demodulation circuit

For FM demodulation, f_c should be adjusted to the centre frequency of the input signal frequency range, which is nothing but the carrier frequency f_c. R_1 should be within the range 2 kΩ to 20 kΩ and. C_1 can be of any value. FM input can be directly connected to the input if its DC content is not much and the DC resistances seen from pins 2 and 3 are equal and there is no DC voltage difference between the pins. A short between pins 4 and 5 connects the VCO to the phase comparator. Pin 6 provides a DC reference voltage that is close to the DC potential of the demodulated output pin 7. A small capacitor, typically 0.001 MF, should be connected between pins 7 and 8 to eliminate possible oscillation.

A single loop filter is formed by the capacitor C_2 connected between pin 7 and the positive supply. The R for the filter is internally provided in NE 565, which is about 3.6 k. If we want to decrease the lock range, a resistor should be connected between pins 6 and 7. This reduces the lock range from $\pm$ 60% of f_o to $\pm 20\%$ of f_o at ± 6 V supply. More information on NE 565 PLL can be had from *Signetics Linear Data Manual*, Volume I, Communications. More details on PLL are given at the end of the chapter.

4.14.9 Quadrature Detectors

This type of FM detector extracts the original information signal from the composite IF waveform by multiplying two quadrature – 90° out of phase signals. This method is also called *coincidence detector*.

A quadrature detector uses a 90° phase shifter, a single tuned circuit, and a product detector to demodulate FM signals. The 90° phase shifter produces a signal that is in quadrature with received IF signals. The tuned circuit converts frequency variations to phase variation and the product detector multiplies the received IF signals and the phase shifted IF signal.

Figure 4.36 shows the circuit for an FM quadrature detector. R_o, L_o, and C_o forms a tank circuit. Capacitor C_1 in series with the tank circuit produces a 90° phase shift at the IF centre frequency. The tank circuit is tuned to the IF centre frequency and produces an additional phase shift θ that is proportional to the frequency deviation. The FM input signal V_i is multiplied by the quadrature signal V_o in the product detector. The product detector output is proportional to the frequency deviation. The tank circuit impedance is resistive at resonant frequency. The frequency variation in the IF signal produces an additional positive or negative phase shift. Hence, the product detector output voltage is proportional to the phase difference between the two input signals. This can be expressed mathematically as

$$V_{out} = V_i V_o = [V_i \sin(\omega_c t + \theta)] [V_o \cos(\omega_c t)] \qquad (4.101)$$

using trigonometric identities

$$V_{out} = (V_i V_o/2) [\sin(2\omega_c t + \theta) + \sin\theta] \qquad (4.102)$$

The second harmonic $2\omega_c$ is filtered out by the LPF and the demodulated output will be

$$V_{out} = (V_i V_o/2)\sin\theta \qquad (4.103)$$

Here $\theta = \tan^{-1} KQ$

where $K = 2\pi f/f_c$ (fractional frequency deviation)

Q = quality factor of the tank circuit

For small phase angle θ, V_{out} becomes

$$V_{out} = (V_i V_o/2)\theta \qquad (4.104)$$

since $\sin \theta = \theta$ for small θ

i.e., the output voltage linearly varies as the phase angle θ, which varies almost linearly with frequency.

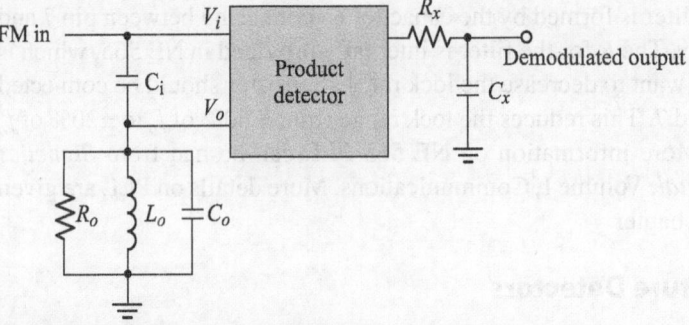

Fig. 4.36 Quadrature FM demodulator

4.14.10 Zero Crossing Detector

In FM wave, the message information is contained in the frequency of the carrier not in the amplitude (that is why the limiter is able to remove all amplitude information from the carrier without destroying any of the information content of the carrier). The frequency of the carrier can be measured by observing the zero crossings of the carrier. The output of the zero crossing detector will be a pulse width modulated square wave. This PWM signal is differentiated and fully rectified to get triggering pulses to a monostable multivibrator, which produces a pulse train whose position depends on the amplitude of the message signal. The output of the monostable multivibrator is passed through a low-pass filter to regenerate the baseband signal.

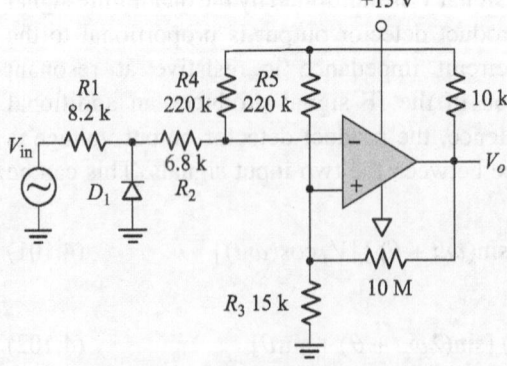

D_1 restricts negative input swing to 0.6 V.
$R_1 + R_2 = R_3$

$R_3 \leq R_5/10$ for small error in zero crossing

Fig. 4.37 Zero crossing detector using an op amp as a comparator

The advantage of zero crossing detection is that no source of the carrier frequency is required to demodulate the signal. A digital signal can be easily recovered from an FM signal in this manner. Decoding an analog signal may be difficult by this method, since the low-pass filter output does not closely resemble the baseband signal.

Figure 4.37 shows a zero crossing detector using an operational amplifier as a comparator. The non-inverting input of the comparator is about 0.0015 V_o, which is quite small and can be considered as zero volts. When the input given to the inverting input is positive, the output will be zero volts and when it crosses the zero volt point and enters the negative portion, the output will be 15 V. Hence, the output of the comparator is a square wave whose width depends on the input signal frequency.

Figure 4.38 shows the block diagram of FM demodulation using zero crossing detector, which is nothing but a limiter followed by a differentiator, rectifier, pulse generator (monostable multivibrator) and a lowpass filter. The waveforms are also shown. Figure 4.39 shows a complete zero crossing detector.

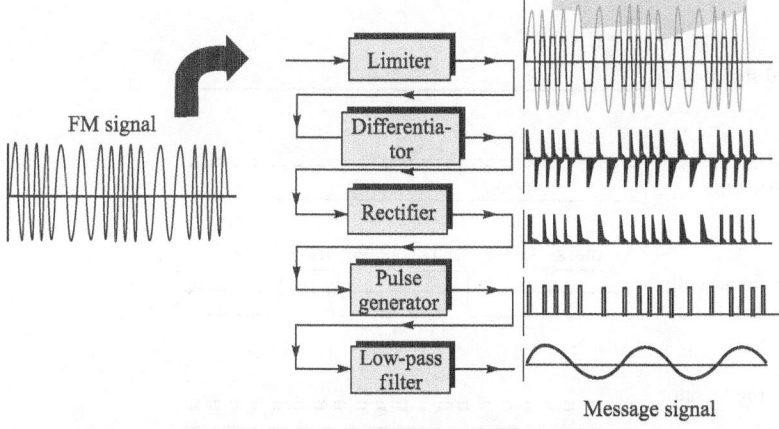

Fig. 4.38   Block diagram of FM demodulation using zero crossing detector

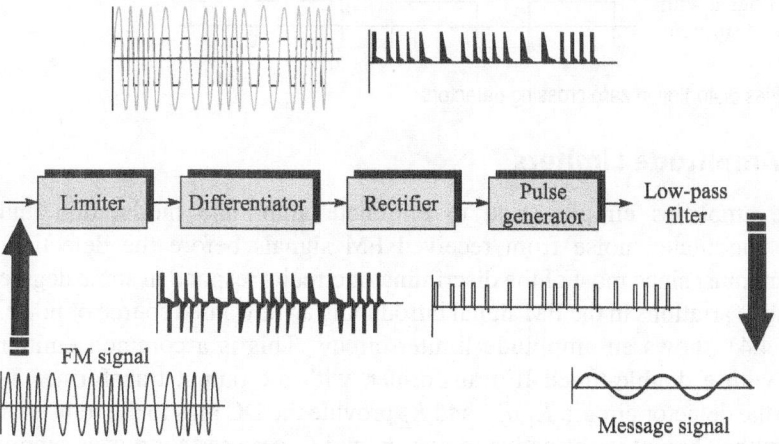

Fig. 4.39   Complete zero crossing detector

4.14.11 Bias Distortion in FM Demodulation Using Zero Crossing Detectors

The design of low-pass filter is quite critical in these types of demodulators. A poorly adjusted threshold can cause bias distortion, where the digital signal produced is not identical to the original signal. The bias distortion is given by

$$\text{Bias distortion} = 1 - \frac{\tau_{\text{actual}}}{\tau_{\text{original}}}$$

If the threshold is too high, we get regenerated signal with positive distortion and if it is too low, we get negative bias distortion. Figure 4.40 shows both positive and negative distortion.

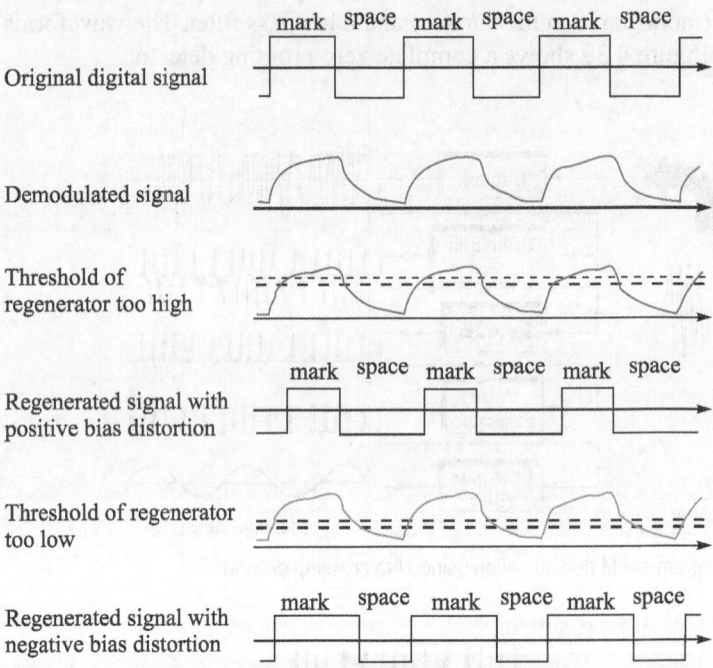

Fig. 4.40 Bias distortion in zero crossing detectors

4.14.12 Amplitude Limiters

These are amplifies circuits used to eliminate amplitude modulation and amplitude modulated noise from received FM signals before the detection. Limiters are must since most of the discriminator circuits respond in some degree to amplitude variations in the FM signal introducing an unwanted source of noise.

Figure 4.41 shows an amplitude limiter circuit. This is a common emitter amplifier with a double-tuned IF transformer, with the output transformer T_2 supplying the detector circuit. R_1, R_2, and R_E provide the DC bias to the transistor to maintain the Q point in the active region. R_c and C_c are used for power supply decoupling.

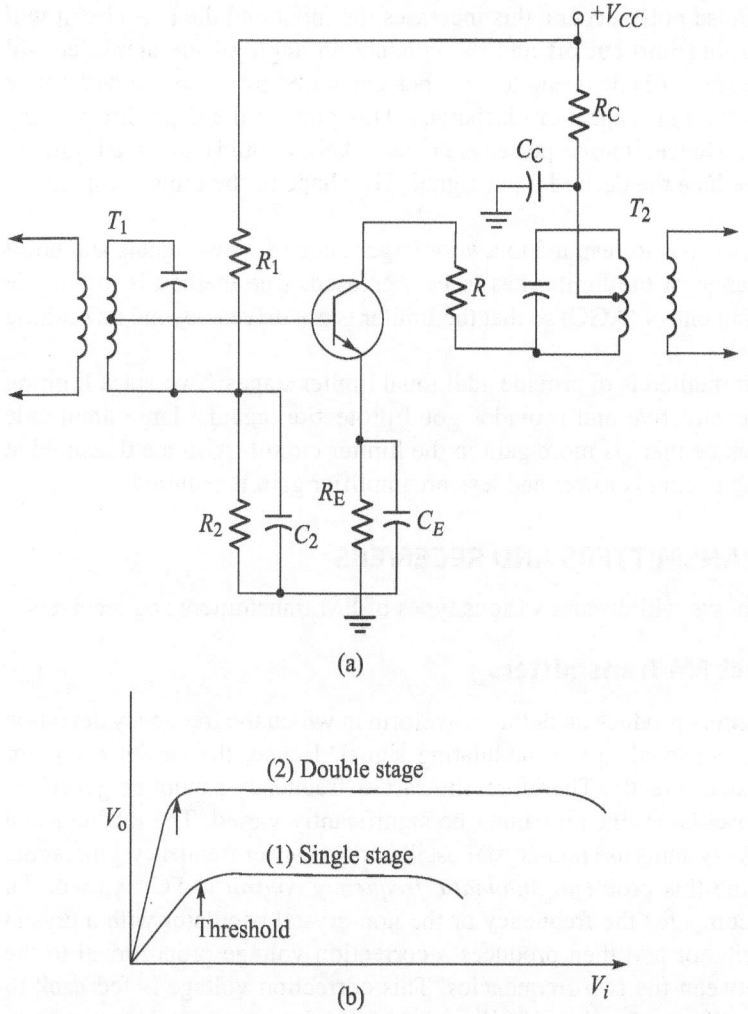

Fig. 4.41 Amplitude limiter: (a) schematic diagram and (b) transfer characteristic

For very low signal levels, the circuit behaves like a normal class A amplifier. The transistor is driven into cut off for large signals by the base leak circuit formed by the combination of C_2 and R_2. The positive peak drives the transistor into saturation and hence the collector current waveform has its top and bottom clipped off resulting in the output that is more or less a rectangular waveform. The fundamental of this rectangular waveform is tuned by the output transformer like in class C amplifier. The point at which the signal input range exceeds the active range of the transistor forms the threshold of the limiter, beyond which the limiting action takes place. Further, increase of the input signal does not significantly increase the collector current magnitude and the fundamental component remains almost constant.

If a large noise pulse occurs, this increases the input and the bias circuit will drive even further into cut off and the conduction angle of the amplifier will decrease. The limiter is then said to have been captured by noise and will not be released until the bias capacitor discharges. This puts on the upper limit on the limiting range. Hence, if noise pulses are received, they could cause the limiter to respond and reduce the desired input signal. The shape of the limiter response is indicated.

If the receiver has to respond to a very large range of input signals and noise signals, the range of the limiter has to be increased. One method is to provide automatic gain control (AGC) so that the limiter is not driven beyond its limiting range.

The second method is to provide additional limiter stages. Two-stage limiting is much more effective and provides good protection against large amplitude noise pulses since there is more gain in the limiter circuit. Also the threshold at which limiting occurs is lower and less preamplifier gain is required.

4.15 FM TRANSMITTERS AND RECEIVERS

In this section, we will discuss various types of FM transmitters and receivers.

4.15.1 Direct FM Transmitters

These transmitters produce an output waveform in which the frequency deviation is directly proportional to the modulating signal. Hence, the carrier oscillator must be deviated directly. Therefore, the carrier frequency cannot be generated by a crystal oscillator since it cannot be significantly varied. The medium- and high-index FM systems use non-crystal oscillator for carrier frequency generation.

To overcome this problem, *automatic frequency control* (AFC) is used. An AFC circuit compares the frequency of the non-crystal oscillator with a crystal reference oscillator and then produces a correction voltage proportional to the difference between the two frequencies. This correction voltage is fed back to the carrier oscillator to automatically compensate for any drift that may have occurred.

Generally, class C power amplifiers are used in FM systems since amplitude distortion introduced by class C amplifier has no effect on FM signals. This makes more efficient transmitter compared to AM.

Commercial direct FM transmitter

Figure 4.42 shows the block diagram for a commercial broadcast band transmitter. This configuration is called Crosby direct FM transmitter. It includes an AFC loop. The frequency modulator can be of either VCO or reactance type. The carrier wave is generated from a master oscillator. The centre frequency of the master oscillator $f_c = 5.1$ MHz, which is multiplied by 18 in three steps ($3 \times 2 \times 3$) to produce a final transmit carrier frequency $f_t = 91.8$ MHz. In a frequency multiplier, when the frequency of a frequency modulated carrier is multiplied, its phase and frequency deviations are also multiplied. The rate at which the carrier is deviated, i.e., the modulating frequency, is unaffected by the multiplication process. Hence, the modulation index is also multiplied.

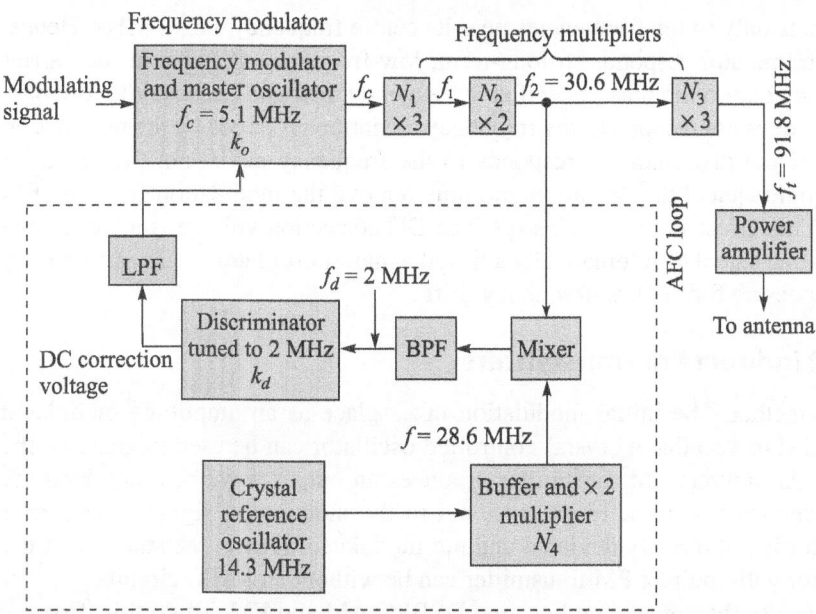

Fig. 4.42 Crosby FM transmitter: direct method

Also when an FM modulated wave is heterodyned with another frequency in a non-linear mixer, the carrier can be either upconverted or downconverted depending on the passband of the output filter. But the frequency deviation, phase deviation, and the rate of change are unaffected by the heterodyning process. So, for the transmitter shown in Fig. 4.42, the frequency and phase deviations at the output of the modulator are also multiplied by 18. To achieve the maximum frequency deviation of 75 kHz allowed for FM, the deviation at the output of the modulator should be

$$\Delta f = 75 \text{ kHz}/18 = 4166.7 \text{ H}$$

and the modulation index should be

$$\beta = 4166.7/f_m$$

For the maximum modulating frequency, $f_m = 15$ kHz,

$$\beta = 4166.7 /15,000 = 0.2778$$

Thus, the modulation index at the output of power amplifier is

$$\beta = 0.2778 \times 18 = 5$$

This is nothing but the deviation ratio for a commercial FM broadcast transmitter with a 15 kHz modulating signal. AFC is used to automatically correct the carrier frequency drift. Thus, a near-crystal stability is achieved for the transmitter carrier frequency without using a crystal oscillator with AFC. The carrier signal is mixed with the output signal from a crystal reference oscillator in a non-linear device, downconverted in frequency, and then fed back to the input of a frequency discriminator. The output from the doubler f_2 is 30.6 MHz, which is mixed with a crystal-controlled reference frequency $f_r = 28.6$ MHz to produce a difference frequency $f_d = 2$ MHz. The discriminator is a high Q tuned circuit that reacts only to the frequencies near its centre frequency, i.e., 2 MHz. Hence, the discriminator responds to long-term, low-frequency changes in the carrier frequency due to the master oscillator frequency drift and because the low-pass filtering does not respond to the frequency deviation produced by the modulating signal. If the discriminator responds to the frequency deviation, the feedback loop would cancel the deviation and thus remove the modulation from the FM wave. This effect is called *wipe off*. The DC correction voltage is added to the modulating signal to automatically adjust the master oscillator's carrier frequency to compensate for the low-frequency drift.

4.15.2 Indirect FM Transmitters

In this method, the initial modulation takes place as an amplitude modulated DSBSC signal so that a crystal-controlled oscillator can be used to generate the carrier. An indirect FM transmitter produces an output waveform in which the phase deviation is directly proportional to the modulating signal. The carrier oscillator is not directly deviated and not modulated. Hence, the stability of the oscillator with indirect FM transmitter can be without any AFC circuit.

Let us see the connection between DSBSC AM and FM. To do this, first we have to analyse a low-level phase modulated signal. The equation for the FM modulated wave is

$$f(t) = V_c \cos(\omega_c t + \beta \sin \omega_m t)$$
$$= V_c [\cos \omega_c t \cos(\beta \sin \omega_m t) - \sin \omega_c t \sin (\beta \sin \omega_m t)] \qquad (4.105)$$

For narrowband FM, since the frequency deviation is quite small, the modulation index β is quite small. Hence, substituting for $\cos(\beta \sin \omega_m t) = 1$ and $\sin (\beta \sin \omega_m t) = \beta \sin \omega_m t$ in Eq. (4.105), we get

$$f(t) = V_c (\cos \omega_c t - \beta \sin \omega_c t \sin \omega_m t) \qquad (4.106)$$

When this expression for $f(t)$ is compared with that of narrowband AM, we get

$$f(t) = V_c (\cos \omega_c t + m \cos \omega_c t \sin \omega_m t) \qquad (4.107)$$

We observe that in narrowband FM, the carrier is shifted by $\pi/2$ and modulated. Thus, by shifting the carrier by $\pi/2$ and giving it to a balanced modulator along with the modulating signal, we can achieve phase modulation through AM. Since the modulated signal is passed through an integrator before a balanced modulator, we get frequency modulation.

Figure 4.43 shows a very popular method of achieving FM, *Armstrong method*, named after its inventor.

The crystal oscillator generates a low-frequency subcarrier, which can be as low as 100 kHz. This subcarrier is phase shifted and fed to a balanced modulator, where it is mixed with the input modulating signal after integration for FM generation.

The output from the balanced modulator is a double sideband suppressed carrier wave that is combined with the original carrier in a summing circuit to produce the phase modulated waveform.

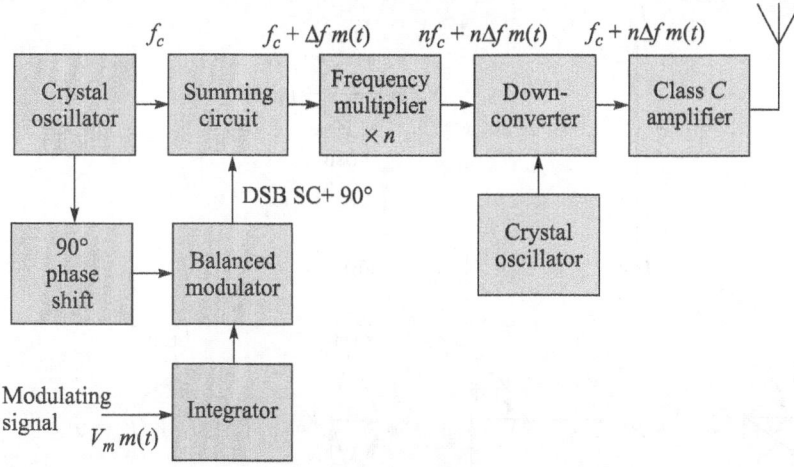

Fig. 4.43 FM transmitter: Armstrong indirect method

Figure 4.44(a) shows the phasor for the original carrier V_c and Fig. 4.44(b) shows the phasor for the two sidebands of the suppressed carrier from the balanced modulator V_{USB} and V_{LSB}. Since the suppressed carrier voltage V_c is 90° out of phase with V_c, the resultant of upper and lower sidebands produces a component V_m, which is always at right angle with V_c. Figure 4.44(c) through (f) shows the progressive phasor addition of V_c, V_{USB}, and V_{LSB}. It can be observed that the output from the combining network is a signal whose phase is varied at a rate equal to f_m and whose magnitude is directly proportional to the magnitude of V_m. The modulation index, i.e., the peak deviation, is given by

$$\theta = \tan^{-1} V_m / V_c \tag{4.108}$$

For very small angles, $\tan \theta = \theta$. Therefore,

$$\theta = V_m / V_c \tag{4.109}$$

From the phasor diagram, it can be seen that $V_{c\,\max}$ occurs when V_{USB} and V_{LSB} are in phase with each other. The maximum phase deviation that can be produced with this modulator is quite low. Hence, for wideband FM, multipliers are used. Multiplication alone will be inadequate and a combination of multiplication and mixing is necessary to develop the desired carrier frequency with standard 75 kHz deviation.

In the Armstrong transmitter, the phase of the carrier is directly modulated through the balanced modulator and the summing network producing indirect frequency modulation. The magnitude of the phase modulation is directly proportional to the amplitude of the modulating signal but independent of its frequency. Hence, the modulation index remains constant for all modulating signal frequencies of a given amplitude.

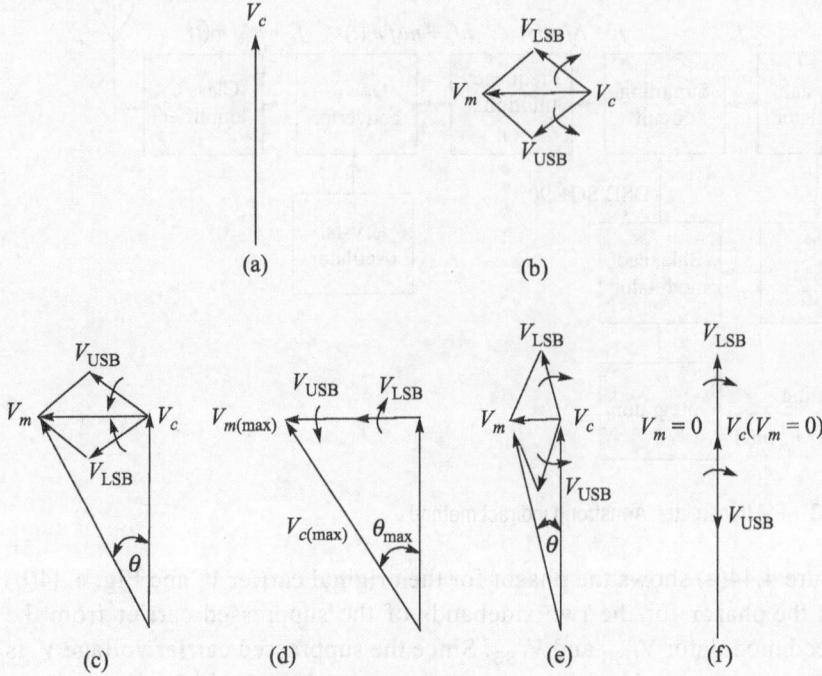

Fig. 4.44 Phasor diagrams: (a) carrier phasor, (b) sideband phasor, and (c)–(f) progressive phase addition, with the peak phase shift at (d)

4.15.3 FM Stereo Broadcasting

Because of excellent fidelity provided by FM systems, commercial radio broadcasting uses FM, since music is the chief programme material, frequency modulation acts in several ways to improve the fidelity. Since FM broadcasting takes place in VHF band 88–108 MHz, a much wider baseband can be used. The base bandwidth presently in use is 50 Hz to 15 kHz. The carrier frequencies are separated by 200 kHz, i.e., the channel spacing is 200 kHz. The power output of the transmitter is about 100 kW.

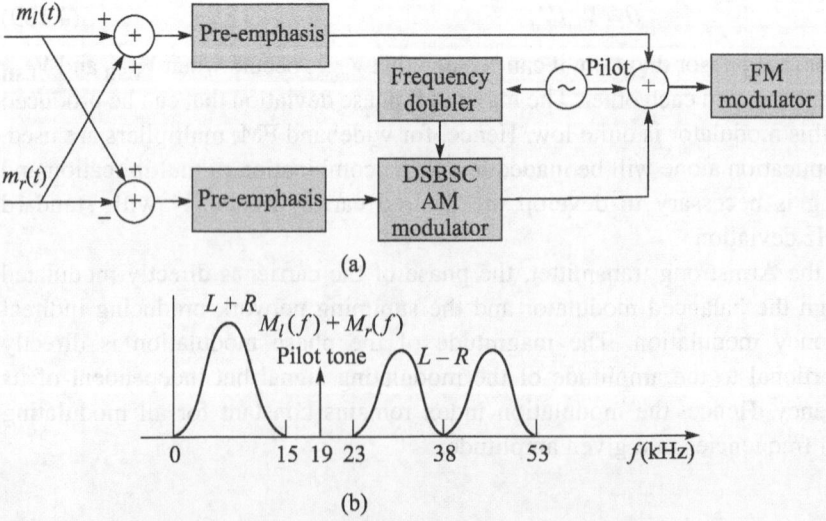

Fig. 4.45 (a) FM stereo transmitter and (b) baseband signal spectrum

Modern FM radio stations transmit music programmes in stereo by using the outputs of two microphones placed in two different parts of the stage. Figure 4.45 shows the block diagram of an FM stereo transmitter with baseband signal spectrum. The signals from the left and right microphones $m_r(t)$ and $m_l(t)$ are added and subtracted as shown. The sum signal $m_r(t) + m_l(t)$ is left unchanged and occupies the frequency band $0 - 15$ kHz. The difference signal $m_r(t) - m_l(t)$ is used to AM modulate a 38 kHz carrier, which is generated from a 19 kHz carrier. To this DSBSC AM signal, a pilot carrier is added for the purpose of demodulating the DSBSC AM signal at the receiver end. A pilot carrier of 19 kHz is used instead of 38 kHz because the pilot is more easily separated from the composite signal at the receiver. The sum channel is modulated directly by baseband 0–15 kHz, whereas the difference signal DSBSC is modulated in the 23–53 kHz slot about a carrier at 38 kHz.

4.15.4 FM in TV Broadcasting

FM is used in TV broadcasting to modulate the audio signal. Picture uses AM. Sound signal is transmitted by frequency modulating carrier at $f_c = 4.5$ MHz. The audio signal bandwidth is limited to 10 kHz. Frequency deviation is limited to 25 kHz and the FM signal bandwidth is 70 kHz.

4.15.5 FM Receivers

Receivers used for PM and FM signals are very similar to those used for conventional AM or SSB reception. The only difference is that in FM receivers, the voltage at the output of the detector is directly proportional to the frequency deviation at its output, whereas in PM receivers, the voltage at the output of the detector is directly proportional to the phase deviation at its input. FM signals can be demodulated by PM receivers and vice versa. FM receivers like their AM counterpart are superheterodyne receivers.

Superheterodyne FM receiver

Figure 4.46 shows the block diagram of a superheterodyne FM radio receiver.

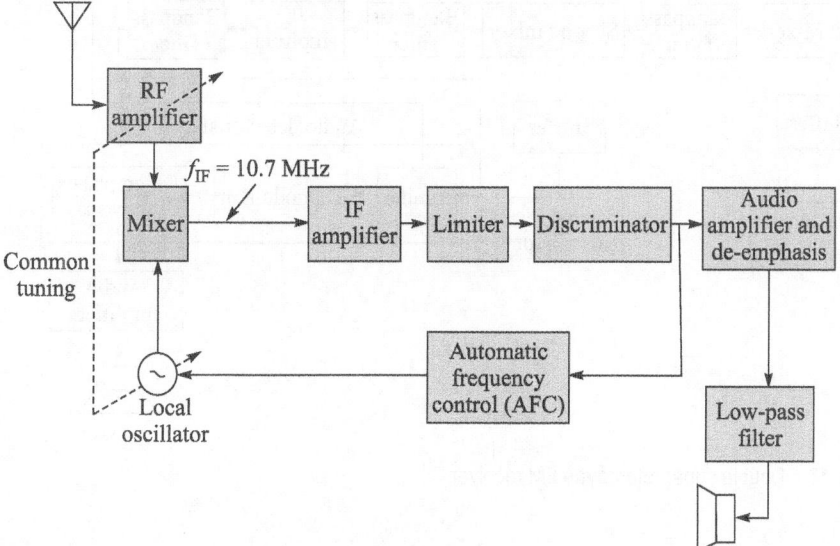

Fig. 4.46   Block diagram of a superheterodyne FM radio receiver

As in an AM radio reception, common tuning between the RF amplifier and the local oscillator allows the mixer to bring all FM signals to a common IF bandwidth of 200 kHz centred at $f_{1F} = 10.7$ MHz. The limiter removes any amplitude variation in the received signal as a result of additive noise and interference by bandlimiting the signal. A BPF with a bandwidth of 200 kHz centred at 10.7 MHz is included in the limiter to remove higher-order frequency components introduced by the limiter's non-linearity. A balanced frequency discriminator is used for frequency demodulation. The resulting message signal is then passed to the audio frequency amplifier, which includes a de-emphasis circuit. The output of the audio amplifier is further filtered by an LPF and given to the loudspeaker. AFC is used to improve the stability of the local oscillator.

Double superheterodyne FM receiver

Figure 4.47 shows a double superheterodyne FM receiver. The FM receiver is similar to the AM receivers. The pre-selector rejects the image frequency, the mixer converter section downconverts RF to IF, the RF amplifier takes care of a signal-to-noise ratio, the IF stage provides most of the gain and selectivity of the receiver, and the detector gets the information from the modulated wave. AGC prevents mixer saturation when strong RF signals are received.

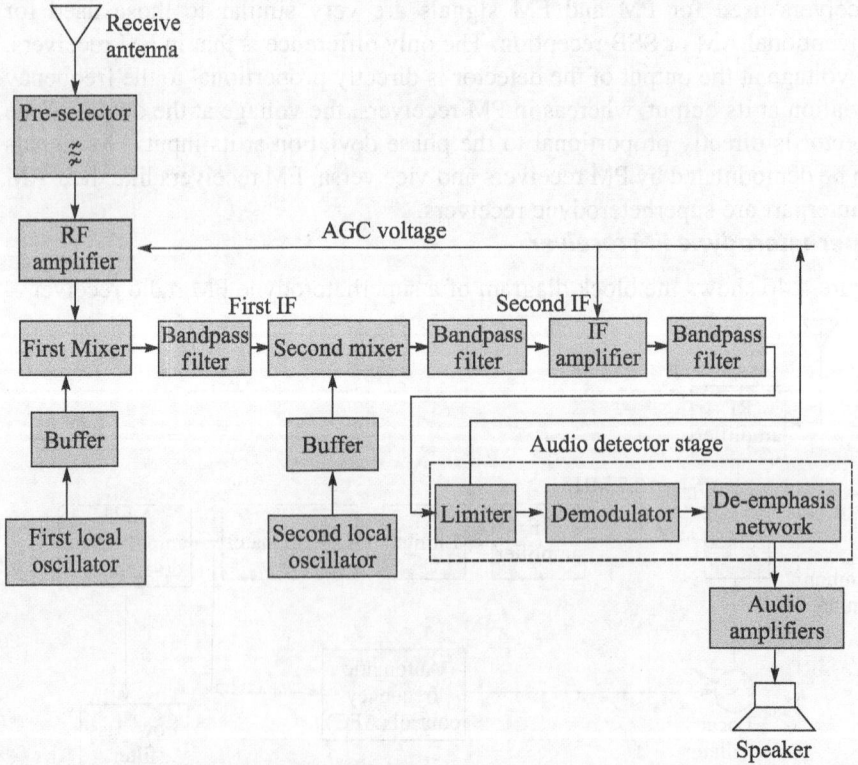

Fig. 4.47 Double superheterodyne FM receiver

For FM receivers, constant amplitude IF signal into the demodulator is desirable. Hence, FM receivers generally have much more IF gain than AM receivers and in most of the cases final IF amplifier is saturated. BPF rejects the harmonics caused by final IF amplifier saturation and passes only the minimum bandwidth necessary to preserve the information signal.

The envelope detector in AM receiver is replaced by a limiter, frequency discriminator, and de-emphasis network in FM receiver. Frequency discriminator extracts the information from the modulated wave. The limiter circuit and de-emphasis network contribute to an improvement in the signal-to-noise ratio that is achieved in the FM receiver.

For broadcasting, FM receiver's first IF is relatively high, often 10.7 MHz for good image frequency rejection, and second IF is relatively low (455 kHz), which allows IF amplifiers to have relatively high gain. With a first IF of 10.7 MHz, the image frequency for even the lowest frequency FM station (88.1 MHz) is 109.5 MHz, which is beyond the FM broadcast band.

The output from the de-emphasis network goes to the audio amplifiers, which amplify the message signal and finally given to the speaker.

FM stereo receiver

Figure 4.48 shows an FM stereo receiver. FM demodulator for FM stereo is basically the same as conventional FM demodulator down to the limiter/discriminator. The baseband signal thus obtained is separated into two signals $m_r(t) + m_l(t)$ and $m_l(t) - m_r(t)$ and passed through a de-emphasis filter. Synchronous demodulator using the pilot carrier generates the difference signal by taking the sum and difference of the two composite signals as shown in Fig. 4.48.

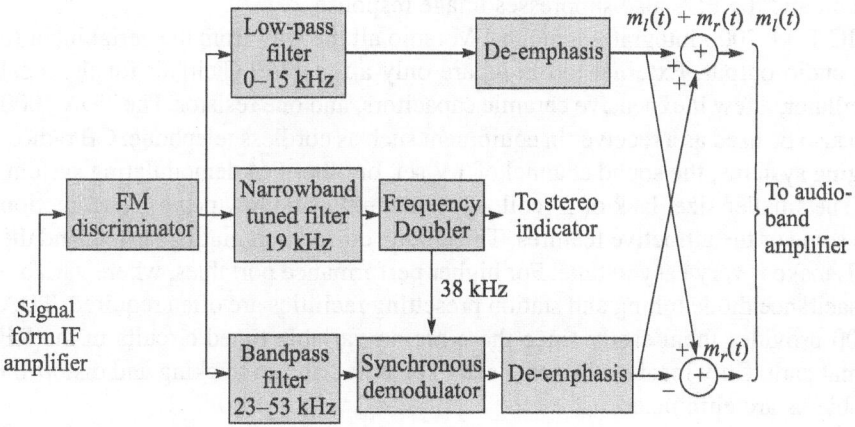

Fig. 4.48 FM stereo receiver

We recover the two signals $m_l(t)$ and $m_r(t)$. The audio signals are amplified and the two outputs drive their respective speakers. FM receivers not configured to receive FM stereo signal see only the baseband signal $(m_l + m_r)$ in the frequency range 0–15 kHz and produces a monophonic output signal that consists of the sum signals at the two microphones.

4.15.6 Single-Chip FM Radio Circuit

The IC TDA 7000 is a monolithic IC for mono-FM portable radios. The IC has a frequency locked loop system with an IF of 70 kHz. IF selectivity is obtained by active RC filters. The oscillator, which selects the reception frequency, requires external resonant circuit. Spurious reception is avoided by means of mute circuit, which also eliminates weak and noisy input signals.

The IC avoids the need for LC tuned circuit in the RF, IF, local oscillator, and demodulator stages by reducing the normally used IF of 10.7 MHz to a frequency that can be tuned by active RC filters, operational amplifiers, and resistors, which can be integrated. An IF of zero is deemed to be ideal because it eliminates spurious signals, such as repeat spots and image response, but it would not allow the IF signals to be limited prior to the demodulation resulting in poor signal-to-noise ratio and no AM suppression.

With an IF of 70 kHz, these problems are overcome and the image frequency occurs about halfway between the desired signal and the centre of the adjacent channel. However, the IF image signal must be suppressed and, in common with conventional FM radios, there is also a need to suppress inter-station noise and noise when tuned to weak signal. Spurious response above and below the centre frequency of the desired station and harmonic distortion in the event of very accurate tuning must also be eliminated.

The IC uses an active 70 kHz IF filter and a unique correlation filter circuit for suppressing spurious signals, such as side response caused by the flanks of demodulated S curve. With such a low IF, distortion would occur with the ± 75 kHz IF swing due to the received signal with maximum modulation. The maximum IF swing is, therefore, compressed to ± 15 kHz by controlling the local oscillator in a frequency locked loop (FLL). The combined action of the muting circuit and the FLL also suppresses image response.

IC TDA 7000 integrates a mono-FM radio all the way from the aerial input to the audio output. External to the IC are only a tunable LC circuit for the local oscillator, a few inexpensive ceramic capacitors, and one resistor. The TDA 7000 can also be used as a receiver in equipment such as cordless telephone, CB radios, paging systems, the sound channel of TV set, or other FM demodulating system.

The smaller size, lack of IF coil, easy assembly, and low power consumption are some of the attractive features. The unique correlation muting system and the FLL make it very easy to tune. For higher performance portables, where variable capacitance diode tuning and station presetting facilities are often required, TDA 7000 provides them easily since there are no variable tuned circuits in the RF signal path. Only local oscillators need to be tuned and so tracking and distortion problems are eliminated.

Figure 4.49 shows a narrowband FM receiver with a crystal-controlled local oscillator. Hence, there is hardly a compression of the IF swing by the FLL. The deviation of the transmitter carrier frequency due to the modulation must, therefore, be limited to prevent severe distortion of the demodulated audio signal. The component value chosen in Fig. 4.45 results in an IF of 4.5 kHz and an IF bandwidth of 5 kHz. If the IF is multiplied by N, the values of capacitors C_{17} and C_{18} in all pass filters and the values of filter capacitors C_7, C_8, C_{10}, C_{11}, and C_{12}

must be multiplied by $1/N$. Audio output is the detector output that can be amplified by an external audio amplifier.

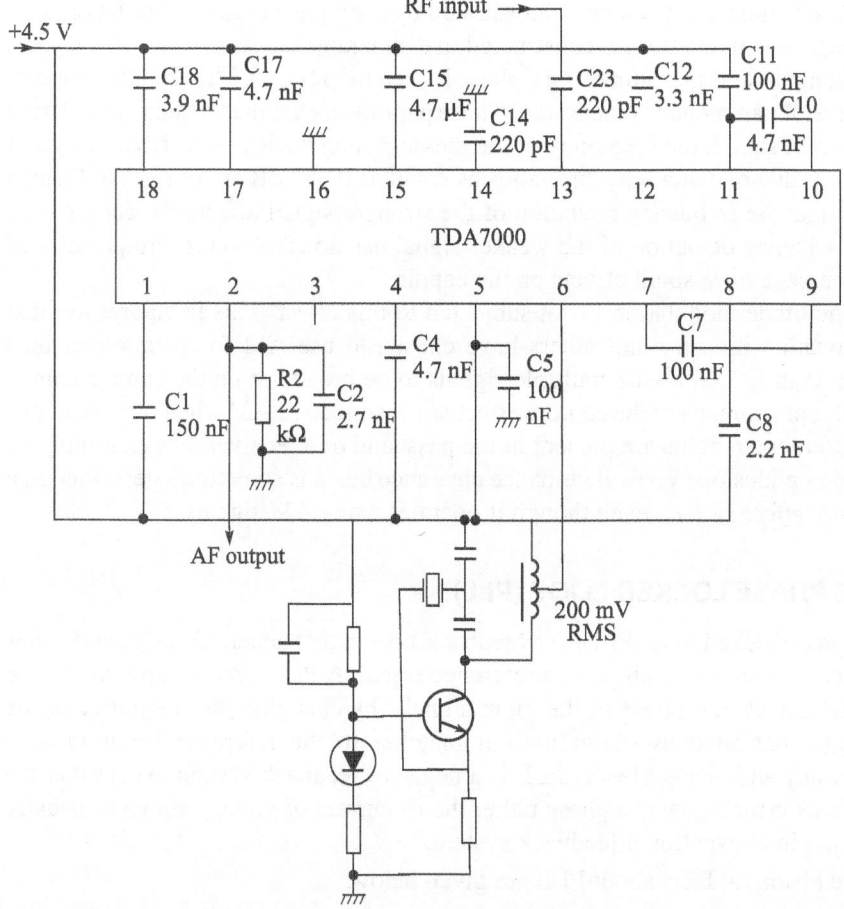

Fig. 4.49 Narrowband FM receiver with crystal-controlled local oscillator

More information and details on this IC can be obtained from *Signetics Linear Data Manual*, Volume 1, Communications.

4.15.7 Capture Effect

It is a phenomenon associated with frequency modulation (FM). The capture effect relates the ability of the receiver demodulator to recover the message of the dominant carrier when two or more FM carriers of unequal power levels are present. The FM receiver tuner is able to clearly receive the stronger of the two stations being broadcast on the same frequency. Hence, one can define the capture effect as the complete suppression of the weaker signal at the receiver limiter (if it has one) where the weaker signal is not amplified but is attenuated. But when both signals are nearly equal in strength or are fading independently, the receiver may switch from one to the other and exhibit picket fencing. However, frequency drift or lack of selectivity may cause one station or signal to be suddenly overtaken by another on an adjacent channel frequency drift typically constituted a problem on very old or inexpensive receivers, while inadequate selectivity may plague any tuner.

This phenomenon can be demonstrated by a simple experiment in laboratory. An experimental system is constructed that generates the sum of two FM signals and demodulates that sum using a phase locked loop (PLL). The effect on capture by several parameters is measured. These parameters are the frequency deviation of the FM signal, the frequency of the message, and the low-pass filter design of the PLL demodulator. Capture ratios as small as 0.387 dB are observed. Results show that the frequency deviation of the stronger signal affects the capture and the frequency deviation of the weaker signal has no effect on it. Frequencies of the message have small effects on the capture.

Amplitude modulation is not subjected to this effect. This is one reason that the aviation industry and others have chosen to use AM for communications rather than FM, allowing multiple signals to be broadcast on the same channel. Similar phenomena to the capture effect are described in AM when offset carriers of different strengths are present in the passband of a receiver. For example, the aviation glideslope vertical guidance clearance beam is sometimes described as a *capture effect system*, even though it operates using AM signals.

4.16 PHASE LOCKED LOOP (PLL)

The phase locked loop (PLL) is a feedback system that generates a signal that has a fixed relation to the phase of a reference signal. A PLL circuit responds to both the frequency and phase of the input signals, by changing the frequency of the voltage-controlled oscillator until it matches to the reference input in both frequency and phase. Hence, PLL is a negative feedback system except that the feedback error signal is a phase rather than a current or voltage signal as usually the case in conventional feedback system.

Some historical facts about PLL are given below.

- It coincides with coherent communication invented in the year 1932.
- Earliest widespread use was in television for synchronizing horizontal and vertical sweeps.
- The first PLL IC was manufactured in the year 1965.
- Signetics Corporation introduced a monolithic IC in 1969.
- RCA introduced few years later CD4046 CMOS PLL.
- The first digital PLL appeared around 1970.

PLLs today have the following features.

- TV, radio, computer, telephone, pager, mobile phone, and digital camera use PLL.
- Optical PLLs are used in clock recovery for very high speed data of the order of 160 Gbps and more.
- At the lower end, PLLs implement entire PLL functionality on sampled data.

4.16.1 PLL Basics

As discussed earlier, the PLL is basically a negative feedback system comprising a bandpass filter to bandlimit the input signal, a phase comparator, a high frequency low-pass filter, a loop filter, which amplifies the error signal, and a voltage-controlled oscillator (VCO) in the feedback path. The block diagram of this basic PLL is shown in Fig. 4.50. The phase detector is a non-linear device whose output contains the phase difference between the two inputs, the reference signal, and the VCO output. The error voltage from the phase detector is then filtered by an LPF, also called a loop filter. The filtered error voltage is given as the correction voltage to the voltage-controlled oscillator. The error voltage forces the VCO frequency to vary in a direction that reduces the frequency difference between the VCO frequency and the reference signal.

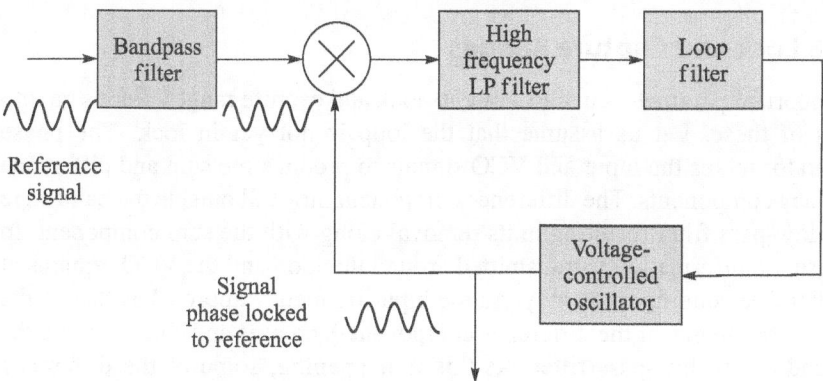

Fig. 4.50 Block diagram of a phase locked loop

4.16.2 PLL Operation

With no reference signal applied to the system, the VCO control voltage is equal to zero. The VCO operates at the set frequency ω_o, called its free-running frequency. When a reference signal is applied to the system, the phase comparator compares the phase and the frequency of the reference signal with that of the VCO and generates an error voltage that is proportional to the difference between the two signals.

The error voltage is then filtered and applied to the control terminal of the VCO. This error voltage forces the VCO frequency to vary in a direction that reduces the frequency difference between the VCO frequency and the reference signal. If the input frequency ω_i is sufficiently close to ω_o, the feedback nature of the PLL causes the VCO to synchronize or lock with the incoming signal. Once in lock, the VCO frequency is identically equal to the reference signal except for a finite phase difference. This net phase difference of θ_d, where $\theta_d = \theta_i - \theta_o$, is necessary to generate the required corrective error voltage to shift the VCO frequency from its free-running value to the input reference signal frequency ω_i and thus keep the PLL in lock. Thus, the self-correcting ability of the system allows PLL to track the frequency changes of the reference signal once it is locked. The range of frequencies over which the PLL can maintain lock with an input reference signal is called *lock range*. The band of frequencies over which the PLL can acquire lock with an incoming reference signal is called its *capture range*. However, it must be noted that the capture range can never be greater than the lock range.

We can observe that the phase comparator is actually a multiplier circuit that mixes the input reference signal with the VCO signal . This results in generating sum and difference frequency components $\omega_i \pm \omega_o$. When the loop is in lock, the VCO duplicates the input reference frequency so that the difference between ω_i and ω_0 is zero. Hence, the output of the phase comparator contains only a DC component. The low pass loop filter removes the sum component $\omega_i + \omega_0$, but passes the DC component which is fed to the VCO. Hence, when the loop is in lock, the difference frequency component is always DC and the lock range is independent of the band edge of the lowpass filter.

4.16.3 Lock and Capture Ranges

The important parameters in a PLL are its lock and capture ranges. Let us go into details of these. Let us assume that the loop is not yet in lock. The phase comparator mixes the input and VCO signals to produce the sum and difference frequency components. The difference component may fall outside the band edge of the low-pass filter resulting in its removal along with the sum component. In this case, no information is transmitted around the loop and the VCO remains at its initial free-running frequency. As the input frequency approaches that of the VCO, the frequency of the difference component decreases and comes within the passband of the low-pass filter. As this is happening, some of the difference component is passed, which tends to drive the VCO towards the frequency of the input signal. This in turn decreases the frequency of the difference component and allows more information to be transmitted through the low-pass filter to the VCO. This is essentially a positive feedback mechanism, which causes the VCO to snap into lock with the input signal. With this mechanism in mind, once again the *capture range* can be defined as the frequency range centred about the VCO initial free-running frequency over which the loop can acquire lock with the input signal. The capture range is a measure of how close the input signal must be in frequency to that of the VCO to acquire lock. The capture range can assume any value within the lock range and depends primarily on the low-pass filter characteristic and closed loop gain of the system. It is this signal capturing phenomenon that gives the loop its frequency selective properties. It is important to distinguish the *capture range* from the *lock range*, which can again be defined as the frequency range usually centred about the VCO initial free that running frequency over which the loop can track the input signal once the lock has been achieved.

When the loop is in lock, the difference frequency component at the output of the phase comparator is DC and will always be passed by the low-pass filter. Hence, the lock range is limited by the range of the error voltage that can be generated and the corresponding VCO frequency deviation produced. The lock range is essentially a DC parameter and is not affected by the characteristic of the low-pass filter.

4.16.4 Mathematical Analysis of PLL

Let us consider Fig. 4.51, which once again shows the block diagram of the PLL. Let the reference input be $A\sin(\omega_i t + \theta_i)$ and the VCO output be $\cos(\omega_o t + \theta_o)$,

i.e.,
$$v_i(t) = A\sin(\omega_i t + \theta_i) \tag{4.110}$$

$$v_o(t) = \cos(\omega_o t + \theta_o) \tag{4.111}$$

With these two inputs, the phase detector output is

$$v_d = AK_m \sin(\omega_i t + \theta_i)\cos(\omega_o t + \theta_o) \tag{4.112}$$

where K_m is the gain of the phase comparator.

Using the trigonometric identities,

$$2\sin(\omega_i t + \theta_i)\cos(\omega_o t + \theta_o) = \sin((\omega_i + \omega_o)t + \theta_i + \theta_o) + \sin((\omega_i - \omega_o)t + \theta_i - \theta_o)$$

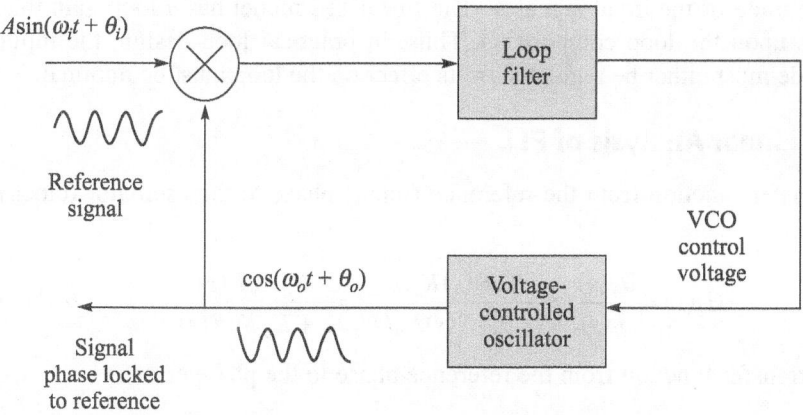

Fig. 4.51 Block diagram of PLL

Let $\theta_d = \theta_i - \theta_o$. With two fundamental assumptions, common PLL analog model can be arrived at.

(a) The first term, i.e., the sum term, is attenuated by the low-pass filter and by the low-pass nature of the PLL itself.

(b) Let $\omega_i \approx \omega_o$ so that the difference can be incorporated into θ_d. This means that the VCO can be assumed to be an integrator.

The baseband phase comparator output is then

$$v_d(t) = \frac{AK_m}{2}\sin\{(\theta_i(t) - \theta_o(t))\} = \frac{AK_m}{2}\sin\{\theta_d(t)\} \tag{4.113}$$

This is still a non-linear system and the analysis to make it a linear system is to assume for very small θ_d, $\sin\theta_d \approx \theta_d$ and $\cos\theta_d \approx 1$.

These assumptions help us to analyse the PLL as a linear system. The linear model is shown in Fig. 4.52. The linear model is useful for studying loops that are in lock, i.e., when θ_d is quite small.

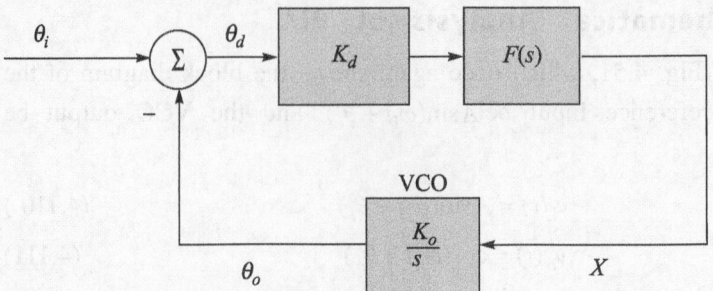

Fig. 4.52 Linear model for analog PLL

The linear model omits the input bandpass filter, which is generally used to bandlimit the input. It is also assumed that there is some idea of the frequency range of the signal and the loop filter is optimized for the stability and performance of the baseband. Amplitude of the phase error is dependent upon the amplitude of the input signal A. The linearized model has a loop gain that depends upon the loop components. Thus, in practical loop design, the input amplitude must either be regulated or its effect on the loop must be minimal.

4.16.5 Linear Analysis of PLL

The transfer function from the reference (input) phase to the oscillator (clock) phase is

$$H_i(s) = \frac{\theta_o(s)}{\theta_i(s)} = \frac{K_d F(s) K_v / s}{1 + K_d F(s) K_v / s} = \frac{K_d K_v F(s)}{s + K_d K_v F(s)} \qquad (4.114)$$

The transfer function from the reference phase to the phase error is

$$H_\phi(s) = \frac{\theta_d(s)}{\theta_i(s)} = \frac{1}{1 + K_d F(s) K_v / s} = \frac{s}{s + K_d K_v F(s)} \qquad (4.115)$$

The steady state error is given by

$$\lim_{t \to \infty} \theta_d(t) = \lim_{s \to 0} s\theta_i(s) H_\phi(s)$$

Now we can define the capture and lock ranges as well as pull-in and pull-out ranges.

Capture or hold range It is the frequency range over which the PLL is able to statically maintain phase tracking. The range is

$$\Delta\omega_c = K_o K_d F(0)$$

Lock range It is the frequency range within which the PLL locks within one single cycle between the reference frequency and the output frequency. The range is

$$\Delta\omega_L \approx \pm K_o K_d F(\infty)$$

Pull-in and pull-out ranges   The pull-in range, $\Delta\omega_{PI}$, is defined as the frequency range in which the PLL will always become locked. The pull-out range, $\Delta\omega_{PO}$, is defined as the limit of dynamic stability for the PLL.

The assumption that a VCO is an integrator makes every PLL at least type 1, i.e., zero steady state error to a phase step. But most PLL designs are second order, with integrator and minimum phase zero in the filter. This is a type 2 system, i.e., zero steady state error to a phase ramp. Third-order and higher PLL designs are not common. Few applications need zero steady state error for an accelerating phase as in the case of deep space communications with Doppler shift. Stability bounds for non-linear third-order systems are not the same as stability bounds for third-order linear systems.

4.16.6 Standard Non-linear Model

Figure 4.53 shows the non-linear model of PLL. The analysis of this model is quite complex and beyond the scope of this book. However, the equations are as given below:

$$\frac{d^2\phi}{dt^2} + K_d\frac{d\phi}{dt} - \cos\phi = -v \tag{4.116}$$

This is a second-order differential equation and limited to first-and second-order systems. It can be seen from the block diagram that the difference between non-linear and the linear models is that in a non-linear model, the transfer characteristic of the high frequency low-pass filter is taken as $K_d \sin(\ )$ instead of K_d as in the case of a linear model. The VCO like in the linear model is assumed as an integrator.

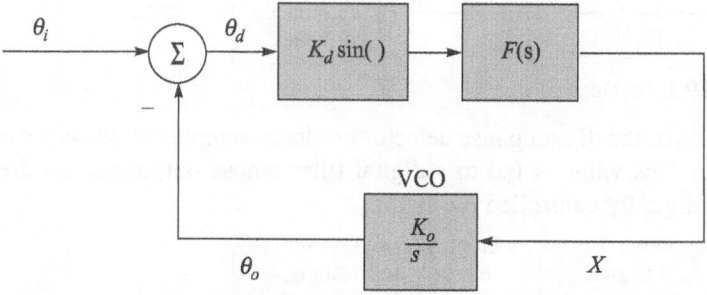

Fig. 4.53   Non-linear model

4.16.7 Digital PLL

As shown in Fig. 4.54, in a digital PLL, the analog input to the conventional PLL is replaced with digital logic levels and a digital phase detector is used. The phase detector is digital working with binary values. Still the filter and the VCO are analog, i.e., the signal space is digital but the signal phase space is analog.

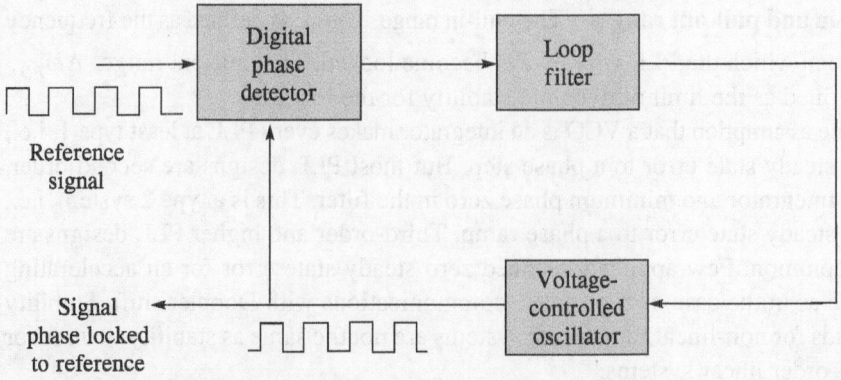

Fig. 4.54 Classical digital PLL

All digital PLLs

Figures 4.55 and 4.56 show two types of all digital PLLs, where all the building blocks of PLL are digital. In the first type, the digital phase detector produces pulses that go into the count up or count down inputs of the counter, which acts as a loop filter. The counter then adjusts the frequency of the digitally controlled oscillator.

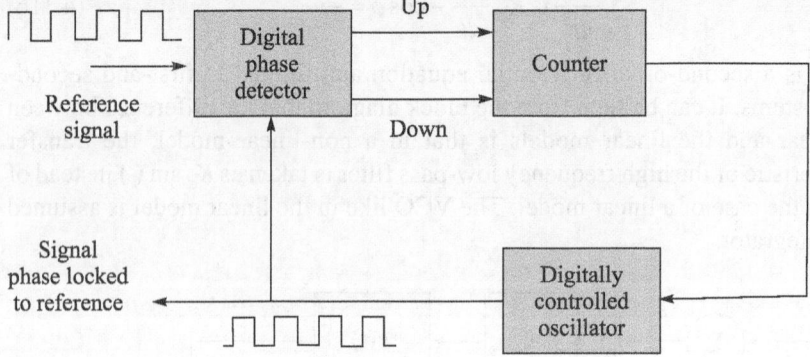

Fig. 4.55 All digital PLL: counter type

In the second type, the digital phase detector produces samples of phase error as an *n*-bit value. This value is fed to a digital filter whose output adjusts the frequency of the digitally controlled oscillator.

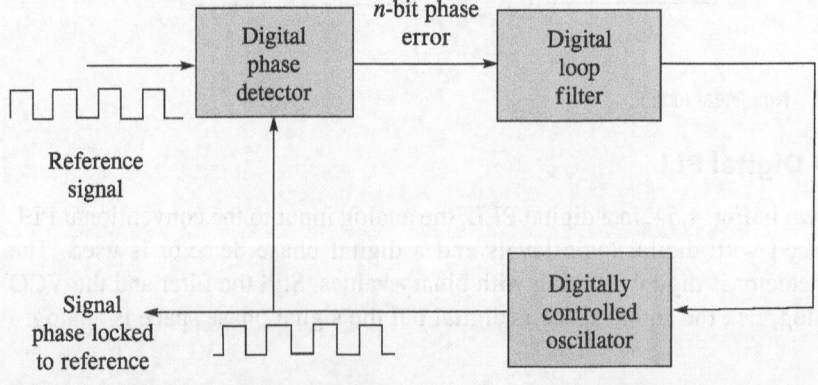

Fig. 4.56 All digital PLL using phase errors

4.16.8 Software PLLs

An example of a software PLL that does clock data recovery is shown in Fig. 4.57. This heuristic loop uses a zero crossing detector on the sampled input. The effective sample rate is derived from the average bit zero crossing rate.

When data can be sampled at a rate substantially faster than the loop centre frequency the entire loop operation can be implemented in software. This has an advantage of flexibility. If the sample rate is high enough, any type of PLL can be implemented in software.

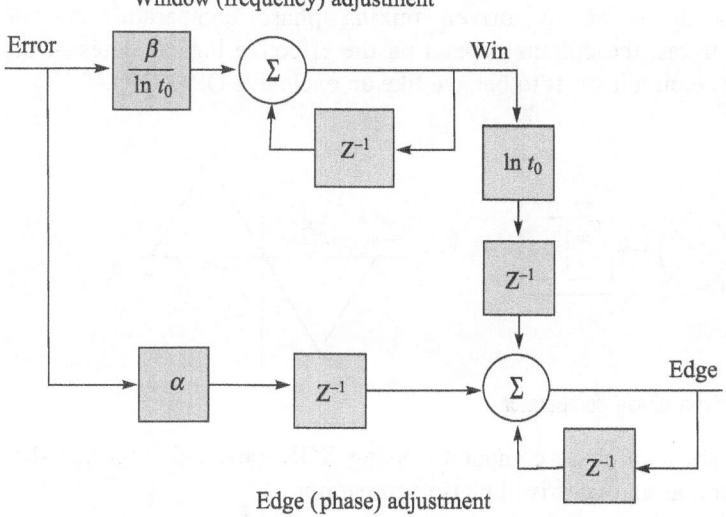

Fig. 4.57   Software PLL

Software loops deal with real data and hence they operate in real time. However, they can also be used in post processing of measured data. Certain operations which are highly effective in hardware, for example limiters which have lot of high frequency component create sampling problems for software loop.

The building blocks of a PLL are
(a) Phase comparator
(b) Loop filters
(c) Voltage-controlled oscillator
These are also sometimes called loop components.

4.16.9 Phase Comparator

The ability of a PLL to track a signal is largely dependent on the phase comparator. Hence, this is usually the starting point of any PLL design. A few types of phase comparators are discussed here.

Phase comparators (memoryless)

Figure 4.58 shows a memoryless phase comparator.

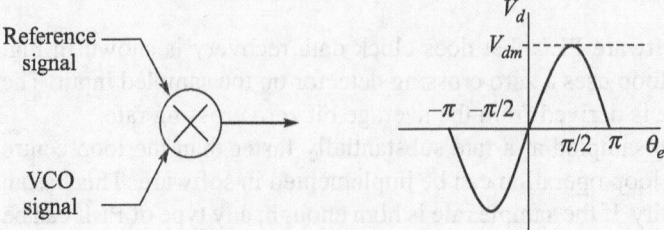

Fig. 4.58 Classical mixing phase comparator

Figure 4.59 shows an overdriven mixing phase comparator. As the comparator saturates, the outputs depend on the effective logical states of the input signals. The circuit starts to behave like an exclusive OR(XOR).

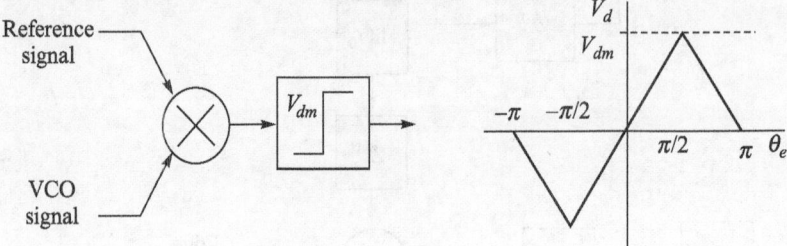

Fig. 4.59 Overdriven phase comparator

Figure 4.60 shows a phase comparator using XOR gate. This accomplishes the same function as an overdriven phase comparator.

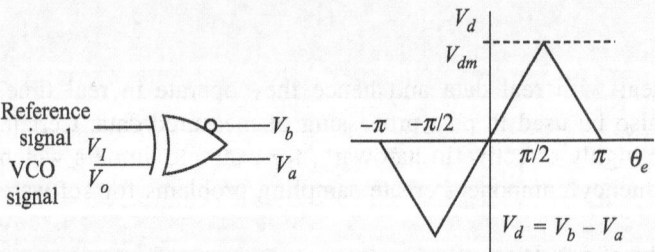

Fig. 4.60 XOR phase detector

Phase detectors

Let us have a look at the basic phase detectors. There are actually two basic types, sometimes referred to as type I and type II. The type I phase detector is designed to be driven by analog signals or digital square-wave signals, whereas the type II phase detector is driven by digital transitions (edges). They are typified by the most common used 565 (linear type I) and the CMOS 4046, which contains both type I and type II. The simplest phase detector is the type I (digital), which

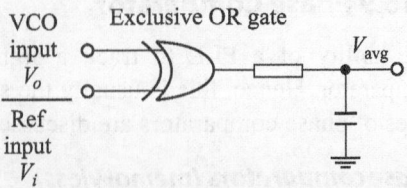

Fig. 4.61 XOR phase comparator

is simply an exclusive-OR gate (Fig. 4.61). With low-pass filtering, the graph of the output voltage versus phase difference is as shown, for input square waves of 50% duty cycle. The type I (linear) phase detector has similar output voltage versus phase characteristics, although its internal circuitry is actually a four-quadrant multiplier, also known as a balanced mixer. Highly linear phase detectors of this type are essential for lock-in detection, which is a fine technique.

More recently, switching speed of PLLs has become a critical parameter in today's design of synthesizers, and especially for our modern networks such as WCDMA, 3G, WLANs, and Bluetooth technology. The switching speed is emerging as a challenging requirement for single-loop, single-chip PLL designs. Speed is mainly a function of loop bandwidth, but in many cases the loop bandwidth cannot be made large because of phase noise considerations. Speed-up techniques have been devised to improve PLL transient time, but most of them have limited efficiency. In addition, speed-up techniques will have to be improved. For the WCDMA and 3G markets, a reasonable value is 100 to 150 ms for a $\Delta f = 60$ MHz excursion and a convergence to $\Delta f = 250$ Hz. To meet this stringent requirement, highly complex sigma–delta fractional PLL architectures are used, which allow a high reference frequency and wide loop bandwidth, while maintaining resolution and a good phase noise profile. This technique is already being implemented nowadays.

The type II phase comparator is sensitive only to the relative timing of edges between the signal and the VCO input. The phase comparator circuit generates either lead or lag output pulses, depending on whether the VCO output transitions occur before or after the transitions of the reference signal, respectively. The width of these pulses is equal to the time between the respective edges.

The output circuitry then either sinks or sources current, respectively, during those pulses and is otherwise open circuited, generating an average output voltage versus phase difference like that in Fig. 4.62. This is completely independent of the duty-cycle of the input signals, unlike as in the case type I phase comparator. Another nice feature of this phase detector is the fact that the output pulses disappear entirely when the two signals are in lock. This means that there is no *ripple* present at the output to generate periodic phase modulation in the loop, as there is with the type I phase comparator. Also, there is an additional difference between the two kinds phase detectors. The type I comparator is always generating an output wave, which must then be filtered by the loop filter. Thus, in a PLL with type I phase comparator, the loop filter acts as a low-pass filter, smoothing this full-swing logic-output signal.

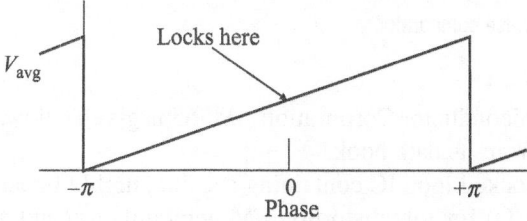

Fig. 4.62   Plot of output voltage vs phase difference

There will always be residual ripple and consequent periodic phase variations in such a loop. In circuits where phase locked loops are used for frequency multiplication or synthesis, this adds phase modulation sidebands to the output signal. By contrast, the type II phase comparator generates output pulses only when there is a phase error between the reference and the VCO signal. Since the phase comparator output otherwise looks like an open circuit, the loop filter capacitor then acts as a voltage storage device, holding the voltage that gives right VCO frequency. If the reference signal moves away in frequency, the phase detector generates a train of short pulses, charging or discharging the capacitor to the new voltage needed to put the VCO back into lock. The second-order PLL serves as the basis for all PLL synthesizer designs and technology. Most PLL designs, especially for synthesizers where third- and fourth-order loops are common, use a different terminology and deal mainly with the open loop gain and phase.

XOR phase detector analysis

Figure 4.63 shows the timing diagram of an XOR phase comparator. The response of this comparator can be split into a 2X term and a residual term. The average of the residual term is the baseband we are looking at. The 2X term gets eliminated by integrating the output over n clock periods. A phase shift of p/2 produces a residual of zero. A phase shift of p/4 produces a non-zero residual. The analog voltages may contain many harmonics. Sufficient care has to be taken to ensure that these harmonics do not modulate the VCO.

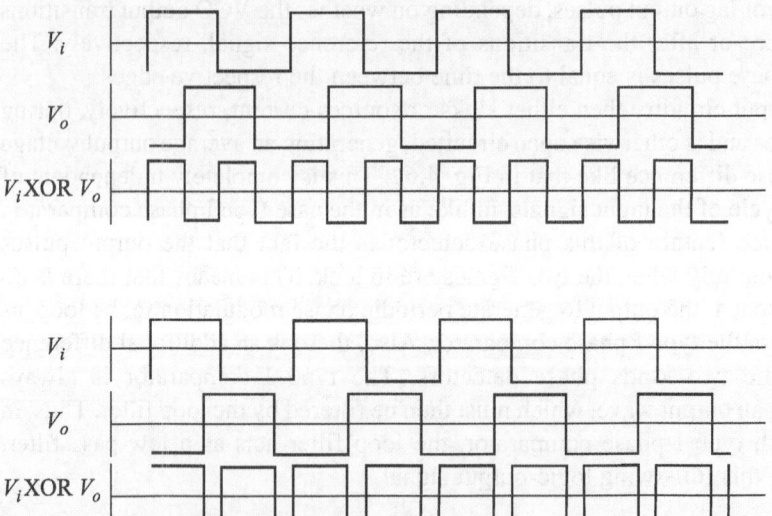

Fig. 4.63 Timing diagram of an XOR phase comparator

Phase comparator ICs

A brief write up on National Semiconductor Corporation LM565 is given below. Further details can be obtained from its data book.

It is a general purpose phase locked loop IC containing a stable, highly linear voltage-controlled oscillator (VCO) for low distortion FM demodulation and a double balanced phase detector with good carrier suppression. The VCO frequency is set with an external resistor and capacitor and a tuning range of

10:1 can be obtained with the same capacitor. The characteristics of the closed loop system—bandwidth, response speed, capture, and pull-in range—may be adjusted over a wide range with an external resistor and capacitor. The loop may be broken between the VCO and the phase detector for insertion of a digital frequency divider to obtain frequency multiplication. The pin diagram and the functional diagram are shown in Fig. 4.64.

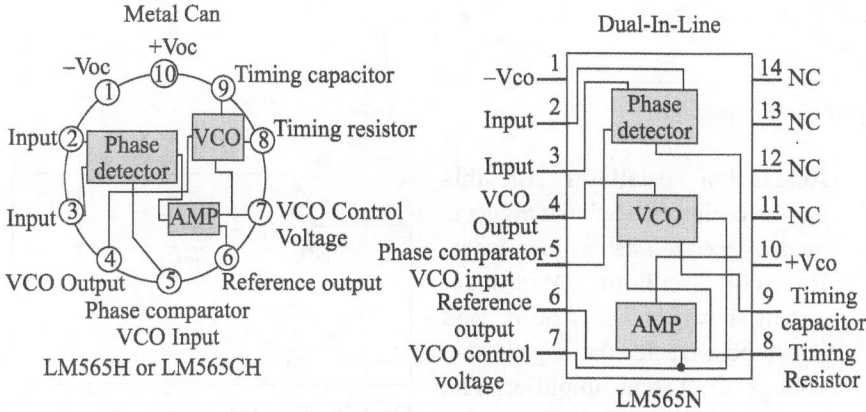

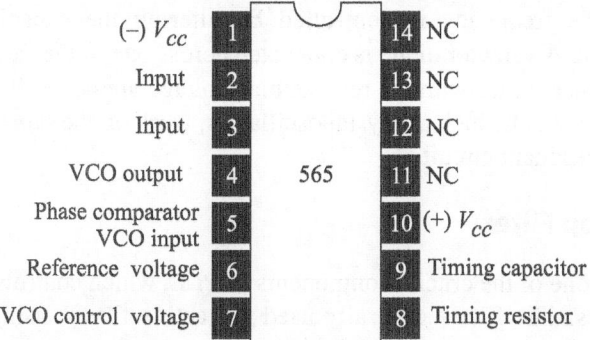

Fig. 4.64 Pin diagram: DIP package

General features

- 200 ppm per °C frequency stability (drifting) of the VCO
- Power supply range of 5 to 12 volts with 100 ppm per % typical 0.2% linearity of demodulated output
- Frequency range of 0.001 Hz to 500 kHz
- Highly linear triangle wave output
- Linear triangular wave with in phase zero-crossings available
- TTL and DTL compatible phase detector input and square wave output
- Adjustable hold in range from 1% to more than 60%

Some applications

Data and tape synchronization, modems, FSK modulation, FM demodulation, frequency synthesizer, tone decoding, frequency multiplication and division, SCA demodulators ('hidden' radio), telemetry receivers, signal regeneration, coherent demodulators, satellite, and robotics and radio control.

4.16.10 Voltage-Controlled Oscillators (VCOs)

There are different types of VCOs. A few of them are discussed here.

Ring oscillator It is generally used in monolithic circuits and uses an odd number of inverters connected in a feedback loop as shown in Fig. 4.65.

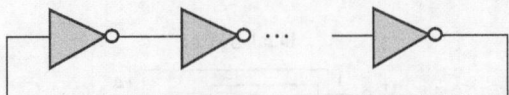

Fig. 4.65 Ring oscillator

Relaxation oscillator In this type of oscillator, a Schmitt trigger is used to generate a stable square wave.
Resonant oscillator A resonant circuit shown in Fig. 4.66 is used in the positive feedback path of a voltage to current amplifier. The amplifier has a gain close to unity.

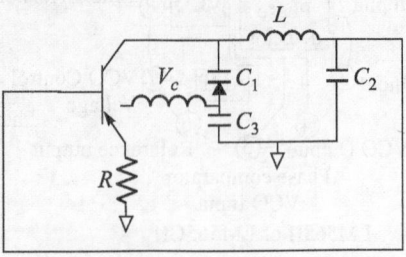

Fig. 4.66 Resonant oscillator

The resonant circuit in the positive feedback path has poles close to the $j\omega$ axis. The frequency is controlled by altering the capacitance of the resonator. A varactor diode is connected across one of the capacitor, whose capacitance varies with the reverse bias voltage applied to it.

Other forms of VCOs, such as crystal oscillators, work on the same principle but modifies the resonant circuit.

4.16.11 Loop Filter

Loop filter is one of the crucial components of PLL, which controls the lock and capture ranges. The filters generally used are active filters, using operational amplifiers for better impedance matching and gain unlike passive filters. Figures 4.67 and 4.68 show the single-ended and differential filters.

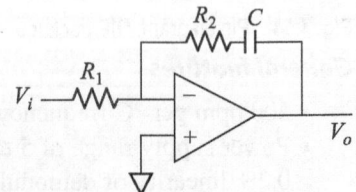

Single-ended input The filters are first order. The filter transfer function is

Fig. 4.67 Single ended filter

$$\frac{V_o}{V_i} = -\frac{sR_2C + 1}{sR_1C}$$

Differential input These filters can be converted as a second-order filter by making C_3 zero. This circuit can be made single ended by connecting the non-inverting terminal to ground. The transfer function of this filter is

$$\frac{V_o}{V_a - V_b} = -\frac{sR_2(C_2 + C_3) + 1}{sR_1C_2(sR_2C_3 + 1)} \tag{4.117}$$

4.16.12 Applications of PLL

PLL finds applications in many fields. Two of the applications are discussed here.

Carrier recovery

PLLs are used to recover the carrier signal in the presence of noise. The free-running frequency of the VCO is chosen as the carrier frequency that has to be recovered. When the carrier has a strong component, the PLL can lock. Figure 4.69 shows the block diagram.

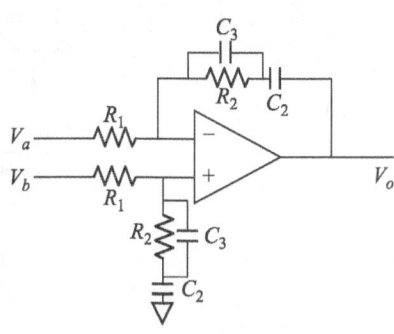

Fig. 4.68 Differential filter

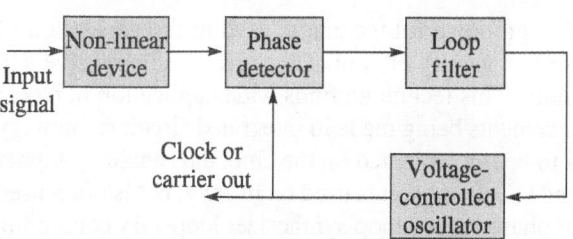

Fig. 4.69 Clock recovery circuit for a modulated wave

For a BPSK modulated wave, the carrier is recovered from a squaring loop as shown in Fig. 4.70. The BPSK signal can be represented as $f(t) = m(t)\sin\omega_i t$ and $m(t) = \pm 1$. The spectrum of $f(t)$ has no component at ω_i for equally probable +1 and −1 bits and the squaring loop locks to $2\omega_i$ and divide by 2 circuit recover ω_i.

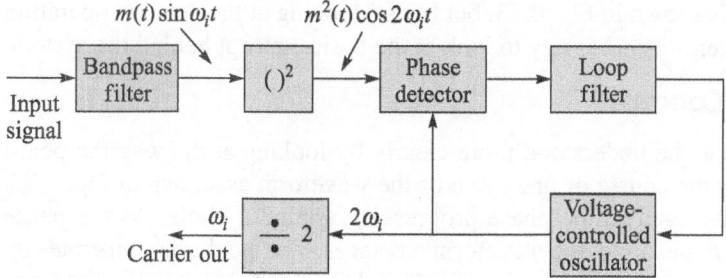

Fig. 4.70 Carrier recovery circuit for BPSK signal

Frequency synthesizer

PLL can be used as a frequency synthesizer to generate various frequencies as low as 0.1 Hz to as high as a few MHz. Figure 4.71 shows the block diagram of a typical synthesizer. The output of the VCO is given to the phase comparator through a divide by N counter. The phase comparator locks a clock with an input signal of different frequency. Harmonics locking loop generates a clock at N times input frequency. For example, if the input reference frequency is 1 MHz, then we can generate frequency as low as 0.1 Hz to as high as 1 MHz, by having a divide by N counter. The N value ranges from 10^7 to 1, i.e., $f' = Nf_i$.

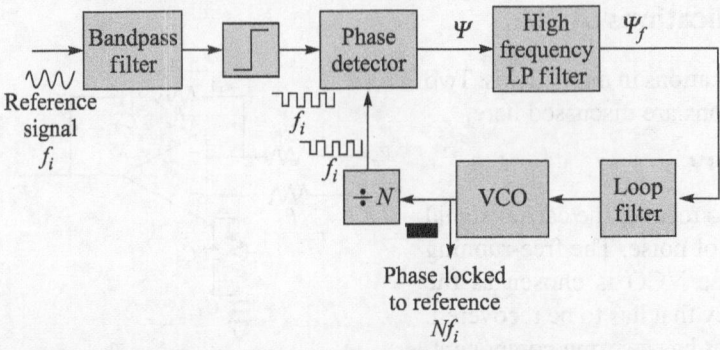

Fig. 4.71 Block diagram of a frequency synthesizer using PLL

4.17 DIRECT DIGITAL SYNTHESIS (DDS)

Direct digital synthesis (DDS) is a powerful technique used in the generation of radio frequency signals for use in a variety of applications from radio receivers to signal generators and many more. This technique finds wide application in recent years with the advent of advancements being made in integrated circuit technology that allow much faster speeds to be implemented on the chip, thus enabling higher frequency DDS chips to be made. Generally, it is used on its own, but is often used in conjunction with indirect or phase locked loop synthesizer loops. By combining both technologies, it is possible to take advantage of the best aspects of each.

As the name suggests, this form of synthesis generates the waveform directly using digital techniques. This is different from the more familiar indirect synthesizers that use a phase locked loop as the basis of their operation. A direct digital synthesizer operates by storing the points of a waveform in digital format and then recalling them to generate the waveform. The rate at which the synthesizer completes one waveform then determines the frequency. The overall block diagram is shown in Fig. 4.73, but before looking at the detailed operation of the synthesizer, it is necessary to look at the basic concept behind the system.

4.17.1 Basic Concept

The operation can be understood more clearly by looking at the way the phase progresses over the course of one cycle of the waveform as shown in Fig. 4.72. This can be envisaged as the phase progressing around a circle. As the phase advances around the circle, the waveform advances. The synthesizer operates by storing various points in the waveform in digital form and then recalling them to

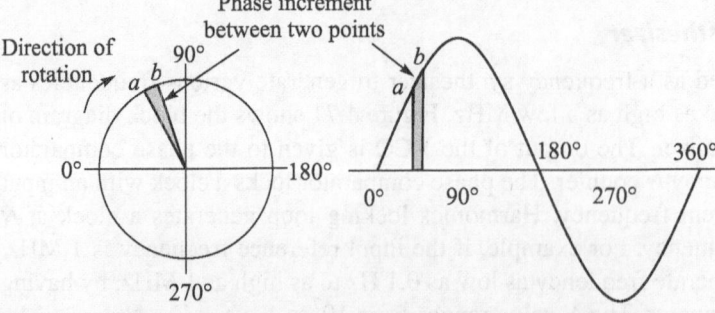

Fig. 4.72 Operation of phase accumulator in a direct digital synthesizer

to generate the waveform. As the phase advances around the circle, which corresponds to the advances in the waveform, i.e., the greater the number corresponding to the phase, the greater the point is along the waveform. By successively advancing the number corresponding to the phase, it is possible to move further along the waveform cycle.

Figure 4.73 shows a block diagram of a basic direct digital synthesizer. The digital number representing the phase is held in the phase accumulator. The number held here corresponds to the phase and gets increased at regular intervals. In this way, it can be seen that the phase accumulator is basically a form of counter. When it is clocked, it adds a preset number to the one already held. When it fills up, it resets and starts counting from zero again. In other words, this corresponds to reaching one complete circle on the phase diagram and restarting again. Once the phase has been determined, it is necessary to convert this into a digital representation of the waveform. This is accomplished using a waveform map. This is a memory that stores a number corresponding to the voltage required for each value of phase on the waveform. In the case of a synthesizer of this nature, it is a sine look up table as a sine wave is required. In most cases, the memory is either a read only memory (ROM) or programmable read only memory (PROM). This contains a vast number of points on the waveform, many more than are accessed in each cycle. A large number of points are required so that the phase accumulator can increment by a certain number of points to set the required frequency.

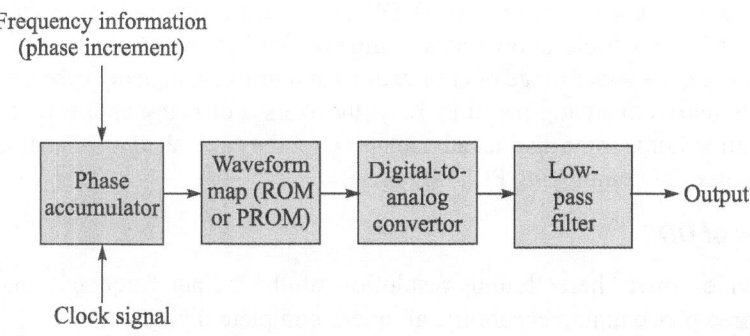

Fig. 4.73   Basic direct digital synthesizer

The next stage in the process is to convert the digital numbers coming from the sine look up table into an analog voltage. This is achieved using a digital-to-analog converter (DAC). This signal is filtered to remove any unwanted signals and amplified to give the required level as necessary.

Tuning is accomplished by increasing or decreasing the size of the step or phase increment between different sample points. A larger increment at each update to the phase accumulator will mean that the phase reaches the full cycle value faster and the frequency is correspondingly high. Smaller increments to the phase accumulator value mean that it takes longer to reach the full cycle value and a correspondingly low value of frequency. In this way, it is possible to control the frequency. It can also be seen that frequency changes can be made instantly by simply changing the increment value. There is no need for a settling time as in the case of phase locked loop based synthesizer.

It can be seen that there is a finite difference between one frequency and the next and that the minimum frequency difference or frequency resolution is determined by the total number of points available in the phase accumulator. A 24 bit phase accumulator provides just over 16 million points and gives a frequency resolution of about 0.25 Hz when used with a 5 MHz clock. This is more than adequate for most purposes.

These synthesizers do have some disadvantages. There are a number of spurious signals generated by a direct digital synthesizer. The most important of these is one called an alias signal. Here images of the signal are generated on either side of the clock frequency and its multiples. For example, if the required signal had a frequency of 3 MHz and the clock was at 10 MHz, then alias signals would appear at 7 MHz and 13 MHz as well as 17 MHz and 23 MHz, etc. These can be removed by the use of a low-pass filter. Also, some low level spurious signals are produced close into the required signal. These are normally acceptable in level, although for some applications they can cause problems.

4.17.2 Need for Direct Digital Synthesis

Today's cost-competitive, high-performance, functionally integrated, and small package-sized DDS products are fast becoming an alternative to traditional frequency-agile analog synthesizer solutions. The integration of a high-speed, high-performance, D/A converter, and DDS architecture onto a single chip (forming what is commonly known as a complete-DDS solution) enabled this technology to target a wider range of applications and provide, in many cases, an attractive alternative to analog-based PLL synthesizers. For many applications, the DDS solution holds some distinct advantages over the equivalent agile analog frequency synthesizer employing PLL circuitry.

Advantages of DDS

- It provides micro-hertz tuning resolution of the output frequency and subdegree phase tuning capability, all under complete digital control.
- It has extremely fast hopping speed in tuning output frequency (or phase). The phase-continuous frequency hops with no overshoot, undershoot, or analog-related loop settling time anomalies.
- The DDS digital architecture eliminates the need for the manual system tuning and tweaking associated with component aging and temperature drift in analog synthesizer solutions.
- The digital control interface of the DDS architecture facilitates an environment where systems can be remotely controlled and minutely optimized under processor control.
- When utilized as a quadrature synthesizer, DDS affords unparalleled matching and control of I and Q synthesized outputs.
- The tuning word is typically 24 to 48 bits long, which enables a DDS implementation to provide superior output frequency tuning resolution.

A case study in designing a function generator using DDS is given below.

4.17.3 DDS Application in Function Generator Design: A Case Study

Function generators have found widespread applications in electronics laboratory for a long time. These instruments have accumulated a long list of features, over a time starting with just a few knobs for setting the amplitude and frequency of a sinusoidal output. Function generators now provide wider frequency ranges, calibrated output levels, a variety of waveforms, modulation modes, computer interfaces, and, in some cases, arbitrary functions. Thus, the many features added to function generators have complicated their design and increased their cost. Now there is an opportunity for a radical redesign of the familiar function generator using direct digital synthesis (DDS). DDS provides remarkable frequency resolution and allows direct implementation of frequency, phase, and amplitude modulation.

Many of the concepts of DDS are illustrated by the way in which a sine wave is generated. Figure 4.74 shows a block diagram of a simple DDS function generator. The sine function is stored in a RAM table. The RAM's digital sine output is converted to an analog sine wave by a DAC. The steps at the DAC output are filtered by a low-pass filter to provide a clean sine wave output. The frequency of the sine wave depends on the rate at which addresses to the RAM table are changed. The addresses are generated by adding a constant stored in the phase increment register (PIR) to the phase accumulator. Usually, the rate of additions is constant, and the frequency is changed by changing the number in the PIR.

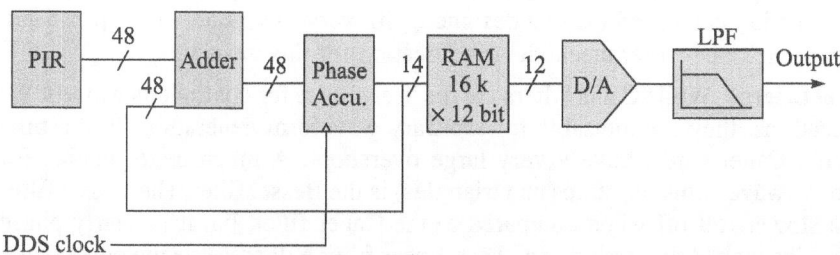

Fig. 4.74   DDS function generator

The frequency resolution depends on the number of bits in the PIR. If the PIR, adder, and phase accumulator support 48 bit additions, then the fractional frequency resolution is one part in 2^{47}. That means a 48 bit DDS generator can provide better than 1 mHz resolution on a 10 MHz output. There are a few more details that need to be addressed in order to understand DDS in this application, such as the sample rate, RAM size, DAC resolution, filter characteristics, and spectral purity of the output.

Sample rate

To achieve good spectral purity of a sine wave, a large number of samples are required for each cycle of the sine wave. A sine wave that is approximated by a small number of samples per cycle hardly looks like a sine wave. In fact, if one could make an arbitrarily sharp, low-pass filter, we would need only two samples per cycle. For example, consider the case where we have four samples per sine cycle. The Fourier spectrum of this sampled output has components at f, $2f$, $3f$, and so on. If one can design a low-pass filter to eliminate the harmonic components of the sampled output, then we are left with the fundamental (a pure sine wave at frequency f), i.e., generating an output at f by sampling at a rate of f_s results in the lowest frequency Fourier component at a frequency of $(f_s - f)$. This simple result becomes the basis of the low-pass filter specification that the filter should pass f but stop $(f_s - f)$.

Filters

The graph in Fig. 4.75 shows a low-pass filter transfer function. As discussed earlier, the filter must pass the highest frequency f_{max} but must begin its stopband at $f_s - f_{max}$. As steep roll-off filters with high stopband attenuation are hard to build, a reasonable compromise in this trade-off occurs when $f_{max} = f_s/3$. This allows the filter a one octave transition band. The stopband attenuation depends on the spurious component specification of the output. A typical specification for a function generator application would be -70 dB. Figure 4.75 shows a typical low-pass filter for DDS outputs.

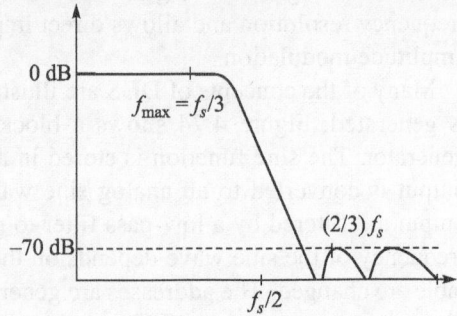

Fig. 4.75 Low-pass filter for DDS outputs

Cauer (elliptic) filters These are a good choice for this application. They have fast transition bands and may be designed with very low ripple in the passband. Ninth-order Cauer filter meets the specification for this example.

Bessel filters While Cauer filters are the best choice for continuous wave (CW) applications, they are unusable for arbitrary waveform generation. In the time domain, Cauer filters have a very large overshoot. A much better choice for arbitrary waveforms (or ramps and triangles) is the Bessel filter. The Bessel filter has a slower roll-off when compared to the Cauer filter, but it is nearly phase linear. The lack of dispersion in a phase-linear filter will preserve the pulse shape and prevent any ringing in the time domain. A seventh-degree Bessel filter, with a -3 dB cut-off of $f_c = f_s/4$, is a good choice for filtering arbitrary waveforms. This filter will exhibit an output rise time of $0.35/f_c$.

DAC and RAM requirements

High-density and fast RAMs and high-speed, high-resolution DACs have made DDS a viable technology for function generator applications. As we have seen, a maximum practical output frequency is $f_s/3$. So, the DDS phase accumulator, RAMs, and DACs must run at three times the maximum desired output frequency. The DAC resolution depends on the spurious component specification for the output (or the desired arbitrary waveform resolution). The DAC's quantization error and non-linearities lead to spurious outputs. The source of these spurious outputs is due to the difference between the actual output of the DAC and the desired sine value. So, a 12 bit DAC, which is linear and monotonic to 2 LSBs, will have output errors of the order of one part in 2048, or about -66 dB. A short RAM table is another way to get the wrong value out of the DAC. To avoid phase quantization noise, there should be two more bits of address to the RAM than bits in the DAC.

Extending frequency range

The frequency range of the DDS output may be extended by a variety of techniques. Depending on which technique is used, some of the advantages of DDS may be lost. Just as with more conventional frequencies synthesizers, the DDS output may be doubled, mixed with other fixed sources, or used as a reference inside of a phase locked loop.

Modulation technique

The power and elegance of DDS are most apparent when a modulated source is required. The frequency of the output may be changed instantly to any frequency from DC to f_{max} by simply changing the number in the phase increment register.

Figure 4.76 shows the block diagram of a DDS phase accumulator with programmable modulation capabilities. This phase accumulator, which has been optimized for function generator applications, has two phase increment registers: PIR-A and PIR-B. A 48 bit wide multiplexer can switch between the PIRs in a single clock cycle. The modulation processor can modify the PIRs at a rate of up to 10 million bytes/s, filling one PIR while the other is used as an input to the adder. Complex modulation programs may be stored in the modulation RAM. This RAM contains op-codes and data for the modulation processor.

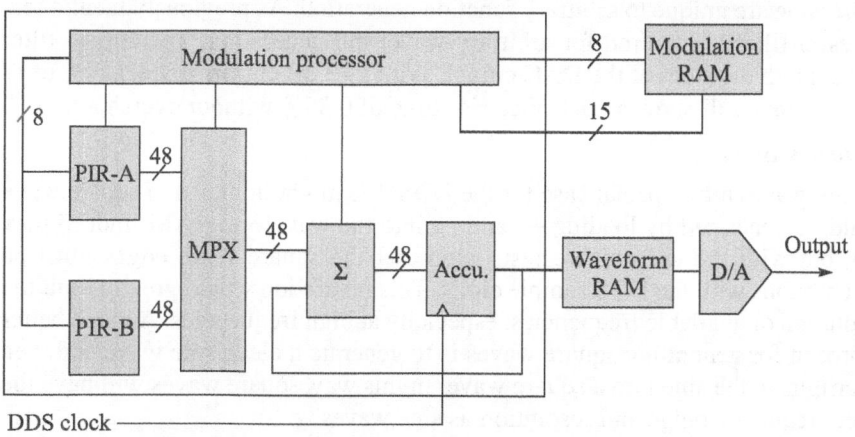

Fig. 4.76   Modulation processor with DDS phase accumulator

Frequency scans illustrate the operation of this processor. When programmed for a log frequency sweeps, a list of up to 4000 discrete frequencies is stored in the modulation RAM by the host system. The modulation processor modifies PIR-A while the adder is using PIR-B and vice versa. More complex modulation programs may be stored, such as frequency modulation by any arbitrary function, linear or log sweeps, frequency hopping, etc. Phase modulation is easily done by programming PIR-A with the nominal frequency and using PIR-B, which contains the nominal phase increment plus any desired phase shift, for a single clock cycle. Wide frequency or phase deviations are no problem. Also phase or frequency hop may be programmed and executed in a single clock cycle. Also since the PIRs may be modified very quickly, modulation frequencies of up to several hundred kilohertz are possible. In fact, arbitrary modulation programs may be stored. This feature allows the function generator to be used for modem testing, communications, bit error rate determination, etc.

Amplitude modulation

There are two approaches for amplitude modulation of the output waveform. Either the digital outputs from the RAM or the analog output from the DAC may be multiplied by the desired amplitude. The later approach is better for function generators, so that either an internal or external source may be used for amplitude modulation.

Arbitrary function

One of the immediate benefits of the DDS architecture is that arbitrary waveform generation comes along for free. Instead of storing a sine table in the waveform RAM, a list of arbitrary values is saved. The phase accumulator is programmed to step through the stored values, one at a time, to play back the desired waveform through the output DAC. The DDS's arbitrary waveform capability simplifies the task of generating the other standard waveforms found in function generators. Ramp, sawtooth, and even Gaussian white noise may be generated by changing the list of values in the waveform RAM. The phase accumulator must be designed to support certain modes that are required for arbitrary waveforms. The rate at which RAM values are retrieved may be changed by simply using a different PIR value. However, variable record lengths, triggering functions, and wrap-around addressing are unique to arbitrary function generation. As previously mentioned, a Bessel filter is required for arbitrary waveform generation. The Bessel filter will smooth the steps at the DAC output. With a -3 dB cut-off frequency f_c of $f_s/4$, the output will show a controlled rise time of $0.35/f_c$ without overshoot.

Square waves

Square waves are a special case for the DDS. We might think that a square wave could be generated by loading $+1$ and -1 into the waveform RAM. Indeed they can, but with the unfortunate restriction that the square wave edges must be synchronous with the DDS sample clock. This restriction would greatly limit the resolution of available frequencies, especially at high frequencies. A much better approach for generating square waves is to generate a clean sine wave and then discriminate the sine into a square wave. In this way, square waves will have the same frequency range and resolution as sine waves.

Output amplifiers

The output amplifier used in a DDS function generator must meet some stringent requirements. In order to preserve waveforms generated in the arbitrary mode, the amplifier must have a wide and flat passband, and exhibit a phase linear response well past the cut-off frequency of the Bessel filter. The amplifier's bandwidth also determines the rise time of the square wave output. Here again, a sharp (phase linear) roll-off is required to prevent overshoot on the square wave output. Finally, the output amplifier must be able to drive 10 Vpp into a 50 Ω load, meet distortion and settling specifications, and be protected against short circuits or connection to external power supplies. The output amplifier should exhibit a 50 Ω output impedance regardless of the output level setting. To generate low signal levels, most function generators have output attenuators. The attenuators allow the output amplifier to work within a limited range of output levels so that the distortion and signal-to-noise ratios remain constant as the output levels are changed.

Floating generator

Many applications require that the function generators be able to provide a signal to a load that is not ground referenced. Even if the load is nominally ground referenced, a floating generator output will provide a much cleaner signal because the system ground loops are eliminated. It is important that the generator output shield is floating under all circumstances even when the function generator is connected to a GPIB controller, or if an external frequency reference is connected to the instrument.

Application specific integrated circuits

DDS provides a new and clean design approach for function generators. Most of the analog functions required for function generators are handled by digital logic circuits. Unfortunately, these logic circuits are big and complicated and have to run fast. For example, a 15 MHz DDS requires a 48 bit adder operating at 40 MHz with lots of glue logic. Fortunately, application specific integrated circuits (ASICs) provide a low-cost solution to the problem. A TTL prototype of the phase accumulator shown in Fig. 4.76 required about 150 ICs. The prototype was just able to work with a clock of 10 MHz. A CMOS gate array of the same design was fabricated in a 68 pin PLCC plastic package. The gate array operates at 40 MHz (worst case), uses about 0.25 watts of power, and has a recurring cost.

Conclusion

DDS-based function generators are just beginning to appear in the market. These function generators offer substantial performance improvements at reduced costs over conventional analog function generators. As the cost of ASICs, RAMs, and DACs decline, while their speed and resolution increase, we can expect to see DDS-based function generators soon replace their analog counterparts. The design can be made still simpler by using FPGAs.

4.17.4 PLL Frequency Synthesizer: A Case Study

When we require a range of frequencies from a single fixed source like an oscillator, frequency synthesizers come to the picture. A frequency synthesizer is an electronic instrument that meets this requirement. It finds a wide variety of applications in radio receivers, walkie-talkies, CB radios, mobile phones, radio telephones, satellite receivers, and GPS systems. It can combine frequency multiplication, frequency mixing, and frequency division operations to produce the desired output signal. Before the widespread use of frequency synthesizers, manual tuning of local oscillators was relied upon. One of the solution was to use many stable resonators or a set of stable quartz crystal oscillators. This is practical only when the requirement of frequencies is somewhat small. But this method proves quite costly and impractical in many applications, where the requirement of frequencies is large as in the case of FM radio band, which needs about hundred individual frequencies in the range of 88 MHz to 108 MHz.

Use of crystals increases both cost and space. Both coherent and incoherent methods are employed. The coherent method generates frequencies derived from a single, stable master crystal oscillator. The incoherent method derives frequencies from a set of several stable oscillators.

Modern synthesizers use coherent technique because of its simplicity and low cost, are based largely on phase locked loops, and are also available as integrated circuits reducing cost and size.

Design procedure

Before attempting to design a synthesizer, the following points are to be considered.
- Output frequency range, i.e., the tuning range or bandwidth
- Frequency resolution, i.e., frequency increments
- Frequency stability and phase stability
- Spectral purity
- Settling time and switching time
- Size and power consumption
- Last but not least, the cost

These are mutually contradictory requirements. But a trade-off has to be made.

PLL synthesizers principle

Today, most of the synthesizers are based on PLL. They have tremendous advantages such as ease of operation, precise required frequency, and frequency stability as the reference quartz crystal. Use of microprocessors/microcontrollers to control the synthesizers enable the user to have a menu-driven facilities such as keypad entries to select the desired frequency, such as display of selected frequency, and running once in a while diagnostic software.

A PLL needs some additional circuitry for converting it into a frequency synthesizer. A frequency divider between the voltage-controlled oscillator and the phase comparator should be added as shown in Fig. 4.77. This frequency divider will be in the form of divide by N counter (also called programmable divider). When the divider is added into the circuit, the PLL still tries to reduce the phase difference between the two signals entering the phase comparator. Again when the circuit is in lock, both signals entering the phase comparator are exactly the same in frequency. For this to be true, the VCO must be running at a frequency equal to the phase comparison frequency times the division ratio. If the division ratio is altered by one, then the VCO will have to change to the next multiple of the reference frequency. This means that the step frequency of the synthesizer is equal to the frequency entering the phase comparator. But the frequency synthesizer has to step in much smaller increments if they are to be of any use. This necessitates the reduction of comparison frequency. This is usually accomplished by running the reference oscillator at a frequency of MHz or so and then dividing this signal down to the required frequency using a fixed divider. In this way, a low comparison frequency can be achieved. This is shown in Fig. 4.78.

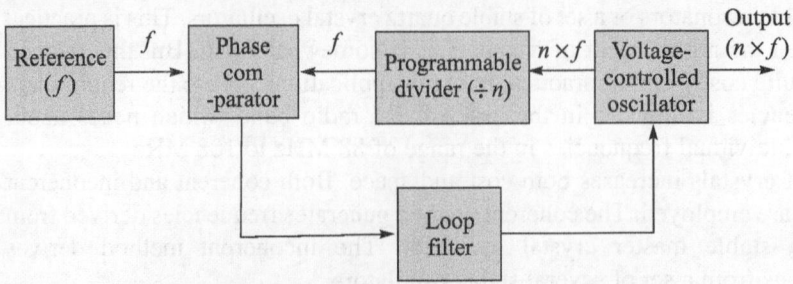

Fig. 4.77 Frequency synthesizer: a block diagram

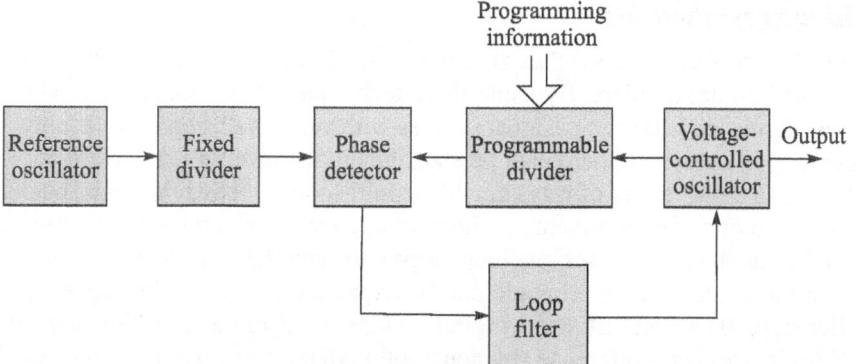

Fig. 4.78 Block diagram showing a fixed divider added to the reference source

Analog method

It is also possible to use a mixer in the loop instead of a digital divider as shown in Fig. 4.79. This technique places an offset into the frequency generated by the loop. When the loop is in lock, the signals entering the phase detector are exactly at the same frequency. The mixer adds an offset equal to the frequency of the signal entering the other port of the mixer. This can be explained by the following example.

If the reference oscillator is operating at a frequency of 10 MHz and the external signal is at 15 MHz, then the VCO must operate at either 5 MHz or 25 MHz. Normally, the loop is set up so that the mixer changes the frequency down and if this is the case, then the VCO will be operating at 25 MHz.

It can be seen that there may be problems with the possibility of two mix products being able to give the correct phase comparison of frequency. This happens as a result of the phasing in the loop, so that only one will enable it to lock. However, to prevent the loop getting into an unwanted state, the range of the VCO is limited. For phase locked loops (PLLs) that need to operate over a wide range, a steering voltage is added to the main tune voltage so that the frequency of the loop is steered into the correct region for required conditions. It is relatively easy to generate a steering voltage by using digital information from a microprocessor and converting this into an analog voltage using a digital-to-analog converter (DAC). The right tune voltage required to pull the loop into the lock is provided by the loop in the normal way.

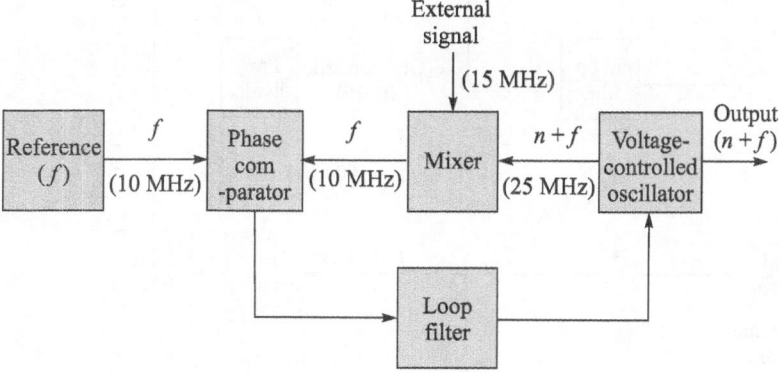

Fig. 4.79 Synthesizer with a mixer

Multi-loop synthesizers

Many high-performance synthesizers use several loops that incorporate both mixers and digital dividers. By using these techniques, it is possible to produce high-performance wide-range signal sources with very small step sizes. If only a single loop is used, then there may be shortfalls in the level of the performance.

There is a large variety of ways in which multi-loop synthesizers can be made, dependent upon the requirements of the individual system. A two-loop system is shown in Fig. 4.81. This uses one loop to give the smaller steps and the other to provide larger steps. This principle can be expanded to give wider ranges and smaller steps. The first PLL has a digital divider and operates over the range 19 to 28 MHz. Having a reference frequency of 1 MHz, it can move in steps of 1 MHz. The signal from this loop is fed into the mixer of the second one. The second loop has division ratios of 10 to 19, but has the reference frequency divided by 10 to 100 kHz to give smaller steps.

The operation of the whole loop can be examined by looking at extremes of the frequency range. With the first loop set to its lowest value, the divider is set to 19 and the output from the loop is at 19 MHz. This feeds into the second loop. Again this is set to the minimum value and the frequency after the mixer must be at 1.0 MHz. With the input from the first loop at 19 MHz, this means that the VCO must operate at 20 MHz if the loop is to remain in lock.

At the other end of the range, the divider of the first loop is set to 28, giving a frequency of 28 MHz. The second PLL has the divider set to 19 giving a frequency of 1.9 MHz between the mixer and the divider. In turn, this means that the frequency of the VCO must operate at 29.9 MHz. As the PLLs can be stepped independently meaning that the whole synthesizer can move in steps of 100 kHz between the two extremes of frequency. As mentioned before, this principle can be extended to give greater ranges and smaller steps, providing for the needs of modern receivers.

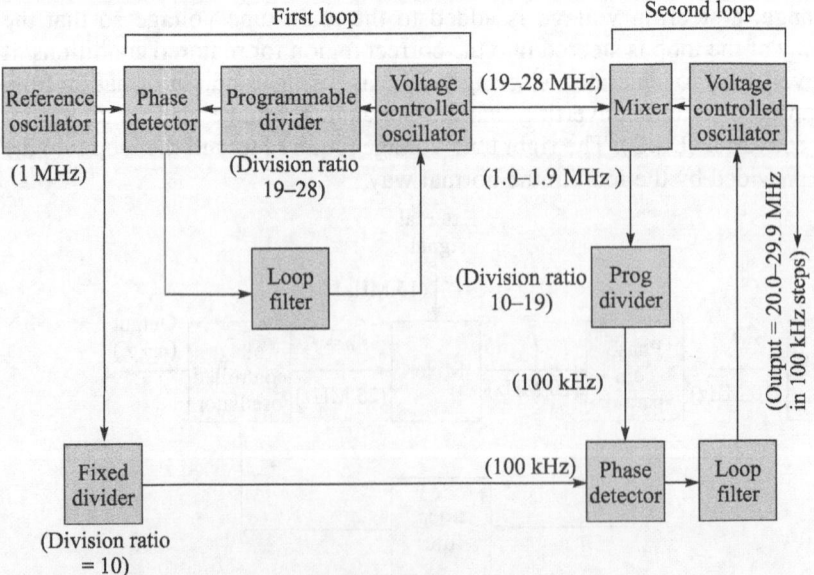

Fig. 4.80 An example of a synthesizer using two loops

Practical frequency synthesizer

A practical synthesizer shown in Fig. 4.81 can be designed with 8051 microcontroller, a NE 565 phase locked Loop and a crystal oscillator. In this design divide by M counter, used for dividing the reference input before feeding to the input of PLL and divide by N counter used for dividing VCO output before feeding to the input of phase comparator is replaced by the two internal timer of the microcontroller. These timers can be configured as a 16 bit divider and wide variety of choice is available for selecting the required frequency. The VCO output after buffering gives the synthesized output. A keypad is included in the design so that required frequencies can be set. A display interface is provided to display the output frequency.

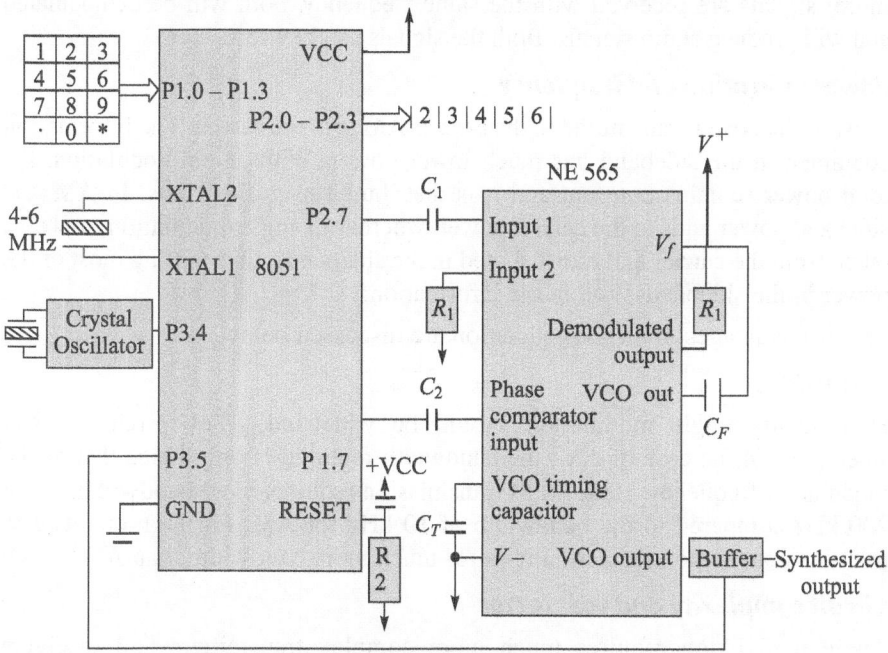

Typical frequency synthesizer practical circuit

Fig. 4.81   A practical frequency synthesizer

4.18 COMPARISON OF ANGLE MODULATION WITH AMPLITUDE MODULATION

Angle modulation has several inherent advantages over amplitude modulation.

Noise immunity

Angle modulation scores over amplitude modulation (AM) as far as noise immunity is concerned. The noise results in unwanted amplitude variations in the modulated wave. These are removed in angle modulated receivers by the limiters employed. Most of the AM noise is removed by the limiters before the final demodulation process occurs. This process cannot be used with AM receivers because the information is also contained in amplitude variations and removing the noise would also remove the information.

Noise performance and signal-to-noise improvement

With the use of limiters, the angle demodulation can actually reduce the noise level and improve the signal-to-noise ratio during the demodulation process. This is called FM *thresholding*. This is not possible in AM once the noise contaminates the signal.

Capture effect

In angle modulation, a phenomenon known as capture effect allows a receiver to differentiate between two signals received with the same frequency. If one signal has high amplitude compared to the other, the receiver will capture the stronger signal and eliminate the weaker signal. With amplitude modulation, if two (or more) signals are received with the same frequency, both will be demodulated and will produce audio signals. Both the signals can be heard.

Power utilization and frequency

Most of the power transmitted in the case of AM is in the carrier. The information contained in the sideband has much lower power. With angle modulation, the total power remains constant and it is distributed over sidebands. In AM, the sideband power adds to the carrier power, whereas in angle modulation, power is taken from the carrier and redistributed in the sidebands, thus putting most of the power in the sidebands, i.e., in the information.

The disadvantages of angle modulation are discussed below.

Bandwidth

High-quality angle modulation should be wideband. They produce many sidebands. In the case of AM, the bandwidth required is only twice that of the modulating frequency. Hence, FM transmission requires more bandwidth, about 200 kHz compared to the bandwidth of 10 kHz for AM. For high-quality FM transmission, angle modulation requires much more bandwidth than AM.

Circuit complexity and cost factor

Angle modulation requires much more complex transmitters and receivers compared to amplitude modulation. This involves extra cost. But for this, we have to pay for quality noise-free transmission. But with the advent of LSI ICS, the cost comparison between angle modulation and AM is getting reduced considerably.

SUMMARY

- In angle modulation, the modulating signal is used to vary either the phase or frequency of the carrier signal.
- If the modulating signal varies the phase of the carrier, it is called phase modulation and if it is the frequency that is varied, then it is called frequency modulation.
- Unlike amplitude modulation (AM), which is linear, both frequency modulation (FM) and phase modulation (PM) are non-linear.
- Because of its high-noise immunity, angle modulation is preferred to AM but at the expense of bandwidth.

- Angle modulation was first introduced by E.H. Armstrong in 1931 and the first successful FM radio system was developed in 1936.
- The instantaneous frequency of a sinusoidal carrier is given by

$$f_i(t) = \lim_{\Delta t \to 0} f_{\Delta t}(t)$$

where $f_{\Delta t}(t) = \dfrac{\theta_i(t + \Delta t) - \theta_i(t)}{2\pi \Delta t}$

So, $\qquad f_i(t) = \lim_{\Delta t \to 0} \dfrac{\theta_i(t + \Delta t) - \theta_i(t)}{2\pi \Delta t}$

$$f_i(t) = \frac{1}{2\pi} \frac{d\theta_i}{dt}$$

The instantaneous phase angle is given by

$$\theta(t) = \int_{-\infty}^{t} \omega_i(\alpha)d\alpha = 2\pi f_c t + \phi_c$$

- If a modulating wave is integrated and then phase modulated, the resultant waveform is an FM wave. If the modulating signal is differentiated and frequency modulated, the resulting wave is a phase modulated wave.
- The modulaton index of an FM wave is $m = \Delta f / f_m$, where Δf is the frequency deviation proportional to the amplitude of the modulating signal but independent of the frequency of the modulating signal and f_m is the modulating frequency.
- The modulation index of a PM wave is $m = K_p V_m$, where K_p is the phase deviation sensitivity in rad/V and V_m is the peak modulating signal amplitude.
- If the modulaton index is < 1, it is called low index or narrowband FM. The bandwidth required will be $2f_m$ Hz same as DSB in AM.
- If the modulation index is > 10, then it is called wideband FM and the bandwidth is approximated by $2\Delta f$ Hz, where Δf is the peak-to-peak deviation. Wideband FM is analysed using the Bessel function.
- The actual bandwidth of wideband FM is given by $2(n \times f_m)$, where f_m is the highest modulating frequency and n is the number of significant sidebands determined from the Bessel function table. But a general expression for the bandwidth of a wideband FM wave was given by Carson as $2(\Delta f + f_m)$ Hz. Using this general formula for narrowband FM, where $\Delta f \ll f_m$, bandwidth = $2f_m$ and for wideband FM, where $\Delta f \gg f_m$, bandwidth = $2\Delta f$.
- Wideband sinusoidal is analysed using the Bessel function.
- The significant sidebands are determined from the Bessel function table.
- In the case of amplitude modulation, the power is wasted in the carrier, whereas in angle modulation, the carrier power is distributed over sidebands and the total power can never exceed the unmodulated carrier power. Also for a certain value of modulation index, the carrier power is zero and the entire power is carried by sidebands.

- Deviation ratio for a non-sinusoidal modulation signal D is given by $\Delta F/f_m$, where Δf is the maximum frequency deviation and f_m is the higher frequency component of the modulating signal.
- The bandwidth required for non-sinusoidal modulation is given by

$$\text{Bandwidth} = 2\left(\frac{\Delta f}{f_m}+1\right)f_m = 2(\Delta f + f_m)$$

which is nothing but the Carson's relation.
- Phase modulation can be considered as if the angular velocity of the phasor $f(t)$ undergoes modulation around the nominal angular velocity ω_c, i.e., the signal $f(t)$ is an angular velocity modulated waveform.
- The peak phase deviation $\Delta\Phi = K_p E_m$ is directly proportional to the peak modulation signal.
- Phase modulation is widely used in digital systems, such as PSK and QPSK. When we compare FM and PM, they are indistinguishable as far as a single modulating frequency is considered.
- If the modulating frequency is changed, the modulation index for PM will remain constant whereas the modulation index for FM will decrease as the modulating frequency is increased and vice versa.
- Various methods are employed to generate FM. They are direct and indirect methods. In the direct method, the frequency of the carrier determined by an LC circuit is varied with the amplitude of the modulating voltage by shunting an inductor, a capacitor, or a resistor whose value changes with the applied voltage. Varactor diodes are used as variable capacitor, PIN diode, and FET used as variable resistor and saturated reactor used as a variable inductor. Linear integrated circuits are also used for direct FM modulators. PLL ICs are also used for generating FM waves. In the indirect method, a narrowband signal is generated initially and then converted to wideband signal.
- Varactor diode, transistor can be used in direct FM modulator.
- FM demodulators are implemented by generating an AM signal whose amplitude is proportional to the instantaneous frequency of the FM signal and then using an AM demodulator to recover the message signal.
- PLLs are also used for demodulation of FM wave.
- Amplitude limiters are used to eliminate amplitude modulation and amplitude modulation noise signal from received FM signal.
- Direct FM transmitters cannot use crystal oscillators for generating carrier frequency. To keep the carrier frequency constant, they use AFC.
- In the indirect method of FM generation, initial modulation takes place as a narrowband FM and then it is multiplied to the required carrier frequency. This circuit does not require AFC since crystal oscillator can be used to generate the carrier as the deviation is quite small for narrowband FM.
- Normally, superheterodyne receiver is used for FM receivers.
- Single-chip FM radio circuits are available to receive FM radio programme.

IMPORTANT FORMULAE

- Instantaneous frequency, $\omega_i(t) = \dfrac{d\theta}{dt}$

- $\theta(t) = \displaystyle\int_{-\infty}^{t} \omega_i(\alpha)d\alpha$

- $\theta(t) = \omega_c t + \theta_0 + K_p m(t) = \omega_c t + K_p m(t)$ with $\theta_0 = 0$
- PM wave, $f_{PM}(t) = A \cos[\omega_c t + K_p m(t)]$
- Instantaneous frequency $\omega_i(t)$ is then given by
  $\omega_i(t) = d\theta(t)/dt = \omega_c + K_p dm(t)/dt$

- $\theta(t) = \displaystyle\int_{-\infty}^{t} [\omega_c + K_f\, m(\alpha)d\alpha] = \omega_c t + K_f \int_{-\infty}^{t} m(\alpha)d\alpha$

- The FM wave is $f_{FM}(t) = A \cos\left[\omega_c t + K_f \displaystyle\int_{-\infty}^{t} m(\alpha)d\alpha\right]$

- Deviation sensitivity:
  For frequency modulation: $K_f = \Delta\omega/\Delta V$
  For phase modulation: $K_\Phi = \Delta\theta/\Delta V$

- The expression for a carrier that is being phase or frequency modulated by a single modulating signal is
  $f(t) = V_c \cos(\omega_c t + m \sin \omega_m t)$
  where m is the modulation index given by

$$m = b = \frac{\Delta f}{f_m} \text{ for FM}$$

$$m = \Delta\theta = K_p V_m$$

  where K_p is the deviation sensitivity in rad/V.

- Percentage modulation $= \dfrac{\Delta f_{actual}}{\Delta f_{max}} \times 100$

- Equation for narrowband FM

$$f(t) = V_c \left[\cos \omega_c t - \frac{\beta}{2} \cos(\omega_c - \omega_m)t + \frac{\beta}{2} \cos(\omega_c + \omega_m)t \right]$$

- Equation for wideband FM

  $f(t) = J_0(\beta) \cos \omega_c t - J_1(\beta)\{\cos(\omega_c - \omega_m)t + \cos(\omega_c + \omega_m)t\} + J_2(\beta)$
  $\{\cos(\omega_c - 2\omega_m)t + \cos(\omega_c + 2\omega_m)t\} - J_3(\beta)\{\cos(\omega_c - 3\omega_m)t$
  $+ \cos(\omega_c + 3\omega_m)t\} + \cdots$

- Average power in sinusoidal FM:

$$P_T = P_c\,[J_o^2(\beta) + 2\{J_1^2(\beta) + J_2^2(\beta) + J_3^2(\beta) + \cdots\}]$$

  where P_c is the unmodulated carrier power V_c^2/R.

- Deviation ratio for non-sinusoidal modulation:

$$\mathrm{BW}_{max} = 2(\Delta f + f_m)$$

- Expression for sinusoidal PM:

$$f(t) = E_{c\,max} \cos[\omega_c t + K_p E_m(t)\,]$$

where $\quad K_p E_m(t) = K_p E_m \sin \omega_m t$

$$= \Delta\Phi \sin \omega_m t$$

So, $\qquad f(t) = E_c \cos(\omega_c t + \Delta\Phi \sin \omega_m t)$

ADDITIONAL EXAMPLES

1. An 800 Hz, 3V modulating signal in an FM system produces a deviation of 6 kHz. If the modulating voltage is increased to 6 V, what is the new deviation?

Solution

Since the deviation is directly proportional to the modulating voltage,

$$\frac{\Delta f}{V_m} = \frac{6}{3} = 2 \text{ kHz/V}$$

Now for $V_m = 6$ V, the new deviation $\Delta f = 2 \times 6 = 12$ kHz

2. Determine the carrier and modulating frequencies and the maximum deviation for an FM wave $f(t) = 15 \sin(8 \times 10^8 + 6 \sin 1300t)$. What is the power dissipated by this FM wave in a 12 Ω resistor? Assume the modulation index β to be 3.

Solution

We have

$$f_c = \frac{8 \times 10^8}{2\pi} = 127.33 \text{ MHz} \quad \text{and} \quad f_m = \frac{1300}{2\pi} = 206.9 \text{ Hz}$$

Maximum deviation $= \beta \cdot f_m = 6 \times 206.9 = 1241 \text{ Hz}$

Power dissipated $= \dfrac{V_{c\,rms}^2}{R} = \dfrac{\left(\dfrac{15}{\sqrt{2}}\right)^2}{12} = 9.375$ W

3. What is the bandwidth required for an FM signal whose modulating frequency is 3 kHz and the maximum deviation is 18 kHz?

Solution

We have

Modulation index, $\beta = \dfrac{\Delta f}{f_m} = \dfrac{18}{3} = 6$

From the Bessel function table (Table 4.1), it can be seen that for $\beta = 6$, the highest J coefficient included is J_{12}. The coefficient whose values lie below 0.01 are ignored. Hence, we include the 12th pair of sidebands.

This gives the bandwidth as BW $= f_m \times$ highest needed sideband $\times$ 2

$$= 3 \text{ kHz} \times 12 \times 2 = 72 \text{ kHz}$$

4. Find the maximum deviation Δf and modulation index β for an FM modulator with a deviation sensitivity of 6 kHz/V and a modulating signal $m(t) = 3\cos(18,850)t$.

Solution

Modulating signal frequency, $f_m = \dfrac{18,850}{2\pi} = 3000$ Hz

Maximum frequency deviation, $\Delta f = 6$ kHz/V $\times 3$ V $= 18$ kHz

Modulation index, $\beta = \dfrac{\Delta f}{f_m} = \dfrac{18\,\text{kHz}}{3\,\text{kHz}} = 6$

5. For an FM modulator with a modulating signal $m(t) = V_m \sin(3000)t$, the carrier signal $V_c(t) = 8\sin(6.5 \times 10^6)t$ and the modulating index $\beta = 2$, find out the number of significant side frequencies and their amplitudes.

Solution

From the Bessel function table (Table 4.1), it can be seen that for $\beta = 2$, the number of significant side frequencies is six. They are J_1, J_2, J_3, J_4, J_5, and J_6. Their amplitudes with the carriers are as follows:

$J_0 = 0.224 \times 8 = 1.792$ V

$J_1 = 0.577 \times 8 = 4.616$ V

$J_2 = 0.353 \times 8 = 2.824$ V

$J_3 = 0.129 \times 8 = 1.032$ V

$J_4 = 0.034 \times 8 = 0.272$ V

$J_5 = 0.007 \times 8 = 0.056$ V

$J_6 = 0.001 \times 8 = 0.008$ V

6. An FM modulator has maximum frequency deviation $\Delta f = 12$ kHz, modulating frequency $f_m = 6$ kHz, and a carrier 500 kHz with amplitude 5 V. Find out the actual minimum bandwidth from the Bessel function table and also the approximate minimum bandwidth using the Carson's rule.

Solution

Modulation index, $\beta = \dfrac{\Delta f}{f_m} = \dfrac{12\,\text{kHz}}{6\,\text{kHz}} = 2$

From the, Bessel function table (Table 4.1), it can be seen that for $\beta = 2$, the number of significant sideband frequencies is six. Hence, the minimum bandwidth, $B = 2(6 \times 6) = 72$ kHz

From the Carson's formula, the approximate minimum bandwidth,

$$B = 2(\Delta f + f_m) = 36 \text{ kHz}$$

7. For Example 5, find the unmodulated carrier power for the FM modulator. Assume the load resistance to be 30 Ω. Also find the total power of the modulated wave. Further, prove that the power in the unmodulated wave is equal to the power in the modulated wave.

Solution

Unmodulated carrier power $= \dfrac{8^2}{2(30)} = 1.066$ W

The total power of the modulated wave,

$$P_t = \frac{1.792^2}{2(30)} + \frac{2(4.616)^2}{2(30)} + \frac{2(2.824)^2}{2(30)} + \frac{2(1.032)^2}{2(30)} + \frac{2(0.272)^2}{2(30)}$$

$$+ \frac{2(0.056)^2}{2(30)} + \frac{2(0.008)^2}{2(30)}$$

$$= 0.053 + 0.710 + 0.264 + 0.058 + 0.002 + 0 + 0$$

$$= 1.087$$

It can be seen that the power in the modulated wave is equal to the power in the unmodulated wave. The small error we observe is due to the fact that we have rounded the values in the Bessel function.

8. A balanced modulator with BPF tuned to the sum frequency and an RF carrier input frequency of 102.8 MHz. The output of the balanced modulator after bandpass filtering is given to a frequency multiplier with multiplying constant equal to 12.

If $\Delta f = 5$ kHz, $\beta = 0.5$, $f_m = 8$ kHz, and $f_c = 800$ kHz, find these parameters at the output of the balanced modulator and at the output of the multiplier.

Solution

The outputs at the balanced modulator for these parameters are same as the inputs. They remain unaltered, i.e., $\Delta f = 5$ kHz, $\beta = 0.5$, $f_m = 10$ kHz, and $f_c = 800$ kHz.

But they get multiplied by the multiplication factor when passed through the multiplier, i.e., all the factors except the modulating frequency get multiplied by 12 (since the rate of deviation remains unaffected). Their new values are as below:

$$\Delta f = 60 \text{ kHz}, \ \beta = 6, f_m = 10 \text{ kHz, and } f_c = 9.6 \text{ MHz}$$

9. For the transmitter shown in Fig. 4.42, if the multiplying factor is 25 and the final carrier frequency is 100.2 MHz, find
   (a) master oscillator centre frequency,
   (b) frequency deviation at the output of the modulator for a frequency deviation of 60 kHz at the power amplifier
   (c) deviation ratio at the output of the modulator for a maximum modulating signal frequency of 10 kHz, and
   (d) deviation ratio at the power amplifier.

Solution

(a) $f_c = \dfrac{f_t}{25} = \dfrac{100.2 \text{ MHz}}{25} = 4.008$ MHz

(b) $\Delta f = \dfrac{\Delta f_t}{25} = \dfrac{60 \text{ kHz}}{25} = 2400$ Hz

(c) $\beta = \dfrac{2.4}{10} = 0.24$

(d) $\beta_t = 0.24 \times 25 = 6$

10. An angle modulated wave with a carrier frequency $\omega_c = \pi \times 10^5$ is

$$f(t) = 5\cos(w_c t + 3\sin 2000t + 5\sin 2000\pi t)$$

Find

(a) frequency deviation Δf,

(b) deviation ratio β,

(c) phase deviation $\Delta\phi$, and

(d) the bandwidth.

Solution

The signal bandwidth is the highest frequency component of the modulating signal, which is $f_m = 2000\pi/2\pi = 1000$ Hz

(a) The frequency deviation Δf is obtained by finding the instantaneous frequency as follows:

$$\omega_i = \frac{d}{dt}\theta(t) = \omega_c + 6000\cos 2000t + 10{,}000\pi\cos 2000\pi t$$

The carrier deviation is $6000\cos 2000t + 10{,}000\pi\cos 2000\pi t$. The maximum value of this deviation occurs when both the sinusoids add in phase. At this instant, the maximum value of the carrier deviation is $6000 + 10000\pi$. This is the maximum carrier deviation $\Delta\omega$. Hence, the frequency deviation

$$\Delta f = \frac{\Delta\omega}{2\pi} = \frac{6000 + 10{,}000\pi}{2\pi} = 5955 \text{ Hz}$$

(b) Deviation ratio, $\beta = \dfrac{\Delta f}{f_m} = \dfrac{5955}{1000} = 5.955$

(c) Angle $\theta(t) = \omega t + (3\sin 2000t + 5\sin 2000\pi t)$. The phase deviation is the maximum value of the angle inside the bracket and is $\Delta\phi = 8$ rad.

(d) Bandwidth, $B = 2(\Delta f + f_m) = 2(5955 + 1000) = 13.9$ kHz

REVIEW QUESTIONS

1. What do you mean by angle modulation?
2. What is the difference between phase modulation and frequency modulation?
3. Define frequency deviation and phase deviation.
4. What do you mean by deviation sensitivity in PM and FM?
5. What is a narrowband FM?
6. Why is pre-emphasis used in FM?
7. Define modulation index for FM.
8. Why is FM preferred to PM?
9. What is the function of AFC?
10. State Carson's rule for determining the bandwidth for an angle modulated

wave.
11. Sketch the waveform of FM and PM.
12. What are the salient features of phase modulation?
13. What is threshold effect?
14. Why is de-emphasis network used in FM receiver?
15. What is meant by capture effect in FM?
16. What property of varactor diode is used to generate FM?
17. Name two circuits used for demodulating FM signal.
18. Why is Armstrong method of generating FM signal called an indirect method?
19. Describe the relationship between the instantaneous carrier phase and the modulating signal for PM.
20. Explain the effect of limiting on the FM waveform.
21. Describe the relationship between the modulation index and the modulating signal for PM and FM.
22. How do you get FM from PM and vice versa?
23. Describe the difference between a direct frequency modulator and a direct phase modulator.
24. Which integrated circuit is used for detection of FM signal?
25. What is the purpose of limiter in an FM receiver?
26. What is the advantage of ratio detector over the slope detector and Foster–Seeley detector?
27. What is the principle of reactance FM modulator?
28. Draw the block diagram of a phase locked loop.
29. Compare FM to PM.
30. What do you mean by instantaneous frequency?

PROBLEMS

1. For an FM, what is the deviation ratio if the modulating frequency is 15 kHz and the maximum frequency deviation is ±75 kHz.
2. Find the frequency of the carrier for an FM wave $f(t)=12 \sin[(6 \times 10^8 t) + \sin 1250t]$.
3. A typical transmitter uses two stages of frequency multipliers in cascade. The first stage is a tripler, while the second stage is a doubler. Find the output frequency if the input frequency is 10 MHz.
4. What is the bandwidth required for an FM signal in which the modulating frequency is 2 kHz and the maximum deviation is 10 kHz?
5. The carrier frequency of a broadcast signal is 100 MHz. The maximum frequency deviation is 75 kHz. If the highest audio frequency modulating the carrier is limited to 15 kHz, calculate the approximate bandwidth.
6. An FM produces a 10 kHz frequency deviation for a 10 V modulating signal. Find the deviation sensitivity. How much frequency deviation is

produced for a 3 V modulating signal?

7. A phase modulator produces 1.5 rad of phase deviation for a 6 V modulating signal. Determine the deviation sensitivity. How much phase deviation will be produced for a 3 V modulating signal?

8. For an FM transmitter with 50 kHz carrier swing, determine the frequency deviation. If the amplitude of the modulating signal gets reduced by half, determine the new frequency deviation.

9. An FM transmitter with a carrier frequency of 80 MHz has deviation sensitivity of 4 kHz/V. Determine the frequency deviation for a modulating signal $f_m(t) = 12 \sin(2\pi\, 2000t)$. Also find the modulating index.

10. For an FM modulator with 30 kHz frequency deviation and a modulating signal frequency of 10 kHz, determine the bandwidth using Carson's formula.

11. From the Bessel table, determine the number of sets of sidebands produced for the modulation indices 0.2, 2.0, and 5.0.

12. What is the bandwidth of a narrowband FM if the carrier frequency is 1 MHz and the modulating frequency is 15 kHz?

13. Consider an angle modulation signal $f(t) = A\cos(\omega_c t + 10\cos\omega_m t)$.
    (a) If $\varphi(t)$ is a PM signal, $K_p = 2$ rad/V, derive the message signal $f(t)$ and maximum frequency deviation of the PM signal.
    (b) if $\varphi(t)$ is an FM signal, $K_f = 2$ kHz/V, derive the message signal $f(t)$ and maximum frequency deviation of the FM signal.

14. Use 10 kHz sinusoid to modulate a 100 MHz carrier by AM, SSB, and FM ($\Delta f = 50$ kHz). Calculate the corresponding bandwidth of the modulated signals.

15. A 30 MHz carrier (amplitude $A = 10$) is frequency modulated by a 10 kHz sinusoid with modulation index $\beta = 3$. For the FM signal, calculate
    (a) the power of carrier frequency component and
    (b) the power of sideband components within Carson's bandwidth.
    The value of the Bessel function is given below:

n	0	1	2	3	4
$J_n(3)$	–0.2601	0.3391	0.4861	0.3091	0.1320

(Assume the load resistance to be 50 ohms.)

16. An f_m Hz sinusoid message is used for AM and FM with the same carrier signal. Suppose peak frequency deviation of the FM signal is four times the bandwidth of the AM signal and the sideband component $f_c \pm f_m$ of AM and FM signals share the same spectral magnitude. Calculate
    (a) modulation index of the FM signal and
    (b) modulation index of the AM signal.

17. Two signals at 100 MHz are tuned in alternately. The carriers are of equal intensity. One is modulated with a 10 kHz signal and has $\beta = 5$ and the other is modulated with a 2 kHz signal and has $\beta = 25$. Which signal requires the larger bandwidth? Explain. Compare the audio amplifier outputs in the two cases.

18. An FM modulator has a modulation index of 1, a modulating signal $f_m(t)$ = $E_m \sin(2\pi 1000t)$ and an unmodulated carrier $E_c \sin(2\pi 50010^3 t)$. Determine
    (a) the number of sets of significant side frequencies and
    (b) their amplitudes.
    Also draw the frequency spectrum showing the relative amplitudes.

19. An FM signal, 2000 sin $(2\pi \times 10^8 t + 2 \sin 2\pi \times 10^4 t)$, is applied to a 50 Ω antenna. Determine
    (a) the carrier frequency,
    (b) the transmitter power,
    (c) the modulating index,
    (d) the intelligence signal frequency,
    (e) the bandwidth (using the two methods), and
    (f) the power in the largest and smallest sidebands.

20. The input frequency deviation of an FM receiver Δf = 30 kHz and the transfer function k = 0.03 V/kHz. Determine the output.

21. An Armstrong transmitter uses a VHF band for transmission, at 180 MHz with a maximum frequency deviation of 25 kHz. At a minimum audio frequency of 100 Hz. The primary oscillator is a 150 kHz crystal oscillator. The initial phase modulation deviation has to be kept to less than 10°, to avoid distortion. Find the multiplying factor to give proper deviation. The multiplier can be a combination of doublers and triplers. Specify the mixer crystal and the multiplier stages needed.

22. An FM carrier is limited to a maximum deviation of 60 kHz. Compare the bandwidth when the modulating signal is sinusoidal of frequency:
    (a) 100 Hz, (b) 400 Hz, (c) 1 kHz, (d) 4 kHz, and (e) 10kHz. The carrier has to reach maximum deviation in each case. How the bandwidth limits are determined?

23. A 1 kHz square wave is used to frequency modulate a carrier of 60 kHz. The square wave harmonics up to and including the eleventh is to be taken into account. Calculate the deviation ratio and using Carson's rule the bandwidth

24. The carrier $f_c(t) = 50\cos 2\pi 10^8 t$ is frequency modulated by the signal $f_m(t) = 5\cos 15,000\pi t$. The peak frequency deviation is 30 kHz. Assuming the load resistor to be 50 ohms, determine:
    (a) the amplitude and frequency of all signal components that have a power level of at least 10
    (b) the power of unmodulated carrier component.

25. A 35 MHz carrier is modulated by a 1 kHz sine wave. If the carrier voltage is 5 V and the maximum deviation is 25 kHz, give the equation of this modulated wave for FM and PM. What will be the equations if the modulating frequency is changed to 5 kHz with other parameters remaining the same?

Answers to problems

1. 5
2. 95.54 MHz
3. 60 MHz
4. 24 kHz
5. 180 kHz
6. 1 kHz/V, 3 kHz
7. 0.25 rad/v, 0.75 rad
8. 12.5 kHz
9. 48 kHz, 24
10. 80 kHz
11. Set of sidebands 2, 6, and 10
12. 30 kHz
13. For PM: message signal = $5\cos\omega_m t$, Frequency deviation = $10 f_m$

    For FM: message signal = $5 f_m \sin\omega_m t$, Frequency deviation = $10 f_m$ kHz
14. 20 kHz, 10 kHz, 120 kHz
15. 0.067W, 0.933W
16. FM modulation index = 8

    AM Modulation index = 0.46
17. Bandwidth is more for the case with $\beta = 5$
18. (a) 3

    (b) Carrier = $0.77\, Ec$

        first sideband = $0.44\, Ec$

        second sideband = $0.11\, Ec$

        third sideband = $0.02\, Ec$

    (c) Frequency spectrum:

Frequency	Amplitude
500 kHz,	$0.77\, Ec$
499 kHz, 501 kHz	$0.44\, Ec$
498 kHz, 502 kHz	$0.11\, Ec$
497 kHz, 503 kHz	$0.02\, Ec$.

19. (a) 100 MHz

    (b) 40 kW

    (c) 2

    (d) 10 kHz

    (e) 26.9 kW

    (f) 1.352 kW.
20. 0.9 V
21. 34.8 MHz, two tripler
22. 120.02 kHz, 120.8 kHz, 122 kHz, 128 kHz and 140 kHz.

    bandwidth determined by number of significant sidebands.
23. 142 kHz
24. Frequency Amplitude

$$f_c \cdots 50 \text{ V}$$

$$f_c \pm f_m \cdots 3.5 \text{ V}$$

$$f_c \pm 2f_m \cdots 18 \text{ V}$$

$$f_c \pm 3f_m \cdots 21.5 \text{ V}$$

$$f_c \pm 4f_m \cdots 14 \text{ V}$$

$$f_c \pm 5f_m \cdots 6.5 \text{ V}$$

$$\text{BW} = 75 \text{ kHz}$$

25. (a) PM: $f(t) = 5\sin(2.19 \times 10^8 t + 25\sin 6.283 \times 10^3 t)$

    FM: $f(t) = 5\sin(2.19 \times 10^8 t + 25\sin 6.283 \times 10^3 t)$

    (b) PM: $f(t) = 5\sin(2.19 \times 10^8 t + 25\sin 31.415 \times 10^3 t)$

    FM: $f(t) = 5\sin(2.19 \times 10^8 t + 5\sin 31.415 \times 10^3 t)$

MATLAB EXAMPLES

```
1.%Frequency modulation and demodulation%

fc=10000;
fs=100000;
f1=200; f2=500;

t=0:1/fs:((2/f1)-(1/fs));
x1=cos(2*pi*f1*t);
x2=cos(2*pi*f1*t)+ cos(2*pi*f2*t);
kf=2*pi*(fc/fs)*(1/max(max(x1)));
kf=kf*(f1/fc);

%modulation%
opt=10*kf;
y1=modulate(x1,fc,fs,'fm',opt);
subplot(521); plot(x1); title('original single tone message');
subplot(522); plot(y1); title('time domain FM single tone');
fx1=abs(fft(y1,1024)); fx1=[fx1(514:1024) fx1(1:513)];
f=(-511*fs/1024):(fs/1024):(512*fs/1024);
subplot(524);plot(f,fx1);title('freq description single tone
dev=10*fm');

%demodulation

x1_recov=demod(y1,fc,fs,'fm',opt);
subplot(523); plot(x1_recov);
title('time domain recovered,singlr tone, dev=10*fm')
```

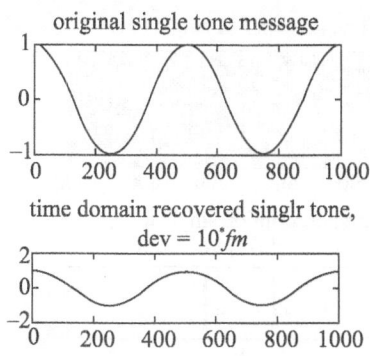

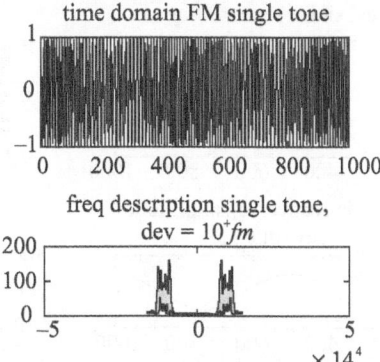

2.%Phase modulation and demodulation%

```
fc=10000;
fs=100000;
f1=200; f2=500;

t=0:1/fs:((2/f1)-(1/fs));
x1=cos(2*pi*f1*t);
x2=cos(2*pi*f1*t)+ cos(2*pi*f2*t);
kp=pi/max(max(x1));

%modulation%

opt=kp/6;
y1=modulate(x1,fc,fs,'pm',opt);
subplot(521); plot(x1); title('original single tone message');
subplot(522); plot(y1); title('time domain PM single tone');
fx1=abs(fft(y1,1024)); fx1=[fx1(514:1024) fx1(1:513)];
f=(-511*fs/1024):(fs/1024):(512*fs/1024);
subplot(524);plot(f,fx1);title('freq description single tone
dev=pi/6');

%demodulation

x1_recov=demod(y1,fc,fs,'pm',opt); subplot(523); plot(x1_recov);
title('time domain recovered,singlr tone, dev=pi/6')

%repeat phase deviation=pi/3

opt=kp/3
y1=modulate(x1,fc,fs,'pm',opt);fx1=abs(fft(y1,1024));
fx1=[fx1(514:1024) fx1(1:513)]; subplot(526); plot(f,fx1);
title('freq description single tone,dev=pi/3');
x1_recov=demod(y1,fc,fs,'pm',opt); subplot(525); plot(x1_recov);
title('time domain recovered,singlr tone, dev=pi/3')
```

original single tone message

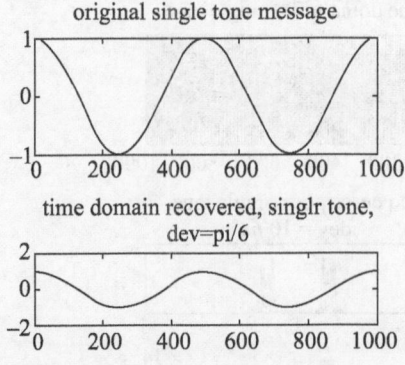

time domain PM single tone

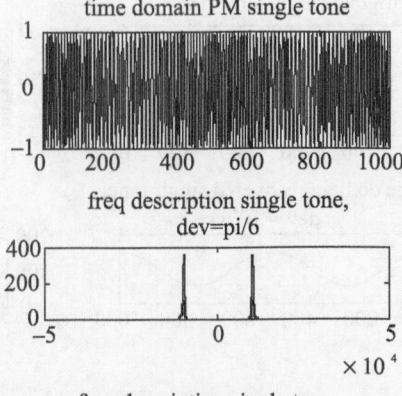

time domain recovered, singlr tone, dev=pi/6

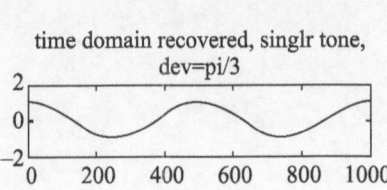

freq description single tone, dev=pi/6

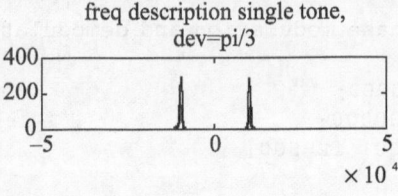

time domain recovered, singlr tone, dev=pi/3

freq description single tone, dev=pi/3

Pulse Modulation

This chapter will enable the students to
- Know what is pulse modulation and how it differs from continuous wave modulation
- Explain how to sample a signal so that it can be reproduced later
- Understand sampling theorem, aliasing error, and Nyquist rate
- Know what are the different types of pulse modulation and how data is encoded in case of pulse code modulation
- Define quantization noise and know how to minimize it and hence improve the resolution
- Know different types of multiplexing: FDM and TDM
- Solve some numerical problems to understand the theoretical concepts

5.1 INTRODUCTION

In Chapters 3 and 4, we studied what is known as continuous wave (CW) modulation, in which parameters like amplitude, phase, and frequency of the carrier wave are varied continuously in accordance with the modulating signal. It is also possible to vary the amplitude, position, and width of a pulse train in accordance with the modulating signal. There are two types of pulse modulation: analog and digital.

In analog pulse modulation, the carrier wave is a periodic pulse train. The amplitude, position, and width of the carrier pulse train are varied in a continuous manner in accordance with the corresponding sample value of message signal. Thus, in this case, information is transmitted basically in analog form, but the transmission takes place at discrete times.

In the case of digital pulse modulation, the message signal is represented in a form that is discrete in both time and amplitude, thus enabling it to be transmitted as a sequence of coded pulse. This type of modulation is also called pulse code modulation (PCM). This type of modulation has no continuous wave counterpart. PCM is the most widely used form in the field of telecommunication. Digital data transmission provides a higher level of noise immunity, more flexibility in the bandwidth, power trade-off, possibility of providing more security to data, and ease of implementation using large scale integrated circuits. Figure 5.1 shows the above types of modulation.

The four predominant methods of pulse modulation are:

- pulse width modulation (PWM)
- pulse position modulation (PPM)
- pulse amplitude modulation (PAM)
- pulse code modulation (PCM)

Pulse modulation consists essentially of sampling analog information signals and then converting those samples into discrete pulses and transporting the pulses from a source to a destination over a physical transmission medium.

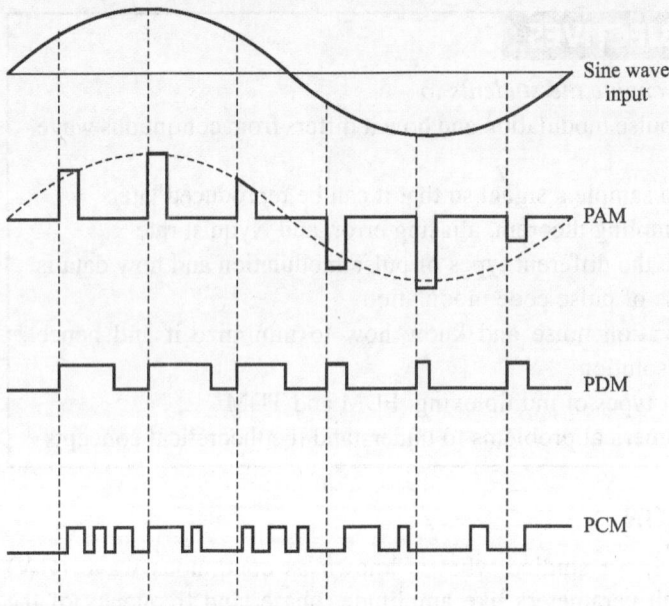

Fig. 5.1 Pulse modulation techniques

To have digital transmission, we have to first convert the analog signals into digital form. This involves three steps. First the analog signal has to be sampled so that we obtain discrete time continuous valued signal from the analog signals. This step is called sampling.

Then these sampled values are quantized, i.e., rounded off to a finite number of values. This is called quantization process. After quantization, we have a discrete time, discrete amplitude signal.

The third step in analog to digital conversion is encoding, i.e., a sequence of bits (ones and zeros) is assigned to different outputs of the quantizer. Since these outputs are finite, each sample of the signal can be represented by a finite number of bits. Since the first step to have digital transmission is sampling, let us go into details.

5.2 SAMPLING THEOREM

The sampling theorem is one of the most important theorems in the analysis of signals. It has widespread applications in signal processing and communications. This theorem and its numerous applications highlight the importance of frequency domain signal analysis. The latest signal processing techniques and the digital communication methods are based on the validity of this theorem and the insight it provides. The idea behind proper sampling of signal is quite simple. Suppose we sample a continuous signal in some manner and able to reconstruct it from the samples then we have done the sampling properly.

Let us see some examples of sampling. Figure 5.2 shows several sinusoids after sampling.

(a) In the first case as shown in Fig. 5.2(a), the analog signal is a DC signal. It can be considered as a cosine wave of zero frequency. Since the analog signal is a series of straight lines between each of the sample, it can be reconstructed easily from the samples. This is regarded as proper sampling.

(b) In the second case as shown in Fig. 5.2(b), about 11.1 samples are taken per cycle, i.e., a 1 kHz sine wave being sampled at approximately 11000 samples/s. It can be seen that from the samples, original analog signal can be reconstructed, i.e., by joining these samples by a smooth curve, original signal can be obtained. This is also a case of proper sampling.

(c) In the third case as shown in Fig.5.2(c), three samples are taken per cycle, i.e., a 1 kHz sine wave is sampled at 3000 samples/s. Although the number of samples are nearly 3.5 times less than the previous case still the signal can be reconstructed from the sample.

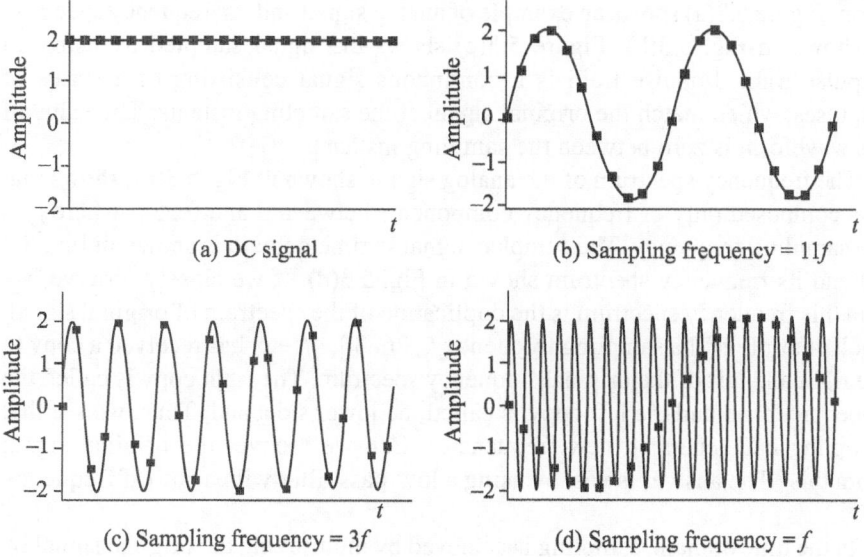

(a) DC signal

(b) Sampling frequency = $11f$

(c) Sampling frequency = $3f$

(d) Sampling frequency = f

Fig. 5.2   Examples of proper and improper sampling: (a) – (c) proper sampling and (d) undersampling resulting in aliasing

(d) In the fourth case as shown in Fig. 5.2(d), only 1.5 samples/cycle are taken. Since the samples are too less, we can infer from the figure that the reconstructed signal from the samples nowhere matches the original signal. A 1 kHz original signal is reconstructed as 52 Hz signal. This phenomenon of sinusoids changing frequency due to insufficient sampling is known as *aliasing*. The original sine wave has hidden its true identity completely. This is an example of improper sampling. The answer to this question is what is known as *sampling theorem* or *Shannon's sampling theorem*. Also it is some times called Nyquist sampling theorem after the paper published by both in the year 1940.

Hence, sampling theorem can be stated as follows: A continuous signal can be reconstructed in full or properly sampled only if it does not contain frequency component above one-half of the sampling rate. For example a sampling rate of 3000 samples/s requires the analog signal to be composed of frequencies below 1500 cycles/s. If frequencies above this limit are present in the signal, they will be aliased to frequencies between 0 and 1500 Hz/s, combining with whatever information that was legitimately there.

While discussing sampling theorem, two terms are widely used: The *Nyquist frequency* and the *Nyquist rate*. The Nyquist frequency is twice the highest frequency present in the signal and Nyquist rate is the minimum sampling rate which is equal to one-half of the sampling rate.

5.2.1 Occurrence of Aliasing Error

Let us go into more detailed analysis of sampling and how aliasing occurs. Let us see what happens to the information when a signal is converted to a discrete form. Figure 5.3(a) shows an example of analog signal and its frequency spectrum is shown in Fig. 5.3(b). Figure 5.3(c) shows the signal sampled by using an impulse train. Impulse train is a continuous signal consisting of a series of impulses, which match the original signal at the sampling instants. The value of the waveform is zero between the sampling instants.

The frequency spectrum of the analog signal, shown in Fig. 5.3(b), shows that it is composed only of frequency components between 0 and $0.35 f_s$ where f_s is the sampling frequency. The sampled signal in time domain is shown in Fig. 5.3 (c) and its frequency spectrum shown in Fig. 5.3(d). If we closely observe, we find this frequency spectrum is the duplication of the spectrum of original signal. Each multiple of the sampling frequency f_s, $2f_s$, $3f_s$, $4f_s$ etc. has received a copy to the right and left of the original frequency spectrum. The right copy is called the upper sideband and the left copy is called the lower sideband. Thus, we see that sampling has generated new frequencies. Can we recover the original signal from this? The answer is yes, by using a low-pass filter with a cut-off frequency $f_s/2$.

In the time domain, sampling is achieved by multiplying the original signal by an impulse train of unity amplitude. When two time domain signals are multiplied their frequency spectra are convolved. This results in the original spectrum being duplicated to the location of each impulse in the impulse train. Viewing the original signal as composed of both positive and negative frequencies accounts for the upper and lower sidebands respectively.

Figure 5.3(e) shows undersampling resulting from too low sampling rate. The analog signal still contains frequencies up to 3.3 kHz, but the sampling rate has been lowered to 5 kHz. The f_s, $2f_s$, $3f_s$, ⋯ along the horizontal axis are spaced closer in Fig. 5.3(f) compared to Fig. 5.3(d).

This figure shows that the duplicated portion of spectrum overlap each other thus, invading the other's territory. Since there is no way to separate the overlapping frequencies, the original signal is lost and cannot be reconstructed. This overlap occurs when the analog signal contains frequencies greater than one-half of the sampling rate. This results in aliasing. Thus, we have indirectly proved the sampling theorem.

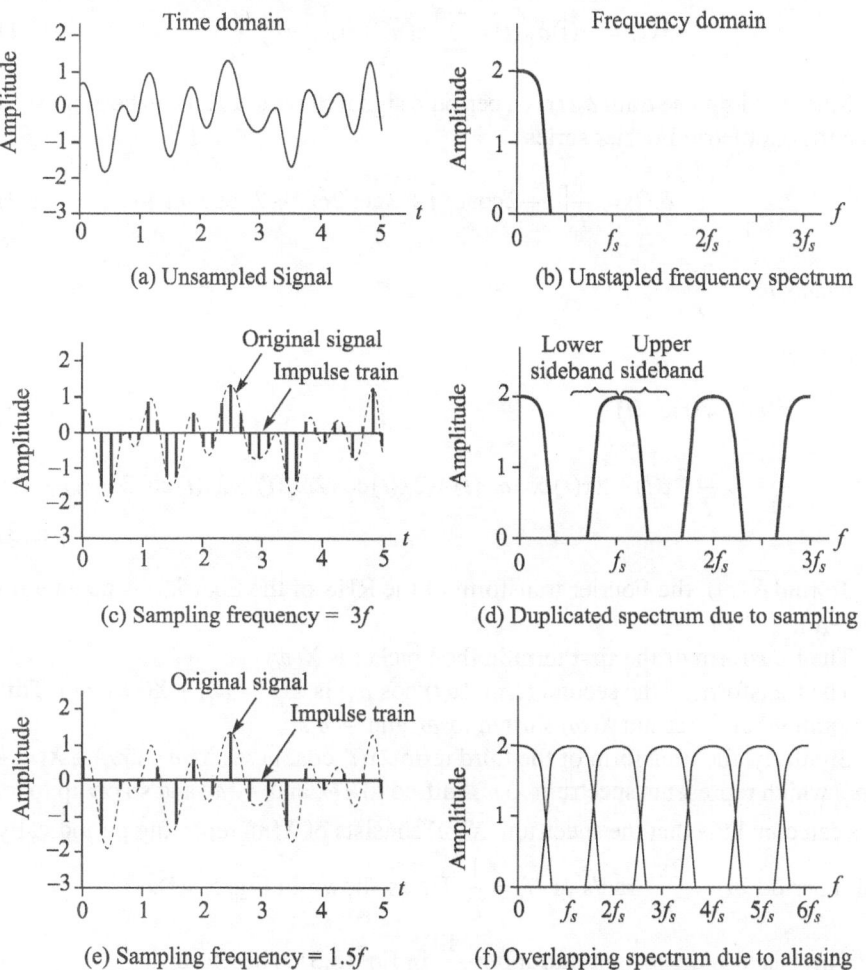

Fig. 5.3   Sampling theorem in time and frequency domain: (a) original analog signal, (b) its frequency spectrum, (c) sampling at 3 times highest frequency by an impulse train, (d) its frequency spectrum showing infinite number of upper and lower sidebands (e) signal sampled at 0.66 of the sampling frequency, and (f) overlapping in the frequency domain due to aliasing, resulting in Infinite number of upper and lower sidebands

5.2.2 Mathematical Proof of Sampling Theorem

Let us take a real signal whose spectrum is bandlimited to B Hz, i.e., [$X(\omega) = 0$ for $|\omega| > 2\pi B$]. This signal can be reconstructed exactly from its samples taken uniformly at a rate $f_s > 2B$ samples/s or the minimum sampling frequency is $f_s = 2B$ Hz.

To prove this theorem, let us consider a signal $x(t)$ whose spectrum is bandlimited to B Hz. Figure 5.4(a) shows $x(t)$ and its spectrum given by Figure. 5.4(b) is bandlimited to B Hz. Now multiplying $x(t)$ by an impulse train $\delta_T(t)$ consisting of unit impulses repeating periodically every T seconds, where $T = 1/f_s$. These are shown in Figures 5.4(c) and 5.4(d).

The resulting sampled signal $\overline{x}(t)$ is shown in Fig. 5.4(e). The sampled signal consists of impulses spaced every T seconds. The nth impulse located at $t = nT$, has a strength $x(nT)$, the value of $x(t)$ at $t = nt$ the sampled signal

$$\overline{x}(t) = x(t)\delta_T(t) = \sum_n x(nT)\delta(t - nT) \tag{5.1}$$

Since the impulse train $\delta_T(t)$ is a periodic signal of period T, it can be expressed as a trigonometric Fourier series

$$\delta_T(t) = \frac{1}{T}\left[1 + 2\cos\omega_s t + 2\cos 2\omega_s t + 2\cos 3\omega_s t + \cdots\right] \tag{5.2}$$

where $\omega_s = \dfrac{2\pi}{T} = 2\pi f_s$

Therefore,

$$\overline{x}(t) = x(t)\delta_T(t)$$

$$= \frac{1}{T}\left[x(t) + 2x(t)\cos\omega_s(t) + 2x(t)\cos 2\omega_s(t) + 2x(t)\cos 3\omega_s(t) + \cdots\right] \tag{5.3}$$

To find $\overline{X}(\omega)$, the Fourier transform of the RHS of the Eq. (5.3) is taken term by term.

The transform of the first term in the bracket is $X(\omega)$.

The transform of the second term $2x(t)\cos\omega_s t$ is $X(\omega - \omega_s) + X(\omega + \omega_s)$. This is nothing but spectrum $X(\omega)$ shifted to ω_s and $-\omega_s$.

Similarly, the transform of the third term $2x(t)\cos 2\omega_s t$ is $X(\omega - 2\omega_s) + X(\omega + 2\omega_s)$ which represents spectrum $X(\omega)$ shifted to $2\omega_s$ and $-2\omega_s$ and so on up to ∞. We can conclude that the spectrum $\overline{X}(\omega)$ consists of $X(\omega)$ repeating periodically with period $\omega_s = \dfrac{2\pi}{T}$ rad/s or $f_s = \dfrac{1}{T}$ Hz as shown in Fig. 5.4(f).

There is also a constant multiplier $\dfrac{1}{T}$ in Eq. (5.3). Therefore,

$$\overline{X}(\omega) = \frac{1}{T}\sum_{n=-\infty}^{\infty} X(\omega - n\omega_s) \tag{5.4}$$

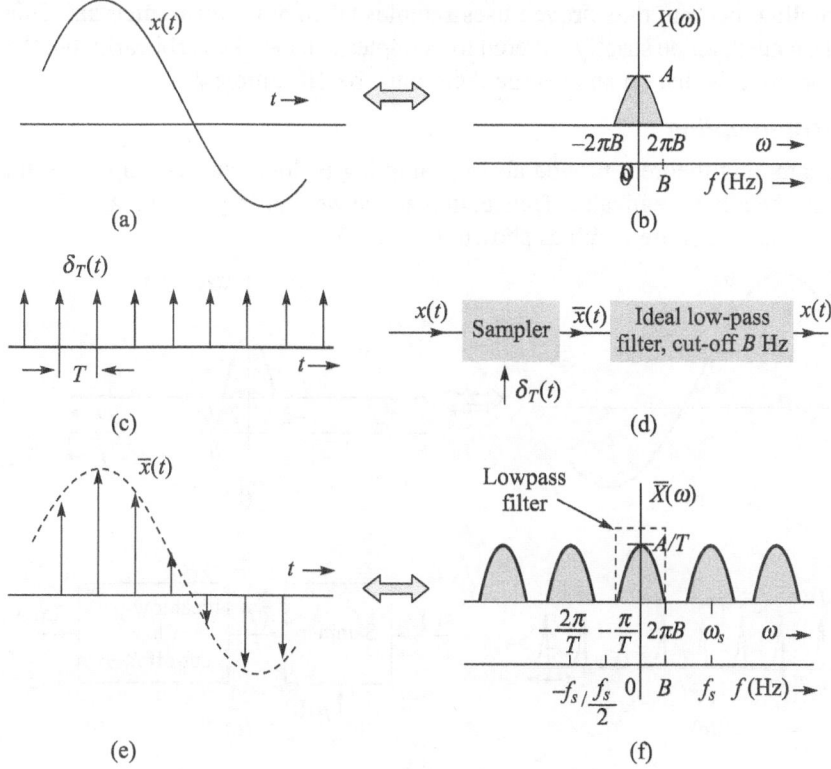

Fig. 5.4   (a) Original signal in time domain, (b) its frequency spectrum, (c) sampling signal– impulse train, (d) block diagram, (e) sampled signal in time domain, and (f) frequency spectrum of sampled signal

If we have to reconstruct $x(t)$ from $\overline{x}(t)$ we must recover $X(\omega)$ from $\overline{X}(\omega)$. This recovery is possible if there is no overlap between successive cycles of $\overline{X}(\omega)$.

This requires $f_s > 2B$ (5.5)

Since the sampling interval $T = \dfrac{1}{f_s}$,

$$T < \frac{1}{2B}$$ (5.6)

Hence, as long as the sampling frequency f_s is greater than twice the signal bandwidth B in Hz, $\overline{X}(\omega)$ consists of non-overlapping repetitions of $X(\omega)$. To recover the original signal we have to just pass this signal through an ideal LPF whose bandwidth is any where between B and $f_s - B$ Hz.

The minimum sampling rate $f_s = 2B$ required to recover $x(t)$ from its samples $\overline{x}(t)$ is called the *Nyquist rate* for $x(t)$ and the corresponding sampling interval $T = \dfrac{1}{2B}$ is called *Nyquist interval* for $x(t)$.

Sampling theorem thus proved uses samples taken at uniform intervals. This condition need not be strictly adhered to. Samples can be taken arbitrarily but the only condition is that on an average there must be $2B$ samples/s.

Practical sampling

In the sampling theorem proved above, sampling is done with an impulse train that is physically unrealizable. Hence, in practice we multiply the signal $x(t)$ by a train of pulses of finite width as shown in Fig. 5.5.

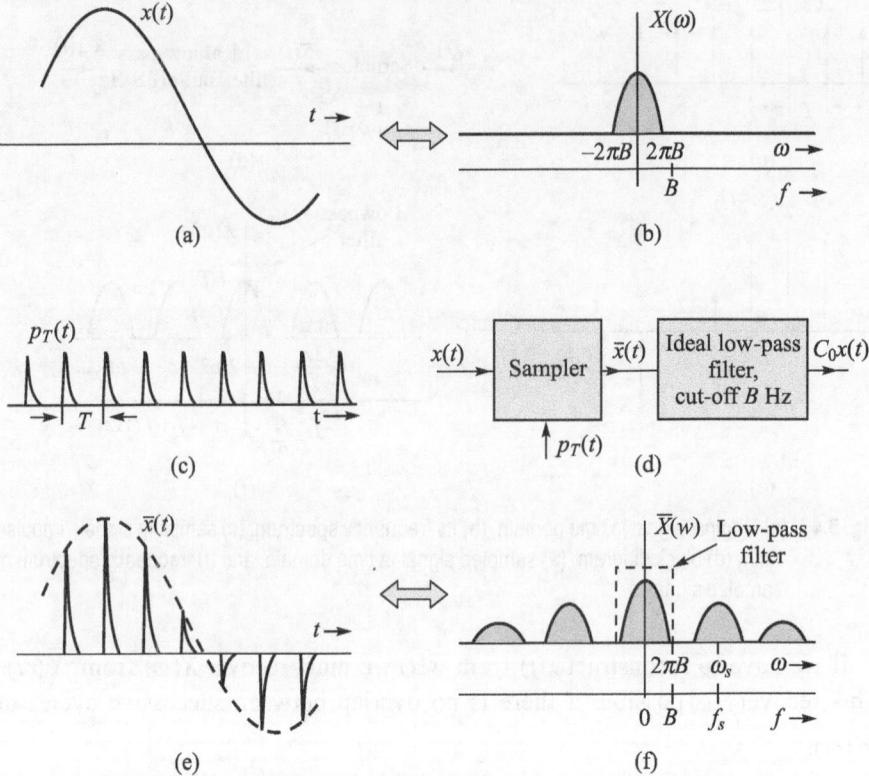

Fig. 5.5 Practical sampling: (a) original signal in time domain, (b) its frequency spectrum, (c) sampling by a train of pulses of finite width, (d) block diagram, (e) sampled signal in time domain, and (f) sampled waveform in frequency domain

The sampler and the LPF to recover back the signal is also shown. The recovery of $x(t)$ requires the knowledge of Nyquist sample values. Since the sampling signal is periodic, Fourier series expansion of this signal gives

$$y_T(t) = C_0 + \sum_{n=1}^{\infty} C_n \cos(n\omega_s t + \theta_n)$$

where

$$\omega_s = \frac{2\pi}{T}$$

and

$$\bar{x}(t) = x(t) y_T(t) = x(t) \left[C_0 + \sum_{n=1}^{\infty} C_n \cos(n\omega_s t + \theta_n) \right]$$

$$= C_0 x(t) + \sum_{n=1}^{\infty} C_n x(t) \cos(n\omega_s t + \theta_n) \quad (5.7)$$

The sampled signal $\bar{x}(t)$ consists of

$$C_0 x(t),\ C_1 x(t) \cos(\omega_s t + \theta_1),\ C_2 x(t) \cos(2\omega_s t + \theta_2), \cdots$$

The first term $C_0 x(t)$ of the desired signal and all other terms are its harmonics centred at $\pm\,\omega_1,\ \pm 2\omega_s,\ \pm 3\omega_s,\ \cdots$ as shown in Fig. 5.5(f). The signal $x(t)$ can be recovered by low pass filtering $\bar{x}(t)$. As before, it is implied that $\omega_s > 4\pi B$, i.e., $f_s > 2B$, where B is the bandwidth of the signal.

We have seen from the sampling theorem that we can replace a continuous time signal by a discrete sequence of numbers. Hence, processing continuous time signal is equivalent to processing discrete sequence of numbers. The continuous time signal is sampled and sampled values are used to modify certain parameters of a periodic pulse train.

We can vary the amplitude, width, or position of the pulses in proportion to the sample values of the continuous signal. Hence, depending on the case, we have pulse amplitude modulation (PAM), pulse width modulation (PWM), or pulse position modulation (PPM). There is also one more type of pulse modulation called pulse code modulation (PCM) in which the amplitude of sampled pulses are converted to a binary code and then transmitted.

In these types of modulation, instead of transmitting the continuous signal, we transmit the corresponding pulse modulated signal. At the receiver, we read the information of pulse modulated signal and reconstruct the analog signal.

There is also an added advantage of pulse modulation. Since a pulse modulated signal occupies only a part of the channel time, we can transmit several pulse modulated signals on the same channel on time sharing basis. This is called time division multiplexing (TDM). By reducing the pulse width several signals can be transmitted on the same channel.

Example 5.1 A single tone 3.5 kHz is sampled with 8 kHz signal and 5 kHz signal. (a) What are the frequencies present after sampling? (b) What will be the output if the sampled signals are passed through a low-pass filter in each case.

Solution
(a) For 8 kHz sampling, the frequencies present will be

$\pm$ 3.5 kHz, $\pm$ (8 – 3.5 kHz) = $\pm$ 4.5 kHz, $\pm$ (8 + 3.5 kHz) = $\pm$ 11.5 kHz, $\pm$
(2 $\times$ 8 – 3.5) kHz = $\pm$ 12.5 kHz, $\pm$ (2 $\times$ 8 + 3.5) kHz = $\pm$ 19.5 kHz, etc.

For 5 kHz sampling, frequencies present will be

$\pm$ 3.5 kHz, $\pm$ (5 – 3.5 kHz) = $\pm$ 1.5 kHz, $\pm$ (5 + 3.5 kHz) = $\pm$ 8.5 kHz, $\pm$
(2 $\times$ 5 – 3.5 kHz) = 6.5 kHz, $\pm$ (2 $\times$ 5 + 3.5 kHz) = $\pm$ 13.5 kHz, etc.

(b) For 8 kHz sampling, the maximum frequency the LPF can be designed to pass is < 8/2 = 4 kHz. Hence, it can pass 3.5 kHz and can reconstruct the original signal.

For 5 kHz sampling, the maximum frequency the LPF can be designed to pass is < 5/2 = 2.5 kHz. Hence, it can pass only 2.5 kHz signal, which means that it cannot pass the original 3.5 kHz signal. This is due to undersampling.

Example 5.2 A signal $x(t) = 2\sin 4000\,\pi t + 3\sin 5000\,\pi t + 4\sin 8000\,\pi t$ has to be truly represented by its samples. Find the minimum sampling rate from low-pass sampling theorem consideration and bandpass consideration.

Solution

(a) Highest frequency component of the signal, $f_h = 8000/2 = 4000$ Hz

Lowest frequency component of the signal, $f_1 = 4000/2 = 2000$ Hz

Minimum sampling frequency from low-pass consideration $= 2 \times f_h = 8000$ Hz

(b) The bandwidth $B = f_h - f_1 = 4000 - 2000 = 2000$ Hz

Integer factor $n = f_h/B = 4000/2000 = 2$

Required sampling frequency in this case $= 2f_h/n = (2 \times 4000)/2 = 4000$ Hz

5.3 PULSE AMPLITUDE MODULATION (PAM)

It is the simplest form of pulse modulation. In this modulation, the signal is sampled at regular interval and each sample is made proportional to the amplitude of the signal at the instant of sampling. These pulses are sent over a distance with or without modulation. Figure 5.6 illustrates the pulse amplitude modulation. This figure shows two types of PAM—double polarity and single polarity. In single polarity system, a fixed DC level is added to the signal to ensure that the pulses are always positive.

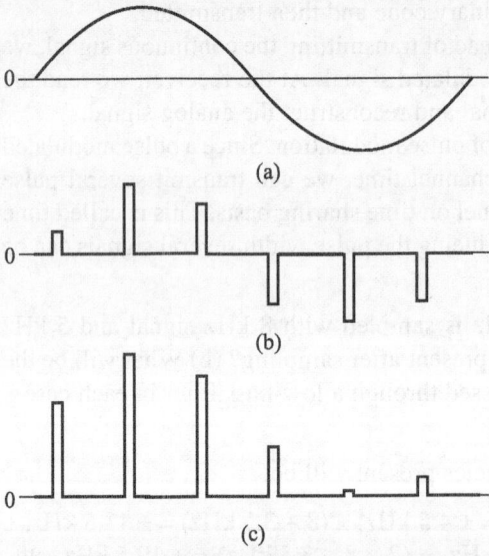

Fig. 5.6   Pulse amplitude modulation: (a) original signal, (b) double polarity, and (c) single polarity

5.3.1 Channel Bandwidth for PAM

Let there be N independent baseband signals $m_1(t)$, $m_2(t)$, $m_3(t)$, $\cdots$. Each of these signals is bandlimited to f_M. The baseband signal $m_1(t)$ must be sampled at intervals not longer than $1/2f_M$, i.e., $T_s = 1/2f_M$. Between successive samples of $m_1(t)$ will appear samples of other $(N - 1)$ signals. Hence, the interval of separation between successive samples of different baseband signals is $1/2f_M N$. The composite signal consists of a sequence of samples that is a sequence of impulses. If the bandwidth of the channel is quite large, then the waveform at the receiving end would be same as that of the transmitting end. If the channel bandwidth is limited, then there will be overlapping of the samples and the output will not be a single baseband signal after filtering through LPF, but instead will be a combination of many baseband signals. This sort of mixing is called cross talk and should be avoided.

Let us take an ideal low-pass filter with a cut-off frequency $\omega_c = 2\pi f_c$, no delay, and unity gain. Let sample be taken of $m_1(t)$ at $t = 0$. Then at $t = 0$ on impulse of strength, $S_1 = m_1(0)\,dt$. The response at the receiving end is $y_1(t)$, which is given by

$$y_1(t) = \frac{S_1\omega_c}{\pi} \frac{\sin \omega_c t}{\omega_c t} \tag{5.8}$$

In Fig. 5.7, the solid curve shows the normalized response $\pi y(t)/\omega_c$. At $t = 0$, the response attains a peak value proportional to the strength of the impulse $S_1 = m_1(0)\,dt$, which in turn is proportional to the value of the sample $m_1(0)$. This response lasts indefinitely. It can be observed that the response passes through zero at intervals that are multiples of $\pi/\omega_c = 1/2f_c$. If a sample $m_2(t)$ is taken and transmitted at $t = 1/2f_c$ and if $S_2 = m_2(t)\,dt$, then the response at the receiving end is

$$y_2(t) = \frac{S_2\omega_c}{\pi} \frac{\sin \omega_c(t - 1/2f_c)}{\omega_c(t - 1/2f_c)} \tag{5.9}$$

This is shown with the dotted curve in Fig. 5.7. If the demultiplexing is also done by instantaneous sampling at the receiving end of the channel for $m_1(t)$ at $t = 0$ and for $m_2(t)$ at $t = 1/2f_c$, in spite of the performance of the channel response, there will be no cross talk and the signals $m_1(t)$ and $m_2(t)$ may be completely separated and individually recovered. Similarly, additional signals may be sampled and multiplexed, provided each new sample is taken synchronously every $1/2f_c$ second. The sequence repeats itself every $1/2f_M$ so that each signal is properly sampled. Hence, with a channel of bandwidth f_c, we need separate samples by intervals $1/2f_c$. The sampling theorem requires that the samples of an individual baseband signal be separated by intervals not larger than $1/2f_M$. Hence, the total number of signals which may be multiplexed is $N = f_c/f_M$ or $f_c = Nf_M$. This implies that multiplexing a number of signals by PAM time division requires no more bandwidth than would be required to multiplex these signals by FDM using SSB transmission.

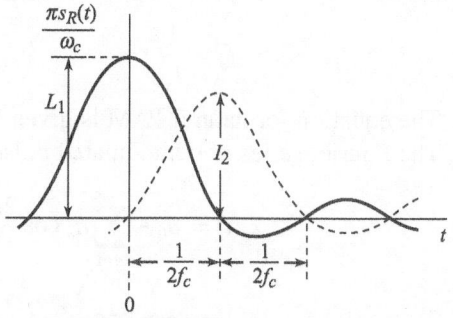

Fig. 5.7 Ideal low-pass filter response to an instantaneous sample at $t = 0$ indicated by solid curve and response at $t = 1/2f_c$ indicated by dashed curve

5.3.2 Natural Sampling

In pulse amplitude modulation, the amplitudes of the pulses are varied in accordance with the modulating signal. Let the modulating signal be $m(t)$. The pulse amplitude modulation is achieved by multiplying the carrier with $m(t)$ signal as shown in Fig. 5.8. The balanced modulators are frequently used as multipliers for this purpose. The output is a series of pulses, the amplitudes of which vary in proportion to the modulating signal. This type of pulse amplitude modulation is called natural PAM, since the top of the pulses is not flat but follows the shape of the modulating signal. The pulse train signal to the modulator (natural PAM signal) is shown in Fig. 5.9, which when switched on allows samples of the modulating signal to pass through the output.

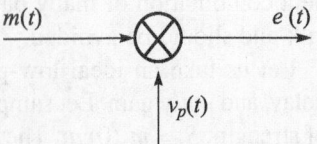

Fig. 5.8 Product modulator: PAM generation

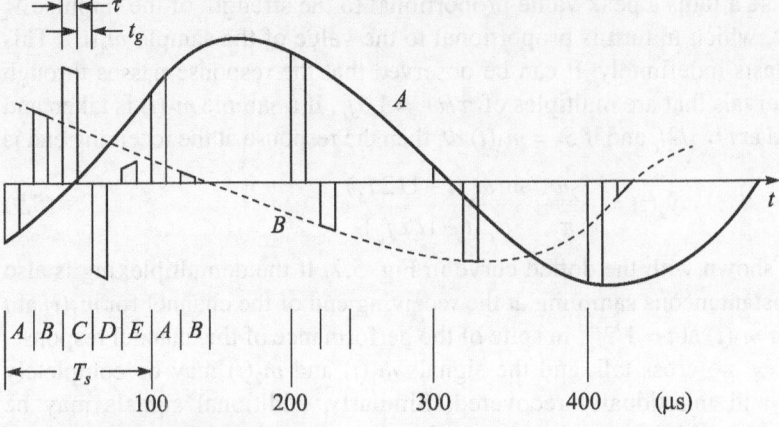

Fig. 5.9 Natural PAM signal

The sampling period is thus the time period of the pulse train T_s and the sampling frequency is

$$f_s = \frac{1}{T_s} \tag{5.10}$$

The equation for natural PAM is given below.
The Fourier series of demodulated pulse train is

$$x_p(t) = a_0 + \sum_{n=1}^{n=\infty} a_n \cos \frac{2\pi nt}{T_s} \tag{5.11}$$

$$= a_0 a_1 \cos \frac{2\pi t}{T_s} + a_2 \cos \frac{4\pi t}{T_s} + \cdots \tag{5.12}$$

The modulating pulse train is then

$$e(t) = m(t) \times (t)$$

$$= a_0 m(t) + a_1 m(t) \cos \frac{2\pi t}{T_s} + a_2 m(t) \cos \frac{4\pi t}{T_s} + \cdots \tag{5.13}$$

The RHS shows that the modulated wave consists of the modulating signal multiplied by the *DC* term a_0 and a series of double sideband suppressed carrier components resulting from the harmonics in the waveform. The modulating signal spectrum $M(f)$ is shown in Fig. 5.10.

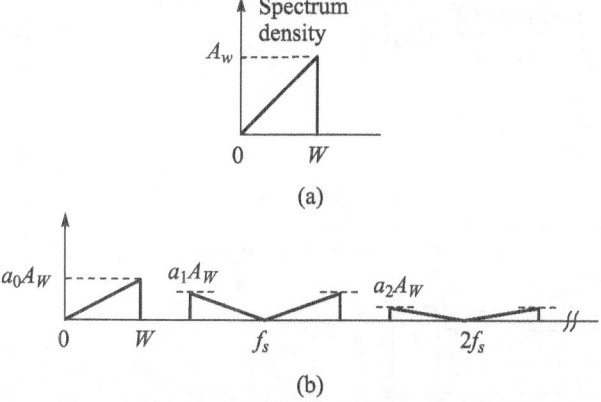

Fig. 5.10   (a) Modulating signal spectrum $M(f)$ and (b) spectrum of natural PAM wave

5.3.3 Flat Top Sampling

The sampled signal with a sequence of pulses of varying amplitude whose tops are not flat but follow the waveform of signal $m(t)$ is seldom used because of the recovery problem; instead the flat top sampling method is employed. If the pulse width of the carrier pulse train used in natural sampling is made very short compared to the pulse period, the natural PAM becomes what is called *instantaneous PAM* . Samples in this case represent the modulating signal at the instant of sampling. This is also called *flat top sampling*. But short pulse means less energy per sample and the magnitudes of the coefficients a_0, a_1, a_2 are proportional to the pulse width. Hence, to maintain reasonable pulse energy, a sample and hold circuit (shown in Fig. 5.11) is employed. A periodic train of short clocking pulses $p(t)$ closes the transistor switch Q_1 allowing instantaneous samples of analog signal to be passed on to the capacitor for a time T. The delayed clocking pulses $p(t)$ operated the transistor switch Q_2 to discharge the capacitor before the next sample arrives.

This is how the flat samples are formed. In the sampling of this type, the baseband signal $m(t)$ cannot be recovered exactly by simply passing the samples through an ideal LPF. But it has a merit that it simplifies the design of electronic circuitry. The extent of distortion can be found out as follows.

Let $M(j\omega)$ be the Fourier transform of $m(t)$. The transform of the sampled signal for flat top sampling is determined by considering that the flat top pulse can be generated by passing instantaneously sampled signal through a network that broadens a pulse of duration dt (an impulse) into a pulse of duration γ. The transform of a pulse of unit amplitude and width dt is

$$\text{Æ [impulse of strength } dt \text{ at } t = 0] = dt \tag{5.14}$$

The transform of a pulse of unit amplitude and width τ is

$$H(j\omega) = \frac{\tau}{dT} \frac{\sin(\omega\tau/2)}{\omega\tau/2} \tag{5.15}$$

Let the signal $m(t)$ with transform $M(j\omega)$ be bandlimited to f_M and be sampled at Nyquist rate or faster. Then in the range 0 to f_M, the transform of the flat topped sampled signal is given by the product

$$H(j\omega)\,M(j\omega) = \frac{\tau}{T_s}\frac{\sin(\omega\tau/2)}{\omega\tau/2}M(j\omega) \qquad (5.16)$$

where $\qquad 0 \le f \le f_M$

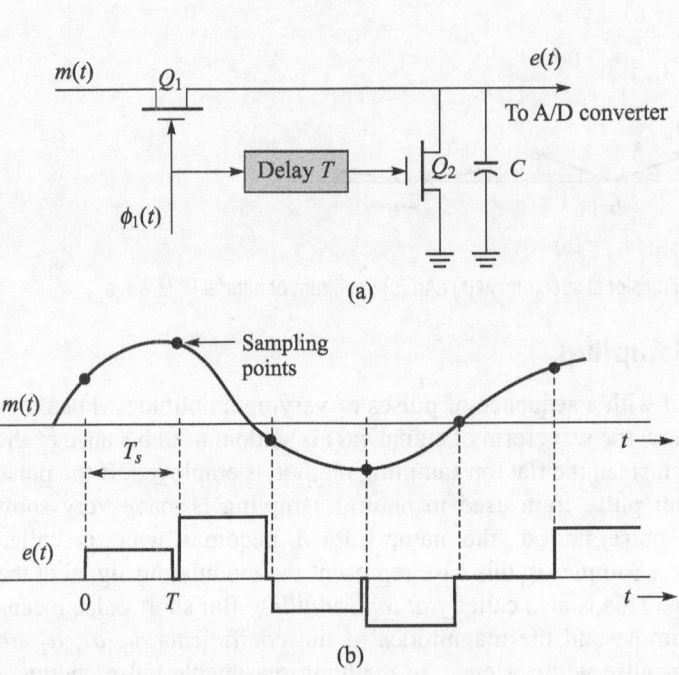

(a)

(b)

Fig. 5.11 Flat topped PAM generation: (a) sample and hold circuit and (b) sampled waveform

Assume that the signal $m(t)$ has a flat spectral density equal to M_0 over its entire range from 0 to f_M as shown in Fig. 5.12(a). Figure 5.12(b) shows the form of transform instantaneous sampled signal. The sampling frequency $f_s = 1/T_s$ is assumed large enough to allow for a guardband between the spectrum of the baseband signal and the DSBSC signal with carrier f_c. The spectra of the flat topped sampled signal is shown in Fig. 5.12. The interested part of the spectrum is in the range 0 to f_M. If in this range, the spectra of the sampled signal and the original signal are identical, then the original signal is recovered by an LPF. But in practice, this is not so. There is always a distortion. This distortion is due to the fact that the original signal was bandlimited and that it is observed through a finite rather than an infinite signal time aperture. Hence, this distortion is called *aperture effect distortion*. This distortion results from the fact that the spectrum is multiplied by the sampling function $\sin x/x$ with $x = \omega\tau/2$. The magnitude of the sampling function falls off slowly with increasing x in the neighbourhood for $x = 0$ and does not fall off sharply until we approach $x = \pi$ at which point the sampling function is zero. If $x = \pi$ corresponds to a frequency very large in comparison with f_M, the distortion is minimized. Since $x = \pi/\tau$, the frequency f_0 corresponding to $x = \pi$ is $f_0 = 1/\tau$.

If $f_0 \geq f_M$ or $1/\tau \ll 1/f_M$, the aperture distortion will be small. The distortion becomes progressively smaller with decreasing τ. As $\tau \to 0$, which is instantaneous sampling, the distortion approaches zero.

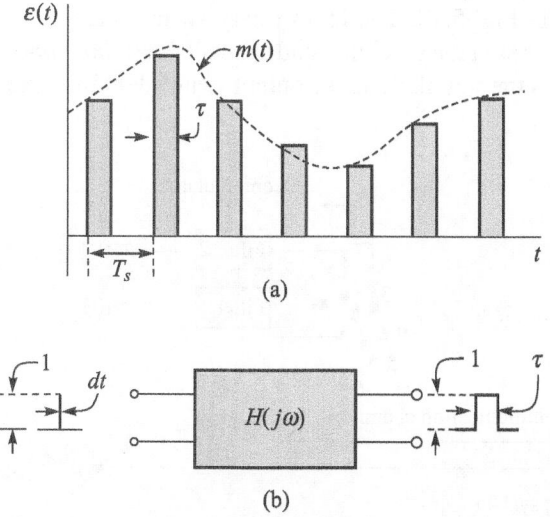

Fig. 5.12 (a) Flat topped sampling and (b) network with transfer function $H(j\omega)$ converting pulse of width dt into a rectangular pulse of same amplitude but with pulse duration τ

Equalization

Making τ as large as possible helps increase the output amplitude. But this causes distortion, which when not acceptable is corrected by using equalizers in cascade with the output low-pass filter. Equalizer is a passive network whose transfer function has a frequency dependence of the form $x/\sin x$ [inverse to the form $H(j\omega)$]. The equalizer in combination with the aperture effect will then yield a flat overall transfer characteristic between the original baseband signal and the output at the receiving end of the system.

5.3.4 Pulse Amplitude Modulation and Time Division Multiplexing (TDM)

Figure 5.13 shows a simple TDM system. The advantage of the sampling principle is that we can time multiplex various bandlimited signals on the same channel.

At the transmitting end, a number of bandlimited signals are connected to the contact point of a rotary switch. The assumption is that the signals are similarly bandlimited like voice signals limited to 3 kHz. As the rotary arm of the switch moves around, it samples each signal sequentially. The rotary switch at the receiving end is in synchronism with the switch at the sending end. The switches make contact simultaneously at similar numbered contacts. Thus, one sample is taken at a time and sent to the receiving end. The train of samples passes through low-pass filters connected to each contact. The output of the filter gives the reconstructed original signal. The switch must make $2f_M$ revolutions/s, where f_M is the highest spectral frequency component present in any of the input signals. Rotary mechanical switches are used when the signals to be multiplexed vary slowly with time so that the sampling rate is correspondingly slow. This rotary switch arrangement is shown in Fig. 5.13. The input rotary switches are called *commutator* and the rotary switches at the receiving end are called *decommutator*, i.e., the input switch multiplexes the samples and the output switch demultiplexes them.

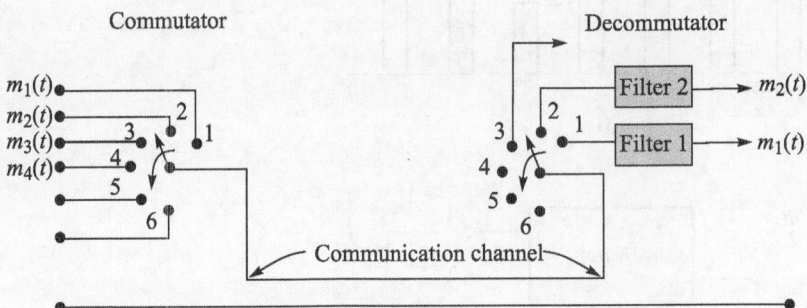

Fig. 5. 13 Time division multiplexing system

Figure 5.14 shows two signals sampled at regular interval of T_s, $m_1(t)$ is sampled first and $m_2(t)$ next. But each is sampled at regular interval of T_s. From the figure, it can be observed that the train of pulses corresponding to the samples of each signal is modulated in amplitude in accordance with the signal itself. This is nothing but the PAM we have explained previously. Since each sample is allotted a time slot, this system is an example of time division multiplexing (TDM).

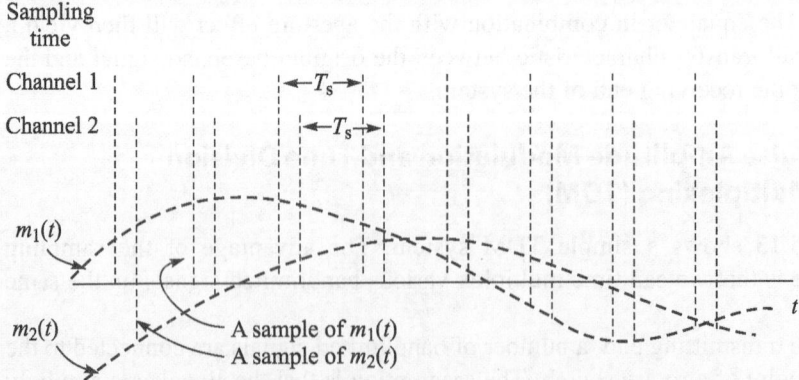

Fig. 5.14 Interlacing of two baseband signals

5.3.5 Signal Recovery

The sampled and TDM signals can be recovered by what is known as holding. A sample and hold circuit is shown in Fig. 5.15. In this figure, the baseband signal $m(t)$ and its flat topped samples are shown. After demultiplexing, the sample pulses are extended at the receiver, i.e., the sample value of each individual baseband signal is held until the occurrence of the next sample of same baseband signal. This results in a staircase waveform with no blank intervals.

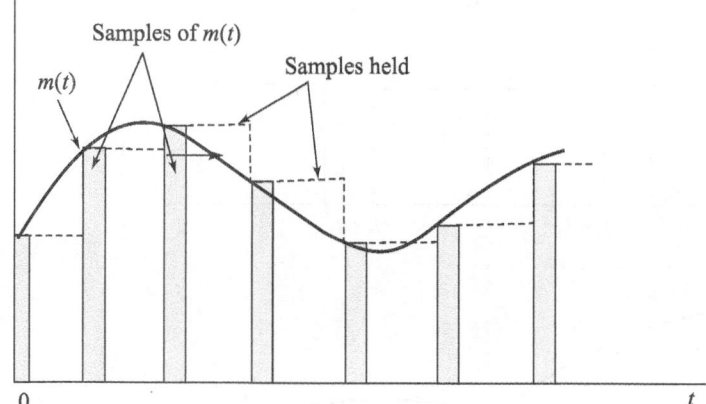

Fig. 5.15   Sample and hold circuit waveforms

A simple holding circuit is shown in Fig. 5.16. Input samples of a baseband signal is given to an amplifier and its output is connected through a switch S to the capacitor C across which the output is taken. Switch S normally opens and closes after occurrence of the leading edge of a sample pulse and opens just before the occurrence of trailing edge. The amplifier should have a low output impedance so that when switch S is closed the capacitor charges rapidly to the voltage proportional to the sample value. The capacitor holds this voltage until the next sample. Although the figure shows abrupt transition in voltage level from one sample to another, in practice these voltage transitions will be somewhat rounded since the capacitor charges and discharges exponentially. Since the sample interval is very small compared to the interval between the samples, the voltage variation of the baseband signal during the sampling interval is small enough and hence can be neglected.

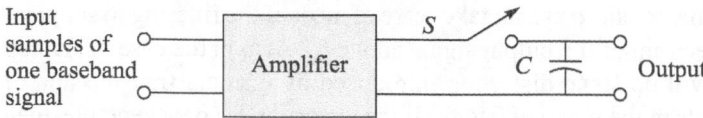

Fig. 5.16   Holding circuit

Let the baseband signal be $m(t)$ with spectral density
$$M(j\omega) = \text{Æ}\,[m(t)] \tag{5.17}$$

Consider that the flat tops have been stretched to encompass the entire interval between instantaneous samples. Then the spectral density is similar to Eq. (5.17) except with τ replaced by the time interval between samples. Thus, we have

$$\text{Æ}[m(t)] = \frac{\sin(\omega T_s/2)}{\omega T_s/2}\,M(j\omega) \tag{5.18}$$

Figure 5.17 shows the spectrum of instantaneously sampled signal, magnitude of the aperture effect factor, and spectrum of sample and held signal.

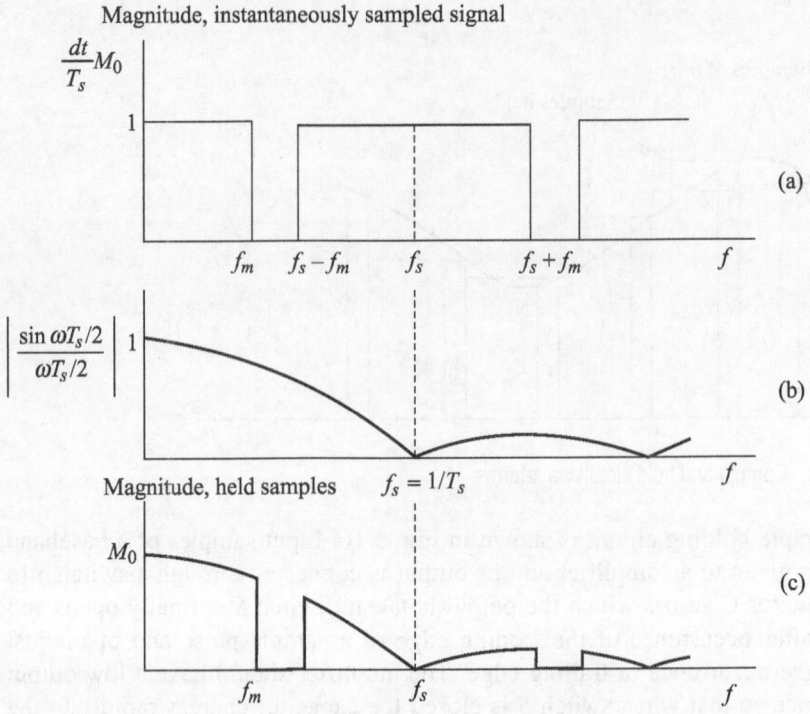

Magnitude, instantaneously sampled signal

Fig. 5.17 (a) Idealized spectrum of sampled signal $m(t)$, (b) aperture effect factor, and (c) sample and held signal's spectrum

In Fig. 5.17(a), we have assumed that the bandlimited signal $m(t)$ has a flat spectral density of magnitude M_0. Figure 5.17(b) indicates the magnitude of the aperture factor and Fig. 5.17(c) shows the magnitude of the spectrum of the sampled and held signal. The nulls of $\sin x/x$ term occur at f_s. The aperture effect, which is responsible for $\sin x/x$ term, takes care of most of the filtering to suppress the part of the spectrum of the output signal above f_M. Also in the case of flat top sampling, there will be some distortion introduced by unequal transmission of special components in the range of 0 to f_M. If the distortion is not acceptable, then an equalizer $x/\sin x$ corrects the distortion.

As in the case of FM modulation, where the amplitude of the carrier remains constant, there are two types of pulse modulation in which the pulse amplitude remains constant and either the pulse width or its position is varied according to modulating signal. As in the case of FM amplitude, limiters can be used to provide good degree of noise immunity.

5.4 PULSE WIDTH MODULATION (PWM)

Pulse width modulation is also sometimes called pulse duration modulation. In this modulation, both the amplitude and the starting time of each pulse are fixed but its width is changed proportional to the amplitude of the signal at that particular instant. Figure 5.18 shows a pulse width modulation system. It can be observed from the figure that the maximum pulse width occurs at E_{max} and the minimum pulse width at $- E_{max}$. The average pulse width occurs at zero of the modulating signal. For a recurrence rate of 8000 pulses/s, the time interval between pulses is 125 µs. This interval is adequate not only to accommodate the varying width but also to permit TDM.

The disadvantage of PWM is that the pulses are of varying width and hence the power content of pulses also vary. Hence, the transmitter has to take care of this.

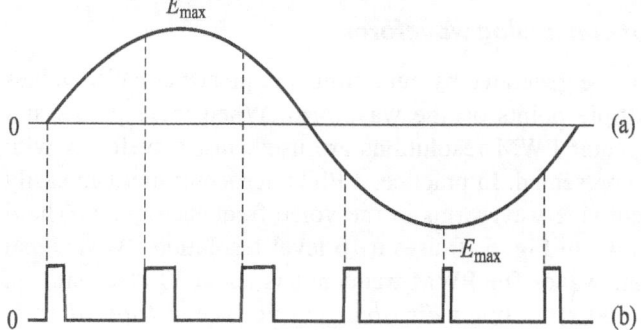

Fig. 5.18   PWM:(a) modulating signal and (b) modulated waveform

Pulse width modulation can be thought of as using digital pulses to create some analog value other than just *high* and *low* signal levels. Many digital systems are powered by a 5 volt power supply. So, if we can filter a signal that has a 50% duty cycle, then an average voltage of 2.5 volt is obtained. Other duty cycles produce any voltage in the range of 0 to 100% of the *high* voltage, depending upon the PWM resolution.

The *duty cycle* is defined as the percentage of digital high to digital low signals present during a PWM period. The *PWM resolution* is defined as the maximum number of pulses that we can pack into a PWM period. The PWM period is an arbitrarily time period in which PWM takes place. It is chosen to give best results for a particular use.

5.4.1 Uses of PWM

PWM is used for the following purposes:

(i) To digitally create an analog output voltage level for control functions and power supplies

(ii) To digitally create analog signals for arbitrary waveforms, sounds, music, and speech.

Using PWM to generate an analog voltage level

A common use is in power supplies. The PWM resolution is selected to be equal to or greater than the resolution requirements of the power supply. A 5 volt power supply that can be adjusted to +/– 1 millivolt should use a PWM resolution of 5000 or greater. The PWM output is then filtered to obtain acceptable ripple. The filter can be a simple low-pass filter.

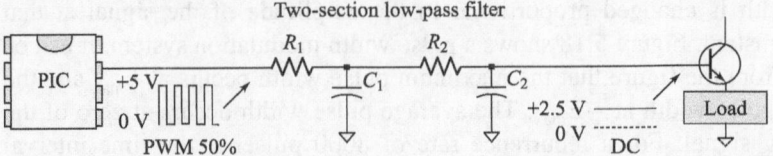

Fig. 5.19 Two-section low-pass filler

Figure 5.19 shows a PIC microcontroller generating a 50% duty cycle PWM signal at 5000 Hz, a two-section 5000 Hz low-pass filter, and a pass transistor with a direct current input of +2.5 V. The filter frequency is $1/2\pi RC$ for each section.

Using PWM to generate an analog waveform

Any shape waveform can be generated by outputting a sequence of PWM values that correspond to multiple points on the waveform. When more points are, heavier filtering and greater PWM resolutions are used, fast waveforms with great accuracy can be represented. In practice, a PIC microcontroller can easily output reasonably decent sine waveforms in the voice frequency range. The 4 kHz sine waveform shown in Fig. 5.20 uses a 16 level resolution PWM signal and 16 points on the sine wave. The PWM frequency is about 30 kHz using an *on/off* cycle, and about 500 kHz using a *distributed* cycle. The *distributed* cycle produces smoother results, as shown in Fig. 5.20.

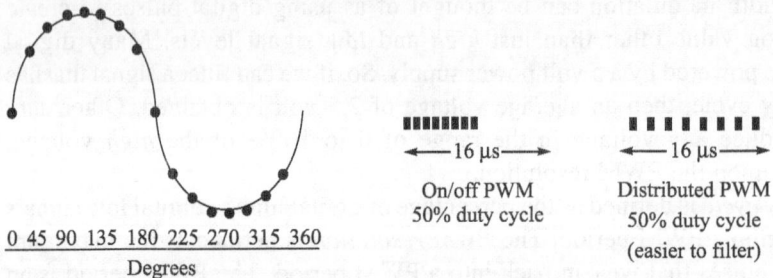

Fig. 5.20 The 4 kHz sine waveform

The on/off PWM duty cycle It starts with eight pulses *high* for the first part of the PWM cycle, then finishes with eight *low* pulses for the remainder of the PWM cycle. The ratio of the percent *high* time to the percent *low* time is called the duty cycle. Note that the PWM frequency of the *on/off* PWM for any duty cycle in the drawing above is: $F = 1/P = 1/16$ µs $= 62.5$ kHz.

The distributed PWM cycle It also uses eight *high* pulses and eight *low* pulses as does the previous example, but spreads them out during the entire PWM cycle. Note that the PWM frequency of the *distributed* PWM 50% duty cycle in the drawing above is: $F = 1/P = 1/2$ µs $= 500$ kHz. This is great if most of our PWM is around 50% duty cycle but the PWM frequency gradually slows to 62.5 kHz as you approach 0% or 100%.

5.4.2 Why the PWM Frequency is Important

The PWM is a large amplitude digital signal that swings from one voltage extreme to the other and this wide voltage swing takes a lot of filtering to smooth out. When the PWM frequency is close to the frequency of the waveform that is being generated, then any PWM filter will also smooth out the generated waveform and drastically reduce its amplitude. So, a rule of thumb is to keep the PWM frequency much higher than the frequency of the waveform being generated. Finally, filtering pulses is not just about the pulse frequency but about the duty cycle and how much energy is in the pulse. The same filter will do better on a low or high duty cycle pulse compared to a 50% duty cycle pulse. Because the wider pulse has more time to integrate to a stable filter voltage and the smaller pulse has less time to disturb it.

5.5 PULSE POSITION MODULATION (PPM)

In this type of modulation, the amplitude and the width of the pulse are kept constant. The position of each pulse with respect to the position of the reference pulse is varied proportional to the modulating signal. To recover the signal back at the receiving end, the transmitter must send synchronizing pulses to operate timing circuits in the receiver. Unlike the PWM, where the power varies depending on the pulse width, PPM requires constant transmitter power output. Both PWM and PPM are also called pulse time modulation since in these types of modulation one of the timing parameter of the pulse is changed.

5.6 GENERATION OF PAM

The circuit to generate PAM is shown in Fig. 5.21. As shown in the figure, the carrier pulse is connected to the base circuit and the message signal to the emitter of the transistor.

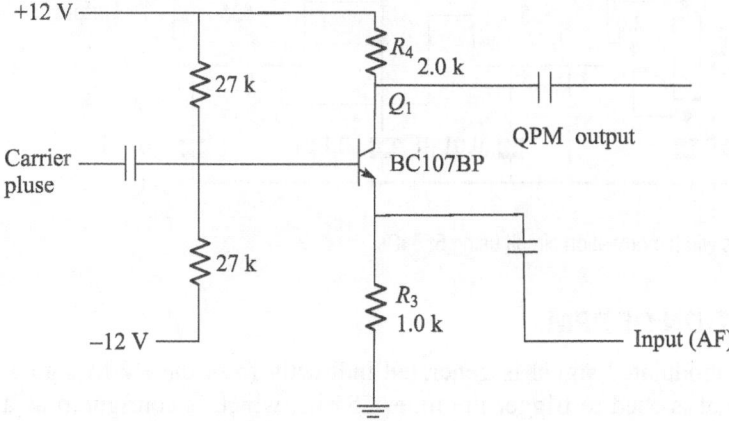

Fig. 5.21   PAM generation

The transistor is held in a cut-off state by the voltage divider bias circuit. The amplitude of the message signal given to the base has to be selected to make the amplifier still in a cut-off state. The amplifier is switched on by the large amplitude square wave pulse at the input. The amplifier is never hard driven to saturate the transistor and the base drive combined with the instantaneous level of the emitter signal will result in a pulsed voltage at collector having pulse amplitude that is proportional to the level of the message signal at the emitter.

The amplifier is class D type and the power dissipation is quite less. The frequency of the square wave should be at least five times higher than the highest frequency of the message signal. This takes care of the Nyquist rate.

5.7 GENERATION OF PWM

Pulse width modulated signal can be generated using 555 timer IC. The circuit is shown in Fig. 5.22. The timer is connected in monostable mode. It is triggered with a continuous train of pulses.

The circuit given below uses two 555 ICs and is actually a combination of two types of circuit. In the example shown, the first is a free-running multivibrator (astable) with an adjustable frequency around 30 Hz. The output of this circuit then triggers a pulse shaping (monostable) circuit, which adjusts the width of the pulse. The circuit produces a duty cycle in the range of approximately 0.3% to 97%. This is a test circuit where 30 Hz generated by first 555 IC is given as a modulating signal. In actual case, only the second 555 IC is used and the modulating signal is given as a trigger input to the second 555 IC.

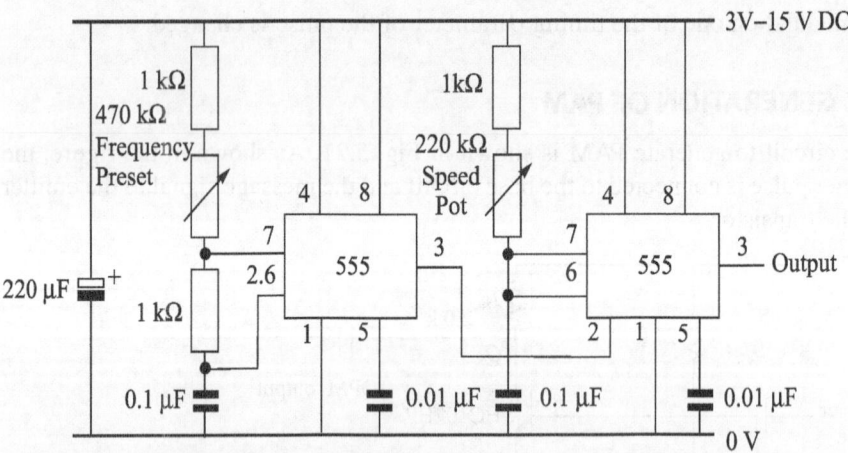

Fig. 5.22 A pulse width modulation circuit using 555 ICs

5.8 GENERATION OF PPM

Pulse position modulated signal is generated indirectly from the PWM signal. The PWM signal is used to trigger the timer 555 IC, which is configured as a monostable circuit. The pulse width of the one shot is fixed by the timing resistor and capacitor. The differentiated PWM signal acts as a trigger to this monostable circuit. The trigger pulse interval depends on the PWM signal. Since the width of the PWM signal varies proportional to the message sign, the trigger interval is not constant and it depends on the width of PWM signal, hence, proportional to the message signal. The circuit diagram is shown in Fig. 5.23.

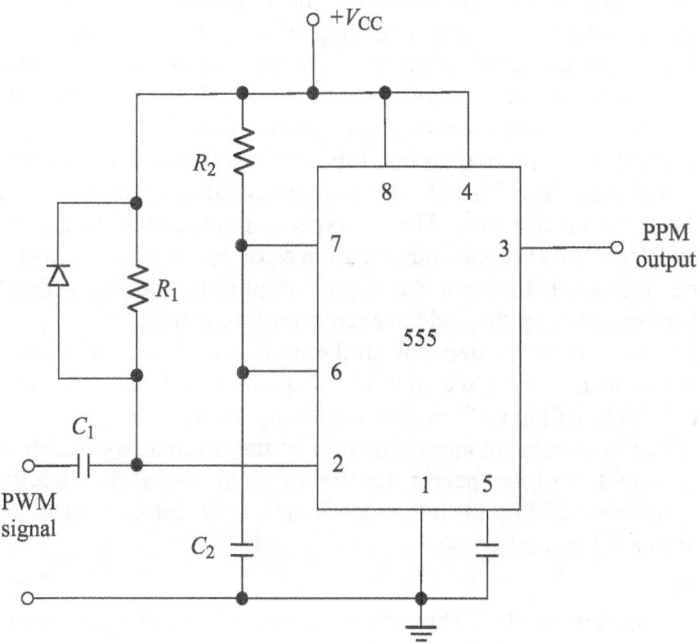

Fig. 5.23 PPM generation from PWM signal

5.9 PULSE CODE MODULATION (PCM)

So far we have discussed PAM, PWM, and PPM. Although all these methods of modulation score over conventional analog modulation, they have their own shortcomings.

The disadvantage of the PAM is that any noise riding on the signal changes the pulse height which introduces distortion. To avoid this sort of problem, PWM is used, where the information is carried by the pulse duration or length rather than the amplitude or height as in the case of PAM. The PWM is less susceptible to noise than PAM. However, any distortion of the pulse shape may change the apparent pulse duration, thereby producing a distorted output signal.

Pulse code modulation (PCM) offers a method of overcoming some of the disadvantages of other types of pulse modulation. In the PCM, the instantaneous amplitude of the sample is represented by a binary code resulting in a series of ones and zeros or mark and space. All pulses have the same height and same shape since only ones and zeros are sent. The receiver has only to detect the presence or absence of a pulse. A distorted pulse does not degrade the signal as long as the pulse can still be recognized. Hence, PCM is less sensitive to noise than PAM or PWM.

5.9.1 PCM Basics

The PCM technique achieves considerable immunity to noise and other transmission difficulties. The message signal is sampled and then coded before transmission. Before it is coded for transmission, the analog signal is sampled just as in other forms of pulse modulation. The range of the pulse amplitude from zero to full scale is then divided into a number of discrete steps so that each step can be represented by a particular arrangement of binary pulses in the form of ones and zeros. This coded arrangement of binary pulses is the PCM signal.

Since the PCM signal is an approximation of the original signal to the nearest discrete level at each sampling instant, the reconstructed waveform at the receiving end of the system is distorted. The reason is quantization of the signal. The error introduced due to this is called quantization error or quantization noise. It results from the difference between the signal amplitude and the nearest quantization level represented by the code at each sampling instant.

The number of quantizing levels depends on the number of bits in that code, i.e., if there are n bits in the code, the number of quantizing level is 2^n. For example, an eight bit code will have $2^8 = 256$ quantizing levels.

Thus, a PCM signal is always an approximation of the original signal. How far the quantized signal will be nearer to the original signal is directly proportional to the number of bits in that code. Hence, a 16 bit code is more accurate compared to a 10 bit code.

PCM word format

The actual encoding occurs in an ADC. Three different types of encoding processes, viz. successive approximation, dual slope integration, and flash types, are generally used in an ADC. The type of encoding depends on the application, e.g., dual slope conversion is used for industrial application where immunity from noise is more important than speed.

Basic PCM encoder

Figure 5.24 shows a simplified block diagram of a PCM encoder. A number of transducer signals are applied to the input of a multiplexer switch. Signals are sampled at a rate and in any order as defined by the user. Selection of channels can be programmable.

The output of the multiplexer switch is PAM signal, which carries TDM samples of each input channel. The ADC samples this PAM signal one at a time and the serial output from the ADC is formatted into a PCM waveform.

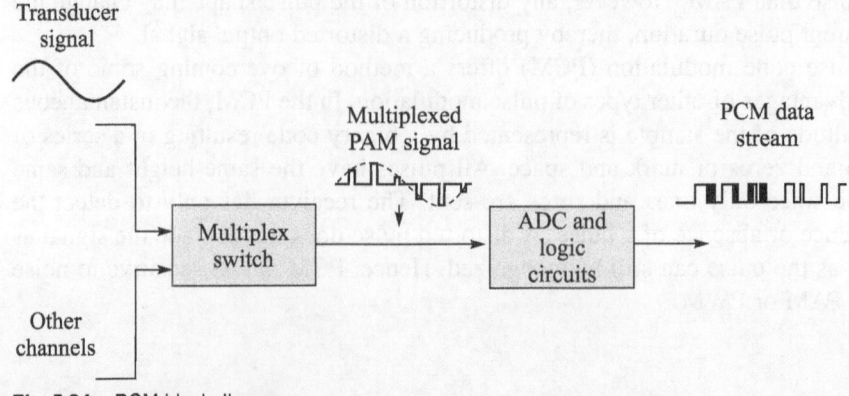

Fig. 5.24 PCM block diagram

5.10 PCM TRANSMITTER AND RECEIVER

Figure 5.25 shows a typical PCM transmitter and receiver. The input analog signal to the transmitter is first passed through a bandpass filter that bandlimits the signal to avoid aliasing. This analog signal is converted to a PAM signal by the sample and hold circuit. The analog-to-digital converter converts the PAM signal to a digital signal and the parallel output from the ADC is converted to a serial PCM code stream and transmitted through the communication channel. Regenerative repeaters are used to improve the signal-to-noise ratio along the channel.

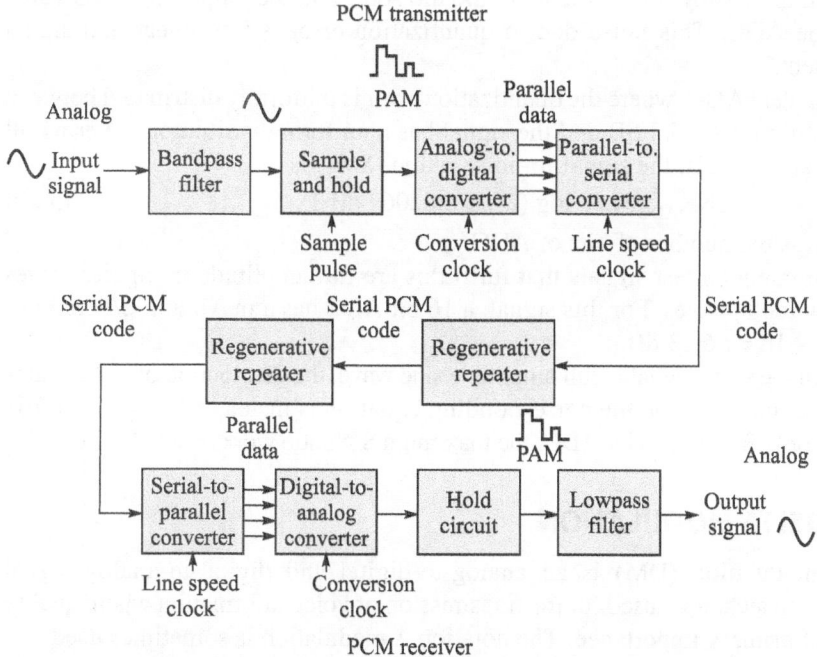

Fig. 5.25 PCM transmitter and receiver

The received signal enters a serial-to-parallel converter. The converted parallel data is given to a digital-to-analog converter and the original PAM waveform is recovered by a hold circuit. This PAM signal is passed through a low-pass filter to get back the message signal in analog form.

5.10.1 Quantization

Quantization is the process of approximating a continuous range of values or a very large set of possible discrete values by relatively a small set of discrete symbols or integer values.

A typical application of quantization is in the conversion of discrete signal or a sampled continuous signal into a digital signal. Generally, sampling or multiplexing and quantizing are done in an ADC. One such application is compact disk (DC) in which audio signal is sampled at 44,100 Hz and quantized with 16 bits, which can be one of 65536 (2^{16}) possible values.

The simplest best known form of quantization is scalar quantization. It operates on scalar input data as compared to multidimensional vector. The scalar quantization operator can be represented as

$$Q(x) = g\ [f(x)] \tag{5.19}$$

where x is a real number to be quantized.

Quantization error

The difference between the actual analog value and the quantized digital value is called *quantization error*. This error can be due to rounding or truncation in the PCM and is introduced by quantization in the ADC. It is the rounding error between the analog input voltage to the the ADC and the output digitized value from the ADC. This noise due to quantization error is non-linear and signal dependent.

In an ideal ADC, where the quantization error is uniformly distributed between $-1/2$ LSB and $+1/2$ LSB and the signal has a uniform distribution covering all quantization levels, the signal-to-noise ration (*SNR*) is

$$\text{SNR}_{\text{ADC}} = 20 \log\ (2^n) \approx 6.0206n \text{ dB} \tag{5.20}$$

where n is the number of bits of ADC.

Most common test signals that fulfil this are full amplitude triangular waves and saw tooth waves. For this signal, a 16 bit ADC has a maximum S/N ratio of $6.0206 \times 16 = 96.33$ dB.

When the input signal is full amplitude sine wave, the distribution of the signal is no longer uniform and the corresponding equation is instead $\text{SNR}_{\text{ADC}} = (1.761 + 6.0206n)$ dB. For a 16 bit ADC, the maximum S/N ratio works out to be 98.08 dB.

5.11 DELTA MODULATION

Delta modulation (DM) is an analog-to-digital and digital-to-analog signal conversion technique used for the transmission of voice information where quality is not of primary importance. The notation Δ modulation is sometimes used.

Delta modulation is the simplest form of the differential pulse code modulation (DPCM), where the difference between successive samples is encoded into n bit data streams. In this type of modulation, the transmitted data is reduced to a 1 bit data stream. Its main features are given below.

- The analog signal is approximated with a series of segments.
- These segments are compared to the original analog signal to determine the increase or decrease in the relative amplitude.
- The comparator determines this increase or decrease.
- If there is any change in the amplitude level, i.e., if there is any change in the information, then this change of information is only sent.
- That is to say, only an increase or decrease of the signal amplitude from the previous sample is sent. If there is no change in the signal, then the modulated signal remains at the same 0 or 1 of the previous sample.
- To achieve high signal-to-noise ratio, delta modulation uses oversampling, i.e., the analog signal is sampled at a rate several times the Nyquist rate.

5.11.1 Principle

The differential pulse code modulation is the other name for delta modulation. In this method, the derivative of the signal is quantized. When signal changes between the sample periods are small, the quantized word length can be reduced. When the sampling rates are very high, the changes between the sample periods are made very small and hence the quantizer can be reduced to low bit. A 1 bit DPCM encoder is known as a delta modulation. The delta modulator codes the differences in the signal amplitude instead of the signal amplitude itself.

Figure 5.26 shows a delta modulation encoder. It is also called a single integration modulator. In this modulator, the input signal is compared to the integrated output pulses and the difference signal (delta) is applied to the quantizer. The quantizer generates a positive pulse when the difference signal is negative and a negative pulse when the difference signal is positive. The integrator moves step by step closer to the present value due to the difference signal fed to the quantizer, thus tracking the derivative of the input signal.

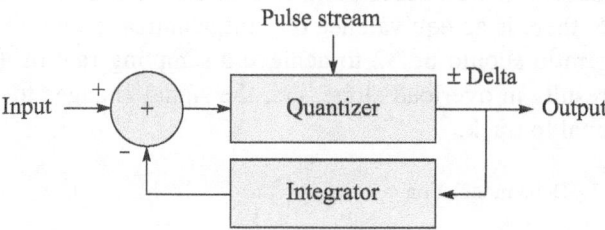

Fig. 5.26 Delta modulation encoder

Let us consider a 1.5 kHz input signal with a maximum amplitude 1 V and delay to be 0.0625. Sampling is done with 4 bit quantization, i.e., 16 levels. To achieve a bit rate equivalent to 4 bit quantization with 4 kHz sampling rate, an oversampling ratio of 16 is needed, i.e., 64 kHz is required. Figure 5.27 shows a

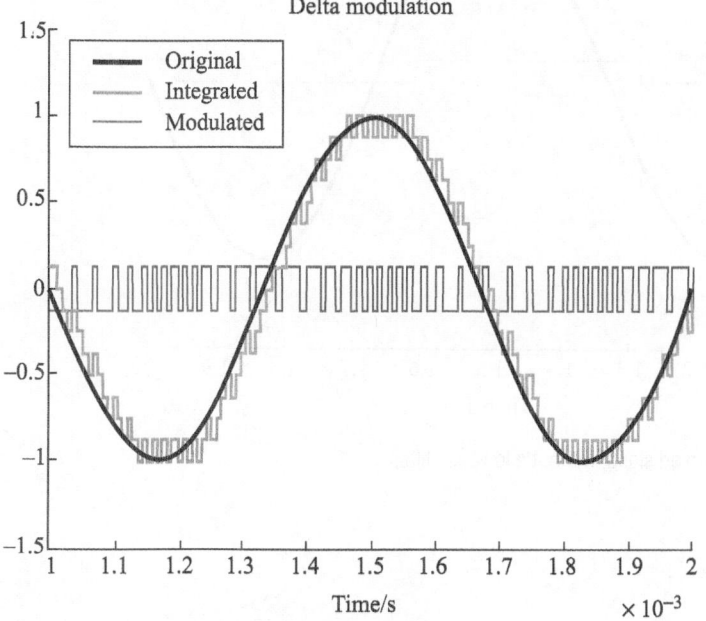

Fig. 5.27 Signal thirty two times oversampled

MATLAB simulation at 32 times oversampling. It can be seen that the output of the integrator tracks nearly the output signal.

Filtering is critical in oversampling in order to avoid the unwanted effects of aliasing and distortion. Output of the delta modulator is passed through a low-pass filter in order to attenuate high frequency unwanted signals. Figure 5.28 shows the MATLAB simulation of not using an LPF. In this example, oversampling is 16 times. It can be seen that modulating signal hardly tracks the actual input signal, as the integrated signal is shifted due to the effect of aliasing. It can be clearly seen from the figure that a lot of distortion is introduced and the reproduced signal at the demodulator will not resemble the original input signal.

Figure 5.29 shows the importance of choosing the right oversampling ratio or the size of delta V, which determines the number of quantizing levels. If the size of the delta V or the oversampling ratio is relatively low to the required values, a slope overload occurs. If we choose delta V to be 0.125, doubles the previous delta $V = 0.0625$, there is an equivalence of 3 bit quantization with 8 levels. The oversampling ratio should be 32 to achieve a sampling rate of 4 kHz. The oversampling results in overload slope, i.e., the signal changes too fast for the modulated signal to track.

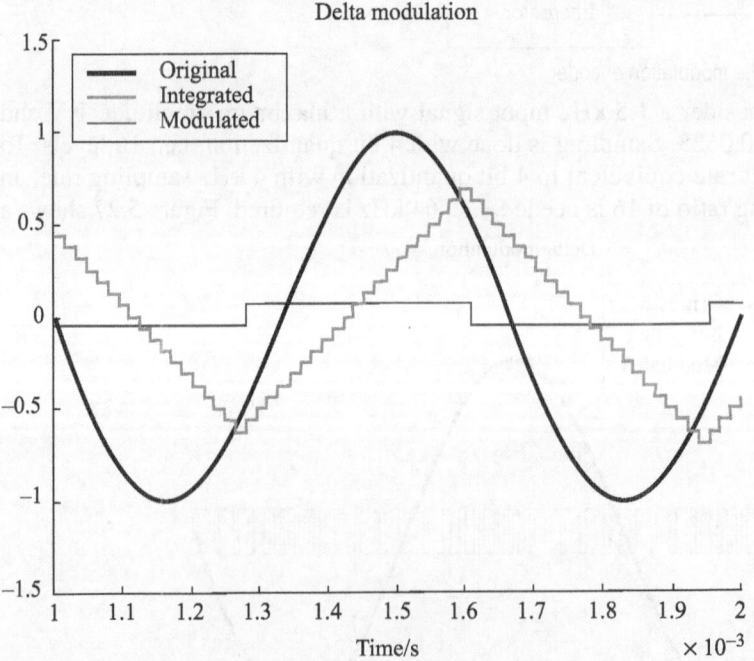

Fig. 5.28 Over sampled signal without a lowpass filter

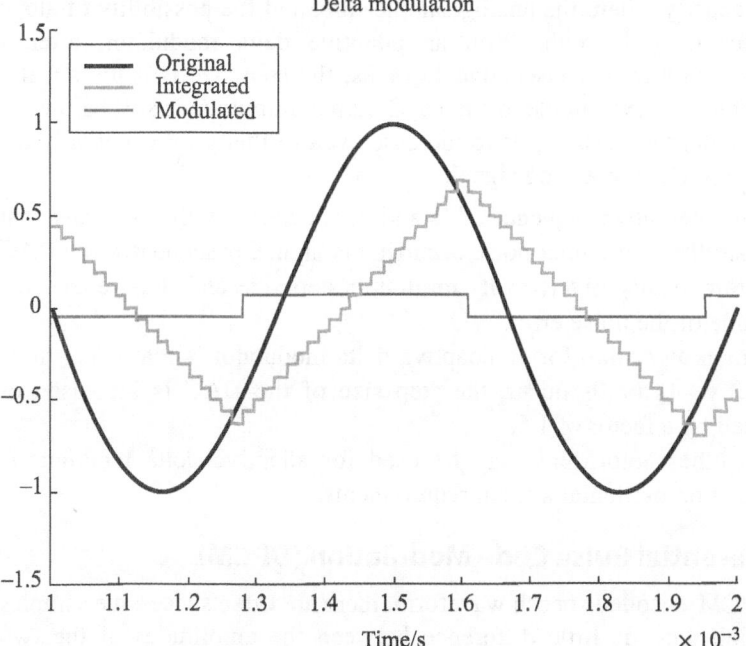

Fig. 5.29 Signal with right oversampling ratio

5.11.2 Adaptive DM

Adaptive delta modulation is a delta modulation system where the step size of the DAC is automatically varied, depending on the amplitude characteristics of the analog input signal. Figure 5.30 shows how an adaptive delta modulator works.

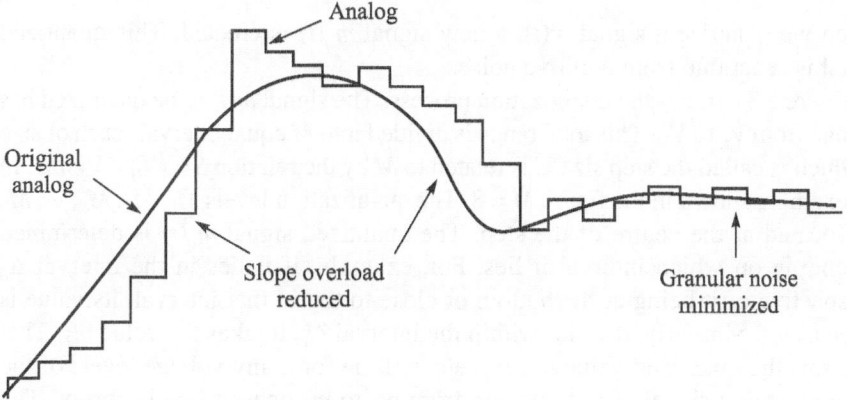

Fig. 5.30 Adaptive delta modulation

When the output of the transmitter is a string of consecutive 1s or 0s, this indicates that the slope of the DAC output is less than the slope of the analog signal in either the positive or the negative direction. Essentially, the DAC has lost track of exactly where the analog samples are, and the possibility of slope over load occurring is high. With an adaptive delta modulator, after a predetermined number of consecutive 1s or 0s, the step size is automatically increased. After the next sample, if the DAC output amplitude is still below the sample amplitude, the next step is to increase even further until eventually the DAC catches up with the analog signal.

- When an alternative sequence of 1s and 0s is occurring, this indicates that the possibility of granular noise occurring is high. Consequently, the DAC will automatically revert to its minimum step size and thus reduce the magnitude of the noise error.
- A common algorithm for an adaptive delta modulator is that when three consecutive 1s or 0s occur, the step size of the DAC is increased or decreased by a factor of 1.5.
- Various other algorithms may be used for adaptive delta modulators, depending on particular system requirements.

5.11.3 Differential Pulse Code Modulation (DPCM)

In a typical PCM-encoded speech waveform, there are often successive samples taken in which there is little difference between the amplitudes of the two samples. This necessitates transmitting several identical PCM codes, which is redundant. Differential pulse code modulation (DPCM) is designed specifically to take advantage of the sample-to-sample redundancies in typical speech waveforms. With DPCM, the difference in the amplitude of two successive samples is transmitted rather than the actual sample. Because the range of sample differences is typically less than the range of individual samples, fewer bits are required for DPCM than conventional PCM.

5.11.4 Quantization of Signals

When we quantize a signal $m(t)$, a new signal $m_q(t)$ is created. This quantized signal is separable from additive noise.

Figure 5.31 shows the quantization process. The signal $m(t)$ to be quantized has a range from V_L to V_H. This total range is divided into M equal intervals each of size S, which is called the step size. S is related to M by the relation $S = (V_H - V_L)/M$. In the example shown in the figure $M = 8$. The quantization levels $M_0, M_1, M_2, \cdots, M_7$ are located at the centre of the step. The quantized signal $m_q(t)$ is determined depending on which interval it lies. For, example, if it lies in the interval Δ_0 irrespective of it being at the bottom or close to top in that interval, its value is taken as m_0. Similarly, if it lies within the interval Δ_1, it takes the value m_1. That is to say, the quantized value m_0, m_1, etc. will be for many voltage levels of the original analog signal. The transition from m_0 to m_1 or m_1 to m_2 is abrupt. The quantization error $m(t) - m_q(t)$ will be equal to or less than $S/2$. Thus, we can see that the quantized signal is an approximation to the original signal. To reduce the quantization error, the number of steps has to be more or the difference between each step should be very small. Hence, the more number of steps, i.e., more number of bits used to quantize the signal, the more accurate will be the quantized signal to original signal.

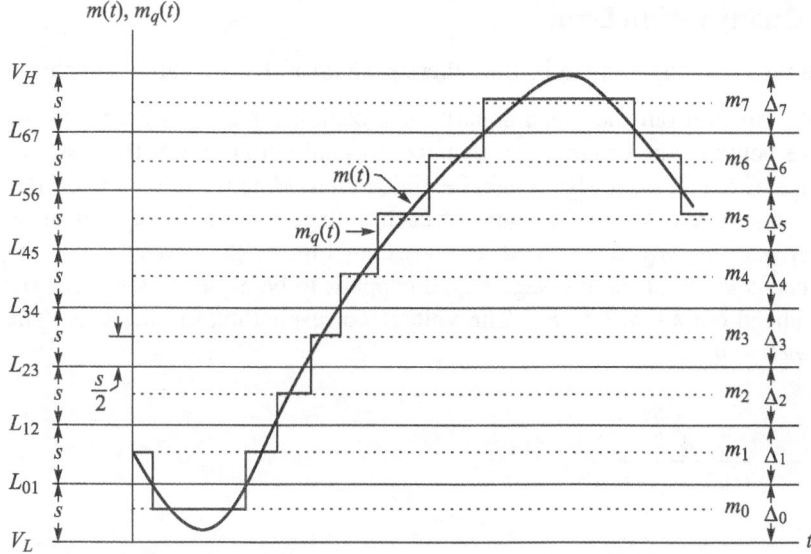

Fig. 5.31 Quantization process

Let us consider that a quantized signal has reached a distortion with attenuation and corrupted with noise. As shown in Fig. 5.32, as long as the noise level does not exceed $S/2$, it does not affect the signal, but if the level exceeds $S/2$, then an error results as shown. The statistical nature of noise is such that even if the average noise magnitude is much less than $S/2$, there is always a finite probability that from time to time the noise magnitude will exceed $S/2$. Hence, it is generally not possible to suppress completely the effect of noise on the signal. One method to reduce the effect of noise is to increase the step size, i.e., the number of levels should be less. No doubt this reduces the noise but at the expense of quantization error. The noise introduced by the channel is called additive noise and the error introduced by quantization is called *quantization noise*. Due to increase in the step size, i.e., having less number of levels, no doubt the contribution due to additive noise is reduced but the difference $m(t) - m_q(t)$ increases and this results in quantization error, which is called *quantization noise*. Therefore, the received signal is not a perfect reproduction of the transmitted signal $m(t)$. This is due to the effect of both additive and quantization noise.

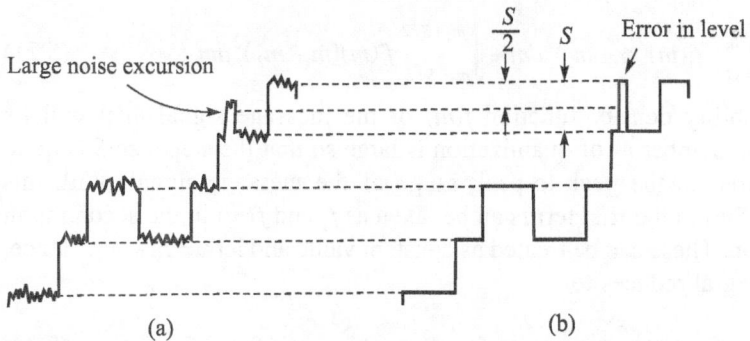

Fig. 5.32 (a) Quantized signal with noise added and (b) error after requantization

5.11.5 Quantization Error

We have seen that due to quantization, there is an error that we call quantization error. We can find out the mean square quantization error $\overline{e^2}$, where e is the difference between the original and quantized signal voltages. Let the peak-to-peak range of the message signal $m(t)$ be divided into M equal voltage intervals, each of magnitude S V. At the centre of each voltage interval the quantization levels m_1, m_2, $\cdots$, m_M are located as shown in Fig. 5.33. In the figure, the instantaneous value of the message signal happens to be closer to the level m_k. The quantized output will be m_k. The voltage corresponding to that level. The error is $m(t) - m_k$.

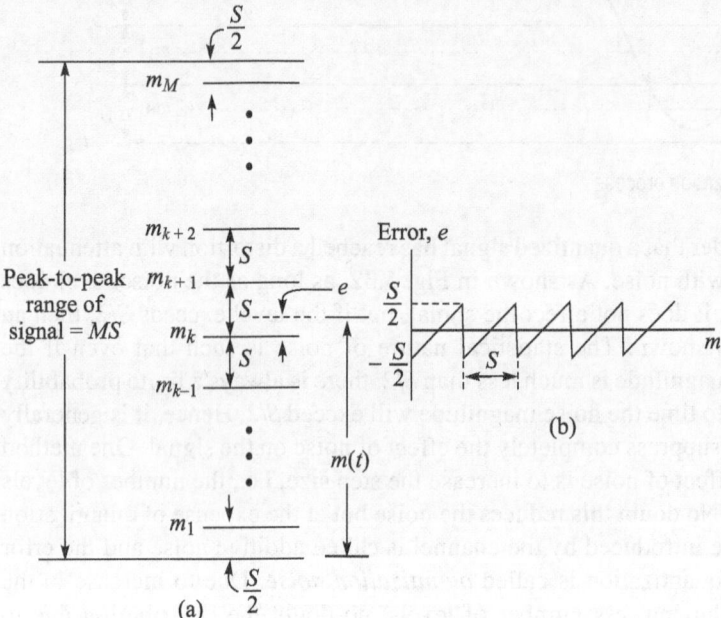

Fig. 5.33 (a) Signal $m(t)$ divided into M quantization range each of size S and (b) the error voltage $e(t)$ as a function of the instantaneous value of signal $m(t)$

Let $f(m)\,dm$ be the probability that $m(t)$ lies in the voltage range $m - dm/2$ to $m + dm/2$. Then the mean square quantization error is

$$\overline{e^2} = \int_{m_1-S/2}^{m_1+S/2} f(m)(m-m_1)^2\,dm + \int_{m_2-S/2}^{m_2+S/2} f(m)(m-m_2)^2\,dm + \cdots \quad (5.21)$$

The probability density function $f(m)$ of the message signal $m(t)$ will be constant if the number M of quantization is large so that the step size S is quite small compared to the peak-to-peak range of the message signal. With this assumption, $f(m)$ in the first term can be taken as f_1 and $f(m)$ in the second term as f_2, and so on. These can be treated as constant value and let $x = m - m_k$. Hence, the above integral reduces to

$$\overline{e^2} = (f_1 + f_2 + \cdots)\int_{-S/2}^{S/2} x^2\,dx = (f_1 + f_2 + \cdots)\frac{S^3}{12} = (f_1 S + f_2 S + \cdots)\frac{S^2}{12} \quad (5.22)$$

Now $f_1 S$ is the probability that the signal voltage $m(t)$ will be in the first quantization range, $f_2 S$ is the probability that $m(t)$ is in the second quantization range, and so on. Hence, $f_1 S + f_2 S + \cdots + f_n S$ is equal to unity and then

$$\overline{e^2} = \frac{S^2}{12} \tag{5.23}$$

This is called mean square quantization noise voltage. For a total number of M levels, the peak-to-peak signal range is $\pm(MS)/2$. For a signal having uniform probability density distribution within this range, the mean square signal voltage is

$$\overline{e^2}_s = \frac{(MS)^2}{12} \tag{5.24}$$

It follows that the signal to quantization noise ratio is

$$\left(\frac{S}{N}\right)_q = \frac{\overline{e^2_s}}{\overline{e^2}} = M^2 \tag{5.25}$$

Hence, to maintain a high $(S/N)_q$, ratio the number of steps should be high. For example, if $M = 256$, $(S/N)_q = 48$ dB. In terms of the number of bits per code word, n $M = 2^n$ and hence,

$$\left(\frac{S}{N}\right)_q = 2^{2n} \tag{5.26}$$

It can be shown that for a sine wave

$$\left(\frac{S}{N}\right)_q = 1.5 M^2 \tag{5.27}$$

Thus, the ratio between the peak and RMS values of the signal voltage will be some value $k = e^2_{RMS} / \overline{e^2}$. If distortion is to be avoided, the maximum peak signal level must not be allowed to exceed half the total input voltage range of $MS/2$. In this case, the signal to quantization ratio becomes.

$$\left(\frac{S}{N}\right)_q = \frac{e^2_{RMS}}{\overline{e^2}} = \left(\frac{kMS}{2}\right)^2 \times \frac{12}{S^2} = 3k^2 M^2 \tag{5.28}$$

5.12 NOISE CONSIDERATION IN PCM SYSTEM

The performance of a PCM system depends on the influence of two major sources of noise. One is the *channel noise,* which is introduced anywhere between the transmitter output and the receiver input and the other is the *quantization noise,* which is introduced in the transmitter due to the quantization and is carried all along to receiver output. This noise unlike channel noise is signal dependent and vanishes once the signal is switched off. Although these two noise sources appear simultaneously, they are considered separately.

The main effect of channel noise is that bit errors are introduced into the received signal. This causes symbol 1 to be mistaken for symbol zero or vice versa. If the bit error occurs frequently, the transmission becomes poor and unreliable. The average probability of symbol error is a measure of fidelity of information transmission in PCM in the presence of noise. It is defined as the probability that the reconstructed symbol at the receiver output differs from the transmitted binary symbol. This is also referred to as the bit error rate (BER). To optimize the system performance in the presence of channel noise, we have to minimize the average probability of symbol error. For this, the channel noise is modelled as additive, white, and Gaussian. Adequate signal-to-noise ratio can minimize this problem.

The quantization noise is essentially under designer's control. It can be minimized by increasing the number of quantizing levels and using the proper companding technique, matching to the characteristics of the message being transmitted. Hence, we can find that a PCM system is more rugged one compared to its analog counterpart, the CW modulation.

5.13 FDM AND TDM

Both use the concept of multiplexing. Multiplexing means sharing a physical channel when more than one application or connection share the capacity of one link. This results in better utilization of resources. A typical example is conversations over telephone line, trunk line, wireless channel etc. Some examples of multiplexing are TDM (time division multiplexing), FDM (frequency division multiplexing), WDM (wavelength division multiplexing), and CDMA (code division multiple access). Figure 5.34 shows N channels sharing a link.

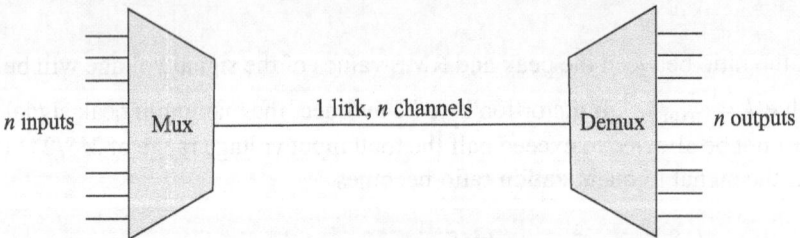

n inputs — Mux — 1 link, n channels — Demux — n outputs

Fig. 5.34 A typical multiplexed channel: N channels sharing the capacity of a link

FDM has the following features.

- Each channel gets separate frequency band.
- The frequency bands are nonoverlapping.
- There is one band per channel, e.g., radio, TV, and trunk lines in telephony.
- Signals are frequency shifted and then combined.

The features of TDM are given below.

- Users are allocated entire bandwidth.
- Each channel is allotted a time slot, i.e., each channel is sampled during the time slot allotted to it.
- It is widely used in telephony.

5.14 FREQUENCY DIVISION MULTIPLEXING TRANSMITTER

Figure 5.35 shows an FDM transmitter. In this transmitter, analog or digital information signals $m_1(t), m_2(t), \cdots, m_n(t)$ are modulated with subcarrier frequencies $f_1, f_2, \cdots, f_n$, respectively, and the composite baseband signal $m_b(t)$ modulate the carrier f_c. The resultant signal is the overall FDM signal $s(t)$. The figure shows the transmitter and spectrum function of composite baseband modulating signal $m_b(t)$.

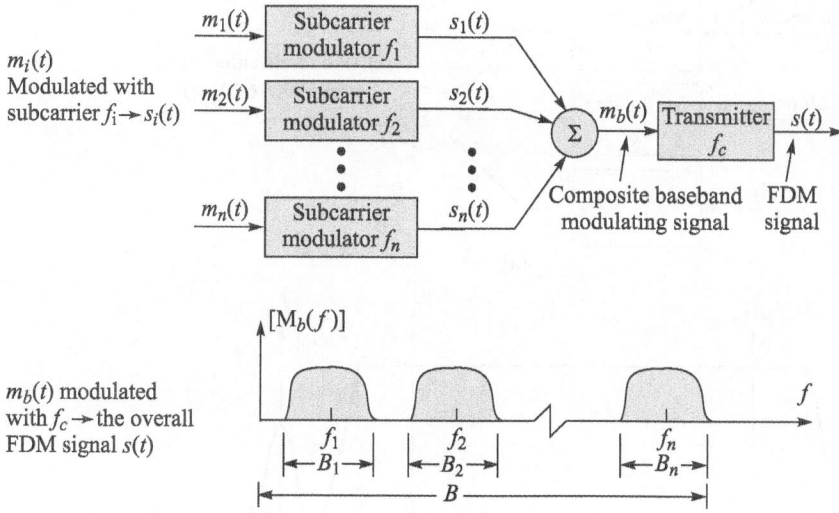

Fig. 5.35 Frequency division multiplexing transmitter

5.14.1 Frequency Division Multiplexing Receiver

Figure 5.36 shows a typical FDM receiver. Composite baseband modulating signal $m_b(t)$ is retrieved by demodulating the FDM signal $s(t)$ using carrier $f_c \cdots m_b(t)$ is passed through a parallel bank of bandpass filters centred around f_i (i) = 1, 2, 3, $\cdots$, I, $\cdots$, n). For example, the output of the ith filter is the ith signal $s_i(t)$ and $m_i(t)$ is retrieved by demodulating $s_i(t)$ using subcarrier f_i.

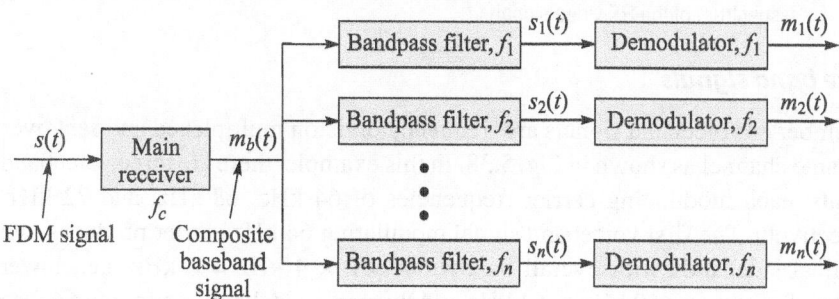

Fig. 5.36 FDM receiver

Cable TV transmission

An example of transmitting cable TV signal using FDM is shown in Fig. 5.37(a). There are three separate modulators, each for video (black and white), color signal, and audio signal. The modulated signals are sent as an overall video signal in the same channel. The bandwidth of this video signal is 6 MHz. The magnitude spectrum of the signal is shown in Fig. 5.37(b).

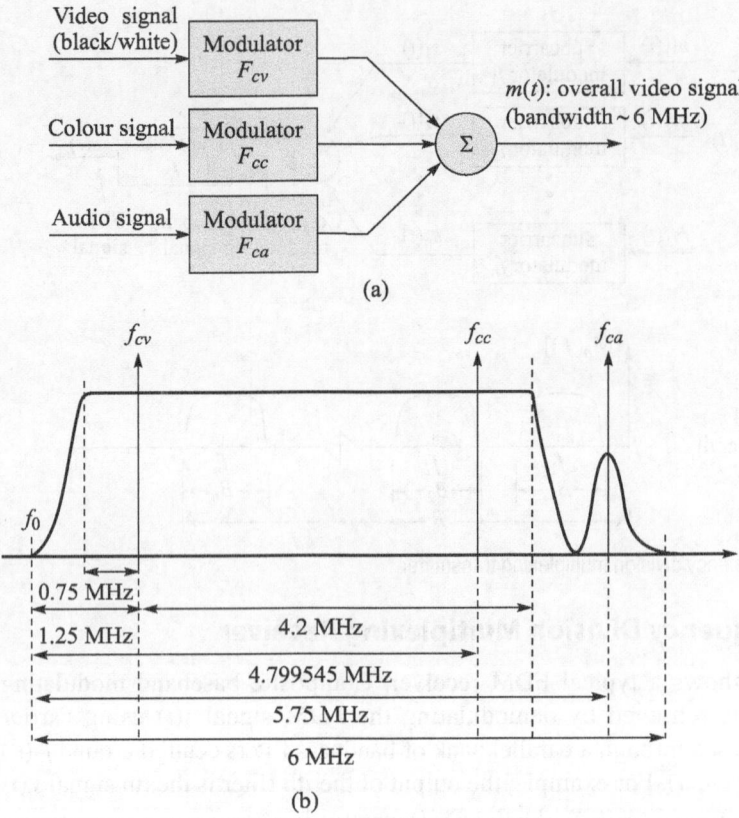

(a)

(b)

Fig. 5.37 Cable TV transmission: (a) transmitting cable TV signal using FDM and (b) magnitude spectrum of the RF video signal

Voice band signals

A number of voiceband signals are frequency division multiplexed and sent over the same channel as shown in Fig. 5.38. In this example, there are three voiceband signals, each modulating carrier frequencies of 64 kHz, 68 kHz, and 72 kHz, respectively. The first voiceband signal modulating 64 kHz carrier produces two identical sidebands with overall bandwidth = 2×4 kHz = 8 kHz, i.e., lower sideband occupying 60 kHz to 64 kHz and the upper sideband occupying 64 kHz to 68 kHz. Similarly, other two voice signals after modulating their respective carriers produce upper and lower sidebands.

Filtering the upper sidebands in each case, we get the spectrum for the composite signals using three subcarriers as shown in the figure.

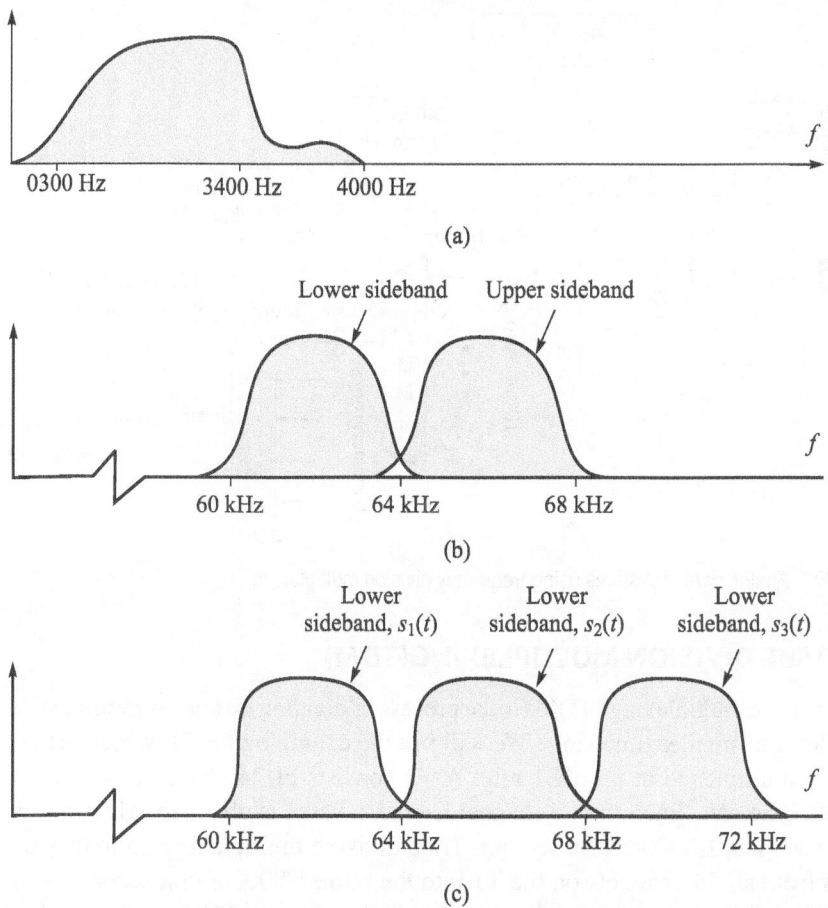

Fig. 5.38 Voiceband signals employing FDM: (a) spectrum of $m_1(t)$, positive f, (b) spectrum of $s_1(t)$ for $f_1 = 64$ kHz, and (c) spectrum of composite signal using subcarrier at 64 kHz, 68 kHz, and 72 kHz

5.15 ANALOG CARRIER SYSTEM

Figure 5.39 shows a frequency division multiplexed analog carrier system. Here 12 voice channels, each with a bandwidth of 4 kHz, are multiplexed in the first stage resulting in 12 channels with an overall bandwidth of 48 kHz, called a group. In the second stage of multiplexing, five more such groups each having 12 channels, are added to the input of second stage resulting in overall output at the second stage to be 60 channels with an overall bandwidth of 240 kHz called a supergroup. Such ten 60 channel supergroups are given to the third stage resulting in over all output at the third stage to be 600 channels with an overall bandwidth of 2.52 MHz, called master group. Final stage has got six master groups, each having 600 channels resulting at the output 3600 channels with an overall bandwidth of 16.984 MHz, called jumbo group. Depending on the bandwidth of the channel, some more stages can be added.

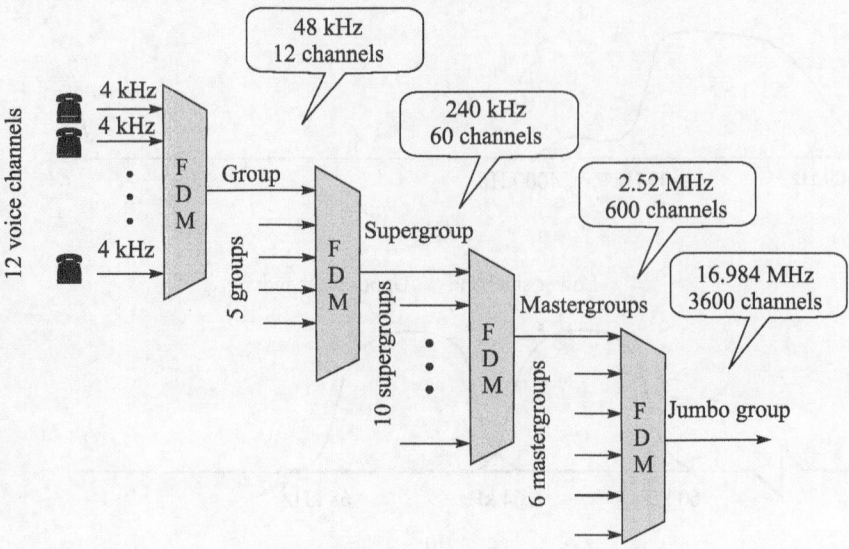

Fig. 5.39   Analog carrier systems using frequency division multiplexing

5.16 TIME DIVISION MULTIPLEXING(TDM)

Time division multiplexing (TDM) is the process of dividing up one communication time slot into smaller time slots. We will use the example of a T1, which is time division multiplexed at the DS1 rate. A T1 consists of 24 channels, which are read 8000 times/s. Each time a channel is read, a value is obtained. Thus, a time slot for a T1 is 1/8000th of a second. Time division multiplexing combines the values from all 24 channels on the T1 into the same 1/8000th of a second.

In Fig. 5.40, we can see the 24 channels (CH 1 through CH 24) of a standard T1. At a given point in time (t), every channel provides an input value (not really, but that is a more complicated topic to be discussed later). These input values are represented in the output diagram, which is shown for a time frame. In this diagram, each channel is shown with its allotted time slot. This frame repeats itself as long as the scanning continues at the input.

Figure 5.41 shows a typical TDM system. All channels of a time division multiplex system use the same portion of the transmission links' frequency spectrum but not at the same time. Each channel is sampled in a regular sequence by a multiplexer (or less commonly by a scanner) as shown in this figure. When all the channels have been sampled, the sequence starts over with the first channel. Thus, samples from a particular channel are interleaved in time between the samples from all other channels.

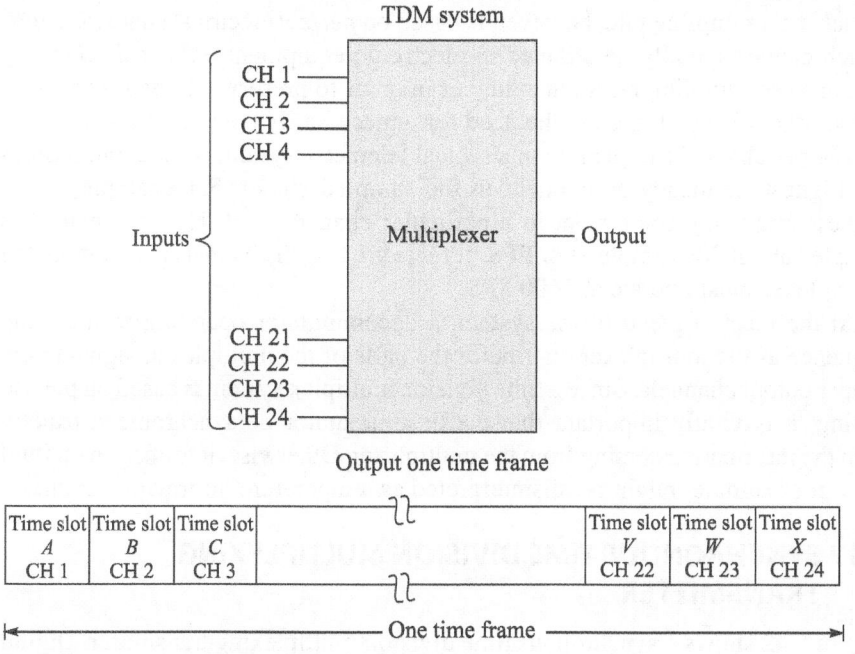

Time slot	Time slot	Time slot	⌇	Time slot	Time slot	Time slot
A	*B*	*C*		*V*	*W*	*X*
CH 1	CH 2	CH 3	⌇	CH 22	CH 23	CH 24

|← ————————————————— One time frame ————————————————— →|

Fig. 5.40   Time slot of TDM system

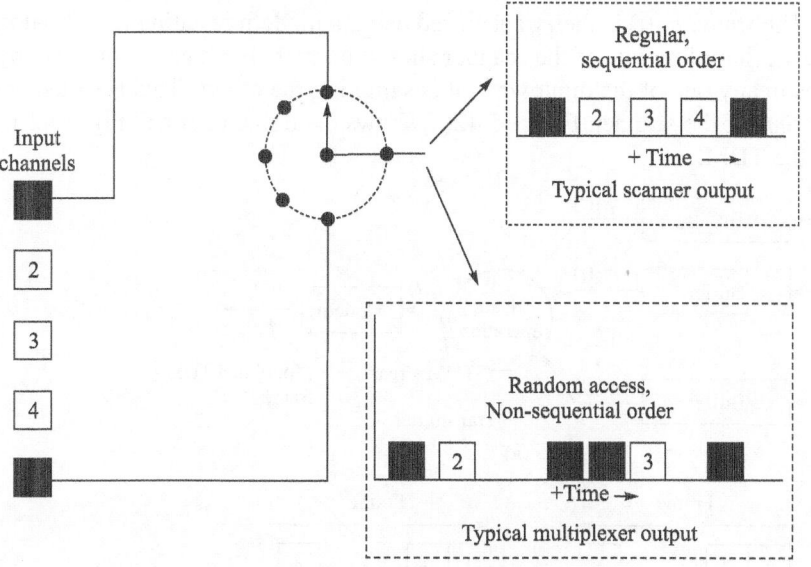

Time division multiplexing techniques

Fig. 5.41   A typical time division multiplex system

Since no channel is monitored continuously in a time division multiplex system, the sampling must be rapid enough so that the signal amplitude in a particular channel does not change too much between the samples. Theoretical studies, based on idealized conditions, have shown that no information is lost if the sampling rate is at least twice the highest frequency component in the sampled signal. This sampling rate, however, is based on *perfect* electrical characteristics, which cannot actually be attained in electronic equipment. Practical telemetry systems use sampling rates much higher in order to preserve all the information in the original signal without the need for unnecessarily complex circuitry.

The per channel sample rate in a typical telemetry system is set about 5 times the highest frequency component in the sampled signal. For example, if the highest frequency component in a particular channel is 40 Hz, the channel is sampled about 200 samples/s or SPS. If there are 8 such channels in a system, the multiplexer must operate at 1600 SPS.

At the receiving end of the system, a decommutator operating in the same sequence as the multiplexer distributes the parts of the multiplexed signal to the proper output channels. Since a time division multiplex system is based on precise timing, it is vitally important that the decommutator be synchronized exactly with the information coming from the multiplexer. Otherwise, information on fluid flow, for example, might be misinterpreted as temperature information frames.

5.17 SYNCHRONOUS TIME DIVISION MULTIPLEXING TRANSMITTER

Figure 5.42 shows a synchronous time division multiplexing transmitter. Digital sources $m_1(t)$, $m_2(t)$, $\cdots$, $m_n(t)$ are buffered and a scanner samples these inputs in a cyclic manner to form a frame. The source $m_c(t)$ is called a TDM stream or a frame. The frame $m_c(t)$ is then transmitted using a modem resulting in an analog signal $s(t)$. Sampling rate of the scanner should be such that it must not miss any changes in any one of the input when it is sampling the others. This is called the *update time* of the scanner. Figure 5.42(a) shows the transmitter and Fig. 5.42(b) shows the TDM.

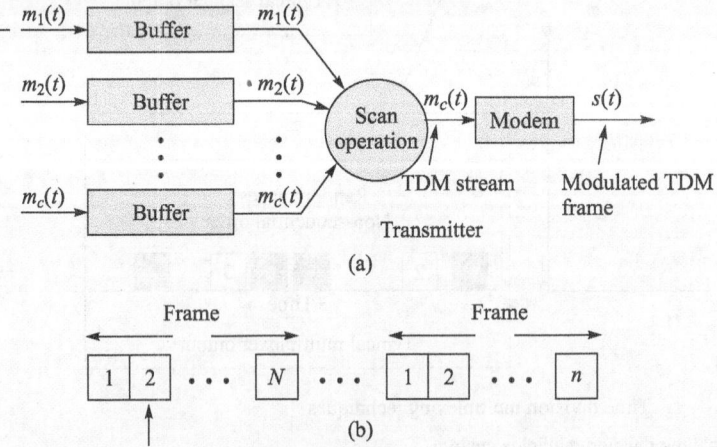

Fig. 5.42 Synchronous TDM transmitter

5.18 SYNCHRONOUS TIME DIVISION MULTIPLEXING RECEIVER

Figure 5.43 shows the block diagram of synchronous TDM receiver. The modulated TDM stream is demodulated in the modem and the demodulated frame $m_c(t)$ is obtained. This frame is scanned into n parallel buffers. For example, the

ith buffer corresponds to the original $m_i(t)$ digital information. The frequency of the scanner should match with the transmitter scanner.

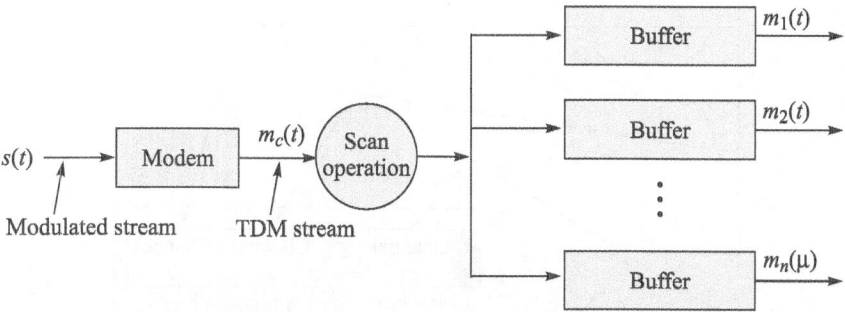

Fig. 5.43   Synchronous TDM receiver

Time slots are pre-assigned to sources and are fixed. There is no header and error control for the frame. One or more slot per digital source is provided. The order of the slots is determined by the scanner control. The slot length equals the transmitter buffer length. The disadvantage of synchronous TDM is that if there is no data for a particular time slot, the slot is left unoccupied. This is a cause of inefficiency.

5.19 TDM DIGITAL CARRIER SYSTEM

PCM-coded voice calls are used in digital carrier system. Voice signal is PCM coded as 8 bit per sample. Each 64 kbps PCM digitized voice call is called DS-0.

Such 24 digitized voice calls are called DS-1, 96 digitized voice calls are called DS-2, and 672 digitized voice calls are called DS-3. It can be observed that channel 1 has a digitized sample from first call, channel 2 has a digitized sample from second call etc. Figure 5.44 gives the DS-1 transmission format.

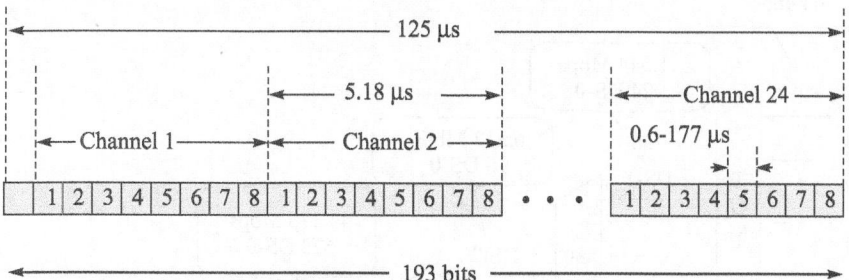

Fig. 5.44   DS-1 transmission format

The first is a framing bit used for synchronization. As far as the voice channels are concerned, 8 bit PCM is used on five of six frames and 7 bit PCM is used on every sixth frame. Bit 8 of each channel is a signalling bit. In the case of data channels, channel 24 is used for signalling in some schemes. Bits 1–7 are used for 56 kbps service and bits 2–7 are used for 9.6, 4.8 and 2.4 kbps service.

A typical TDM frame, called T-1 frame, is shown in Fig. 5.45. The figure shows the arrangement of sending 8000 frames, each containing 24 channels (D-1). Each frame consists of 193 bits as shown and hence the T-1 frame consists of $8000 \times 193 = 1.544$ Mbps.

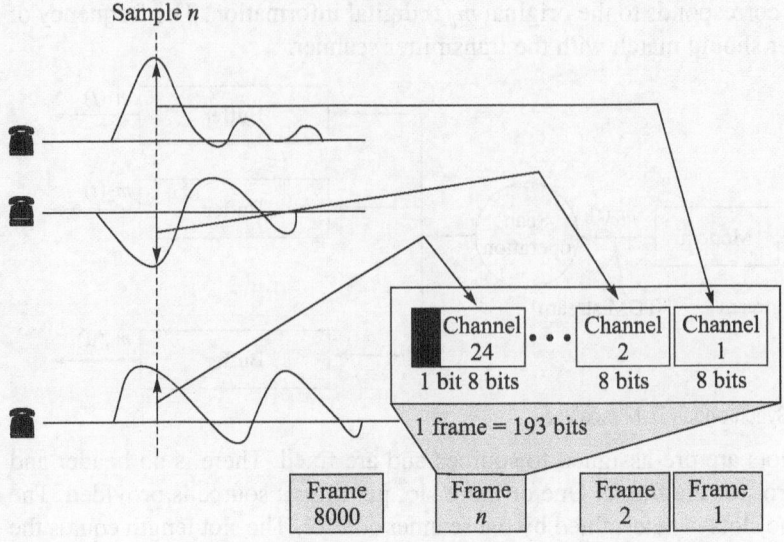

T-1 = 8000 frames/s = 8000 × 193 bps = 1.544 Mbps

Fig. 5.45 T-1 frame of a TDM system

Figure 5.46 shows a typical digital carrier system employing TDM. The first stage consists of twenty four DS-0 channels, each of 64 kbps as input, the second stage consists of four DS-1 channel (96 DS-0), each of 1.544 Mbps as input, the third stage consists of seven DS-2 channels (672 DS-0), each of 6.312 Mbps, and the fourth stage consists of six DS-3 channel (672 DS-0), each of 44.376 Mbps. The final stage output consists of one DS-4 channels (4032) of 274.176 Mbps.

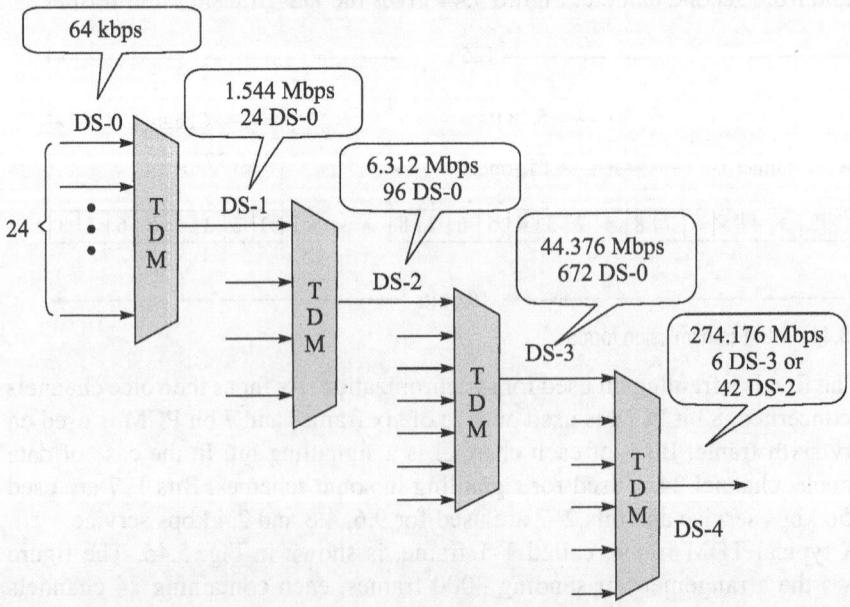

Fig. 5.46 TDM digital carrier system

SUMMARY

This chapter has explained

- how to represent messages in the form of pulses,
- the sampling theorem and the Nyquist rate, which determines the minimum sampling rate to reproduce the signal,
- various types of pulse modulation like modulating its amplitude, width, and position,
- pulse code modulation where the sampled amplitude is encoded and sent as digital data,
- concept of quantization noise and how to minimize it, and
- frequency division multiplexing and time division multiplexing with examples.

IMPORTANT FORMULAE

- Sampled signal: $\bar{x}(t) = x(t)\delta_T(t) = \sum_n x(nT)\delta(t - nT)$

  where $\delta_T(t) = \dfrac{1}{T}\left[1 + 2\cos\omega_s t + 2\cos 2\omega_s t + 2\cos 3\omega_s t + \cdots\right]$

  Hence, $\bar{x}(t) = x(t)\delta_T(t)$

  $$= \frac{1}{T}\left[x(t) + 2x(t)\cos\omega_s(t) + 2x(t)\cos 2\omega_s(t) + 2x(t)\cos 3\omega_s(t) + \cdots\right]$$

- Fourier transform of sampled signal, $\bar{X}(\omega) = \dfrac{1}{T}\displaystyle\sum_{n=-\infty}^{\infty} X(\omega - n\omega_s)$

- PAM signal $= a_0 m(t) + a_1 m(t)\cos\dfrac{2\pi t}{T_s} + a_2 m(t)\cos\dfrac{4\pi t}{T_s} + \cdots$

- The transform of the flat topped sampled signal

  $$H(j\omega)\,M(j\omega) = \frac{\tau}{T_s}\frac{\sin(\omega\tau/2)}{\omega\tau/2}M(j\omega)$$

- The mean square quantization error is

  $$\overline{e^2} = \int_{m_1-S/2}^{m_1+S/2} f(m)(m - m_1)^2\,dm +$$

  $$= \int_{m_2-S/2}^{m_2+S/2} f(m)(m - m_2)^2\,dm + \cdots$$

- If the number M of quantization is large, then the quantization error is

  $$\overline{e^2} = (f_1 + f_2 + \cdots)\int_{-S/2}^{S/2} x^2\,dx = (f_1 + f_2 + \cdots)\frac{S^3}{12} = (f_1 S + f_2 S + \cdots)\frac{S^2}{12}$$

- The signal to quantization noise ratio:

  $$\left(\frac{S}{N}\right)_q = \frac{e_s^2}{e^2} = S^2$$

  For a sine wave, $\left(\dfrac{S}{N}\right)_q = 1.5M^2$

ADDITIONAL EXAMPLES

1. For the sample and hold circuit shown in Fig. 5.10, find the largest value capacitors that can be used. The on resistance of Q_1 can be assumed to be 5 ohms and the output impedance to be 10 ohms. The acquisition time is 15 μs. The input voltage is 5 V peak to peak, the maximum output current is 5 mA and the accuracy required from the device is 0.1%.

Solution
Current through the capacitor is

$$i = c\frac{dv}{dt}$$

$$\therefore \qquad c = \frac{dt}{dv} \cdot i$$

Substituting the values of c, I, dv (maximum voltage across c), dt (the acquisition time), we get

$$c_{max} = \frac{(5\text{ mA})(15\text{ μs})}{5\text{ V}} = 15\text{ nF}$$

The charging time constant for C when Q_1 is on is $\tau = RC$, here τ is one. Also R is the output impedance plus on resistance of Q_1 in ohms, and C is the capacitance in farads. From the above relation,

$$C_{max} = \tau/R$$

The charge time of capacitor is also dependent on the accuracy desired from the device. For 0.1% accuracy, the charge time is $6.9\,\tau$.

Hence, $\qquad C = \dfrac{15\text{ μs}}{6.9(15)} = 145\text{ nF}$

To satisfy the output current limitation of output impedance, a maximum capacitance of 15 nF was required. But to satisfy the accuracy requirement, 145 nF was required. To satisfy both the requirements, the smaller of the two, i.e., 15 nF has to be chosen.

2. In a PCM system the maximum audio input frequency is 5 kHz. What must be the minimum sampling rate? If a 6 kHz signal enters the sample and hold circuit, what is the aliasing frequency produced?

Solution
The minimum sampling rate = 2×5 kHz = 10 kHz
When the audio frequency 6 kHz enters the sample and hold circuit, it will overlap the audio spectrum and produces alias frequency of 4 kHz.

3. A PCM system has the following parameters: Minimum dynamic range of 35 dB, maximum analog frequency of 5 kHz, and a maximum decoded voltage at the receiver of 3 V. Find out the following:

(a) Minimum sample rate

(b) Minimum number of bits used

(c) Resolution

(d) Quantization error

Solution

(a) Minimum sampling rate = 2×5 kHz = 10 kHz

(b) Since minimum dynamic range in dB $6n$, where n is the number of bits used. Hence, 35 dB = $6n$, i.e., $n = 35/6 = 5.833$. Closest whole number is 6. Hence, 6 bits must be used for amplitude. Since the amplitude range is ± 3 V, a sign bit is also necessary. The total number of bits is then 7 bits. There will be 126 positive codes and 127 negative codes and two zero codes.

(c) Resolution = $3 \text{ V}/2^7 = 0.023$ V

(d) Maximum quantization error = Resolution/2 = 0.023/2 = 0.0115

4. Sixteen telephone channels are time division multiplexed by using PCM. Each channel is bandlimited to 4 kHz. Calculate the bandwidth of PCM system for 256 quantization levels and an 10 kHz sampling frequency.

Solution

We have $n = 16$ and $m = 256 = 2^8$

Therefore, the number of bits = 8

Since $2f_m = 10$ kHz

Bandwidth = $[(16 \times 9)+1] \times 10$ kHz = 1.45 MHz

REVIEW QUESTIONS

1. Compare the power required for transmitting information by analog modulation and pulse modulation.
2. What is sampling theorem and what does it determine?
3. What is Nyquist rate?
4. What do you mean by aliasing error?
5. What is the need for bandlimiting a signal before sampling?
6. What is the difference between natural and flat top sampling?
7. What is a sample and hold circuit?
8. What is the difference between PWM and PPM?
9. How do you generate PPM from PWM?
10. What do you mean by quantization of signals?
11. What is quantization error?
12. How can the quantization error be reduced?
13. What are the advantages of using the PCM over other types of pulse modulations?
14. What are the main features of delta modulation?
15. Explain a delta modulation encoder.

16. Explain the quantization process.
17. What are the two different types of multiplexing techniques employed?
18. Explain a typical cable TV transmission using the FDM.
19. Explain the T-1 frame of a TDM system.
20. What is the difference between FDM and TDM?

PROBLEMS

1. What sampling rate is required for the following?
   (a) A 3.5 kHz telephone channel
   (b) A music channel with maximum frequency of 18 kHz
   (c) A TV channel with 6 MHz bandwidth
2. A bandpass signal has a spectral range that extends from 200 Hz to 6.4 kHz. Determine the acceptable range of sampling frequency.
3. A bandpass signal has centre frequency f_o extending from $f_o - 8$ kHz to $f_o + 8$ kHz. The sampling rate is 30 kHz. If the centre frequency varies from 10 kHz to 70 kHz, find the ranges of f_o for which the sampling rate is employed is adequate.
4. A sinusoidal signal to be sampled has a magnitude of 0.4 V peak at 4 kHz. A natural PAM generator uses a 1.5 V carrier pulse for which the sampling rate is 10 kHz and the duty cycle is 30%. Calculate the magnitude and frequency for the first six components in the spectrum.
5. A PAM TDM system transmits 10 messages, each having pulse width of 120 μs, in 1.4 ms. Find the guard time. If the same guard time is maintained and 15 PAM messages are to be transmitted, how narrow the pulses should be?
6. The discrete samples of an analog signal are uniformly quantized such that the maximum value of the analog signal is to be represented within 0.05% accuracy. Find the minimum number of binary digits required.
7. Three signals m_1, m_2, and m_3 are multiplexed. These signals have different bandwidths of 30 kHz, 40 kHz, and 50 kHz, respectively. Design a TDM system so that each signal is sampled at its Nyquist rate.
8. A sinusoidal signal with maximum input voltage of 4 V is given to a PCM channel using a 12 bit code. Find (a) the number of quantization levels used, (b) the RMS quantization noise level in volts, and (c) the maximum sinusoidal signal to quantization noise ratio in dB.
9. A PCM system is to have a signal-to-noise ratio of 35 dB. For the speech signal, an RMS to peak ratio of −10 dB is allowed. Find the number of bits required for coding.
10. Thirty two voice signals are sampled uniformly and then time division multiplexed. Flat top sampling is used with 2 μs duration. Provision for synchronization is there by adding an extra pulse of sufficient amplitude also of 2 μs duration. The highest frequency component of each voice signal is 3.3 kHz. Calculate the spacing between successive pulses of the multiplexed signal if the sampling rate is 10 kHz. Repeat the calculation with Nyquist sampling rate.

11. A signal bandlimited to 2.5 kHz is sampled at 20% higher than the Nyquist rate. The maximum accepted error in the sampled amplitude due to quantization must not exceed 0.4% of the peak amplitude E. The quantized samples use binary coding. Find the required sampling rate, the number of bits required to encode each sample, and the bit rate of the resulting PCM signal.

12. A PCM system has a maximum input message frequency of 5 kHz, the maximum decoded voltage at the receiver of ± 5 V, and a minimum dynamic range of 35 dB. Find out the minimum sampling rate, minimum number of bits used in the PCM code, resolution, and quantization error.

Answers to problems

1. 7 kHz, 36 kHz, 12 MHz
2. 12.8 kHz
3. 10 kHz to 15 kHz
4.

Harmonics	Amplitude	Frequency
First	0.306 V	10 kHz
Second	0.18 V	20 kHz
Third	0.039 V	30 kHz
Fourth	– 0.054 V	40 kHz
Fifth	– 0.075 V	50 kHz
Sixth	– 0.367 V	60 kHz

5. 0.2 ms, 80 μs
6. 10 bits
7. 100 kHz sampling rate to be used
8. 4096, 281 μV, 74 dB
9. 8 bits
10. 1.06 μs, 2.54 μs
11. 6 kHz, 8 bits, 48,000 bits/s
12. 10 kHz, 9 bits, 0.019 V, 0.0095 V

MATLAB EXAMPLES

```
1.%program to analyze over-sampling%
  %Over-sampling%
  fm=200;
  fs=800; t=0:1/fs:((10/fm));
  x=sin(2*pi*fm*t); fx=fft(x,64); xr=ifft(fx,64);
  f=(-31*fs/64):(fs/64):(32*fs/64); fx=[fx(34:64) fx(1:33)];
  subplot(231),stem(x),title('sampled signal,fm=200,fs=800');
  subplot(232),stem(f,abs(fx)),axis([-300 300 0 30]);
  title('frequency spectrum,fm=200,fs=800');
  subplot(233); stem(xr); title('recovered signal,
  fm=200,fs=800');
```

2.
```
%program to analyze under-sampling%
fm=600;
x=sin(2*pi*fm*t);
fx=fft(x,64); xr=ifft(fx,64); fx=[fx(34:64) fx(1:33)];
subplot(234),stem(x),title('sampled signal,fm=600,fs=800');
subplot(235),stem(f,abs(fx)),axis([-300 300 0 30]);
title('frequency spectrum,fm=600,fs=800');
subplot(236); stem(xr); title('recovered signal,
fm=600,fs=800');
```

sampled signal,fm=200,fs=800

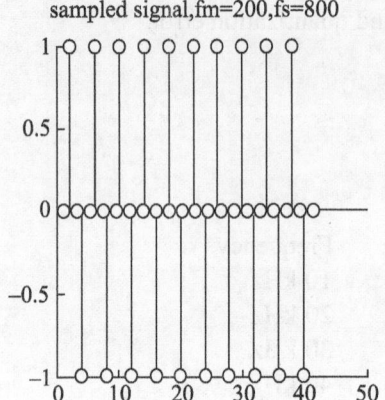

sampled signal,fm=600,fs=800

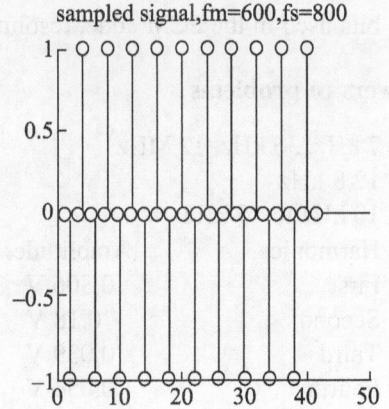

frequency spectrum,fm=200,fs=800

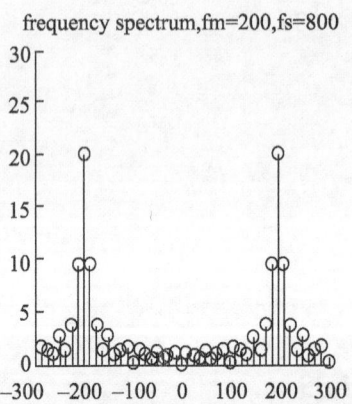

frequency spectrum,fm=600,fs=800

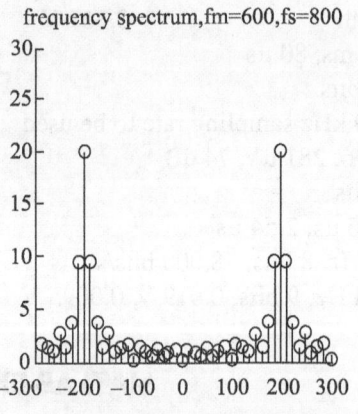

recovered signal, fm=200,fs=800

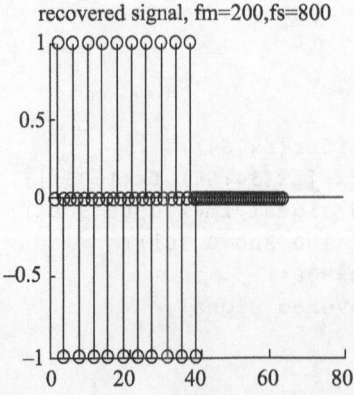

recovered signal, fm=600,fs=800

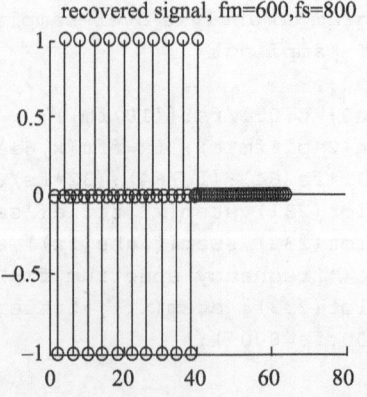

3. ```
%Pulse Width Modulation%
fc=1000;
fs=10000;
f1=200; f2=300;
t=0:1/fs:((2/f1)-(1/fs));
x=0.4*cos(2*pi*f1*t)+0.5;
%modulation%
y=modulate(x1,fc,fs,'pwm');
subplot(421); plot(x); title('original single tone
  message,f1=200,fs=10000');
subplot(422); plot(y); axis([0 500 -0.2 1.2]);
title('PWM,one cycle of f1, fc=1000,f1=200');
fx=abs(fft(y,1024)); fx=[fx(514:1024) fx(1:513)];
f=(-511*fs/1024):(fs/1024):(512*fs/1024);
subplot(424); plot(f,fx); title('frequency description
  PWM,single tone,fc=1000');
%Demodulation%
x_recov=demod(y,fc,fs,'pwm'); subplot(423); plot(x_recov);
title('time domain recovered, single tone, f1=200');
```

original single tone message, f1=200,fs=10000

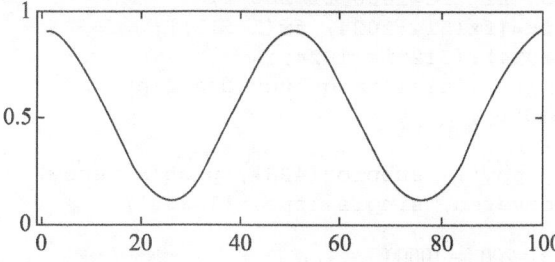

time domain recovered single tone,f1=200

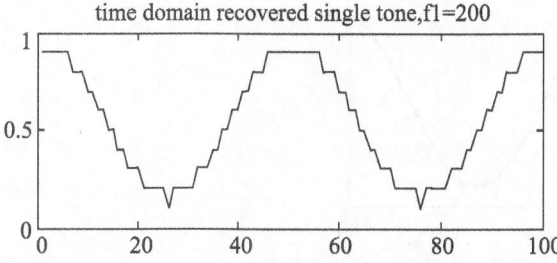

PWM, one cycle of f1, fc=1000,f1=200
```

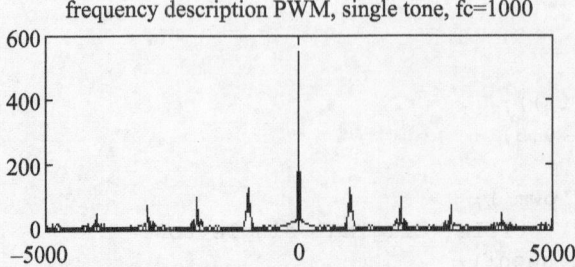

frequency description PWM, single tone, fc=1000

4. ```
%Pulse Position Modulation%
fc=1000;
fs=10000;
f1=200; f2=300;
t=0:1/fs:((2/f1)-(1/fs));
x=0.4*cos(2*pi*f1*t)+0.5;
%modulation%
y=modulate(x1,fc,fs,'ppm');
subplot(421); plot(x); title('original single tone
    message,f1=200,fs=10000');
subplot(422); plot(y); axis([0 500 -0.2 1.2]);
title('PPM,one cycle of f1, fc=1000,f1=200');
fx=abs(fft(y,1024)); fx=[fx(514:1024) fx(1:513)];
f=(-511*fs/1024):(fs/1024):(512*fs/1024);
subplot(424); plot(f,fx); title('frequency description
    PPM,single tone,fc=1000');
%Demodulation%
x_recov=demod(y,fc,fs,'ppm'); subplot(423); plot(x_recov);
title('time domain recovered, single tone, f1=200');
```

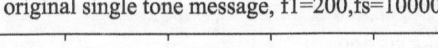

original single tone message, f1=200,fs=10000

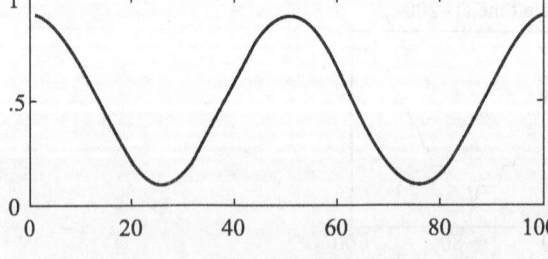

time domain recovered, single tone, f1=200

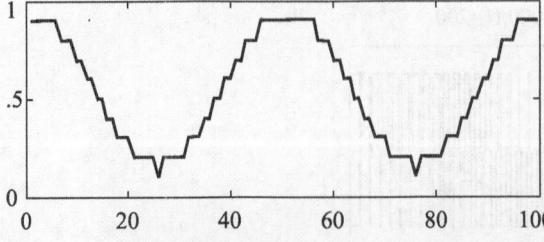

PPM,one cycle of f1, fc=1000,f1=200
```

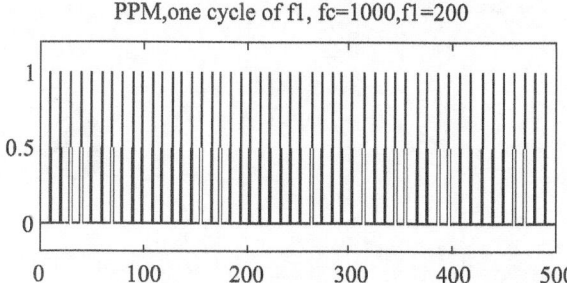

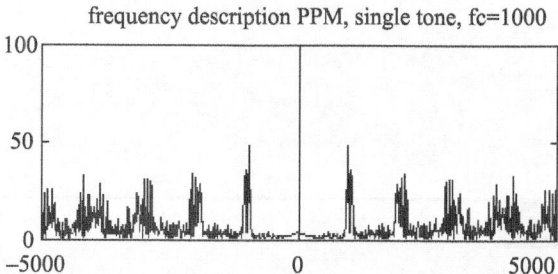

6

Noise

LEARNING OBJECTIVES

This chapter will enable the students to
- know what is noise, its definition, its source, different types of noise, noise power in a thermal noise source, noise voltage for a chain of resistors, and thermal noise power in a reactance circuit
- become familiar with spectral density of noise, noise equivalent bandwidth, noise bandwidth, signal-to-noise (*S/N*) ratio, *S/N* ratio of cascaded system, noise figure, noise factor, noise factor of amplifier in cascade, and noise temperature
- understand effect of noise on various types of modulation systems and pre-emphasis and de-emphasis circuits
- emphasize threshold effect in angle modulation and capture effect
- know mathematical representation of noise, narrowband noise, super-position of noise, quadrature component of noise, and representation of noise using orthogonal components
- solve some numerical problems to understand the theoretical concept

6.1 INTRODUCTION

Noise as commonly understood is a disturbance we hear, but electrical noise is defined as any undesirable electrical energy that falls within the passband of the signal. This gives rise to audible noise in a system. It is also a relative term. Noise to someone is a valid signal to other. In telecommunication, noise is described as the electrical disturbance that gives rise to audible noise in a system. It appears in video systems as white flecks on TV picture when the signal received is weak. Picture in such a case is referred to as *noisy picture*. Noise can be divided into two general categories: *correlated* and *uncorrelated*. Correlation implies a relationship between the signal and noise. Therefore, correlated noise exists only when a signal is present. Uncorrelated noise, on the other hand, is present all the time even if the signal is not present.

Noise can arise in a variety of ways such as due to faulty connection in a piece of equipment and due to the connection carrying the current being broken intermittently, e.g., at the brushes of an electric motor. In principle, these types of noise can be suppressed at the source itself. Noise can limit the range of systems for a given transmitted power. It affects the sensitivity of receivers by placing a limit on the weakest signals that can be amplified. It also sometimes even forces a reduction in the bandwidth of a system. Hence, noise is mainly of concern in receiving systems. Even when precautions are taken to eliminate noise from faulty connections or arising from external sources, it is found that certain fundamental sources of noise are present within electronic equipment that limit the receiver sensitivity. Unfortunately, adding extra amplifier does not improve the receiving system since it also amplifies the noise. Therefore, the study of the fundamental sources of noise within a system is essential if the effects of noise are to be maintained. Natural phenomena like electric storms, solar flares, and certain belt of radiations that exist in space give rise to noise. Noise from these sources is very difficult to suppress. The only way is to reposition the antenna to minimize the noise.

As explained earlier, uncorrelated noise is present regardless of whether there is a signal present or not. These types of noise fall into two general categories: external and internal.

6.2 EXTERNAL NOISE

This is the noise that is generated outside the device or circuit. The three primary kinds of external noise are atmospheric, extraterrestrial, and industrial or man made.

6.2.1 Atmospheric Noise

The noise generated due to the electrical disturbance within the earth's atmosphere is called *atmospheric noise*. This type of noise creates strange sounds like sputtering, crackling, etc., in shortwave receivers. Most of these sounds are a result of spurious radio signals with components distributed over a wide range of frequencies. These signals are propagated over the earth's atmosphere in the same way as ordinary radio waves of same frequencies. Atmospheric noise is commonly called *static electricity*. It is caused by lightening and thunderstorms (local or distant). The static is likely to be more severe but less frequent if the storm is local. It is in the form of impulses that spread energy throughout a wide range of frequencies. The magnitude of this energy is inversely proportional to the frequency. Hence, at frequencies above 30 MHz, atmospheric noise is less relevant. The atmospheric noise interferes more with reception of radio than that of television. The reason for this noise becoming less severe at frequencies above 30 MHz is due to line-of-slight propagation. Also the nature of the mechanism generating this noise is such that very little of it is created in the very high frequency (VHF) range and above.

6.2.2 Extraterrestrial Noise

It consists of electrical signals that originate from outside the earth's atmosphere of and is, therefore, sometimes called *deep space noise*. It originates from the Milky Way, other galaxies, and the sun. Extraterrestrial noise is classified into two subgroups: solar and cosmic.

Solar noise

This noise is generated directly from the sun's heat. This is due to the radiation from the sun, which is of two types. One takes place in a situation called *quiet condition*, when relatively constant radiation intensity exists. In such a condition, the radiation radiates over a very broad frequency spectrum, which includes the frequencies we use for communication. The other type of radiation arises due to the sun spot activity and solar flare-up. This high-intensity radiation occurs approximately every 11 years. Although this high-intensity noise comes from a limited portion of the sun's surface, it may still be of the order of magnitude greater than that received during the period of quiet sun.

Cosmic noise

The sources of cosmic noise are continuously distributed throughout the galaxies. Since distant stars are also suns and have high temperatures, they radiate RF noise in the same manner as the sun. These sources are located much farther away from the sun and their noise intensity is relatively small. The noise received is called *black body noise* and is distributed evenly throughout the sky.

Space noise is observable at frequencies in the range from about 8 MHz to 1.43 GHz. Apart from the man-made noise, it is the strongest component over the range of about 20 to 120 MHz. The noise below 20 MHz does not penetrate much through the ionosphere but it disappears at frequencies in excess of 1.5 GHz.

6.2.3 Industrial Noise (Man-made Noise)

This is simply the noise generated by the human race. It lies between the frequencies of 1 MHz and 600 MHz. The intensity of this noise overtakes the noise created by any other source, internal or external to the receiver. The predominant sources of man-made noise are automobile and aircraft ignition, electric motor, switching equipment, leakage from high-voltage lines, and a multitude of other heavy electric machines. Fluorescent lights are another powerful source of this noise and should be avoided near very sensitive receivers. This noise is impulsive in nature and contains a wide range of frequencies that propagate through space in the same manner as radio waves. They are most intense in densely populated and industrial areas. Hence, it is called industrial noise. The nature of industrial noise is so variable that it is difficult to analyse it on any basis other than static. But the received noise increases as the receiver bandwidth increases.

6.3 INTERNAL NOISE

This is the noise generated within a device or circuit. It can be the noise generated by any of the active or passive devices found in a receiver. Such noise is generally random, impossible to treat on individual voltage basis but easy to describe and observe statistically, since it is distributed randomly over the entire radio spectrum. Random noise power is proportional to the bandwidth over which it is measured. Various types of internal noise are discussed below.

6.3.1 Thermal Noise (Johnson Noise)

Thermal noise is associated with the rapid and random movement of electrons within a conductor due to thermal agitation. Free electrons within an electrical conductor possess kinetic energy as a result of heat exchange between the conductor and its surroundings. The atoms and molecules of all substances vibrate constantly in a minute motion, which causes the sensation of heat—the higher the temperature, the greater the vibration. As they vibrate, they send electromagnetic waves and as there are many atoms, a chaotic jumble of electromagnetic waves of all frequencies is created. These motions form the ultimate, unavoidable noise background to all electronic process. It is a continuous Gaussian noise. Since this noise arises from thermal causes, it is called *thermal noise* or *Johnson noise* after its discoverer.

The average or mean noise voltage across a conductor is zero, but the root mean square (RMS) value is finite and can be easily measured. It has been observed that this RMS value of the noise voltage is proportional to the resistance of the conductor, its absolute temperature, and the frequency bandwidth of the device measuring noise. This noise is also called *white noise* since it contains all spectral frequencies equally on an average. When there is no FM signal, this noise can be heard as a sound 'hiss'. In an FM receiver, we hear this voice when there is no signal and the volume is increased. Squelch circuits are used to suppress this noise. Figure 6.1 shows a plot of instantaneous noise voltage vs time, which varies randomly.

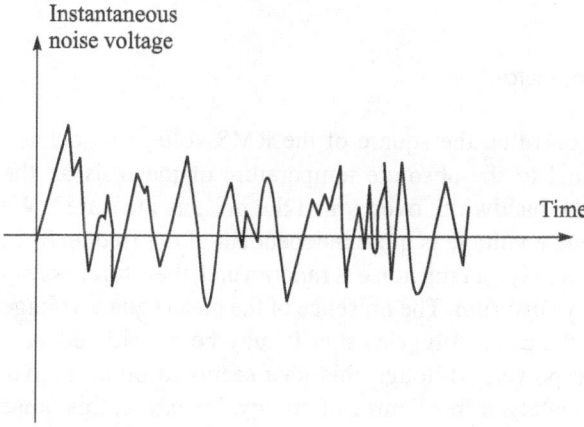

Fig. 6.1   Thermal noise

Johnson proved that the thermal noise power is proportional to the product of the bandwidth and the temperature. Mathematically, the noise power P_n is

$$P_n = kTB \qquad (6.1)$$

where
P_n = noise power in watt
T = absolute temperature in kelvin
B = bandwidth in Hz
k = Boltzmann's constant (1.38×10^{-23} joule/kelvin)

6.3.2 Noise Voltage

Figure 6.2 shows an equivalent circuit for a thermal noise source where the internal resistance of the source R is in series with the RMS voltage V_N. To receive maximum power, the noiseless R_L should be equal to R. Hence, the noise voltage dropped across R_L is equal to $V_N/2$. The noise power P_N developed across the load resistor R_L is

$$P_n = kTB = \left(\frac{V_N^2/2}{R}\right)^2 = \frac{V_N^2}{4R} \qquad (6.2)$$

Thus,
$$V_N^2 = 4RkTB \qquad (6.3)$$

and
$$V_N = \sqrt{4RkTB} \qquad (6.4)$$

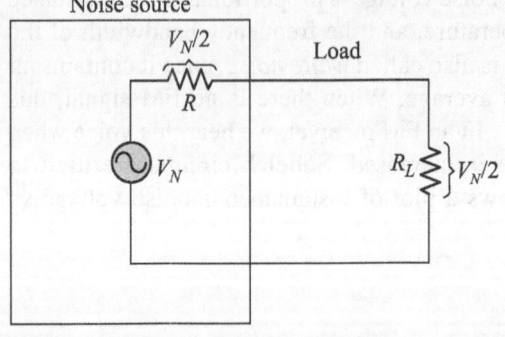

Fig. 6.2 Equivalent circuit for noise source

It is observed from Eq. (6.4) that the square of the RMS voltage associated with a resistor is proportional to the absolute temperature of the resistor, the value of its resistance, and the bandwidth over which the noise is measured. We also see that the generated noise voltage is quite independent of the frequency at which it is measured. The reason is that the noise is random and, therefore, evenly distributed over the frequency spectrum. The presence of the mean square voltage at the terminals of the resistance R_L suggests that it may be considered as a generator of electrical noise power. Although this idea seems to be attractive, thermal noise is not, unfortunately, a free source of energy. To extract this noise power, the resistor R_L should be connected to a resistive load and in thermal equilibrium. The load would supply as much energy to R_L as it receives.

Example 6.1 Calculate the thermal noise power available from any resistor at a room temperature of 290 K for a bandwidth of 1 MHz.

Solution

We get $\qquad P_n = kTB_n = 1.38 \times 10^{-23} \times 290 \times 10^6 = 4 \times 10^{-15}$ watt

6.3.3 Equivalent Sources for Thermal Noise

The thermal noise properties of a resistor R may be represented as an equivalent voltage generator. Norton's theorem can then be used to find the equivalent current generator. These two source representations are shown in Fig. 6.3. Since the mean square noise voltage is $E_n^2 = 4RkTB$, the Norton equivalent mean square noise current will be $I_n^2 = \dfrac{4kTB}{R}$. If we represent $\dfrac{1}{R} = G$, then

$$I_n^2 = 4kTGB \qquad (6.5)$$

Example 6.2 Calculate the corresponding noise voltage for $R = 60\ \Omega$ for Example 6.1.

Solution

We have $\qquad E_n^2 = 4 \times 60 \times 1.38 \times 10^{-23} \times 290 = 960{,}480 \times 10^{-24}$

i.e., $\qquad E_n = \sqrt{960480 \times 10^{-24}} = 0.980 \times 10^{-9}$ volt

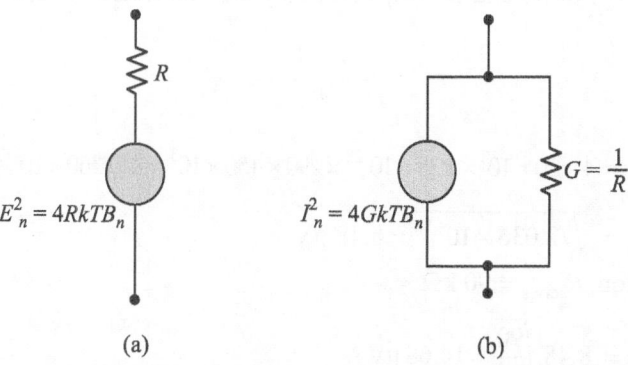

(a)   (b)

Fig. 6.3   Thermal noise equivalent: (a) voltage source and (b) current source

Although the bandwidth of open circuit noise voltage is infinite, this is not so in practical cases. Since all resistance will have lead inductance and self-capacitance, this sets the finite limit on bandwidth. In case of open circuit load, the self-capacitance sets the limit on bandwidth.

6.3.4 Noise Voltage for Resistors Connected in Series

Let R_S represent the equivalent series resistance of a resistance chain connected in series, i.e.,

$$R_S = R_1 + R_2 + R_3 + \cdots$$

Then the noise voltage of the equivalent series resistance is

$$E_n^2 = 4R_S kTB \qquad (6.6)$$
$$= 4(R_1 + R_2 + R_3 + \cdots)kTB$$

or $\qquad E_n^2 = (E_{n1}^2 + E_{n2}^2 + E_{n3}^2 + \cdots)kTB \qquad (6.7)$

This shows that the total noise voltage squared is equal to the sum of the noise voltage squared of individual resistors, i.e., the noise voltage of the series chain is given by

$$E_n = \sqrt{E_{n1}^2 + E_{n2}^2 + E_{n3}^2 + \cdots} \qquad (6.8)$$

6.3.5 Resistors in Parallel

Let G_P be the equivalent conductance of a parallel combination of resistors R_1, R_2, R_3, $\cdots$. Taking the equivalent conductance G_1, G_2, G_3, $\cdots$,

$$G_P = G_1 + G_2 + G_3 + \cdots \qquad (6.9)$$

Then $\qquad I_n^2 = 4G_P kTB \qquad (6.10)$

or $\qquad I_n^2 = 4(G_1 + G_2 + G_3 + \cdots)kTB$

Therefore, $\qquad I_n^2 = I_{n1}^2 + I_{n2}^2 + I_{n3}^2 + \cdots \qquad (6.11)$

Example 6.3 For a bandwidth of 150 kHz, calculate the thermal noise voltage generated by two resistors of 30 and 60 kΩ, when they are connected in series and in parallel.

Solution
For 30 kΩ resistor,

$$E_n^2 = 4 \times 30 \times 10^3 \times 1.38 \times 10^{-23} \times 290 \times 150 \times 10^3 = 720{,}360 \times 10^{-17}$$

i.e., $\qquad E_n = \sqrt{72.036 \times 10^{-12}} = 8.48 \ \mu V$

For series combination, R_{series} = 90 kΩ

$$E_{nseries} = 8.48\sqrt{\frac{90}{30}} = 14.68 \ \mu V$$

For parallel combination, $R_{parallel}$ = 20 kΩ

$$E_{nparallel} = 8.48\sqrt{\frac{20}{30}} = 6.92 \ \mu V \qquad (6.11)$$

6.3.6 Thermal Noise Power in a Reactance Circuit

Reactance does not generate thermal noise. The reason is that it does not dissipate power. Consider the circuit given in Fig. 6.4. A parallel inductor and capacitor circuit is connected to a resistor R as shown. Equal amount of power should be exchanged between them for thermal equilibrium. If the resistor R supplies thermal noise power P_R to the reactance, the reactance must supply thermal noise power P_X. Since for thermal equilibrium, $P_X = P_R$ and the reactance cannot dissipate any power, power $P_X = 0$. Therefore, P_R also must be zero. But the effect of reactance on the noise bandwidth, however, should be taken into account.

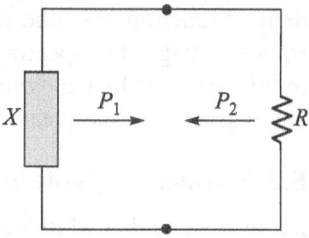

Fig. 6.4 Reactance circuit

6.3.7 Spectral Densities

Thermal noise falls into the category of power signal and has a spectral density. The bandwidth B depends on the external measuring or receiving system and is assumed flat. Hence, from Eq. (6.1), the available power spectral density in watts/Hz or joule is

$$S_P(f) = \frac{P_n}{B_n} = kT \tag{6.12}$$

The spectral power density in terms of mean square voltage is

$$S_V(f) = \frac{E_n^2}{B_n} = 4RkT \tag{6.13}$$

Spectral densities are flat and are independent of frequency. Owing to this, thermal noise is sometimes referred to as *white noise*. Similar to white light, it has a flat spectrum. When white noise is passed through a network, the spectral density will be altered by the shape of the frequency response. The sum of the noise contribution over the complete frequency range gives the total noise power at the output. This takes into account the shape of the frequency response.

6.3.8 Power Spectral Response

Let us consider a power spectral response as shown in Fig. 6.5. The available noise power for infinitesimally small bandwidth δf about the frequency f_1 is

$$\delta P_{n1} = S_P(f_1)\,\delta f \tag{6.14}$$

The assumption made above is that since δf is too small, the bandwidth δf is flat at about f_1. The available power is given as the product of the spectral density (watts/Hz) and the bandwidth (Hz). The area of the shaded strip about f_1 gives the available noise power. Hence, if we consider at frequencies $f_2, f_3, \cdots$, the total power given by the sum of all these contributions is equal to the sum of all these small areas, which is the total area under the curve, i.e., this is equal to the integral of the spectral

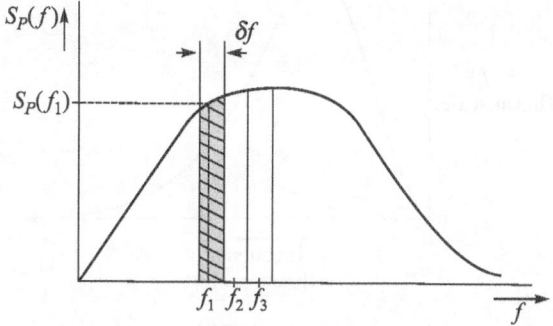

Fig. 6.5 Noise spectral density

density function over the frequency range $f = 0$ to $f = \infty$. Similarly, for mean square voltage, the spectral density curve will have V^2/Hz as its ordinate and multiplying this by bandwidth δf Hz. results in units of V^2. So, the area under the curve gives the total mean square voltage.

6.3.9 Noise Equivalent Bandwidth

Let us assume that white noise is present at the input to a receiver and this white noise is passed through a filter with a transfer function $H(f)$ centred at f_0.

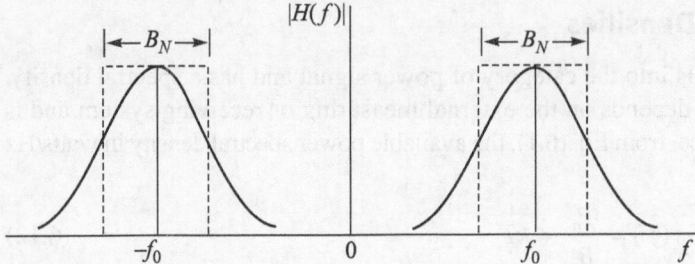

Fig. 6.6 Noise equivalent bandwidth

To restrict the noise power actually passed on to the receiver, the output of the filter is not white anymore, i.e., this filter is being used to restrict the noise power actually passed on to the receiver. Let us take a rectangular filter centred at f_0. Let the rectangular filter bandwidth B_N be adjusted so that the real filter and the rectangular filter transmit the same noise power. This is shown in Fig. 6.6. The bandwidth B_N is called the *noise bandwidth* of the real filter. The noise bandwidth then is the bandwidth of an idealized (rectangular) filter that passes the same noise power as does the real filter.

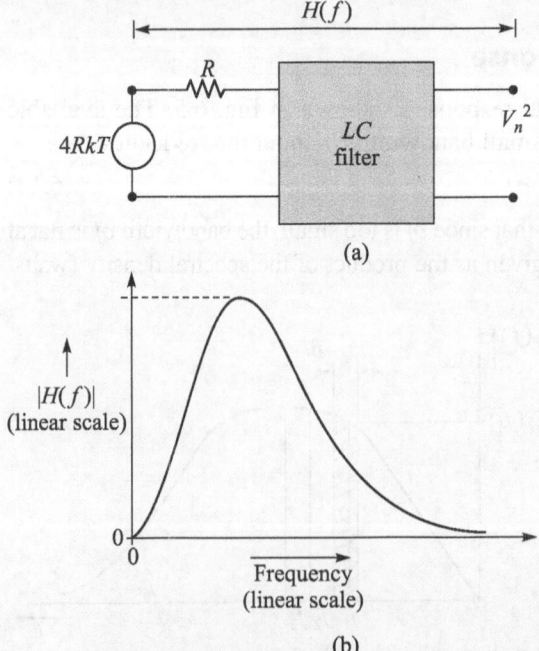

Fig. 6.7 (a) Schematic diagram and (b) transfer functions of the filter

Let us consider a resistor R connected to the input of an LC filter as shown in Fig. 6.7. We can replace R with an input noise generator with mean square voltage spectral density $4RkT$ in series with R. Now we can consider the entire circuit as a noise generator connected to an R and LC filter. Let the transfer function of the network including R be $H(f)$ as shown in Fig. 6.7(b). The spectral density for the mean square output voltage is, therefore, $4RkT|H(f)|^2$.

The total mean square output voltage is given by the area under the output spectral density curve as follows:

$$V_n^2 = \int_o^\infty 4\,kRT\,|H(f)|^2 df$$

$$= 4\,RkT \times \text{area under } |H(f)|^2 \text{ curve} \tag{6.15}$$

Also we have seen earlier that the total mean square voltage at the output can be stated as

$$V_n^2 = 4RkTB_n \tag{6.16}$$

Substituting for V_n^2 from Equations (6.15) and (6.16), we get

$$4RkTB_n = \int_o^\infty 4RkT\,|H(f)|^2$$

$$= 4kRT \int_o^\infty |H(f)|^2$$

Comparing LHS and RHS, we get

$$B_n = \int_o^\infty |H(f)|^2 df$$

i.e., $B_n = \text{area under } |H(f)|^2 \text{ curve}$ \hfill (6.17)

Let us find the noise equivalent bandwidth of a low-pass RC filter. The filter is shown in Fig. 6.8.

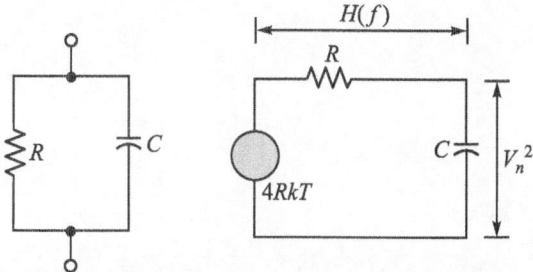

Fig. 6.8 *RC network and its transfer function for determining noise bandwidth*

For this filter, the transfer function

$$H(f) = \frac{1}{1 + j2\pi fRC} \tag{6.18}$$

Let $RC = \tau$, the time constant. The frequency response of the filter is shown in Fig. 6.9.

Now
$$|H(f)| = \frac{1}{\sqrt{1 + 4\pi^2 f^2 \tau^2}}$$

Also
$$B_n = \int_{-\infty}^{\infty} |H(f)|^2 \, df$$

Substituting the value of $|H(f)|^2$, we get

$$B_n = 2 \int_0^{\infty} \frac{1}{\sqrt{1 + 4\pi^2 f^2 \tau^2}} \, df \tag{6.19}$$

Let
$$v = 2\pi f \tau$$

Hence,
$$B_n = \int_0^{\infty} \frac{1}{\sqrt{1 + v^2}} \frac{dv}{2\pi\tau} \tag{6.20}$$

$$= \frac{1}{\pi\tau} \times \frac{\pi}{2}$$

$$= \frac{1}{2\tau}$$

So,
$$B_n \text{ eq} = \frac{1/2\tau}{2} = \frac{1}{4RC} \tag{6.21}$$

Since the mean square noise voltage at the output can be expressed as

$$V_n^2 = 4RkTB_n$$

Substituting the value of B_n from Eq. (6.21) in Eq. (6.16), we get

$$V_n^2 = 4RkT \times \frac{1}{4RC} = \frac{kT}{C}$$

i.e.,
$$V_n^2 = \frac{kT}{C} \tag{6.22}$$

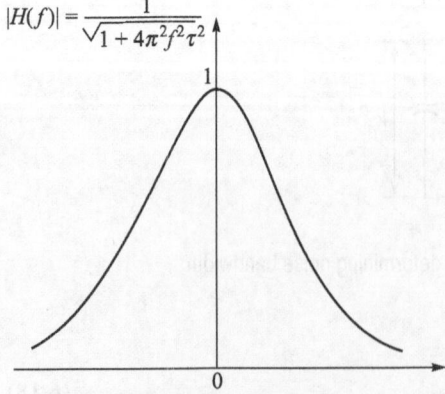

Fig. 6.9 Frequency response

This shows that the mean square output voltage is independent of R although the noise voltage is generated by R. At the same time, it is inversely proportional to C, which never generates any noise.

For most radio receivers, the noise is generated at the front end of the receiver, while the output noise bandwidth is determined by the audio sections of the receiver.

The equivalent noise bandwidth is equal to the area under the normalized power gain/frequency curve for low-frequency section. To normalize, the curve is scaled such that the maximum value is equal to unity. Usually, this information is available in the form of a frequency response curve showing output in decibel relative to maximum and frequency plotted on a logarithmic scale as shown in Fig. 6.10(a). To determine the area under curve, the decibel axis must be converted to a linear power ratio scale and the frequency axis to a linear frequency scale as shown in Fig. 6.10(b). The equivalent noise bandwidth is then equal to the area under this curve for SSB receiver. The noise bandwidth appears on both sides of the carrier for DSB receiver and is effectively doubled. This is shown in Fig. 6.10(c).

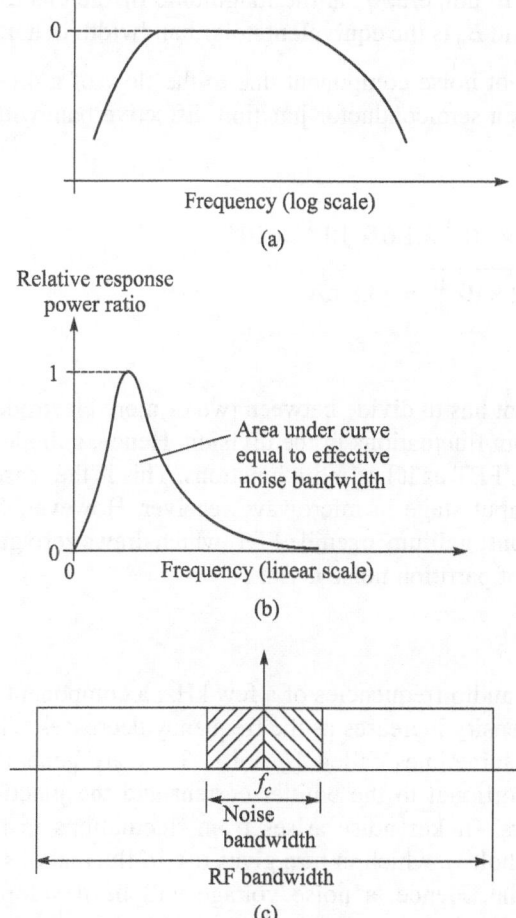

Fig. 6.10 Frequency response curve for amplifier showing noise bandwidth

6.3.10 Shot Noise

Any direct current crossing a potential barrier in a random fashion results in shot noise. This occurs because the carrier holes and electrons do not cross the barrier simultaneously, but with a random distribution in the timing for each carrier. This gives rise to a random component of current superimposed on the steady current. In bipolar transistors, the bias current crossing the forward biased emitter base junction carries shot noise. The name 'shot noise' came from the fact that similarity exists between an electron striking the anode and a lead shot from a gun striking a target. Shot noise is also sometimes called *transistor noise* and is additive with thermal noise.

Shot noise also has a flat spectrum similar to thermal noise except in the range of microwave frequency. The mean square component is proportional to the direct current that is flowing and for most devices the mean square shot noise current is given by

$$I_n^2 = 2I_{dc}q_e B_n \text{ ampere}^2 \qquad (6.23)$$

where I_{dc} is the direct current in ampere, q_e is the magnitude of the electron charge (1.6×10^{-19} columbs), and B_n is the equivalent noise bandwidth in hertz.

Example 6.4 Calculate the shot noise component due to the flow of a direct current of 2 mA flowing across a semiconductor junction. Effective bandwidth of 3 MHz can be assumed.

Solution

We have $\qquad\qquad I_n^2 = 2 \times 2 \times 10^{-3} \times 1.6 \times 10^{-19} \times 10^6$

So, $\qquad\qquad I_n = \sqrt{19.2 \times 10^{-16}} = 43.8 \text{ nA}$

6.3.11 Partition Noise

This occurs whenever the current has to divide between two or more electrodes. The reason for this is the random fluctuations in the division. Hence, a diode is less noisy than a transistor or an FET as it has more junctions. This is the reason for using diode circuit as an input stage in microwave receiver. However, for low-noise microwave applications, gallium arsenide FET, which draws zero gate current, is used. The spectrum of partition noise is flat.

6.3.12 Flicker Noise

At low frequencies, i.e., at low audio frequencies of a few kHz, a component of noise appears whose spectral density increases as the frequency decreases. This is known as *flicker noise* and sometimes called $1/f$ noise. They are generally found in transistors. It is proportional to the emitter current and the junction temperature. In semiconductors, flicker noise arises from fluctuations in the carrier densities (electrons and holes), which in turn gives rise to fluctuations in the conductivity of the material. Hence, a noise voltage will be developed whenever a direct current flows through the semiconductor and the mean square voltage will be proportional to the square of the direct current. This flicker noise limits the sensitivity of microwave diode mixers used for Doppler radar system.

This is because, although the input frequencies to the mixer are in microwave range, the Doppler frequency output is in the low audio frequency range.

6.3.13 Burst Noise

This is also a low-frequency noise. It is observed in bipolar transistor. The noise appears as a series of bursts at two or more levels. When present in an audio system, it produces popping sound and hence called *popcorn noise*. Source for this noise is not clearly understood, but spectral density is known to increase as the frequency decreases.

6.3.14 Transit Time Noise

As the name suggests, this noise is due to transit time. If the time taken by an electron to travel from the emitter to the collector in a transistor becomes significant to the period of signal being amplified, transit time effect takes place. This causes the noise input admittance of the device being increased. The transit time noise in a transistor is determined by carrier mobility, bias voltage, and transistor construction. Carriers travelling from emitter to collector suffer from emitter time delays, base transit time delay, collector recombination time, and propagation time delays. If transit delays are excessive at high frequencies, the device may add more noise than amplification to the signal causing frequency distortion.

6.3.15 Avalanche Noise

The reverse current in a reverse biased diode, which is normally small, increases rapidly with a slight increase in the magnitude of the reverse bias voltage. The region where this phenomenon occurs is known as *avalanche region*. In the avalanche region, both holes and electrons in the diode depletion region gain sufficient energy from the reverse biased field to ionize atoms by collision. This ionization process results in creation of additional holes and electrons, which in turn contribute to further ionization process. The collisions that result in the avalanching occurs at random creating large spikes of noise. In zener diodes, this noise is predominant and has to be avoided.

However, the avalanche noise is put into good use in noise measurements. The spectral density of avalanche noise is flat.

6.3.16 Transistor Noise

Bipolar transistor exhibits all the sources of noise discussed previously. Bulk or extrinsic resistance, especially extrinsic base resistance, generates thermal noise. Bias current in the transistor results in shot noise and partition noise. Flicker and burst noise occur due to base current.

In field effect transistor for both JFETs and MOSFETs, thermal noise is generated by the physical resistance of drain source channel. Gate leakage current attributes to shot noise. Flicker noise is due to the channel. Shot noise develops a noise component of voltage across the signal source impedance, which can be quite high due to very high input impedance of FET.

6.4 SIGNAL-TO-NOISE RATIO

In communication system, we more often talk about signal-to-noise (S/N) ratio rather than the absolute value of noise. Signal-to-noise ratio is defined as a power ratio. It is defined as the ratio of the signal power to the noise power. Mathematically, this is expressed as

$$\frac{S}{N} = \frac{P_s}{P_n} \tag{6.24}$$

The S/N ratio is often expressed as a logarithmic function with the decibel unit, i.e.,

$$\frac{S}{N} = 10 \log \frac{P_s}{P_n} \tag{6.25}$$

It can also be expressed in terms of voltages and resistances. Since $P_s = V_s^2/R_s$, where V_s is the signal voltage in volts, R_s is the input resistance in ohms, and $P_n = V_n^2/R_o$, where V_n is the noise voltage in volts and R_o is the output resistance in ohms.

Substituting the values of P_s and P_n in Eq. (6.25), we get

$$\frac{S}{N} = 10 \log \frac{\left(\dfrac{V_s^2}{R_s}\right)}{\left(\dfrac{V_n^2}{R_o}\right)} \tag{6.26}$$

Assume $R_s = R_o$, i.e., input and output resistances of the amplifier, receiver, or network being evaluated are equal. Then, Eq. (6.26) reduces to

$$\frac{S}{N} = 10 \log \frac{\left(\dfrac{V_s^2}{R_s}\right)}{\left(\dfrac{V_n^2}{R_o}\right)}$$

$$= 10 \log \frac{V_s^2}{V_n^2} \tag{6.27}$$

Therefore, $\dfrac{S}{N} = 20 \log \dfrac{V_s}{V_n}$ in dB $\tag{6.28}$

6.4.1 Signal-to-Noise Ratio of a Cascaded System

Let us consider S/N ratio of a cascaded system. In the case of telephony to compensate for the loss in cables, as the speech travels along, repeaters are added at regular intervals. The repeaters compensate for the loss of the signal (L) and its gain (G) is chosen so that $LG = 1$. Hence, the signal loss is compensated, but noise introduced by each repeater is added and the final noise at the output will depend on the number of repeaters used.

Let us take an example of repeaters connected in cascade or tandem as shown in Fig. 6.11. In this configuration, the input signal power to the first repeater is P_s and at this point, the input noise may be assumed negligible. After travelling along the first section of time, the signal is attenuated by a factor of L. At the output of the first repeater, the signal power is still P_s since the gain G exactly compensates for the loss L. The noise output at the output of the first repeater is P_{n1} and consists of the noise added by the line section and amplifier.

As the signal progresses along the links, the power output at each repeater remains at P_s since the signal loss is compensated by the gain G of the repeater, but the noise powers are additive and the total noise output at the Mth link is

$$P_n = P_{n1} + P_{n2} + P_{n3} + \cdots + P_{nm} \tag{6.29}$$

If the links are identical and each link contributes noise power P_n, then the total output noise for the entire system will be

$$P_{nm} = MP_n \tag{6.30}$$

Then the output S/N ratio for this network will be

$$\left(\frac{S}{N}\right)_{out} = 10 \log \frac{P_s}{MP_n} \quad \text{dB} \tag{6.31}$$

$$= 10 \log \left(\frac{P_s}{P_N}\right)\left(\frac{1}{M}\right) \quad \text{dB}$$

i.e.,

$$\left(\frac{S}{N}\right)_{out} = \left(\frac{S}{N}\right)_1 \text{dB} - M \text{ dB} \tag{6.32}$$

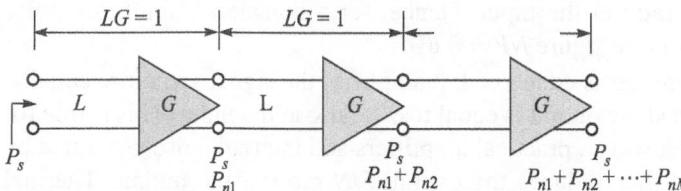

Fig. 6.11 Repeaters connected in tandem

From Eq. (6.32), it can be inferred that $(S/N)_{out}$ ratio for a cascaded connection is equal to the difference between $(S/N)_1$ of any one link in dB and the number of links M expressed in dB.

Example 6.5 Calculate the output signal-to-noise ratio for four systems connected in tandem. The (S/N) ratios of systems are 60 dB, 40 dB, 30 dB, and 50 dB, respectively.

Solution

The noise-to-power ratio of the first system is -60 dB or the power ratio is 10^{-6}. Similarly, for the other systems, the power ratios are 10^{-4}, 10^{-3}, and 10^{-5}, respectively.

The overall $\left(\dfrac{S}{N}\right)_{\text{out}} = \dfrac{1}{10^{-6}+10^{-4}+10^{-3}+10^{-5}} = 10^{-3} = 30$ dB

It is seen that the overall signal-to-noise ratio is equal to that of the worst system.

6.5 NOISE FIGURE

Noise figure is a figure of merit and used to indicate how much the (S/N) ratio gets degraded as a signal passes through a circuit or series of circuits. It is the ratio of the input signal-to-noise power ratio to the output signal-to-noise power ratio expressed in decibels, i.e.,

$$NF \text{ (dB)} = 10 \log \frac{\text{Input signal-to-noise power ratio}}{\text{Output signal-to-noise power ratio}} \qquad (6.33)$$

It can be expressed in terms of noise factor F, where

$$F = \frac{\text{Input singal-to-noise power ratio}}{\text{Output signal-to-noise power ratio}} \qquad (6.34)$$

Hence, $NF \text{ (dB)} = 10 \log F \qquad (6.35)$

Noise figure in effect indicates how much the S/N ratio deteriorates as a waveform progresses from input to the output of a circuit. For example, an amplifier with a noise figure of 3 dB means that the signal-to-noise ratio at the output is 3 dB less than it was at the input. If an amplifier or circuit is perfectly noiseless and adds no additional noise to the signal, the S/N ratio at the output will equal the S/N ratio at the input. Hence, for a noiseless circuit, the noise factor F is 1 and its noise figure $NF = 0$ dB.

As far as the amplifier is concerned, it amplifies the signal and noise equally. Hence, the S/N ratio at the output is equal to S/N ratio at the input. This is true for an ideal amplifier. However, practical amplifiers add internal noise generated by its components and thus reduces the overall S/N ratio at the output. Thermal noise is the major contributor. Therefore, all network, amplifiers, and system add noise to the signal and thus reduce the overall S/N ratio as signals progress through them. This is indicated in Fig. 6.12. Figure 6.12(a) shows an ideal noiseless amplifier with a power gain A_p and an input level (N_i).

Hence, $\dfrac{S_{\text{out}}}{N_{\text{out}}} = \dfrac{A_p}{A_p} \dfrac{S_i}{S_n} = \dfrac{S_i}{S_n} \qquad (6.36)$

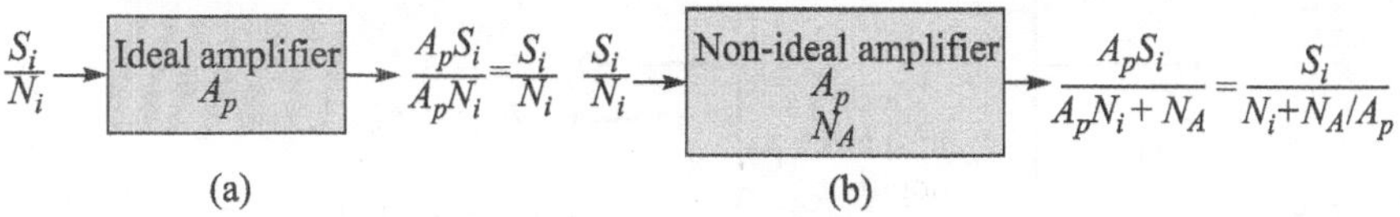

$$\frac{S_i}{N_i} \rightarrow \boxed{\begin{array}{c}\text{Ideal amplifier}\\ A_p\end{array}} \rightarrow \frac{A_p S_i}{A_p N_i} = \frac{S_i}{N_i} \qquad \frac{S_i}{N_i} \rightarrow \boxed{\begin{array}{c}\text{Non-ideal amplifier}\\ A_p \\ N_A\end{array}} \rightarrow \frac{A_p S_i}{A_p N_i + N_A} = \frac{S_i}{N_i + N_A/A_p}$$

(a)   (b)

Fig. 6.12 Noise figure: (a) ideal noiseless amplifier and (b) practical amplifier

This shows that input and output *S/N* ratios are equal. Now let us take the case of practical non-ideal amplifier. This amplifier generates an internal noise N_A, which also gets added to the output. Hence, the output signal-to-noise ratio is less than the input *S/N* ratio by an amount proportional to N_A. This can be expressed mathematically as

$$\frac{S_{\text{out}}}{N_{\text{out}}} = \frac{A_p S_i}{A_p N_i + N_A} = \frac{S_i}{N_i + \dfrac{N_A}{A_p}} \qquad (6.37)$$

where A_p = amplifier power gain

                   N_A = internal noise of the amplifier

Since most of the noise introduced by an amplifier is thermal in nature, let us work out the relation between the noise factor *F* and the available output noise power.

Noise power, $F = \dfrac{\text{Available } S/N \text{ power ratio at the input}}{\text{Available } S/N \text{ power ratio at the output}}$

Thermal noise power = kTB_n

$$F = \frac{P_{si}}{kTB_n} \times \frac{P_{no}}{GP_{si}} \qquad (6.38)$$

where P_{si} = input signal power

         kTB_n = input noise power

            G = gain at the amplifier

         P_{no} = available output noise power

Equation (6.38) can be simplified as

$$F = \frac{P_{no}}{kTB_n G} \qquad (6.39)$$

It follows from this that the available output noise power is

$$P_{no} = FkTB_n G \qquad (6.40)$$

Hence, *F* can be interpreted as the factor by which the amplifier increases the output noise. If the amplifier were noiseless, the output noise would be $GkTB_n$. Figure 6.13 illustrates the noise factor.

Available power gain *G* can be defined unambiguously. This need not depend on the load impedance. Matching of load impedance to source impedance need not be considered since we are interested in the actual power delivered to the amplifier taking into account any mismatch that may be present. Also the noise factor is defined for the source at room temperature $T = 290°$ K. Noise factor is a measured parameter and will be usually specified for a given amplifier or network and specified in decibels. Then it is called noise figure.

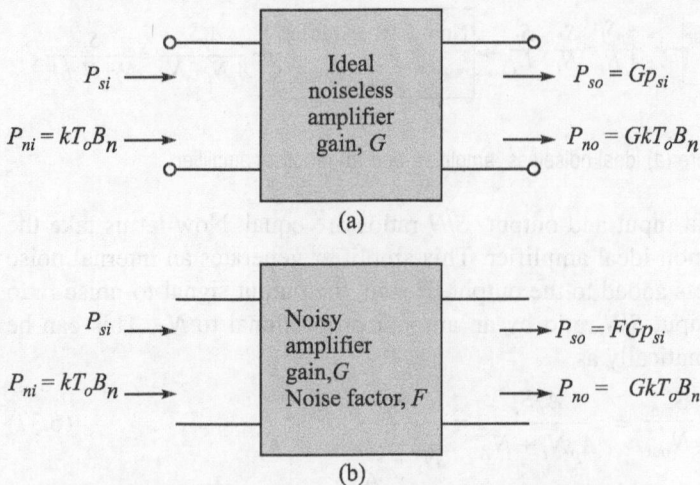

(a)

(b)

Fig. 6.13 Noise factor: (a) noiseless amplifier and (b) practical amplifier

6.5.1 Input Noise of Amplifier in Terms of F

We can imagine the amplifier noise generated by many components throughout the amplifier to be due to some equivalent power source at the input. From Eq. (6.39), since $P_{ni} = P_{no}/G$, we get

$$P_{ni} = FkT_oB_n \tag{6.41}$$

The source contributes an available power kT_oB and hence the amplifier must contribute an amount P_{na}, where $P_{na} = FkT_oB_n - kT_oB_n$

Therefore, $\qquad P_{na} = (F - 1)kT_oB_n \tag{6.42}$

6.5.2 Noise Factor of Amplifiers in Cascade

When two or more amplifiers are in cascade, as shown in Fig. 6.14, the total noise factor is the accumulation of the individual noise factors. The available noise power at the output of the first stage is

$$P_{no1} = F_1G_1kT_oB_n \tag{6.43}$$

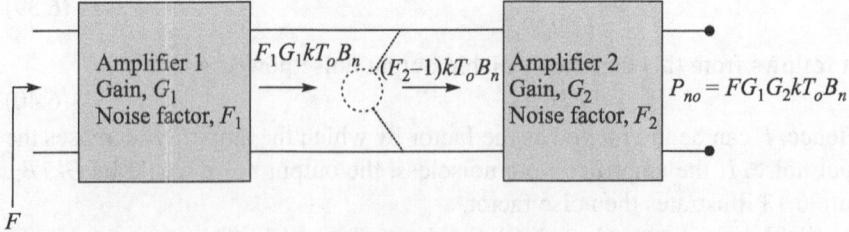

Fig. 6.14 Noise factor: two amplifiers in cascade

This is available to second stage as input. Also stage 2 has a noise $(F_2 - 1)$ kTB_n of its own at its input. Hence, the total input noise to stage 2 is

$$P_{ni2} = F_1G_1kT_oB_n + (F_2 - 1)kT_oB_n \tag{6.44}$$

The noise of stage 2 is represented by its equivalent input source. This stage can be regarded as noiseless and with a gain of G_2, the available noise output of stage is

$$P_{no2} = G_2 P_{ni2} = G_2[F_1 G_1 kT_o B_n + (F_2 - 1)kT_o B_n] \qquad (6.45)$$

The overall power gain of the two stages is $G = G_1 G_2$. If we consider F as the overall noise factor due to both the stages, available at the output of stage 2, then

$$P_{no} = FGkTB_n = FG_1 G_2 kTB_n \qquad (6.46)$$

Comparing Equations (6.45) and (6.46), we observe that

$$F = F_1 + \frac{F_2 - 1}{G_1} \qquad (6.47)$$

Hence, this equation can be generalized for more number of stages, say N, and F is given as

$$F = F_1 + \frac{F_2 - 1}{G_1} + \frac{F_3 - 1}{G_1 G_2} + \cdots + \frac{F_N - 1}{G_1 G_2 \cdots G_{N-1}} \qquad (6.48)$$

This is known as Friss' formula. Equation (6.48) shows the importance of having a high gain low-noise amplifier at the first stage of cascaded system. By having G_1 large, the noise contribution of the second stage can be negligible. Similarly, the noise contribution of the third stage is still small due to the factor $G_1 G_2$ in the denominator and so on. However, care has to be taken to see that the first stage noise is as minimum as possible.

Example 6.5 The noise figure of an amplifier is 8 dB. The input signal-to-noise ratio is 45 dB. Calculate the output signal-to-noise ratio.

Solution

$$\left(\frac{S}{N}\right)_{out} = \left(\frac{S}{N}\right)_{in} - (F) \text{ dB} = 45 - 8 = 37 \text{ dB}$$

6.6 NOISE TEMPERATURE

Since thermal noise plays a major part in determining the noise in a system and it is directly proportional to the temperature, it can be expressed in degrees as well as watts or dBm. The available noise power P_n is

$$P_n = kT_a B_n \qquad (6.49)$$

where T_a is the noise temperature associated only with the available noise power. T_a will not be same as the physical temperature of the noise source. From Eq. (6.49), we can express T_a as

$$T_a = \frac{P_n}{kB_n} \qquad (6.50)$$

If P_{na} is the amplifier noise referred to the input, the equivalent noise temperature of the amplifier referred to the input is

$$T_e = \frac{P_{na}}{kB_n} \qquad (6.51)$$

It was shown earlier that equivalent input power for an amplifier given in terms of noise factor is

$$P_{na} = (F-1)kTB_n$$

Substituting this value of P_{na} in Eq. (6.51), we get

$$T_e = (F-1)T \qquad (6.52)$$

where T_e = equivalent noise temperature in kelvin

T = Ambient temperature 290° K for reference

F = noise factor

This is the relation between T_e and F. Knowing one automatically helps to find the other. In fact, it will be found that noise temperature is a better measure for low-noise devices such as low-noise amplifier used in satellite receiving system, while noise factor is a better measure for the main receiving system.

Friss' formula can be rearranged to give the expression for overall noise temperature T_e for a cascaded system as

$$T_e = T_{e1} + \frac{T_{e2}}{G_1} + \frac{T_{e3}}{G_1 G_2} + \cdots + \frac{T_{en}}{G_1 G_2 \cdots G_{n-1}} \qquad (6.53)$$

where T_{e1}, T_{e2}, $\cdots$, T_{en} are the noise temperatures of the individual stages.

Also the noise factor F can be represented as a function of equivalent noise temperature as shown below:

$$F = 1 + \frac{T_e}{T} \qquad (6.54)$$

6.7 MEASUREMENT OF NOISE FACTOR AND NOISE TEMPERATURE

The methods to measure these parameters depend largely on the range of values expected. For a normal receiving system, an avalanche diode noise source is commonly employed.

A diode generates a large amount of noise when operated in avalanche mode and hence can be considered as a source of noise power at some equivalent temperature T_h, which is called *hot temperature*. The diode reverts to normal noise output once reverse bias is removed at some equivalent cold temperature T_c. The excess noise ratio (*ENR*) is defined as

$$ENR = 10 \log \frac{T_h - T_c}{T_c} \qquad (6.55)$$

Generally, cold temperature is taken as room temperature at 290°K. Normally, the manufacturer gives the *ENR* for a range of frequencies. Hot temperature T_h can be found once we know the values of *ENR* and T_c.

The diode is matched to the input of the amplifier under test. Let T_e be the unknown equivalent noise temperature of the amplifier. The amplifier output noise is measured for two conditions, one with the diode in the avalanche mode denoted by P_h and one with the reverse bias removed denoted by P_c. The two equations for the noise output are

$$P_h = G_k(T_h + T_e)B_n \tag{6.56}$$

$$P_c = G_k(T_c + T_e)B_n \tag{6.57}$$

where G_k is the power gain of the amplifier under test. The power ratio, called Y factor, is

$$Y = \frac{P_h}{P_c} \tag{6.58}$$

Solving the three equations for T_e, we get

$$T_e = \frac{T_h - YT_c}{Y - 1} \tag{6.59}$$

It can be observed from the above equation that the gain and noise bandwidth do not enter into the final equation. Y is a measured quantity. It does not require an absolute measure of power. It can be shown that noise factor in terms of *ENR* is given by the relation

$$F = \frac{ENR}{Y - 1} \tag{6.60}$$

where *ENR* and Y are expressed as power ratios.

Modern noise measurements are made over a range of frequencies and noise factor and noise temperature are automatically calculated and displayed as a function of frequency with the help of microprocessors/microcontrollers.

6.8 NOISE IN A BANDPASS SYSTEM

In communication system, we come across bandpass filters. These filters have equivalent noise bandwidth B_n and centre frequency f_c. If the centre frequency is much greater than the bandwidth, we call it a narrowband system. A narrowband system is shown in Fig. 6.15. Signal source has an internal resistance R_s. As usual, we take the thermal noise as the major contributor.

Fig. 6.15 Bandpass system

System noise is assumed to be at noise temperature T_S. The power spectral density for this noise is given by

$$S_p(f) = kT_S \tag{6.61}$$

For the ideal bandpass system, the spectral density is not changed when the signal is passed through the filter. But the filter bandwidth determines the available noise power as $kT_S B_n$.

In the case of modulation system, noise also acts as modulating signal and results in a modulated wave at the output. The output waveform can be represented as

$$n(t) = A_n(t) \cos[\omega_c t + \phi_n(t)] \tag{6.62}$$

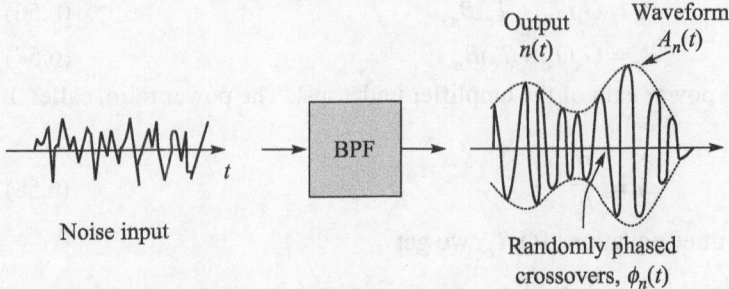

Fig. 6.16 Bandpass system with noise

Figure 6.16 shows the noise input, which is a randomly varying voltage, the modulated envelope $n(t)$, and the random phase angle $\phi_n(t)$.

Equation (6.62) can be represented using trigonometric identities as

$$n(t) = n_I(t) \cos \omega_c t - n_Q(t) \sin \omega_c t \tag{6.63}$$

where $n_I(t)$ is a random noise voltage termed as *in-phase component* since it multiplies a cosine term used as a reference phasor and $n_Q(t)$ is a random noise voltage termed the quadrature component since it multiplies the sine term, which is 90° out of phase or in quadrature with the reference phasor. The noise voltages $n_i(t)$ and $n_q(t)$ appear to modulate a carrier at frequency f_c and are known as the *low-pass equivalent noise voltages*. Carrier f_c may be anywhere within the pass band. There exists a number of relations among $n(t)$, $n_I(t)$, and $n_Q(t)$. All of them have similar characteristic and $n_I(t)$ and $n_Q(t)$ are uncorrelated. When the power spectral density of $n(t)$ is $S_p(t)$, the power spectral densities for $n_I(t)$ and $n_Q(t)$ are

$$S_I(t) = S_Q(t) = 2kT_S \tag{6.64}$$

These are shown in Fig. 6.17.

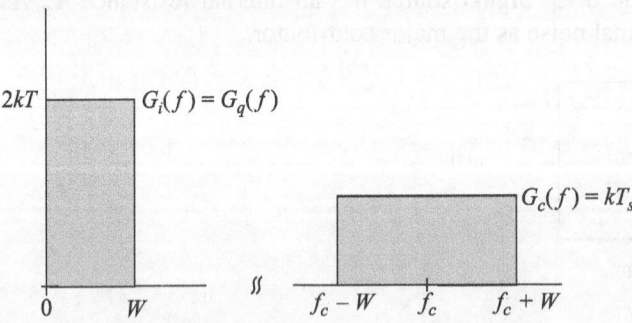

Fig. 6.17 Noise spectral densities

6.9 NOISE IN AM SYSTEMS

The performance of an AM system is generally judged from its signal-to-noise ratio. All the noise generated within the receiver can be referred to the receiver input. This will make it easier to compare receiver noise, antenna noise, and received signal. The receiver and antenna power can be added and the receiver system can be modelled as shown in Fig. 6.18.

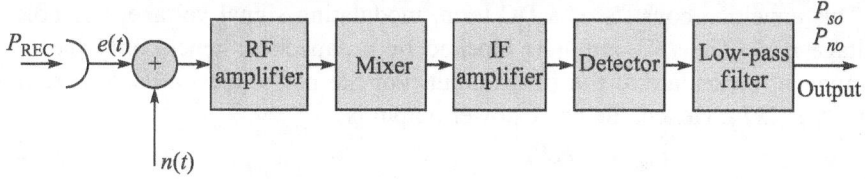

Fig. 6.18 AM receiver with noise

To calculate the signal-to-noise ratio, we also need the bandwidth of the system. From the figure, it can be seen that there are two bandwidths to be considered—the RF bandwidth and the IF bandwidth. The IF bandwidth is much smaller than the RF bandwidth and hence the IF bandwidth will determine the noise reaching the detector. After the detector, baseband comes into picture. Let the baseband be B_B and IF bandwidth be B_{IF}. For AM systems, $B_{IF} = 2B_B$. We can consider them as equivalent noise bandwidth. Hence, noise output from the band-pass system is

$$n_{IF}(t) = n_I(t) \cos 2\pi f_{IF} t - n_Q(t) \sin \omega_{IF}(t) \qquad (6.65)$$

When this noise waveform is passed through the detector, the resulting noise output is very much dependent on whether or not the carrier is present and on the strength of the carrier. Let the modulated carrier be

$$f(t) = E_c(1 + m \cos \pi f_m t) \cos 2\pi f_{IF} t$$

$$= A_c(t) \cos 2\pi f_{IF} t \qquad (6.66)$$

where $\qquad A_c(t) = E_c(1 + m \cos 2\pi f_m t)$

The input to the detector is hence

$$l_{det}(t) = f(t) + n_{IF}(t)$$

$$= [A_c(t) + n_I(t)] \cos 2\pi f_{IF} t - n_Q(t) \sin 2\pi f_{IF} t \qquad (6.67)$$

AM envelope detector recovers the envelope of this waveform and is expressed as $R(t) \cos[\omega_{IF} t + f(t)]$.

The amplitude term, $R(t) = \sqrt{[A_c(t) + n_I(t)]^2 + [n_Q(t)]^2} \qquad (6.68)$

Since AM carrier is much greater than the noise voltage for most of the time, $R(t)$ can be simplified as

$$R(t) = \sqrt{A_c(t)^2 + 2A_c(t)n_I(t) + n_I(t)^2 + [n_Q(t)]^2}$$

$$= \sqrt{A_c(t)^2 + 2A_c(t)n_I(t)} \qquad (6.69)$$

The square root term can be expanded further as

$$R(t) = A_c(t) \left\{ 1 + 2\frac{n_I(t)}{A_c(t)} \right\}^{1/2}$$

$$= A_c(t) + n_I(t) \qquad (6.70)$$

For sinusoidal modulation, this becomes

$$R(t) = E_c + mE_c \cos 2\pi f_m t + n_I(t) \qquad (6.71)$$

This envelope consists of a DC term, modulating signal voltage, and noise voltage $n_i(t)$. The *DC* output is blocked by a capacitor, hence, only the *AC* component contribute to the final output. For the noise, the available spectral density is $2kT_S$. Hence, the noise power output is

$$P_{no} = 2kT_S B_B \qquad (6.72)$$

The peak signal voltage at the output is mE_c. Let E_{rms} be the rms voltage. Then mE_{rms} is the rms value of the output voltage. The available signal power output is, therefore,

$$P_{so} = \frac{m^2 E^2{}_{rms}}{4R_{out}} \qquad (6.73)$$

The output signal-to-noise ratio is

$$\left(\frac{S}{N}\right)_{out} = \frac{P_{so}}{P_{no}}$$

$$= \frac{m^2 E_c^2}{4R_o \times 2kT_S B_B}$$

$$= \frac{m^2 E_c^2}{8R_{out} kT_S B_B} \qquad (6.74)$$

Generally, the output S/N ratio is compared with a reference ratio, which is the S/N ratio at the detector input but with the noise calculated for the base bandwidth B_B. The noise power spectral density at the detector input is kT_S and hence the reference noise power is

$$P_{nref} = kT_S B_B \qquad (6.75)$$

The available signal power from a source with internal resistance R_S is

$$P_R = \frac{E_{rms}^2}{4R_s}\left(1 + \frac{m^2}{2}\right) \qquad (6.76)$$

Hence, the reference S/N ratio is

$$\left(\frac{S}{N}\right)_{ref} = \frac{P_R}{P_{nref}}$$

$$= \frac{E_{rms}^2\left(1 + \frac{m^2}{2}\right)}{4R_s kT_s B_B} \qquad (6.77)$$

The figure of merit (F_m) is the ratio of these two signals-to-noise ratios.

$$F_m = \frac{\left(\frac{S}{N}\right)_o}{\left(\frac{S}{N}\right)_{ref}} = \frac{m^2}{2 + m^2}\frac{R_s}{R_{out}} \qquad (6.78)$$

The higher the figure of merit, the better is the system when $R_{out} = R_S$. For 100% modulation, $m = 1$ and then $F_m = 1/3$.

This is the highest value of the figure of merit we can achieve.

For sinusoidal DSBC, the received signal is of the form

$$f(t) = E_c \cos \omega_m t \cos 2\pi f_{IF} t \qquad (6.79)$$

where E_c is the peak value of the received signal. The input of the detector is, therefore,

$$f_d(t) = f(t) + n(t)$$
$$= E_c \cos \omega_m t + n_I(t) \cos 2\pi f_{IF} t - n_Q(t) \sin \omega_{IF} t$$
$$= A(t) \cos \omega_{IF} - n_Q(t) \sin \omega_{IF} t \qquad (6.80)$$

where $\qquad A(t) = [E_c \cos \omega_m t + n_I(t)]$

Coherent detection is used for DSBSC. Demodulation is done through a balanced mixer. For coherent detection, a locally generated carrier is required that is exactly locked on to the incoming carrier $\cos \omega_{IF} t$ and the two signals are fed into the balanced mixer. The output of the balanced demodulator is

$$f_{out}(t) = k f_{det}(t) \cos \omega_{IF} t \qquad (6.81)$$

$$f_{out}(t) = \frac{k}{2} [E_c \cos \omega_m t + n_I(t) + \text{high-frequency terms}] \qquad (6.82)$$

LPF following the balanced demodulator removes the high-frequency terms and the baseband output is

$$f_{BB}(t) = \frac{k}{2} [E_c \cos \omega_m t + n_I(t)] \qquad (6.83)$$

This is the same term as shown earlier for AM except that the DC term is missing. The $k/2$ factor is common to signal and noise and can be ignored. Also as in the AM case, the condition that the carrier must be greater than noise is not required. Hence, the S/N ratio is given by

$$\left(\frac{S}{N}\right)_o = \frac{\left(\frac{E_c}{\sqrt{2}}\right)^2}{8 R_{out} k T_s B_B}$$

$$= \frac{E_c^2}{16 R_{out} k T_s B_B} \qquad (6.84)$$

The reference noise is $P_{nref} = k T_S B_B$. The rms voltage of the received DSBSC signal

$$E_c \cos \omega_m t \cos \omega_{IF} t = \frac{E_c}{2}$$

Hence, the available signal power at the input is

$$P_R = \frac{\left(\frac{E_c}{2}\right)^2}{4 R_s} = \frac{E_c^2}{16 R_s} \qquad (6.85)$$

The figure of merit is, therefore,

$$F_m = \frac{\left(\dfrac{S}{N}\right)_{\text{out}}}{\left(\dfrac{S}{N}\right)_{\text{ref}}} = \frac{R_s}{R_{\text{out}}} \tag{6.86}$$

For $R_s \approx R_{\text{out}}$, the figure of merit is unity, which is three times better than that of AM.

6.9.1 Signal-to-Noise Ratio for SSB

Let $f(t) = E_c \cos(\omega_{\text{IF}} + \omega_m)t$ be the received SSB signal. In this SSB system upper sideband is used. The noise reaching the detector is narrow pass bandwidth limited noise. The centre frequency $\omega_c = 2\pi f_c = 2\pi\left(\text{IF} + \dfrac{B_B}{2}\right)$. The noise input to the detector is, therefore, πB_B.

$$n(t) = n_I(t) \cos(\omega_{\text{IF}} + \pi B_B)t - n_Q(t) \sin(\omega_{\text{IF}} + \pi B_B)t \tag{6.87}$$

Signal plus noise input to the detector is, therefore,

$$f_{\text{det}}(t) = f(t) + n(t) \tag{6.88}$$

Coherent detection takes place and the baseband signal output is

$$e_{SBB}(t) = \frac{kE_c}{2} \cos\omega_m t$$

$$= AE_c \cos\omega_m t \tag{6.89}$$

Also the baseband noise is given by

$$e_{nBB}(t) = A[n_I(t) \cos \pi B_B t - n_Q(t) \sin \pi\omega t] \tag{6.90}$$

Let A_m is the demodulator multiplier coefficient and A_A is the receiver gain from the antenna. The available noise power detector output is

$$P_{no} = (A_A A_m)kT_S B_B \tag{6.91}$$

The available signal power at the detector output is

$$P_{so} = (A_A A_m) \frac{\left(\dfrac{E_c}{\sqrt{2}}\right)^2}{4R_{\text{out}}} \tag{6.92}$$

Hence, the output signal-to-noise ratio at the detector output is

$$\left(\frac{S}{N}\right)_o = \frac{P_{so}}{P_{no}} = \frac{E_c^2}{8R_{\text{out}}kT_S B_B} \tag{6.93}$$

Let E_r be the maximum value of the received sinusoidal signal voltage at the detector input. Then the available received power is

$$P_R = (A_A) \frac{\left(\dfrac{E_r}{\sqrt{2}}\right)^2}{4R_s} = (A_A) \frac{E_r^2}{8R_s} \tag{6.94}$$

The noise spectral density is kT_S over an IF bandwidth $B_{\text{IF}} = B_B$, which is the same as the base bandwidth.

Hence, the available noise power at the input to the detector is

$$P_{n\text{ref}} = (A_A)\, kT_S B_B \tag{6.95}$$

Hence, the reference S/N ratio

$$\left(\frac{S}{N}\right)_{\text{ref}} = \frac{P_R}{P_{n\text{ref}}} = \frac{E_r^2}{8R_s kT_s B_B} \tag{6.96}$$

The figure of merit,

$$F_m = \frac{\left(\dfrac{S}{N}\right)_{\text{out}}}{\left(\dfrac{S}{N}\right)_{\text{ref}}} = \frac{R_s}{R_{\text{out}}} \tag{6.97}$$

is the same as that of DSBSC but the bandwidth required for SSB is half that of DSBSC.

6.9.2 Single Sideband Companding

Another method to reduce the effect of channel noise is a technique called companding. In this method, the speech signal is compressed in volume range relative to a fixed level typically at –10 dBm. Typical compression ratio is 1 to 2 applied to the decibel difference between the reference level and the average signal level. At the output, the signal is expanded by the same ratio referred to the same fixed level.

The main advantage of companding is that it reduces the *idle noise* on the channel, which allows for an increase in the total number of channels on a multiplexed carrier system. (Idle noise is the background noise.)

Noise reduction is the result of the expander action at the output. Let P_{ns} be the ide noise power from the source. This is increased by a factor x in the compressor so that the noise power presented to the channel input is xP_{ns}. The total noise power at the output before the expansion is $(xP_{ns} + P_{nch})$, where P_{nch} is the noise added by the channel. Expansion reduces the total noise by the factor x. Hence, the expanded output noise is

$$P_{n\text{out}} = \frac{xP_{ns} + P_{nch}}{x} = P_{ns} + \frac{P_{nch}}{x} \tag{6.98}$$

Channel noise comprises channel thermal noise, intermodulation distortion, and adjacent and cochannel interference. All these components except thermal noise increase with the number of channels in a multiplexed carrier system. Hence, for a given level of performance, the noise contribution from the additional channel can be offset by companding action, i.e., the use of companding allows the number of channels in a multiplex system to be increased.

6.10 EFFECT OF NOISE ON ANGLE MODULATION

The performance of angle modulated signal in the presence of noise will be discussed in this section. This performance will be compared with the performance of amplitude modulated signal. In amplitude modulation, the message information is contained in the amplitude of the modulated signal. Since noise is additive, it is directly added to the signal, whereas in the case of frequency and phase modulated signal, the noise is added to the amplitude but the message information is contained in the frequency of the modulated signal. Hence, the message is contaminated by the noise to the extent that the added noise changes the frequency of the modulated signal. The frequency of the signal can be described by its zero crossing. Therefore, the effect of additive noise on the demodulated FM/PM signal can be described by the changes that it produces in the zero crossing of the modulated signal. The effect of noise on the zero crossing of two modulated signals, one with low power and the other with high power, is shown in Fig. 6.19.

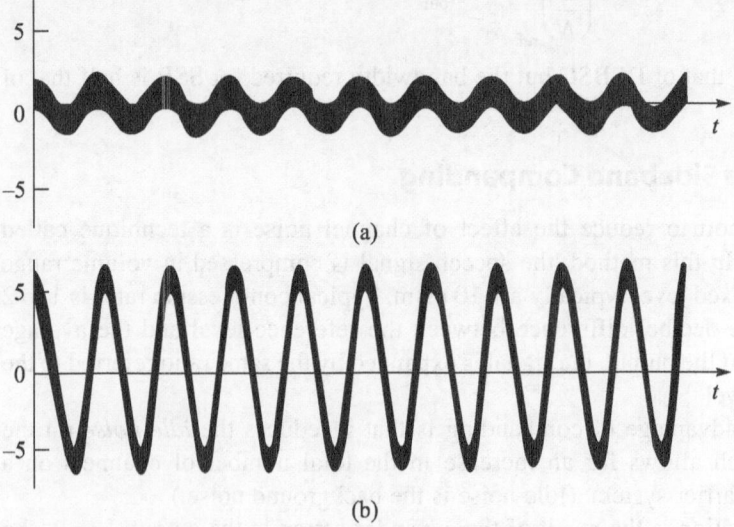

Fig. 6.19 Frequency modulated signal in the presence of noise

From the figure, it can be seen that in low-power signal, noise causes more changes in zero crossing than in the case of high-power signal. Figure 6.20 shows the block diagram of the receiver for a general angle modulated signal.

The angle modulated signal as described earlier can be represented as

$$f(t) = E_c \cos[2\pi f_c + \Phi(t)] \tag{6.99}$$

Then the FM signal will be

$$f(t) = E_c \cos\left[2\pi f_c\, t + 2\pi k_f \int_{-\infty}^{t} m(\tau)d\tau\right] \tag{6.100}$$

and the PM signal will be

$$f(t) = E_c \cos\left[2\pi f_c t + k_p\, m(t)\right] \tag{6.101}$$

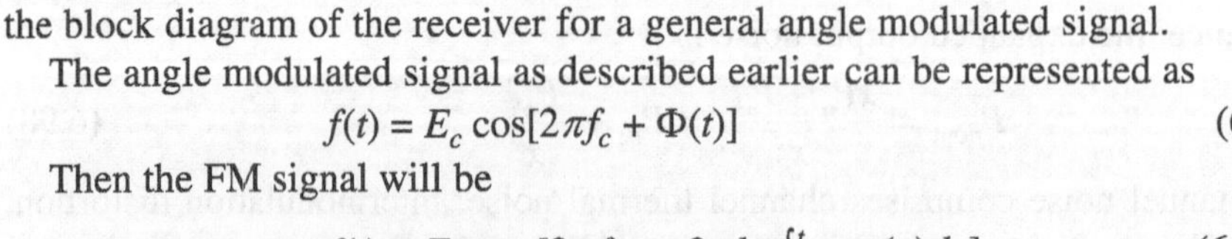

Fig. 6.20 Block diagram of an angle demodulator

The noise $n_\omega(t)$ is added to $f(t)$ and the combined signal is passed through a noise limiting filter, which removes out of band noise. The bandwidth of this filter is equal to the bandwidth of the modulated signal. This filter hence passes the modulated signal without distortion, but eliminates out of band noise. The output of this filter is

$$f_o(t) = f(t) + n_I(t)\cos\,\omega_c(t) - n_Q(t)\sin\,\omega_c(t) \tag{6.102}$$

Since a precise analysis is quite involved due to non-linearity of demodulation process, the signal power is assumed to be quite large compared to the noise power. With this assumption, the bandpass noise is given by

$$n(t) = \sqrt{n_I^2(t) + n_Q^2(t)}\;\cos\left(2\pi f_c t + \arctan\frac{n_I(t)}{n_Q(t)}\right)$$

$$= E_n(t)\cos[2\pi f_c t + \Phi_n(t)] \tag{6.103}$$

where $E_n(t)$ and $\Phi_n(t)$ represent the envelope and the phase of bandpass noise, respectively. Because of the assumption that the signal is much larger than the noise,

$$P[E_n(t)] << E_c = 1 \tag{6.104}$$

The phasor diagram of the signal and the noise is shown in Fig. 6.21. From this figure, we get

$$r(t) = [E_c + E_n\cos\{\Phi_n(t) - \Phi(t)\}] \times \cos\{2\pi f_c t + \Phi(t)\}$$

$$+ \arctan\frac{E_n(t)\sin\{\Phi_n(t) - \Phi(t)\}}{E_c + E_n(t)\cos\{\Phi n(t) - \Phi(t)\}}$$

$$= [E_c + E_n(t)\cos\{\Phi_n(t) - \Phi(t)\}] \times \cos[2\pi f_c(t) + \Phi(t)]$$

$$+ \frac{E_n(t)}{E_c}\;\sin\{\Phi_n(t) - \Phi(t)\}] \tag{6.105}$$

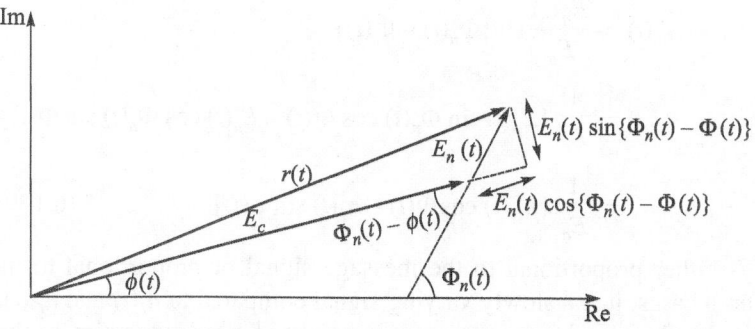

Fig. 6.21   Angle modulated signal: phasor diagram (The signal is much stronger than the noise.)

The demodulator processes this signal and depending on whether it is a phase or frequency demodulation, its output will be the phase or the instantaneous frequency. Now

$$\Phi(t) = k_p\,m(t)\quad\text{for PM}$$

$$= 2\pi k_f \int_{-\infty}^{t}\,m(\tau)d\tau\quad\text{for FM} \tag{6.106}$$

The output of the demodulator is

$$y(t) = \Phi(t) + \frac{E_n(t)}{E_c} \sin\{\Phi_n(t) - \Phi(t)\} \quad \text{for PM}$$

$$= \frac{1}{2\pi} \frac{d}{dt} \left[\Phi(t) + \frac{E_n(t)}{E_c} \sin\{\Phi_n(t) - \Phi(t)\} \right] \quad \text{for FM}$$

$$= k_p \, m(t) + \frac{E_n(t)}{E_c} \sin\{\Phi_n(t) - \Phi(t)\} \quad \text{for PM}$$

$$= k_f m(t) + \frac{1}{2\pi} \frac{d}{dt} \frac{E_n(t)}{E_c} \sin\{\Phi_n(t) - \Phi(t)\} \quad \text{for FM}$$

$$= k_p \, m(t) + Y_n(t) \quad \text{for PM}$$

$$= k_f \, m(t) + \frac{1}{2\pi} \frac{d}{dt} Y_n(t) \quad \text{for FM} \tag{6.107}$$

where $\qquad Y_n(t) = \dfrac{E_n(t)}{E_c} \sin\{\Phi_n(t) - \Phi(t)\}$ $\qquad\qquad$ (6.108)

The first term in Eq. (6.107) is the desired signal component and the second term is the noise component. We can infer from this expression that the noise component is inversely proportional to the signal amplitude E_c. Hence, the higher the signal level, the lower the noise level. In the case of AM, the noise component is independent of the signal component and the scaling of the signal power does not affect the received noise power. Let us consider

$$Y_n(t) = \frac{E_n(t)}{E_c} \sin\{\Phi_n(t) - \Phi(t)\}$$

$$= \frac{1}{E_c} \{E_n(t) \sin \Phi_n(t) \cos \Phi(t) - E_n(t) \cos \Phi_n(t) \sin \Phi(t)\}$$

$$= \frac{1}{E_c} \{n_Q(t) \cos \Phi(t) - n_I(t) \sin\Phi(t)\} \tag{6.109}$$

where $\Phi(t)$ is either proportional to the message signal or proportional to its integral. In both cases, it is a slowly varying signal compared to $n_I(t)$ and $n_Q(t)$, which are the in-phase and quadrature components of bandpass noise at the receiver and have a much higher bandwidth than $\Phi(t)$. In fact, the bandwidth of the filtered noise at the demodulator input is half of the bandwidth of the modulated signal, which is many times the bandwidth of the message signal. Hence, $\Phi(t)$ can be regarded as almost constant when compared to the variations in $n_I(t)$ and $n_Q(t)$. Therefore,

$$Y_n(t) = \frac{1}{E_c} \{n_Q(t) \cos \Phi - n_I(t) \sin\Phi\} \tag{6.110}$$

Let $\qquad\qquad\qquad a = \dfrac{\cos \Phi}{E_c} \quad \text{and} \quad b = \dfrac{-\sin \Phi}{E_c}$

Then $\quad SY_n(f) = (a^2 + b^2)\, S_{nc}(f) = \dfrac{S_{nc}(t)}{E_c^2}$ (6.111)

where $S_{nc}(f)$ is the power spectral density of the in-phase component of the filtered noise. The bandwidth of the filtered noise process extends from $f_c - B_c/2$ to $f_c + B_c/2$. Hence, the spectrum of $n_I(t)$ extends from $B_c/2$ to $B_c/2$. Therefore,

$$Sx_{1i}(f) = Sx_{1q}(f) = kT, \quad |f| < B_c/2$$
$$= 0, \quad \text{otherwise} \tag{6.112}$$

Substituting Eq. (6.112) into Eq. (6.110), we get
$$S_{Yn}(f) = [kT/E_c^2, \quad |f| \le B_c/2$$
$$= 0, \quad \text{otherwise} \tag{6.113}$$

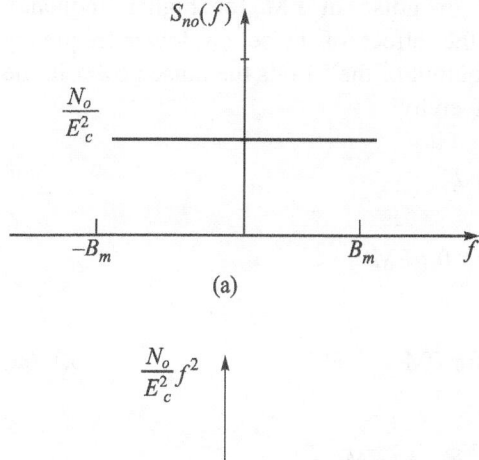

(a)

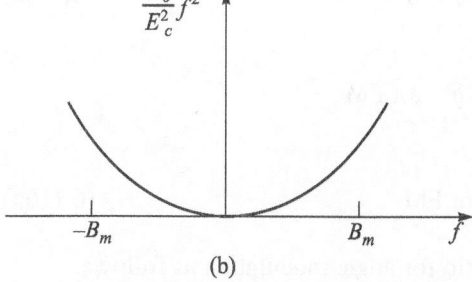

(b)

Fig. 6.22 Noise power spectrum at demodulated output: (a) PM and (b) FM for $|f| < B_m$

This equation provides an expression for the power spectral density of the filtered noise at the front end of the receiver. After demodulation, due to another filtering, the noise bandwidth is reduced to B_m, which is the bandwidth of the message signal. In the case of FM modulation, the process $Y_n(t)$ is differentiated and scaled by $1/2\pi$. The power spectral density in this case is

$$\dfrac{4\pi^2 f^2}{4\pi^2}\, S_{Yn}(f) = f^2 S_{Yn}(f) = \dfrac{kT}{E_c^2}\, f^2, \quad |f| \le \dfrac{B_c}{2}$$
$$= 0, \quad \text{otherwise} \tag{6.114}$$

This shows that in PM, the demodulated noise power spectral density is independent of f, whereas in the case of FM, it is directly proportional to f^2. In both cases, $B_c/2$ must be replaced by B_m to account for post-demodulation filtering. In other words, for $|f| < B_m$, we have

$$S_{no}(f) = \frac{kT}{E_c^2} \quad \text{for PM}$$

$$= \frac{kT}{E_c^2} f^2 \quad \text{for FM} \tag{6.115}$$

The plot of power spectrum of the noise component at the output of the demodulator in the frequency interval $|f| < B_m$ is shown in Fig. 6.22.

It can be observed that PM has a flat noise spectrum and FM has a parabolic noise spectrum. Hence, the effect of noise in FM for higher-frequency components is much higher than the effect of noise on lower-frequency components. The noise power at the output of the LPF is the noise power in the frequency range $-B_m$ to $+B_m$ and is given by

$$P_{no} = \int_{-B_m}^{B_m} S_{no}(f)\, df$$

$$= \int_{-B_m}^{B_m} \frac{kT}{E_c^2}\, df \quad \text{for PM}$$

$$= \frac{2B_m kT}{E_c^2} \quad \text{for PM} \tag{6.116a}$$

and

$$P_{no} = \int_{-B_m}^{B_m} \frac{kT}{E_c^2} f^2 df \quad \text{for FM}$$

$$= \frac{2kT}{E_c^2} \frac{B_M^3}{3} \quad \text{for FM} \tag{6.116b}$$

Now we can find the output S/N ratio for angle modulation as follows. Output signal power is given by

$$P_{so} = k_p^2 P_M \quad \text{for PM}$$

$$= k_f^2 P_M \quad \text{for FM} \tag{6.117}$$

Hence,

$$\left(\frac{S}{N}\right)_{\text{out}} = \frac{P_{so}}{P_{no}}$$

Substituting the values for P_{so} and P_{no} from the above equations, we get

$$\left(\frac{S}{N}\right)_{\text{out}} = \frac{k_p^2 E_c^2}{2} \frac{P_M}{kTB_m} \quad \text{for PM}$$

$$= \frac{3k_f^2 E_c^2}{2B_m^2} \frac{P_M}{kTB_m} \tag{6.118}$$

We know that $\dfrac{E_c^2}{2}$ is the received signal power P_R and

$$\beta_p = k_p \max |m(t)| \quad \text{for PM} \tag{6.119a}$$

$$\beta_f = \frac{k_f \max |m(t)|}{B_m} \quad \text{for FM} \tag{6.119b}$$

Then the output S/N can be expressed as

$$\left(\frac{S}{N}\right)_{\text{out}} = P_R \left(\frac{B_p}{\max |m(t)|}\right)^2 \frac{P_M}{kTB_m} \quad \text{for PM}$$

$$= 3P_R \left(\frac{B_f}{\max |m(t)|}\right)^2 \frac{P_M}{kTB_m} \quad \text{for PM}$$

Since $\dfrac{P_M}{kTB_m} = \left(\dfrac{S}{N}\right)_b$ is the signal-to-noise ratio of a baseband system and kTB_m is the noise power, then we have

$$\left(\frac{S}{N}\right)_{\text{out}} = \frac{P_m B_p^2}{\{\max |m(t)|\}^2}\left(\frac{S}{N}\right)_b \quad \text{for PM}$$

$$= 3\frac{P_M B_f^2}{\{\max |m(t)|\}^2}\left(\frac{S}{N}\right)_b \quad \text{for FM} \tag{6.120}$$

Now $\dfrac{P_M}{\max |m(t)|^2}$ is the average to peak power ratio of the message signal, i.e., the power content of the normalized message P_{Mn}. Therefore,

$$\left(\frac{S}{N}\right)_{\text{out}} = \beta_p^2 P_{Mn}\left(\frac{S}{N}\right)_b \quad \text{for PM}$$

$$= 3\,\beta_f^2 P_{Mn}\left(\frac{S}{N}\right)_b \quad \text{for FM} \tag{6.121}$$

Using Carson's relation $B_{\max} = 2(\beta + 1) B_m$, we can express the output S/N ratio in terms of the bandwidth expansion factor, which is defined as the ratio of the channel bandwidth to the message bandwidth and is denoted by

$$\frac{B_{\max}}{B_m} = 2(\beta + 1) \tag{6.122}$$

Let $\qquad 2(\beta + 1) = D$

Then $\qquad\qquad \beta = \dfrac{D}{2} - 1 \tag{6.123}$

Therefore,

$$\left(\frac{S}{N}\right)_{\text{out}} = P_M \left(\frac{\frac{D}{2}-1}{\max|m(t)|}\right)^2 \left(\frac{S}{N}\right)_b \quad \text{for PM}$$

$$= 3p_M \left(\frac{\frac{D}{2}-1}{\max|m(t)|}\right)^2 \left(\frac{S}{N}\right)_b \quad \text{for FM} \tag{6.124}$$

From Equations (6.120) and (6.124), the following inferences can be concluded.

The output S/N ratio for both PM and FM is proportional to the square of the modulation index β. Hence, increasing β increases the output S/N ratio even with low received power. This is not possible in the case of amplitude modulation.

But this increase in the S/N ratio is at the expense of increased bandwidth. Therefore, angle modulation provides a way to trade-off bandwidth for transmitted power.

The relation between D and the output S/N ratio is quadratic. But for optimum relation between D and the output S/N ratio, this relation must be exponential.

There is a limit to β. Having large β means having large $B_{\max}$. This results in having large noise power at the input of the demodulator. Our approximation $P\{E_n(t)\} \ll E_c\} \approx 1$ in Eq. (6.104) is no longer valid and the preceding analysis will not hold. But as we start increasing β to improve the output S/N ratio, a stage is reached where preceding approximation does not hold, a phenomenon known as the *threshold effect* will occur and the signal will be lost in noise. This means that increasing β helps to increase the output S/N ratio only to a certain extent and beyond this any increase in β will be harmful and deteriorates the performance of the system.

One method to circumvent the above-mentioned problem is to increase the transmitted power, which in turn increases the received power, thus improving the output S/N ratio. In the case of AM, any increase in the received power directly increases the signal power at the output of demodulator. This is due to the fact that the message signal is in the amplitude of the transmitted signal and increase in the transmitted power directly affects the demodulated signal power. However, in angle modulation, the message is in the phase of the modulating signal and hence increasing the transmitter power does not increase the demodulated message power. In angle modulation, the output S/N ratio is increased by a decrease in the noise power.

From the plot of power spectrum of noise component at the output of the demodulator in the frequency interval $|f| < B_m$ as shown in Fig. 6.22, it can be seen that noise affects FM at higher frequencies. This results in signal components at higher frequencies suffering more from noise compared to signal components at lower frequencies. To compensate, this signal components of higher frequencies are amplified so that they have much higher level compared to their lower frequency counterpart. The quadratic characteristic of the demodulated noise spectrum in FM is the basis of pre-emphasis and de-emphasis circuits.

6.11 PRE-EMPHASIS AND DE-EMPHASIS CIRCUITS

It was observed that the noise power spectral density at the output of the demodulator in PM is flat within the message bandwidth. But for FM, the noise power spectrum has a parabolic shape. Hence, we can conclude that FM performs better in low-frequency components of the message signal and PM performs better in high-frequency components. Thus, we need a system that performs frequency modulation for low-frequency components of the message signal and works as a phase modulator for high-frequency components of the message signal. This system results in a better overall performance compared to each system alone. This is the idea behind pre-emphasis and de-emphasis circuits.

Hence, the objective is to design a pre-emphasis and de-emphasis filter circuit that behaves like an ordinary frequency modulator-demodulator pair in the low-frequency band of the message signal and like a phase modulator-demodulator pair in the high-frequency band of the message signal. We can get a phase modulator by cascading a differentiator with a frequency modulator circuit. The differentiator should not affect low-frequency components but differentiates only high-frequency components. A simple high-pass filter is a very good approximation. Such a filter has a constant gain at low frequencies and at high frequencies it has a frequency characteristic approximated by $k|f|$, which is the frequency characteristic of differentiator.

At the receiver end, the demodulator needs a filter that has a constant gain at low frequencies and behaves as an integrator at high frequencies. A good approximation to such a filter is a simple low-pass filter. The modulator filter, which emphasizes the high frequencies is called the pre-emphasis filter, and the demodulator filter, which is the inverse of the modulator filter, is called the de-emphasis filter.

We can also look at pre-emphasis and de-emphasis in another way. The high-frequency components of the message signal in FM is prone to much more noise than the low-frequency components. Hence, we attenuate the high-frequency components of the demodulated signal. No doubt this reduces the noise level but it also attenuates the high-frequency component of the message signal. To compensate this, we amplify these high-frequency components at the transmitter before modulation. Hence, we need a high-pass filter at the transmitter and a low-pass filter at the receiver. The net effect of these filters should be a flat frequency response. Therefore, the characteristic of the receiver filter should be the inverse of the transmitter filter.

Proper care has to be taken on the time constants of these filters. *RC* filters are generally used, which can be passive or active. Figure 6.23 shows typical pre-emphasis and de-emphasis networks along with their relative responses. The characteristic of pre-emphasis and de-emphasis filters depend largely on the power spectral density of the message. For example, in commercial FM broadcasting of music and voice, first-order low-pass and high-pass *RC* filters with a time constant of 75 μs are employed.

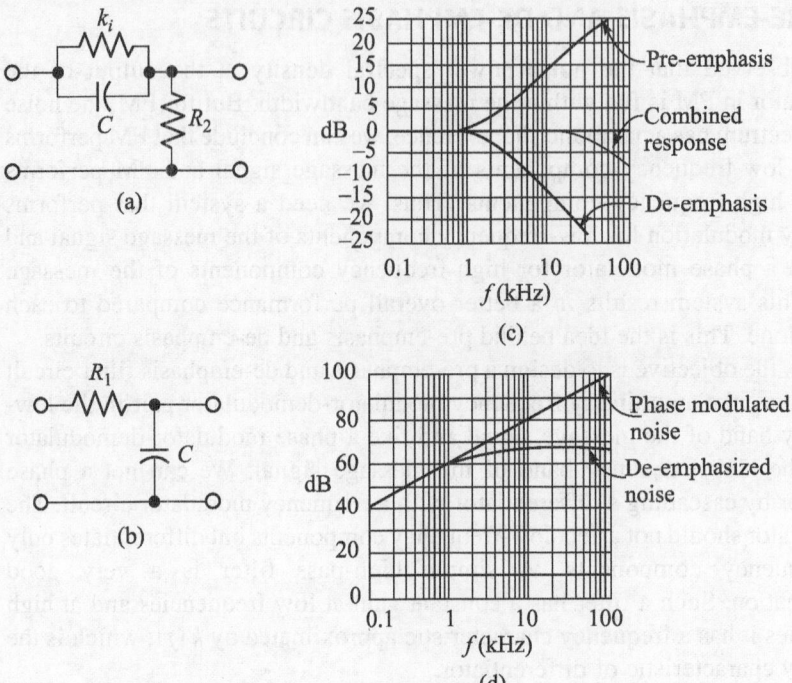

Fig. 6.23 (a) Pre-emphasis network, (b) de-emphasis network, (c) pre-emphasis response, and (d) de-emphasis response

For this, the frequency response of the de-emphasis circuit at the receiver is given by

$$Hd(f) = \frac{1}{1 + j\dfrac{f}{f_o}} \qquad (6.125)$$

where the cut-off frequency f_o of this low-pass de-emphasis circuit is

$$f_o = \frac{1}{2\pi \times 75 \times 10^{-6}} \approx 2100 \text{ Hz}$$

which is the 3 dB cut-off frequency of the filter.

To analyse the overall improvement in the S/N ratio, we note that since the transmitter and the receiver filter cancel the effect of each other, the received power in the message signals remains unchanged. We only have to consider the effect of filtering on the received noise in the receiver filter, i.e., the de-emphasis filter. The noise component before filtering has a parabolic power spectrum. Hence, the noise component after the de-emphasis filter has a power spectral density given by

$$S_{npd}(f) = S_{nc}(f) |Hd(f)|^2$$

$$= \frac{kT}{E_c^2} f^2 \frac{1}{1 + \dfrac{f^2}{f_o^2}} \qquad (6.126)$$

The noise power at the output of the demodulator is P_{npd} and is

$$\int_{-B_m}^{B_m} S_{npd}(f)df$$

$$P_{npd} = \frac{kT}{E_c^2} \int_{-B_m}^{B_m} \frac{f^2}{1+\frac{f^2}{f_o^2}} df$$

$$= \frac{2kTf_o^3}{E_c^2} \left(\frac{B_m}{f_o} - \arctan\frac{B_m}{f_o} \right) \qquad (6.127)$$

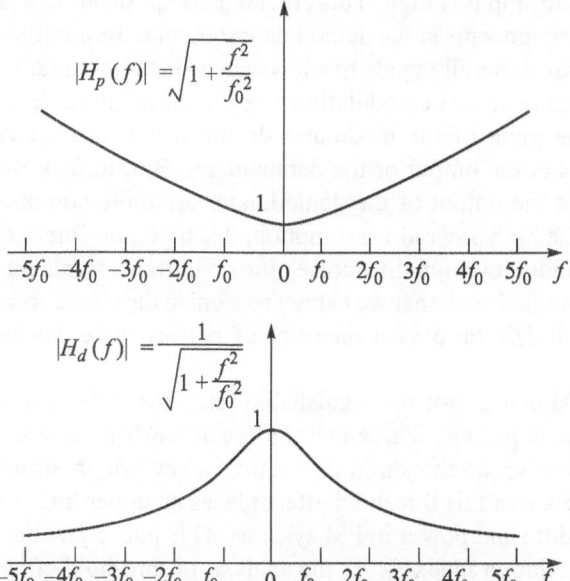

$$|H_p(f)| = \sqrt{1+\frac{f^2}{f_0^2}}$$

$$|H_d(f)| = \frac{1}{\sqrt{1+\frac{f^2}{f_0^2}}}$$

Fig. 6.24   The pre-emphasis and de-emphasis filter characteristics

Since the demodulated message signal power is nothing but that of a simple FM system, with no pre-emphasis and de-emphasis filtering, the ratio of output signal-to-noise ratios in these two cases is inversely proportional to the noise power ratios, i.e.,

$$\frac{\left(\frac{S}{N}\right)_{opd}}{\left(\frac{S}{N}\right)_{out}} = \frac{P_{no}}{P_{npd}}$$

$$= \frac{\frac{2kTB_M^3}{3E_c^2}}{\frac{2kTf_o^3}{E_c^2}\left(\frac{B_m}{f_o} - \arctan\frac{B_m}{f_o}\right)}$$

$$= \frac{1}{3} \frac{\dfrac{B_m}{f_o}}{\left(\dfrac{B_m}{f_o} - \arctan \dfrac{B_m}{f_o} \right)} \tag{6.128}$$

This equation gives the improvement due to pre-emphasis and de-emphasis circuits. The pre-emphasis and de-emphasis filter characteristics are shown in Fig. 6.24.

6.12 THRESHOLD EFFECT IN ANGLE MODULATION

Noise analysis of angle demodulation schemes is based on the assumption that the S/N ratio at the demodulator input is high. This crucial assumption allows us to consider the signal noise components at the demodulator output to be additive. These types of assumptions are generally made to analyse non-linear modulation systems. Due to non-linear nature of the demodulation process, the additive signal and noise components at the input of the modulator do not result in additive signal and noise components at the output of the demodulator. But, in fact, the signal and noise processes at the output of the demodulator are quite complex and non-linear. Only under the high S/N ratio assumption, this high non-linearity is assumed as additive. But under low signal-to-noise ratio conditions, signal and noise components are intermingled such that we cannot recognize the signal from noise. Hence, no meaningful S/N ratio as a measure of performance can be defined.

Under these conditions, signal is not distinguishable from the noise and a mutilation or threshold effect is present. There exists a specific S/N ratio at the input of the demodulator, known as threshold S/N ratio, below which signal mutilation occurs. The existence of this threshold effect places an upper limit on the trade-off between bandwidth and power in FM systems. This puts a practical limit on the value of the modulation index β_f. As the analysis of threshold effect is quite complex, only the results on the threshold effect in FM will be considered.

At the threshold, the following approximate relation between $\dfrac{P_R}{kTB_m} = \left(\dfrac{S}{N} \right)_b$ and β_f holds in an FM system:

$$\left(\frac{S}{N} \right)_{b,\text{th}} = 20(\beta + 1) \tag{6.129}$$

From this relation, given a received power P_R, we can calculate the maximum allowed β to make sure that the system works above threshold. Given the bandwidth allocation B_{max}, we can find an appropriate β using Carson's rule $B_{\text{max}} = 2(\beta + 1)B_m$. Then using the preceding threshold relation, we determine the required minimum received power to make the whole allocated bandwidth usable. Normally, there are two factors that limit the value of the modulation index β. The first is the limitation on the channel bandwidth, which affects β through Carson's rule. The second is the limitation on the received power, which limits the value of β to less than the value derived from Eq. (6.129). Figure 6.25 shows plots of the output S/N ratio in an FM system as a function of the baseband S/N ratio. Different curves correspond to different values of β. The effect of threshold is apparent from the sudden drops in the output S/N ratio.

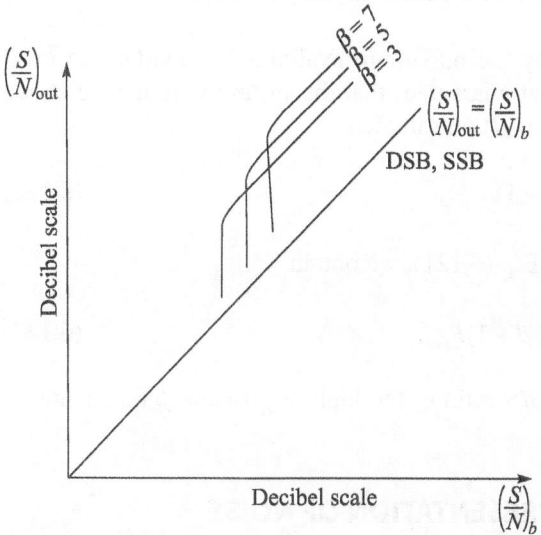

Fig. 6.25 Plot of output SNR vs baseband SNR for various values of β for an FM system

These plots are drawn for sinusoidal message for which

$$\frac{P_M}{\{\max|m(t)|\}^2} = \frac{1}{2} \tag{6.130}$$

In such a case $\left(\dfrac{S}{N}\right)_{\text{out}} = \dfrac{3}{2}\beta^2\left(\dfrac{S}{N}\right)_b$ (6.131)

For $\beta = 6$, we have

$$\left(\frac{S}{N}\right)_{\text{out}} \text{ in dB} = \left[17.3 + \left(\frac{S}{N}\right)_b\right]\text{dB} \tag{6.132}$$

and $\qquad \left(\dfrac{S}{N}\right)_{b,\text{th}} = 20(\beta + 1) = 20(6 + 1) = 140$

i.e., $\qquad \left(\dfrac{S}{N}\right)_{b,\text{th}} = 21.46\text{ dB}$ (6.133)

For $\qquad \beta = 2, \left(\dfrac{S}{N}\right)_{\text{out}} \text{ in dB} = \left[7.8 + \left(\dfrac{S}{N}\right)_b\right] \text{in dB}$ (6.134)

We have

$$\left(\frac{S}{N}\right)_{b,\text{th}} = 60 = 17.8\text{ dB} \tag{6.135}$$

It is apparent that if, for example, $(S/N)_b = 20$ dB, then regardless of the available bandwidth, we can not use $\beta = 5$ for such a system because the system will operate below the threshold. For this case, we use $\beta = 2$. This yields an S/N ratio equal to 27.8 dB at the output of the receiver. This is an improvement of 6.4 dB over a baseband system.

Hence, if we want to employ the maximum available bandwidth, we must choose the largest possible β that guarantees that the system will operate above the threshold. This is the value of β that satisfies

$$\left(\frac{S}{N}\right)_{b,\text{th}} = 20(\beta + 1) \tag{6.136}$$

By substituting this value in Eq. (6.121), we obtain

$$\left(\frac{S}{N}\right)_{\text{out}} = 60\beta^2(\beta + 1)P_{mn} \tag{6.137}$$

which relates a desired output S/N ratio to the highest possible β that achieves that signal-to-noise ratio.

6.13 MATHEMATICAL REPRESENTATION OF NOISE

First we will discuss frequency domain representation of noise.

6.13.1 Frequency Domain Representation of Noise

The frequency domain characteristics of noise can be analysed by first considering a sample function of noise in an interval of length T. Then a periodic function is generated by repeating the sample function every T seconds. This periodic function is next expanded in a Fourier series. The true characteristics of the sample function from $t = -T/2$ to $T/2$ can be obtained in the limit as T goes to infinity. Now, the noise can be represented as

$$n(t) = \lim_{\Delta f \to 0} \sum_{k=1}^{\infty} (a_k \cos 2\pi k \Delta ft + b_k \sin 2\pi k \Delta ft) \tag{6.138}$$

where
$$\Delta f = \frac{1}{T}$$

The exponential form of the above equation is

$$n(t) = \lim_{\Delta f \to 0} \sum_{k=1}^{\infty} C_k \cos(2\pi k \Delta ft + \theta_k) \tag{6.139}$$

where
$$C_k^2 = a_k^2 + b_k^2 \text{ and } \theta_k = \arctan\frac{b_k}{a_k} \tag{6.140}$$

Also a_k, b_k, c_k, and θ_k are all random variables, not constants. Further, as $\Delta f \to 0$, the discrete spectral lines in the Fourier series representation get closer and finally form a continuous spectrum and the power spectral density of noise $n(t)$ becomes

$$G_n(f) = \lim_{\Delta f \to 0} \frac{\overline{c_k^2}}{4\Delta f} = \frac{\overline{a_k^2} + \overline{b_k^2}}{4\Delta f} \tag{6.141}$$

where C_k^2 is replaced by $\overline{C_k^2}$, the expected value or ensemble average of the square of the random variable C_k.

The power in the frequency range f_1 to f_2 can be obtained by integrating the power spectral density function

$$P_{f_1 \to f_2} = 2 \int_{f_1}^{f_2} G_n(f) df \tag{6.142}$$

and the total power $P_t = 2 \int_0^\infty G_n(f) df \tag{6.143}$

since $\qquad G_n(f) = G_n(-f)$

6.13.2 Spectral Component of Noise

Noise $n(t)$ was represented as a superposition of noise spectral components. The coefficient associated with the kth frequency component interval as $\Delta f \to 0$ can be written as

$$n_k(t) = a_k \cos 2\pi k \Delta ft + b_k \sin 2\pi k \Delta ft \tag{6.144}$$

or in exponential form as

$$n_k(t) = C_k \cos(2\pi k \Delta ft + \theta_k) \tag{6.145}$$

Since the components a_k and b_k are random variables, the normalized power P_k, variance of $n_k(t)$, is determined by taking the average over the ensemble of $\overline{\left((n_k(t))^2 \right)}$, i.e.,

$$P_k = \overline{\left((n_k(t))^2 \right)} = \overline{a_k^2} \cos^2 2\pi k \Delta ft + \overline{b_k^2} \sin^2 2\pi k \Delta ft$$

$$+ \overline{2a_k b_k} \sin 2\pi k \Delta ft \cos 2\pi k \Delta ft \tag{6.146}$$

It can be seen that $n_k(t)$ is a stationary process, so that $\overline{(n_k(t)^2)}$ does not depend on the time selected for determining it. Hence, to evaluate P_k, we can substitute in the above equation $t = t_1$ for which $\cos 2\pi k \Delta ft_1 = 1$ and hence

$$\sin 2\pi k \Delta ft_1 = 0$$

Then we also have $P_k = \overline{a_k^2}$

Similarly, we can show that $P_k = \overline{b_k^2}$ and hence $\overline{a_k^2} = \overline{b_k^2}$

Also $P_k = 2G_n(k\Delta f)\Delta f = 2G_n(-k\Delta f)\Delta f = \overline{a_k^2} = \overline{b_k^2} = \dfrac{\overline{a_k^2}}{2} + \dfrac{\overline{b_k^2}}{2} = \dfrac{\overline{c_k^2}}{2}$

Since $\overline{a_k^2} = \overline{b_k^2}$, Eq. (6.146) can be rewritten as

$$P_k = \overline{a_k^2}(\cos^2 2\pi k\Delta ft + \sin^2 2\pi k\Delta f)$$
$$+ 2\overline{a_k b_k} \sin 2\pi k\Delta ft \cos 2\pi k\Delta ft \qquad (6.147)$$

i.e., $\qquad P_k = \overline{a_k^2} + 2\overline{a_k b_k} \sin 2\pi k\Delta ft \cos 2\pi k\Delta ft \qquad (6.148)$

Since $P_k = \overline{a_k^2}$ independent of time, the term $2\overline{a_k b_k} \sin 2\pi k\Delta ft \cos 2\pi k\Delta ft$ in the above equation must be zero, i.e., $\overline{a_k b_k} = 0$, thus the coefficients a_k and b_k are uncorrelated. It can be shown that both a_k and b_k are Gaussian. In Eq. (6.144) by substituting for a value $t = t_1$ and for $\cos 2\pi k\Delta ft_1 = 1$ and $\sin 2\pi k\Delta ft_1 = 0$, we obtain

$$n_k(t) = a_k \qquad (6.149)$$

where $n_k(t)$ is a Gaussian random variable. This after assuming that $n_k(t)$ can be viewed as the output of a very narrowband filter whose input is Gaussian noise, the output is also Gaussian and the noise voltage at any time $t = t_1$ has a Gaussian probability density. Hence, we can conclude from Eq. (6.149) that a_k is also Gaussian variable. Similarly, b_k is also Gaussian variable. Also $n_k(t)$ is the output of a narrowband filter at frequency $k\Delta f$ and has no DC component. So, a_k also has no DC component, i.e., $\overline{a_k} = 0$ and, similarly, $\overline{b_k} = 0$.

Let us consider two spectral components of noise, one at frequency $k\Delta f$ and the other at $l\Delta f$ with

$$n_k(t) = a_k \cos 2\pi k\Delta ft + b_k \sin 2\pi k\Delta ft \qquad (6.150)$$

and $\qquad n_l(t) = a_l \cos 2\pi k\Delta ft + b_l \sin 2\pi k\Delta ft \qquad (6.151)$

and form the product $n_k(t)n_l(t)$, which results in four terms These terms are time dependent provided that $k \neq 1$, the resulting coefficient terms are $a_k a_l$, $a_k b_l$, $b_k a_l$, and $b_k b_l$. Because of the stationary character of the random process involved, the ensemble average of the product $\overline{n_k(t)n_l(t)}$ must be time independent. Hence

$$\overline{a_k a_l} = \overline{a_k b_l} = \overline{b_k a_l} = \overline{b_k b_l} = 0 \qquad (6.152)$$

i.e., each of the coefficients a_k and b_k is uncorrelated with each of the coefficients a_l and b_l.

Some statistical characteristics of interest concerning c_k and θ_k can be inferred as follows.

- The c_k and θ_k are related to the a_k's and b_k's precisely in the same manner in which the random variables R and θ are related to Gaussian variables X and Y.
- c_k's have a Rayleigh probability density

$$f(c_k) = \frac{c_k}{P_k} e^{-c_k^2/2P_k} \qquad (6.153)$$

where P_k is the normalized power in the spectral range Δf at the frequency $k\Delta f$.

- Similarly, the angle θ_k has a uniform probability density

$$f(\theta_k) = \frac{1}{2\pi}, \quad -\pi \le \theta_k \le \pi \tag{6.154}$$

- Also the amplitude c_k and phase θ_k are independent of one another as well as of the amplitude and phase of a spectral component at a different frequency.

6.13.3 Superposition of Noise

The concept of a power spectrum is useful because it allows us to resolve a deterministic waveform or a random process into a sum

$$f(t) = f_1(t) + f_2(t) + f_3(t) + \cdots \tag{6.155}$$

where we can apply superposition theorem. This is possible because of the orthogonality of spectral components of different frequencies. The noise waveform can be represented as a superposition of spectral components, all of which are harmonics of some fundamental frequency Δf, which in the limit approaches zero. But there are instances when there are two noise processes $n_1(t)$ and $n_2(t)$, whose spectral ranges overlap in part or entirely. In this case, the power P_{12} of the sum $n_1(t) + n_2(t)$ will be

$$P_{12} = E[\{n_1(t) + n_2(t)\}]^2 = E[n_1^2(t) + n_2^2(t)] + 2E[n_1(t)n_2(t)] \tag{6.156}$$

or $\qquad\qquad P_{12} = P_1 + P_2 + 2E[n_1(t)n_2(t)] \tag{6.157}$

where P_1 and P_2 are the power, respectively, of the noise processes $n_1(t)$ and $n_2(t)$ and $E[n_1(t)n_2(t))]$ is nothing but the cross correlation of the processes. Thus, one can apply the superposition of power $P_{12} = P_1 + P_2$ provided the processes are uncorrelated.

6.13.4 Mixing Noise with Sinusoid

Let us consider a situation where noise is mixed with a deterministic sinusoidal waveform $\cos 2\pi f_c t$. The product of this sinusoidal waveform with a spectral noise component is given by

$$n_k(t) \cos 2\pi f_c t = \frac{a_k}{2} \cos 2\pi (k\Delta f + f_c)t + \frac{b_k}{2} \sin 2\pi (k\Delta f + f_c)t$$

$$+ \frac{a_k}{2} \cos 2\pi (k\Delta f - f_c)t + \frac{b_k}{2} \sin 2\pi (k\Delta f - f_c)t \tag{6.158}$$

Hence, mixing gives rise to two noise spectral components, one at the sum frequency $f_c + k\Delta f$ and another at the difference frequency $f_c - k\Delta f$, and the amplitude of each of the two noise spectral components has been reduced by two with respect to the original noise spectral component. The variance, i.e., normalized power of the two new noise components is smaller by a factor of four. As the power spectral density of the original component at frequency $k\Delta f$ is $G_n(k\Delta f)$, the spectral densities of the new components is given by

$$G_n(k\Delta f + f_c) = G_n(k\Delta f - f_c) = \frac{G_n(k\Delta f)}{4} \tag{6.159}$$

As the limit reaches zero, i.e., $\Delta f \to 0$, $k\Delta f$ can be replaced by continuous variable f and the above equation becomes

$$G_n(f + f_c) = G_n(f - f_c) = \frac{G_n(f)}{4} \tag{6.160}$$

Hence, it can be inferred that given the power spectral density plot of G_n of a noise waveform $n(t)$, the power spectral density of $n(t) \cos 2\pi f_c t$ is arrived at by dividing $G_n(f)$ by 4 and shifting the divided plot to the left by an amount f_c, to the right by an amount f_c, and then adding both of them.

When we have noise $n(t)$ from which we get two spectral components, one at a frequency $k\Delta f$ and the other at a frequency $l \Delta f$ and mix with a sinusoidal frequency f_c, with f_c selected to be midway between $k\Delta f$ and $l\Delta f$, i.e.,

$f_c = (1/2)(k+l)\Delta f$. We get four spectral components, two sum frequency components and two difference frequency components. The two difference frequency components will be at the same frequency $p\Delta f = f_c - k\Delta f = l\Delta f - f_c$ and they are uncorrelated.

Representing the spectral components at $k\Delta f$ and $l\Delta f$, the difference frequency components are

$$n_{p1}(t) = \frac{a_k}{2} \cos 2\pi p\Delta ft - \frac{b_k}{2} \sin 2\pi p\Delta f$$

and

$$n_{p2}(t) = \frac{a_l}{2} \cos 2\pi p\Delta ft + \frac{b_l}{2} \sin 2\pi p\Delta ft \tag{6.161}$$

where $n_{p1}(t)$ is the difference component due to the mixing of frequencies f_c and $k\Delta f$ and $np_2(t)$ is the difference component due to the mixing of frequencies f_c and $l\Delta f$.

It has been proved already that $\overline{a_k a_l} = \overline{a_k b_l} = \overline{b_k a_l} = \overline{b_k b_l} = 0$, which implies that

$$E\{n_{p1}(t)n_{p2}(t)\} = 0 \tag{6.162}$$

Applying the superposition theorem, the power at the difference frequency due to the superposition of $n_{p1}(t)$ and $n_{p2}(t)$ is

$$E\{(n_{p1}(t) + n_{p2}(t))\}^2 = E\{n_{p1}(t)\}^2 + E\{n_{p2}(t)\}^2 \tag{6.163}$$

Hence, mixing noise with sinusoidal signal results in the frequency shifting of the original noise by f_c. The variance of the shifted noise is determined by adding the variance of new noise component.

6.13.5 Mixing Noise with Noise

Let us consider two spectral components of noise. When the two spectral components are multiplied, one at the frequency $k\Delta f$ and the other at the frequency $l\Delta f$, the resulting expression is

$$n_k(t)n_l(t) = \frac{1}{2}c_k c_l \cos\{2\pi(k+1)\Delta ft + \theta_k + \theta_l\}$$

$$+ \frac{1}{2}c_k c_l \cos\{2\pi(k-1)\Delta ft + \theta_k - \theta_l\} \tag{6.164}$$

This results in two new spectral components of noise, one at the sum frequency $(k+1)$ Δf and the other at the difference frequency $(k-l)$ Δf, and the power equation is given by

$$P_{k+l} = P_{k-l} = \frac{1}{2}\overline{\left(\frac{1}{2}c_k c_l\right)^2} \tag{6.165}$$

Since c_k and c_l are independent random variables,

$$P_{k+l} = P_{k-l} = \frac{1}{8}\overline{c_k^2}\;\overline{c_l^2} = \frac{1}{2}P_k P_l \tag{6.166}$$

6.13.6 Linear Filtering of Noise

The discussion about the noise included different types of noise such as thermal noise, shot noise, and flicker noise. Most of them have wide spectral range. But for the effect of these types of noise on communication system, it is enough if we consider white noise. Its power spectral density is uniform over the entire range of frequency spectrum. This noise power spectral density is given by

$$G_n(f) = \frac{\eta}{2} \tag{6.167}$$

where η is a constant as defined earlier. To minimize the noise power generally we employ a filter before the demodulating stage. The filter bandwidth B is made as small as possible so as to avoid unwanted noise to the demodulator at the same time allowing the required signal. The bandwidth of the filter depends on the nature of modulation. For an AM system, it will be $2f_m$ and for a wideband FM system, it is $2\Delta f$, where Δf is the frequency deviation. Effect of certain types of filters on noise will be considered.

RC low-pass filter

This filter has a transfer function

$$H(f) = \frac{1}{1 + j\dfrac{f}{f_c}} \tag{6.168}$$

where f_c is the 3 dB frequency of the filter. If the input noise to the filter has a power spectral density G_{ni} and the power spectral density of the output noise is G_{no}, then

$$G_{no}(f) = G_{ni}(f)|H(f)|^2 \tag{6.169}$$

For white noise, $G_{ni}(f) = \eta/2$ for all frequencies. Hence, the expression for output noise becomes

$$G_{no}(f) = \frac{\eta}{2}\frac{1}{1 + \left(\dfrac{f}{f_c}\right)^2} \tag{6.170}$$

and the noise power at the filter output is

$$N_o = \int_{-\infty}^{\infty} G_{no}(f)df = \frac{\eta}{2}\left[\int_{-\infty}^{\infty} \frac{df}{1+\left(f/f_c\right)^2}\right] \tag{6.171}$$

Changing the variable to $x = f/f_c$ and noting that $\int_{-\infty}^{\infty} \frac{dx}{1+x^2} = \pi$, the above equation becomes

$$N_o = \frac{\pi}{2}\eta f_c \tag{6.172}$$

Rectangular low-pass filter

This filter is also called ideal filter. This filter has a transfer function
$$H(f) = 1, \quad |f| \le B \tag{6.173}$$
$$= 0, \quad \text{elsewhere}$$
Assuming that the noise input is white, the output power spectral density is

$$G_n(f) = \frac{\eta}{2}, \quad -B \le B \tag{6.174}$$
$$= 0, \quad \text{elsewhere}$$
The output noise power is
$$N_o = \eta B \tag{6.175}$$

6.13.7 Quadrature Component of Noise

Detailed explanation on this subject, including the power spectral density of noise and its in-phase component and quadrature component, is given under narrow-band noise.

6.13.8 Representation of Noise Using Orthogonal Representation

Noise is represented as a function of sines and cosines, which are called orthonormal functions. Let us consider below some features of representing noise in terms of orthonormal function. If $s_i(t)$ is a set of orthonormal functions in the interval T, the noise $n(t)$ in this interval is

$$n(t) = \sum_{i=0}^{\infty} n_i s_i(t) \tag{6.176}$$

where n_i is the coefficient of the ith component and is given by

$$n_i = \int_0^T n(t)s_i(t)dt \tag{6.177}$$

If the noise $n(t)$ is a Gaussian random process with a mean value of zero, then n_i is also a Gaussian random variable with zero mean value.

We can find the correlation between coefficients n_i and n_j as follows.

$$n_i n_j = \int_0^T n(t)s_i(t)dt \int_0^T n(\partial)s_j(\partial)d\partial \qquad (6.178)$$

or $\qquad n_i n_j = \int_0^T dt \int_0^T d\partial \{n(t)n(\partial)\}s_i(t)s_j(\partial) \qquad (6.179)$

where t and ∂ are dummy variables of integration. Taking the ensemble average of both sides of the above integral, we get

$$E(n_i n_j) = \int_0^T dt \int_0^t d\partial E\{n(t)n(\partial)\}s_i(t)s_j(\partial) \qquad (6.180)$$

As the noise process is ergodic, the autocorrelation of the process is

$$R(t-\partial) = E\{n(t)n(\partial)\} \qquad (6.181)$$

Assuming the power spectral density of white noise as $G(f) = \dfrac{\eta}{2}$, we get

$$R(t-\partial) = \frac{\eta}{2}\delta(t-\partial) \qquad (6.182)$$

From Equations (6.178) – (6.181), we can obtain

$$E(n_i n_j) = \int_0^T dt \int_0^T d\partial \frac{\eta}{2}\delta(t-\partial)s_i(t)s_j(\partial) \qquad$$

or $\qquad E(n_i n_j) = \dfrac{\eta}{2}\displaystyle\int_0^T s_i(t)s_j(t)dt = \dfrac{\eta}{2}, \quad$ if $i=j$

$$= 0, \quad \text{if } i \neq j \qquad (6.183)$$

6.14 NARROWBAND NOISE

Most communication systems often deal with bandpass filtering of signals. This filter's bandwidth is just large enough to pass the modulated component of the received signal without distortion, but not large enough to allow excessive noise through the receivers. Thus, wideband is shaped into bandlimited noise. If the bandwidth of this bandlimited noise is relatively small compared to the carrier frequency, then it is called *narrowband noise*. Figure 6.26 shows the method of generating narrowband noise from wide band noise.

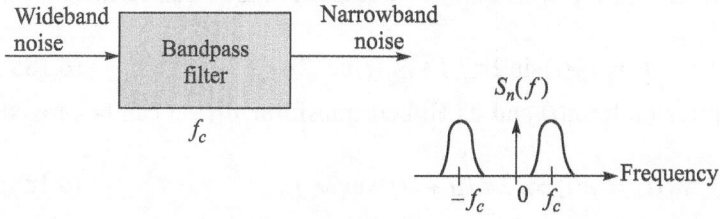

Fig. 6.26   Generation of narrowband noise

The spectral components of narrowband noise is around some midband frequency $\pm f_c$, where f_c is the carrier frequency. The sample function $n(t)$ appears as somewhat similar to a sine wave of frequency f_c, which undulates slowly in both amplitude and phase as shown in Fig. 6.27.

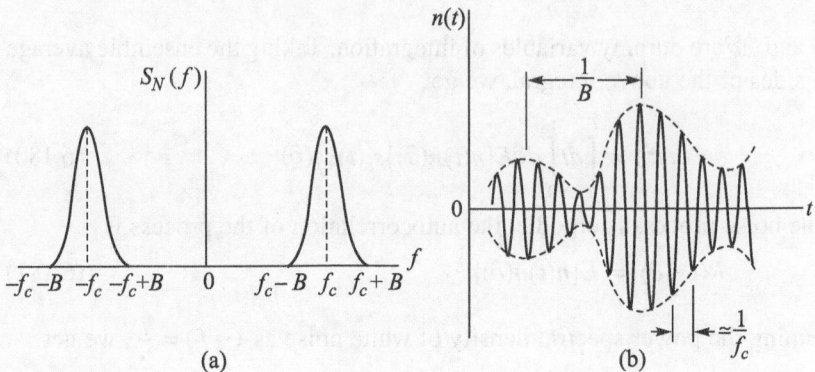

Fig. 6.27 (a) Power spectral density of narrowband noise and (b) its sample function

The power spectral density of the narrowband noise can be derived and used to analyse the performance of linear systems. One method is to deal with mixing, which is nothing but multiplication. But this is a non-linear operation and the system analysis becomes difficult. Hence, depending on the applications, there are two specific representations of narrowband noise:

(a) In terms of a pair of components called in-phase and quadrature components and

(b) In terms of two other components called the envelope and phase

6.14.1 Representation of Narrowband Noise in Terms of In-Phase and Quadrature Components

Let us consider a narrowband noise of bandwidth $2B$ centred around f_c, which is represented as

$$n(t) = n_I(t)\cos 2\pi f_c t - n_Q(t)\sin 2\pi f_c t \qquad (6.184)$$

where f_c is the carrier frequency within the band occupied by the noise and $n_i(t)$ and $n_q(t)$ are known to be the in-phase component and quadrature component of $n(t)$, respectively. Both $n_I(t)$ and $n_Q(t)$ are low-pass signals. Except for the midband frequency, these two components fully represent the narrowband noise $n(t)$. To find $n_I(t)$ and $n_Q(t)$ in terms of $n(t)$, we find the Hilbert transform of $\hat{n}(t)$, which is

$$\hat{n} = n_I(t)\sin 2\pi f_c t + n_Q(t)\cos 2\pi f_c t \qquad (6.185)$$

From the expression for $n(t)$ and its Hilbert transform $\hat{n}(t)$, it can be proved that

$$n_I(t) = n(t)\cos 2\pi f_c t + \hat{n}(t)\sin 2\pi f_c t \qquad (6.186)$$

and

$$n_Q(t) = n(t)\cos 2\pi f_c t - \hat{n}(t)\sin 2\pi f_c t \qquad (6.187)$$

Given $n(t)$, its in-phase and quadrature components can be obtained. Similarly, if the in-phase and the quadratue components are given, $n(t)$ can be obtained. Figure 6.28 shows the generation for both the cases.

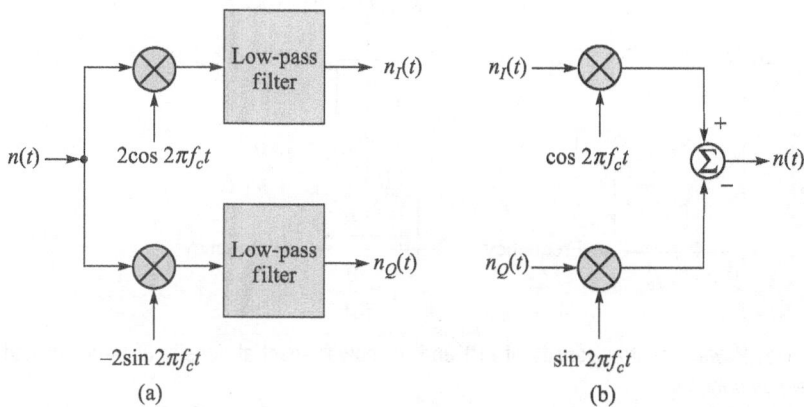

Fig. 6.28 (a) Generation of in-phase and quadrature components of $n(t)$ and (b) generation of $n(t)$ from its in-phase and quadrature components

The in-phase and quadrature components have some important properties, which are given below.

(a) $E[n_I(t)\, n_Q(t)] = 0$, i.e., $n_I(t)$ and $n_Q(t)$ are uncorrelated with each other.
(b) They have zero mean.
(c) They have the same mean and variance as $n(t)$.
(d) If $n(t)$ is Gaussian, then both the in-phase and quadrature components of $n(t)$ are jointly Gaussian.
(e) If $n(t)$ is stationary, then both the in-phase and quadrature components are jointly stationary.
(f) Both the in-phase and quadrature components have the same power spectral densities related to the power spectral density $S_N(f)$ of the narrowband noise $n(t)$ by

$$S_{NI}(f) = S_{NQ}(f) = S_N(f-f_c) + S_N(f+f_c), \quad -B \le f \le B$$
$$= 0, \quad \text{otherwise} \tag{6.188}$$

where it is assumed that $S_N(f)$ occupies the frequency interval $f_c - B \le |f| \le f_c + B$ and $f_c > B$. This equation will enable us to calculate the effect of noise on AM and FM systems. It implies that the power spectral density of in-phase and quadrature components can be found by shifting the positive and negative portions of $S_N(f)$ to zero frequency and adding to give S_{NI} and S_{NQ} as shown in Fig. 6.29.

(g) The in-phase and quadrature components have the same variance as the narrowband noise $n(t)$.
(h) The cross spectral density of the in-phase and quadrature components of narrowband noise $n(t)$ is purely imaginary and is given as

$$S_{NINQ}(f) = -S_{NQNI}(f)$$

$$= j[S_N(f + f_c) - S_N(f - f_c)], \quad -B \le f \le B$$

$$= 0, \quad \text{otherwise} \tag{6.189}$$

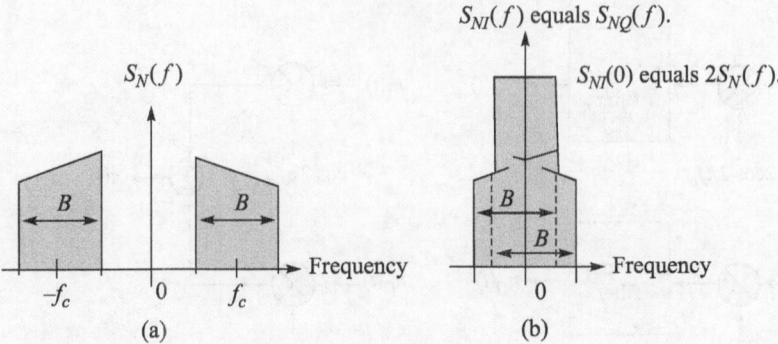

Fig. 6.29 (a) Power spectral density of $n(t)$ and (b) power spectral density of in-phase and quadrature components

(i) The in-phase and quadrature components are statistically independent if the narrowband noise $n(t)$ is Gaussian and its power spectral density $S_N(t)$ is symmetric about the midband frequency f_c.

6.14.2 Representation of Narrowband Noise in Terms of Envelope and Phase Components

The narrowband noise $n(t)$ can also be represented in terms of its envelope and phase components as

$$n(t) = r(t) \cos[2\pi f_c t + \psi(t)] \tag{6.190}$$

where

$$r(t) = \sqrt{n_I^2(t) + n_Q^2(t)} \tag{6.191}$$

and

$$\psi(t) = \tan^{-1} \frac{n_Q(t)}{n_I(t)} \tag{6.192}$$

The function $r(t)$ is called the envelope of $n(t)$ and the function $\psi(t)$ is called the phase of $n(t)$. These are both sample functions of low-pass random process. The time interval between two successive peaks of the envelope $r(t)$ is approximately $1/B$, where $2B$ is the bandwidth of the narrowband noise $n(t)$. Let N_I and N_Q denote the random variables obtained by observing the random process represented by the sample function $n_I(t)$ and $n_Q(t)$, respectively. Then it is easy to obtain the probability distribution of $r(t)$ and $\psi(t)$. N_I and N_Q are independent Gaussian random variables of zero mean and variance σ^2. The joint probability density function of $n_I(t)$ and $n_Q(t)$ is given by

$$f_{N_I N_Q}(n_I, n_Q) = \frac{1}{2\pi\sigma^2} \exp\left(-\frac{n_I^2 + n_Q^2}{2\sigma^2}\right) \tag{6.193}$$

The probability of the joint event that N_I lies between $n_I + dn_I$ and that N_Q lies between $n_Q + dn_Q$ is given by

$$f_{N_I N_Q}(n_I, n_Q) dn_I dn_Q = \frac{1}{2\pi\sigma^2} \exp\left(-\frac{n_I^2 + n_Q^2}{2\sigma^2}\right) dn_I dn_Q \qquad (6.194)$$

The coordinate system for representation of narrowband noise is shown in Fig. 6.30. From the figure, it can be seen that the pair of random variables N_I and N_Q lies jointly inside the shaded area.

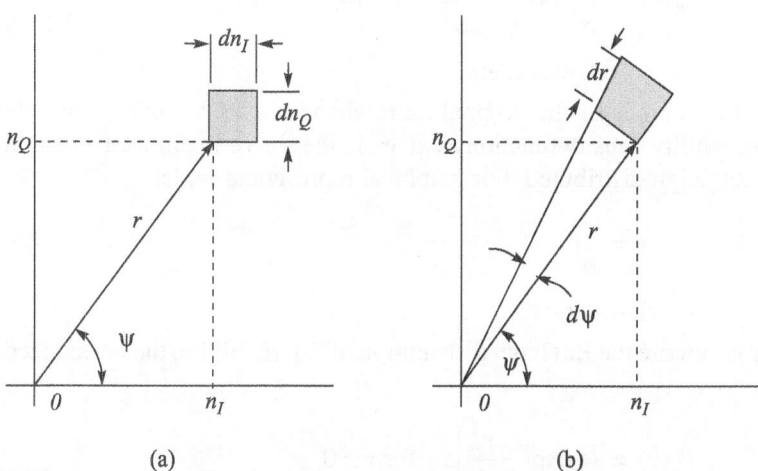

(a)   (b)

Fig. 6.30 Coordinate system representation for narrowband noise: (a) in terms of in-phase and quadrature components and (b) in terms of envelope and phase

Let $\qquad\qquad n_I = r \cos \psi \quad$ and $\quad n_Q = r \sin \psi \qquad$ (6.195a)

In the limiting case, the two incremental areas shown shaded in the figure and can be written as

$$dn_I dn_Q = r dr d\psi \qquad (6.195b)$$

Since R and ψ denote the random variable and observing them at some time t, the random processes are represented by the envelope $r(t)$ and phase $\psi(t)$ respectively. By substituting Equations. (6.195a) and (6.195b) into Eq. (6.194), we get the probability of the random variables R and ψ lying jointly in the shaded area of Fig. 6.30(b) as

$$\frac{r}{2\pi\sigma^2} \exp\left(-\frac{r^2}{2\sigma^2}\right) dr \, d\psi \qquad (6.196)$$

i.e., the joint probability density function of R and ψ is

$$f_{R,\psi}(r,\psi) = \frac{r}{2\pi\sigma^2} \exp\left(-\frac{r^2}{2\sigma^2}\right) \qquad (6.197)$$

This probability density function is independent of the angle ψ. This shows that the random variables R and ψ are statistically independent. Thus, one can express $f_{R,\psi}(r,\psi)$ as the product of $f_R(r)$ and $f_\psi(\psi)$. In effect, the random variable ψ representing phase is uniformly distributed inside the range 0 to 2π, as shown by

$$f_\psi(\psi) = \frac{1}{2\pi}, \quad 0 \le \psi \le 2\pi \tag{6.198}$$

$$= 0, \text{ elsewhere}$$

Hence, the probability density function of the random variable is

$$f_R(r) = \frac{r}{\sigma^2} \exp\left(-\frac{r^2}{2\sigma^2}\right), \quad r \ge 0 \tag{6.199}$$

$$= 0, \text{ elsewhere}$$

where σ^2 is the variance of the original narrowband noise. A random variable having the probability density function as given in the above-mentioned equation is said to be Rayleigh distributed. For graphical representation, let

$$v = \frac{r}{\sigma}$$

$$f_V(v) = \sigma f_R(r)$$

Then we can rewrite the Rayleigh distribution of Eq. (6.199) in the normalized form as

$$F_v(v) = \left\{ v \exp\left(-\frac{v^2}{2}\right), \quad \text{for } v \ge 0 \right. \tag{6.200}$$

$$= 0, \quad \text{otherwise}$$

The plot of this function is shown in Fig. 6.31. The peak value of the distribution $f_V(v)$ occurs at $v = 1$ and is equal to 0.607. Unlike the Gaussian distribution, the Raleigh distribution is zero for negative values of v. This is because the envelope $r(t)$ can assume only a non-negative value.

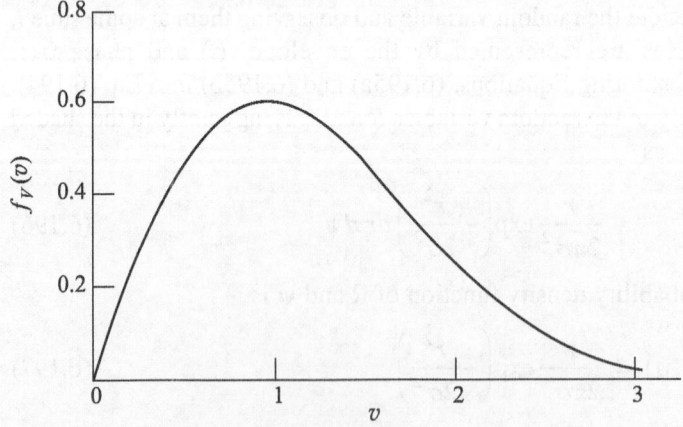

Fig. 6.31 Normalized Rayleigh distribution

6.14.3 Sine Wave Plus Narrowband Noise

Let us add to the narrowband noise $n(t)$, a sinusoidal wave $E_m \cos(2\pi fct)$, where E_m and f_c are constants. The assumption made is that the frequency of the sine wave is the same as the nominal carrier frequency of the noise. Then we can express this combination as

$$x(t) = E_m \cos(2\pi f_c t) + n(t) \qquad (6.201a)$$

We can replace the expression $n(t)$ for narrowband noise with its in-phase and quadrature components. Then the above equation becomes

$$x(t) = n_I(t) \cos(2\pi f_c t) - n_Q(t) \sin(2\pi f_c t) \qquad (6.201b)$$

where

$$n_I(t) = E_m + n_I(t) \qquad (6.202)$$

Here $n(t)$ is assumed to be Gaussian with zero mean and variance σ^2.
With this assumption, we can state the following:

(a) Both $n'_I(t)$ and $n_Q(t)$ are Gaussian in nature and statistically independent.
(b) The mean of $n'_I(t)$ is E_m and that of $n_Q(t)$ is zero.
(c) The variance of both $n'_I(t)$ and $n_Q(t)$ is σ^2.

Hence, we can express the joint probability density function of the random variables N'_I and N_Q corresponding to $n'_I(t)$ and $n_Q(t)$ as follows:

$$f_{N_I N_Q}(n_I, n_Q) = \frac{1}{2\pi\sigma^2} \exp\left(-\frac{(n'_I - E_m)^2 + n_Q^2}{2\sigma^2}\right) \qquad (6.203)$$

Let $r(t)$ denote the envelope of $x(t)$ and $\psi(t)$ denote its phase. From Eq. (6.201b), we can infer that

$$r(t) = \sqrt{[n'_I(t)]^2 + \left[n_Q(t)\right]^2} \qquad (6.204)$$

and

$$\psi(t) = \arctan\left[\frac{n_Q(t)}{n'_I(t)}\right] \qquad (6.205)$$

It can be proved that the joint probability density function of the random variables R and ψ, corresponding to $r(t)$ and $\psi(t)$ for some fixed time t is given by

$$f_{R,\psi}(r, \psi) = \frac{r}{2\pi\sigma^2} \exp\left(-\frac{r^2 + E_m^2 - 2E_m r \cos\psi}{2\sigma^2}\right) \qquad (6.206)$$

From the above expression, it can be seen that we cannot express the joint probability density function $f_{R,\psi}(r, \psi)$ as a product $f_R(r) f_\psi(\psi)$. This is because of the term involving the values of both random variables multiplied together as $r \cos\psi$. Hence, R and ψ are dependent random variables for non-zero values of the amplitude E_m of the sinusoidal wave component. To get the probability density function of R, Eq. (6.206) is integrated over all possible values of ψ, giving the marginal density as

$$f_R(r) = \int_0^{2\pi} f_{R,\psi}(r, \psi) d\psi \qquad (6.207)$$

$$= \frac{r}{2\pi\sigma^2} \exp\left(-\frac{r^2 + E_m^2}{2\sigma^2}\right) \int_0^{2\pi} \exp\left(\frac{E_m r}{\sigma^2} \cos\psi\right) d\psi \qquad (6.208)$$

Since the integral for the modified Bessel function of the first kind of zero order is given by

$$I_o(x) = \frac{1}{2\pi} \int_0^{2\pi} \exp(x\cos\psi)d\psi \qquad (6.209)$$

Now letting $x = E_m \dfrac{r}{\sigma^2}$, Eq. (6.208) can be rewritten as

$$f_R(r) = \frac{r}{\sigma^2}\exp\left(-\frac{r^2 + E_m^2}{2\sigma^2}\right)I_o\left(\frac{E_m r}{\sigma^2}\right) \qquad (6.210)$$

This relation is called *Rician distribution.*

The Rician distribution in a normalized form can be obtained by letting

$$v = \frac{r}{\sigma}, a = \frac{E_m}{\sigma}, \text{ and } f_V(v) = \sigma f_R(r)$$

as

$$f_V(v) = v\exp\left(-\frac{v^2 + a^2}{2}\right)I_o(av) \qquad (6.211)$$

This relation is shown in Fig. 6.32 for $a = 0, 1, 2, 3,$ and 5.

From the figure, it can be seen that

- When a is zero, the Rician distribution reduces to the Rayleigh distribution.
- The envelope distribution is approximately Gaussian in the vicinity of $v = a$ when a is large, i.e., when the sine wave amplitude E_m is large compared with σ, the square root of the average power of the noise $n(t)$.

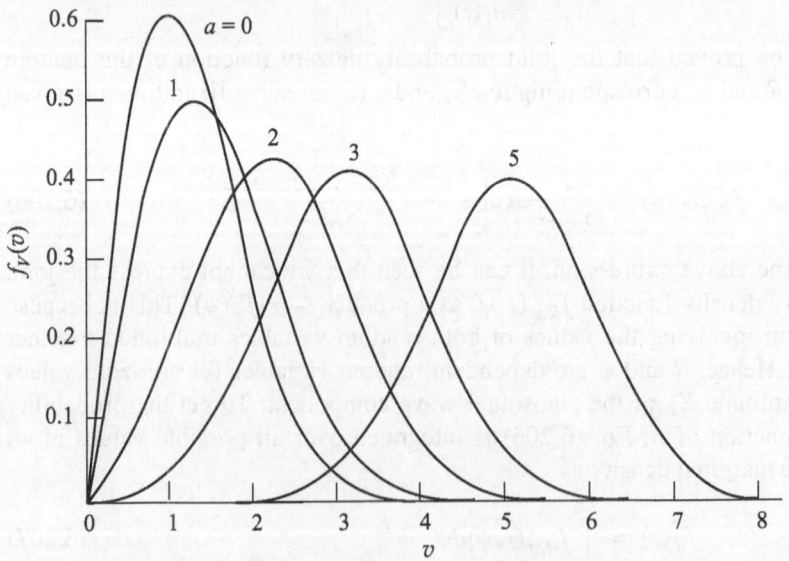

Fig. 6.32 Normalized Rician distribution

6.15 FREQUENCY MODULATION FEEDBACK (FMFB) TECHNIQUE

A communication link must achieve a satisfactory baseband S/N quality at the lowest possible C/N (carrier-to-noise power). The relatively high threshold of carrier-to-noise ratio that an angle modulated signal must exceed for satisfactory demodulation of the signal with conventional FM detectors is, therefore, an obvious disadvantage. However, in recent years, two demodulation techniques have the improvement factor of angle modulation at signal levels above the threshold. One demodulation technique utilizes negative feedback, called the frequency modulation feedback (FMFB). The FMFB circuit is also called a threshold extension circuit and a signal enhancer. The second technique employs a phase locked loop (PLL) detector. The FMFB and PLL circuitry differ considerably, but the performance is essentially the same for both circuits. The received signals are at relatively low power levels. Therefore, the amount of noise that accompanies the received signals is of prime importance. This noise may be reduced by narrowing the receiver's bandpass filter but this method would also introduce distortion if the deviation of the angle modulated signal exceed the bandpass frequency of the receiver. A more acceptable method is to degenerate the signal automatically when the deviation exceeds certain limit. In effect, the bandpass filter appears to be narrower to the higher frequency (more troublesome) noise signals. Both frequency and phase modulations have a carrier that deviates (higher and lower, or ahead and behind, in phase) with the complex wave shape of the baseband signal but, practically, it also contains noise. The greater the deviation, the greater the bandwidth occupied by the spectrum of the modulated carrier.

Hence, in frequency modulation, there is particular interest in reducing the noise threshold in an FM receiver so as to satisfactorily operate the receiver with the minimum signal power possible. The threshold extension is measured with respect to the standard frequency discriminator (i.e., one without feedback).

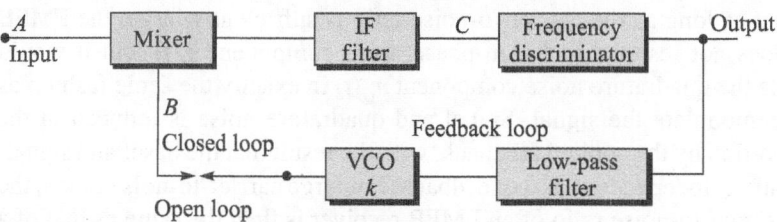

Fig. 6.33 Block diagram of FMFB demodulator

The block diagram of an FMFB demodulator is shown in Fig. 6.33. We see that the local oscillator of the conventional FM receiver has been replaced by a voltage-controlled oscillator (VCO) whose instantaneous output frequency is controlled by the demodulated signal. The operation of this receiver can be understood by assuming for the moment that the VCO is removed from the circuit and the feedback loop is left open. Now a wideband FM signal is applied to the receiver input and a second FM signal from the same source but whose modulation index is a fraction smaller compared to the wideband FM signal, is applied to the VCO terminal of the mixer. The output of the mixer will consist of the difference frequency component, since the sum frequency component is removed by the bandpass filter. The frequency deviation output from the mixer will be small, although the frequency deviation of both input FM waves is large, since the difference between their instantaneous deviations is small, i.e., the modulation indices would subtract and the resulting FM wave at the mixer output would have a smaller modulation index. The bandwidth of the bandpass filter required for this FM wave with reduced modulation index needs only to be a fraction of that required for the wideband FM. It is now apparent that the second wideband FM signal applied to the mixer may be obtained from the output of VCO by feeding the output of the frequency discriminator back to the VCO.

An argument may be put forth, that the signal-to-noise ratio of an FMFB receiver is the same as that of a conventional FM receiver with the same input signal and noise power if the carrier-to-noise ratio is sufficiently large. Let it be assumed that there is no feedback around the demodulator. In the combined presence of an unmodulated carrier $E_c \cos(2\pi f_c t)$ and the expression for the narrowband noise are given by

$$n(t) = n_I(t)\cos 2\pi f_c(t) - n_Q(t)\sin 2\pi f_c(t) \qquad (6.212)$$

The phase of the composite signal $x(t)$ at the limiter-discriminator input is approximately equal to $n_Q(t)/E_c$ assuming that the carrier-to-noise ratio is high. The envelope of $f(t)$ is of no interest to us, because the limiter removes all variations in the envelope. Thus, the composite signal at the frequency discriminator input consists of a small index phase modulated wave with the modulation derived from the component $n_Q(t)$ of noise, i.e., in-phase quadrature with the carrier. When feedback is applied, the VCO generates a frequency modulated signal that reduces the phase modulation index of the wave in the bandpass filter output, i.e., the quadrature component $n_Q(t)$ of noise. Thus, it can be seen that as long as the carrier-to-noise ratio is sufficiently large, the FMFB receiver does not respond to the in-phase noise component $n_I(t)$, but it would demodulate the quadrature noise component $n_Q(t)$ in exactly the same fashion as it would demodulate the signal. Signal and quadrature noise is reduced in the same proportion by the applied feedback, with the result that the baseband signal-to-noise ratio is independent of the feedback. For large carrier-to-noise ratios, the baseband signal-to-noise ratio of an FMFB receiver is then the same as that of a conventional FM receiver.

The reason for an FMFB receiver to extend the threshold is that, unlike a conventional FM receiver, it uses a very important piece of a priori information, i.e., the amount of frequency deviation of the incoming wave whose rate of change is at the baseband rate. An FMFB demodulator is essentially a tracking filter that can track only the slowly varying frequency of a wideband FM signal, and, consequently, it responds only to a narrowband of noise centred around the instantaneous carrier frequency. The bandwidth of noise to which the FMFB receiver responds is precisely the band of noise that the VCO tracks. The end result is that an FMFB receiver is capable of realizing a threshold extension of the order of 5 to 7 dB, which represents a significant improvement in the design of minimum power FM systems.

The phase locked loop like the FMFB demodulator is also a tracking filter and, as such, the noise bandwidth to which it responds is precisely the band of noise tracked by the VCO. Indeed, the phase locked loop demodulator offers a threshold extension capability with a relatively simple circuit. Unfortunately, the amount of threshold extension is not predictable by any existing theory, and it does depend on signal parameters. The threshold achieved in a typical application with a PLL is a few decibels (of the order of 2 to 3), which is not as good as FMFB demodulator.

SUMMARY

- Noise is defined as any undesirable electrical energy that falls within the passband of the signal.
- There are various sources for noise. Generally, they are classified as external and internal. Most common noise is thermal noise or Johnson noise. Sometimes, it is also called white noise.
- Equivalent circuit for noise source can be represented as a noise voltage generator along with a series resistor. Thermal noise is directly proportional to the absolute temperature in Kelvin and the bandwidth B.
- In communication system, we talk about signal-to-noise ratio rather than the absolute value of noise signal and define noise as a relative term. It is define and as a power ratio and generally expressed in decibels. Mathematically, it is expressed as $S/N = P_s/P_n$, where P_s is the power of the signal and P_n is the power of the noise. The S/N ratio of a cascaded system is given by $10 \log(P_s/P_n)(1/M)$, where M is the total number of links. Figure of merit of a communication system is given by a factor called noise figure, which is defined as the ratio of input signal-to-noise ratio to output signal-to-noise ratio.
- We define noise temperature T_e associated with the available noise power. Equivalent noise temperature in kelvin T_e is given by the expression $(F-1)T$, where F is the noise factor and T is the ambient temperature (290 K). The overall noise temperature for a cascaded system is

$$T_e = T_{e1} + \frac{T_{e2}}{G_1} + \frac{T_{e3}}{G_1 G_2} + \cdots + \frac{T_{en}}{G_1 G_2 \cdots G_{n-1}}$$

- In case of modulation system, noise also acts as modulating signal and results in a modulated wave at the output. Pre-emphasis and de-emphasis circuits are employed in FM since FM performs better in low-frequency component of message signals and PM performs better in high-frequency component. Hence, pre-emphasis and de-emphasis circuits are designed in such a way that they behave like an ordinary frequency modulator-demodulator pair in the low-frequency band of message signal and like a phase modulator-demodulator pair in the high-frequency band of message signals. This system results in a better overall performance compared to each system alone.

- Mathematically, noise is analysed as a periodic signal by applying Fourier series. Superposition theorem is applicable to noise. The output from a narrowband filter is called a narrowband noise. It is defined in terms of in-phase and quadrature components. It is also defined in terms of two other components called envelope and phase.

IMPORTANT FORMULAE

- Noise power $P_n = KTB$, where P_n = noise power in watts, T = absolute temperature in kelvin, B = bandwidth in hertz, and K = Boltzmann's constant.

- Noise voltage, $E_n = \sqrt{4RKTB}$

- Mean square noise voltage is $E_n^2 = 4RKTB$

- Mean square noise current is $I_n^2 = 4KTGB$

- Noise voltage for resistors connected in series, $E_n = \sqrt{E_{n1}^2 + E_{n2}^2 + E_{n3}^2 + \cdots}$

- Noise current for resistors connected in parallel, $I_n = \sqrt{I_{n1}^2 + I_{n2}^2 + I_{n3}^2 + \cdots}$

- Available power spectral density in W/HZ or joule is $S_p(f) = KT$

- Special power density in terms of mean square voltage is $S_v(f) = 4RKT$

- Noise equivalent bandwidth, B_n = area under $|H(f)|^2$ curve

- Noise equivalent bandwidth for a low-pass filter, $B_n = \dfrac{1}{4RC}$

- Signal-to-noise ratio in dB, $\dfrac{S}{N} = 10 \log \dfrac{P_s}{P_n} = 20 \log \dfrac{V_s}{V_n}$

- Signal-to-noise ratio for a cascaded system at the Mth link,

$$\left(\frac{S}{N}\right)_{out} = \left(\frac{S}{N}\right)_1 dB - M \; dB$$

- $NF = \dfrac{\left(\dfrac{S}{N}\right)_{in}}{\left(\dfrac{S}{N}\right)_{out}}$

- Noise figure in dB, $NF = 10 \log \dfrac{\text{Input signal to noise power ratio}}{\text{Output signal to noise power ratio}}$

- Noise factor, $F = \dfrac{\text{Input signal to noise power ratio}}{\text{Output signal to noise power ratio}}$

- $NF = 10 \log F$

- Input noise of amplifier in terms of F, $P_{ni} = FKTB_n$

- Noise factor of an amplifier in cascade:

$$F = F_1 + \frac{F_2 - 1}{G_1} + \frac{F_3 - 1}{G_1 G_2} + \cdots + \frac{F_N - 1}{G_1 G_2 \cdots G_{N-1}}$$

This is known as *Friss' formula*.

- Equivalent noise temperature in kelvin, $T_e = (F - 1)T$
  where T is the ambient temperature in Kelvin and F is the noise factor.

- Friss' formula can be rearranged to give the expression for overall noise temperature T_e for a cascaded system as

$$T_e = T_{e1} + \frac{T_{e2}}{G_1} + \frac{T_{e3}}{G_1 G_2} + \cdots + \frac{T_{en}}{G_1 G_2 \cdots G_{n-1}}$$

where $T_{e1}, T_{e2}, \cdots$ are noise temperatures of individual stages

- Signal-to-noise ratio for an AM system, $\left(\dfrac{S}{N}\right)_{ref} = \dfrac{E_{rms}^2 \left(1 + \dfrac{m^2}{2}\right)}{4 R_s K T_s B_B}$

- Figure of merit for an AM system, $F_m = \dfrac{m^2}{2 + m^2} \dfrac{R_s}{R_{out}}$

- For a DSBSC system: $\left(\dfrac{S}{N}\right)_{out} = \dfrac{E_c^2}{16 R_{out} K T_s B_B}$;

  Figure of merit, $F_m = \dfrac{R_s}{R_{out}}$

- For an SSB system, $\left(\dfrac{S}{N}\right)_{ref} = \dfrac{P_R}{P_{N\,ref}} = \dfrac{E_r^2}{8 R_s K T_s B_B}$;

  Figure of merit, $F_m = \dfrac{R_s}{R_{out}}$

- Signal-to-noise ratio for an angle modulation system:

$$\left(\frac{S}{N}\right)_{out} = \beta_p^2 P_{mn} \left(\frac{S}{N}\right)_b \quad \text{for PM}$$

$$= 3\beta_f^2 \left(\frac{S}{N}\right)_b \quad \text{for FM}$$

ADDITIONAL EXAMPLES

1. An amplifier is operating at 17°C with a bandwidth of 15 kHz. Find
   (a) thermal noise power in watts and dBm and
   (b) RMS noise voltage for a 60 Ω internal resistance and a 60Ω load resistance.

Solution

(a) Thermal noise power, $N = KTB = 1.38(1.38 \times 10^{-23}) \times (290) \times (15 \times 10^3)$

$= 6 \times 10^{-17}$ W

Thermal noise in dBm, $N_{\mathrm{dBm}} = 10\log\dfrac{(6 \times 10^{-17})}{0.001} = -132$ dBm

(b) RMS noise voltage, $V_{\mathrm{rms}} = \sqrt{4RKTB} = \sqrt{4 \times 60 \times (6 \times 10^{-17})} = 0.12\,\mu\text{V}$

2. For an amplifier:

Input signal power = 1.5×10^{-9} W

Input noise power = 1.5×10^{-18} W

Power gain = 10, 00,000

Internal noise = 4×10^{-12} W

Find

(a) input S/N ratio in dB,

(b) output S/N ratio in dB, and

(c) noise factor and noise figure.

Solution

(a) Input S/N ratio in dB = $10\log\dfrac{1.5 \times 10^{-9}}{1.5 \times 10^{-18}} = 10\log 10^9 = 90$ dB

(b) To find out the output S/N ratio, we must find the output noise power. This is the sum of the internal noise and the amplified input noise, i.e.,

$$N_{\mathrm{out}} = 1,000,000(1.5 \times 10^{-18}) + 4 \times 10^{-12} = 5.5 \times 10^{-12} \text{ W}$$

The output signal power is the product of input signal power and the amplifier power gain, i.e.,

$$P_{\mathrm{out}} = 1,000,000 \times (1.5 \times 10^{-9}) = 1.5 \times 10^{-3} \text{ W}$$

$$\left(\frac{S}{N}\right)_{\mathrm{out}} = \frac{1.5 \times 10^{-3}}{5.5 \times 10^{-12}} = 273,000,000$$

$$10\log(273,000,000) = 84.4\,\text{dB}$$

(c) The noise factor, $F = \dfrac{10^9}{273 \times 10^6} = 3.66$

The noise figure, $NF = 10\log 3.66 = 5.63$ dB

3. The equivalent noise resistance for a system is 200 Ω and the equivalent shot noise current is 5 μA. The system is fed from a 100 Ω, 10 μV RMS sinusoidal signal source. Find out the individual noise voltages at the input and the input S/N ratio. The noise bandwidth is 8 MHz. Assume 290 K to be the room temperature.

Solution

The shot noise current, $I_n = \sqrt{2q_e I_{eq} B_n} = \sqrt{2(1.6 \times 10^{-19})(5 \times 10^{-6})(8 \times 10^6)}$]

$$= 3.58 \text{ nA}$$

Shot noise voltage $= I_n R_s = (3.58 \times 10^{-9})100 = 0.358 \,\mu\text{V}$

Thermal noise voltage from the source,

$$V_{ns} = \sqrt{4 R_s k T_o B_n} = \sqrt{4 \times 100 \times (1.38 \times 10^{-23}) \times 290 \times (8 \times 10^6)} = 3.58 \,\mu\text{V}$$

The noise voltage generated by equivalent noise resistance,

$$V_{ne} = \sqrt{4 R_n k T_o B_n} = \sqrt{4 \times 200 \times (1.38 \times 10^{-23}) \times 290 \times (8 \times 10^6)} = 5.66 \,\mu\text{V}$$

The total noise voltage at the input of the amplifier, $V_n = \sqrt{0.358^2 + 3.58^2 + 5.06^2}$

$$= 6.10 \,\mu\text{V}$$

The signal-to-noise ratio in dB is

$$\frac{S}{N} = 20 \log \frac{V_s}{V_n} = 20 \log \frac{10 \,\mu\text{V}}{6.10 \,\mu\text{V}} = 4.4 \text{dB}$$

4. Four stages of amplifiers are connected in cascade. Each stage has the same S/N ratio. If S/N ratio is 55 dB, calculate the output S/N ratio of the entire system.

Solution
We have

$$\left(\frac{S}{N}\right)_{out} = 55 - 10 \log 4 = 49.02 \text{ dB}$$

5. The noise figure of an amplifier is 5 dB and its input S/N ratio is 55 dB. Find the output S/N ratio.

Solution
We have

$$\left(\frac{S}{N}\right)_{out} = \left(\frac{S}{N}\right)_{in} - F$$

$$= 55 - 5 = 50 \text{ dB}$$

6. An amplifier has a noise figure of 12 dB. Calculate the equivalent amplifier input noise for a bandwidth of 5 MHz.

Solution

The noise figure of 12 dB is a power ratio of approximately 16:1. Hence

$$P_{no} = (F-1) \, k T_o B_n = (16 - 1)(4 \times 10^{-21})(5 \times 10^6) = 0.3 \times 10^{-12} \text{ W}$$

7. A mixer stage has a noise figure of 25 dB and a stage before it is an amplifier with a noise figure of 7 dB and an available power gain of 15 dB. Find out the overall noise figure referred to input.

Solution

Convert each decibel values to the equivalent power ratios:

$$F_2 = 25 \text{ dB} = 316.22{:}1 \text{ power ratio}$$

$$F_1 = 7 \text{ dB} = 5.01{:}1 \text{ power ratio}$$

$$G_1 = 15 \text{ dB} = 31.62{:}1 \text{ power ratio}$$

$$F = F_1 + \frac{F_2 - 1}{G_1} = 5.01 + \frac{316.22 - 1}{31.62} = 14.97$$

This is the overall noise factor.

Overall noise figure in dB = 10 log 14.97 = 11.75 dB

8. A receiver has a noise figure of 15 dB and a low-noise amplifier with a gain of 60 dB and a noise temperature of 80 K feed it. Determine the noise temperature of the receiver and the overall noise temperature of the receiving system.

Solution

The noise figure of 15 dB represents a power ratio of 31.62:1. Therefore,

$$T_{em} = (31.62 - 1) \times 290 = 8880.6 \text{ K}$$

The 60 dB gain gives the power ratio of 10^6. This gives

$$T_e = 80 + \frac{8880.6}{10^6} = 88.88 \text{ K}$$

9. An avalanche diode is used in the measurement of noise temperature, the *ENR* being 15 dB. The measured *Y* factor is 8 dB. Find out the equivalent noise temperature of the system under test.

Solution

The excess noise ratio as a power ratio is $ENR = 10^{1.5} = 31.62$

The *Y* factor expressed as a power ratio is $Y = 10^{0.8} = 6.30$

The hot temperature is $T_h = T_o(ENR + 1) = 290(31.62+1) = 9459.8$

Then $\qquad T_e = \dfrac{9459.8 - (6.3 \times 290)}{(6.30 - 1)} = 1440 \text{ K}$

REVIEW QUESTIONS

1. What are the various types of noise?
2. List the various types of noise in a modulation system realized with transistors.
3. Briefly comment on the importance of *S/N* ratio in a communication system.
4. Plot the noise distribution curve in the case of AM and FM systems.
5. What is meant by noise figure of a network?
6. What is the formula for the signal-to-noise ratio for FM?
7. Describe electrical noise.
8. Which types of noise are considered external?
9. What is the relationship between thermal noise power, bandwidth, and temperature?

10. Describe white noise.
11. What do you mean by signal-to-noise power ratio?
12. Define equivalent noise temperature.
13. What is the difference between noise factor and noise figure?
14. Find the noise bandwidth of an *RC* low-pass filter.
15. Explain the types, causes, and effects of the various forms of noise that may be created within a receiver of an amplifier.
16. When the value of the resistor creating thermal noise is doubled, what happens to the noise power?
17. What is Friss' formula?
18. How are the noise temperature and noise factor measured?
19. Describe narrow bandpass noise.
20. Give the noise model for an AM system.
21. Derive an expression for the *S/N* ratio for an SSB system.
22. Explain the various types of noise generated by a transistor.
23. How does a pre-emphasis circuit provide noise immunity?
24. How does the random noise affect the output of an FM receiver with an amplitude limiter?
25. Find an expression for the noise figure of cascaded amplifiers. What do you conclude from this?

PROBLEMS

1. Calculate the thermal noise power for an amplifier with a bandwidth of 300 Hz and a temperature of 20°C.
2. Determine the bandwidth necessary to produce 6×10^{-15} watts of thermal noise power at a temperature of 20°C.
3. Determine the overall noise figure and noise factor for four cascaded amplifiers with the following parameters:
   $A_1 = 10$ dB, $A_2 = 8$ dB, $A_3 = 10$ dB, and $A_4 = 20$ dB
   $NF1 = 15$ dB, $NF2 = 10$ dB, $NF3 = 6$ dB, and $NF4 = 8$ dB
4. An amplifier has a bandwidth of 30 kHz with a total noise power of 3×10^{-16}. Find out the total noise power if the bandwidth increases to 50 kHz. Determine the noise power in watts for an amplifier operating at a temperature of 300°C with a 2 MHz bandwidth.
5. Find out the noise figure for an equivalent noise temperature of 1220 K. Use the reference temperature as 290 K.
6. Find the equivalent noise temperature for a noise figure of 15 dB.
7. Find out the noise figure for an amplifier with an input *S/N* ratio of 80 and output *S/N* ratio of 30.
8. Calculate the output *S/N* ratio for an amplifier with an input *S/N* ratio of 30 dB and noise figure of 5 dB.
9. An amplifier operating over the frequency range of 300 to 600 kHz has a 300 kΩ input resistor. What is the RMS noise voltage at the input to this amplifier if the ambient temperature is 18°C?

10. A noiseless amplifier having a gain of 70 and a bandwidth of 30 kHz amplifies the noise output of a resistor. The output of the amplifier is 1 mV RMS. Now the bandwidth of the amplifier is reduced by 10 kHz, its gain remaining constant. What is the output of the amplifier in this case? If the resistor is operated at 65°C, what is the value of the resistor?

11. A receiver has a noise figure of 15 dB. It is fed by a low-noise amplifier, which has a gain of 40 dB and a noise temperature of 25°C. Find out the noise temperature of the receiver and the overall noise temperature of the receiving system.

12. The input resistance of an RF amplifier is 1.5 kΩ for a receiver. It has the gain of 20, equivalent shot noise resistance of 2 kΩ, equivalent shot noise current of 5 μA, and load resistance of 100 kΩ. Calculate the equivalent noise voltage at the input to this amplifier if the bandwidth is 2 MHz and the temperature is 30°C.

13. Define noise factor in terms of input and output signal-to-noise ratio of a network. The noise factor of the amplifier is 4:1. Calculate the output S/N ratio in dB if the input S/N ratio is 40 dB.

14. The *ENR* for an avalanche diode is 10 dB and the cold temperature is 320 K. Determine the hot temperature.

15. An avalanche diode is used for noise measurement. The *ENR* was 15.2 dB. The noise power output with the diode on was 40 dBm and with the diode off was 30 dBm. Find out the noise temperature of the device under test.

16. A DSBSC receiving system has an equivalent noise temperature of 900 K, which includes the antenna noise. It has a baseband bandwidth of 5 kHz. The received signal is 1 μV RMS across 60 Ω. Find out (a) the output S/N ratio, (b) the reference S/N ratio, and (c) the figure of merit. $R_S = R_{out}$ can be assumed.

17. What is the required receiver power in an FM system with $\beta = 5$, bandwidth = 12 kHz, and $N_o = 10^{-15}$ W/Hz? The power of the normalized message signal is assumed to be 0.2 watt and the required SNR after demodulation is 70 dB.

18. Design an FM system that achieves an SNR at the receiver equal to 35 dB and requires the minimum amount of transmitter power. The bandwidth of the channel is 120 kHz and the message bandwidth is 15 kHz. The average to peak power ratio for the message $P_{mn} = P_M/(\max|(t)|)^2$ is 1/2 and the one-sided noise power spectral density is $N_o = 10^{-9}$ W/Hz. What is the required transmitter power if the signal is attenuated by 35 dB in transmission through the channel?

19. In an FM broadcasting, BW = 15 kHz, $f_m = 3100$ Hz, and $\beta = 5$. Assuming that the average to peak power ratio of the message signal is 0.6, find the improvement in the output SNR of FM when we use pre-emphasis and de-emphasis filtering rather than a baseband system.

20. A 3 kHz message signal modulates a carrier of 1 MHz. SSB transmission is employed. The received signal strength is 4 mW. The accompanied noise has a uniform power spectral density of 10^{-9} W/Hz. This is multiplied by a local oscillator of frequency 1 MHz and followed by a filter with a cut-off frequency of 3 kHz to recover the message signal. Calculate the signal and noise energy at the output of the filter and find the S/N ratio here. How does the S/N ratio change if the bandwidth of the message signal is reduced by 30%?

21. A message signal of 3 kHz bandwidth is to be transmitted as an SSB signal through a channel that has 40 dB loss and white noise of spectral density 10^{-10} W/Hz. Calculate the minimum required transmitter power if the required S/N ratio at the output is to be more than 50 dB.

22. An FM system has $\beta = 8$, bandwidth = 45 kHz, and the noise at the output $= 10^{-13}$ W/Hz. The power due to the message signal is 0.15 W. What is the required receiver power if the S/N ratio at the output of the demodulator is 40 dB?

23. Calculate the output S/N ratio of an FM demodulator if the input signal strength is 0.8 W, maximum frequency deviation is 40 kHz, and the message signal is cut off at 12 kHz. The received white Gaussian noise power spectral density is 10^{-8} W/Hz and the average power of message signal is 0.15 W. Find also the required transmitter power if the channel has 30 dB loss and the required output S/N ratio should be greater than 25 dB.

24. The pre-emphasis circuit used in an FM system employs an RC first-order filter with $R = 1.2$ kΩ and $C = 0.1$ μF. Calculate the gain in dB if the message has a bandwidth of 10 kHz. What will be the improvement in gain if the message bandwidth is increased to 25 kHz.

25. For an angle modulated carrier, $f_c(t) = 5 \cos(300\pi \times 10^6 t)$ with 50 kHz deviation due to the message signal and a single frequency interference signal $f_n(t) = 0.5 \cos(299.950 \times 10^6 t)$, find
    (a) the frequency of the demodulated interference signal,
    (b) the peak phase and frequency deviation due to the interfering signal, and
    (c) the voltage S/N ratio improvement at the output of the demodulator.

Answers to problems

1. 1.123×10^{18} W
2. 1.438 MHz
3. 15.13 dB, 32.648
4. 5×10^{-16} W 8.28×10^{-15} W
5. 1.52 dB
6. 9118.8° K
7. 4.24 dB
8. 24.94 dB
9. 4.35 μV
10. 812 μV, 42.025 kΩ
11. 911.8 K, 299 K

12. 11.09 uv
13. 33.9 dB
14. 3520 K
15. 777.5 K
16. 135.4 dB, 135.4 dB, and 1
17. 8 μW
18. 11.45 dB (13.96 W)
19. 26.79 dB
20. 1mw, 0.75 μW, 31.2 dB, and 36.47 dB
21. 600 W
22. 1.56 μW
23. 31.54 dB, 177.1W
24. 6.64 dB, 10.16 dB
25. 25 kHz, 0.1 rad, 2.5 kHz, 3dB

MATLAB EXAMPLES

1. ```
%program to represent power spectrum of random noise%
N=2096;
n=rand(N,1);
fn=fft(n);
Gn=(abs(fn).^2)/N;
w=linspace(0,pi,N);
subplot(121);
plot(w,Gn);
title('power spectrum random noise(0,1), no dc clip');
xlabel('angular frequency in radian');
subplot(122); plot(w,Gn); axis([0 3.5 0 2]);
title('power spectrum random noise(0,1), dc clip');
xlabel('angular frequency in radian');
```

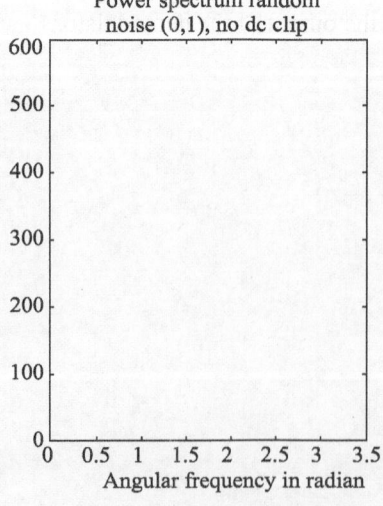

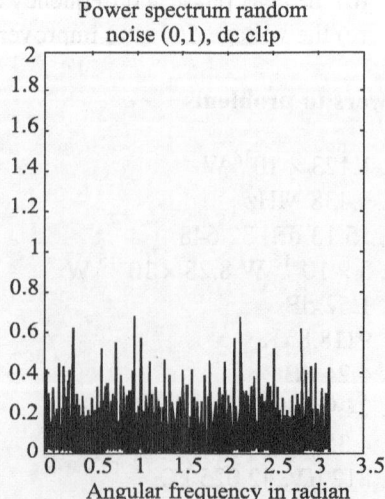
```

2. %program for power spectrum white gaussian noise across different load%

```
N=4096;
n=wgn(N,1,10)
fn=fft(n);
Gn=(abs(fn).^2)/N;
w=linspace(0,pi,N);
subplot(121);
plot(w,Gn);
title('power spectrum real noise, 1 ohm load');
xlabel('angular frequency in radian');
n=wgn(N,1,10,50,'complex');
fn=fft(n);
Gn=(abs(fn).^2)/N;
subplot(122);
plot(w,Gn);
title('power spectrum complex noise, 50 ohm load');
xlabel('angular frequency in radian');
```

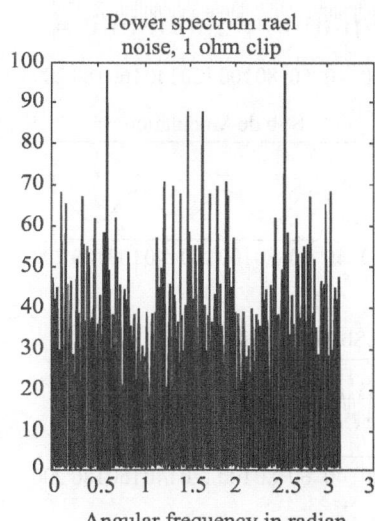

Power spectrum rael noise, 1 ohm clip

Angular frequency in radian

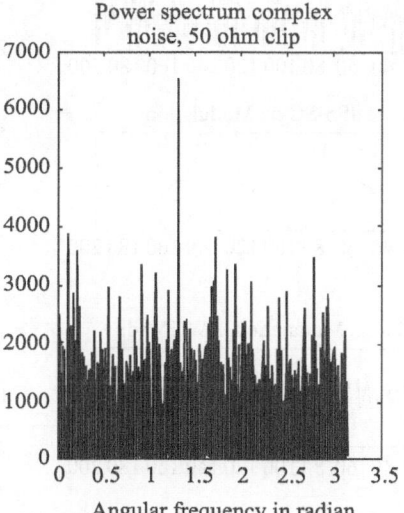

Power spectrum complex noise, 50 ohm clip

Angular frequency in radian

3. %effect of AWGN noise in DSB-SC and SSB%

```
fc=2000;
fs=10000;
f=200;
t=0:1/fs:((4/f)-(1/fs));
x=cos(2*pi*f*t);
y_dsb_sc=modulate(x,fc,fs,'amdsb-sc');
y_ssb=modulate(x,fc,fs,'amssb');
y_dsb_sc_n=awgn(y_dsb_sc,20,'measured')
subplot(421);plot(y_dsb_sc_n); title('DSB-SC modulation,
AWGN  20 db');
y_ssb_n=awgn(y_ssb,20,'measured');
subplot(422); plot(y_ssb_n); title('ssb modulation');
x1=demod(y_dsb_sc_n,fc,fs,'amdsb-sc');
```

```
subplot(423); plot(x1); title('dsb-sc demodulation');
x2=demod(y_ssb,fc,fs,'amssb');
subplot(424); plot(x2); title('ssb demodulation');
y_dsb_sc_n=awgn(y_dsb_sc,5,'measured');
subplot(425);plot(y_dsb_sc_n); title('dsb-sc modulation,AWGN
5 db');
y_ssb_n=awgn(y_ssb,5,'measured');
subplot(426); plot(y_ssb_n); title('ssb modulation,AWGN 5
db');
x1=demod(y_dsb_sc_n,fc,fs,'amdsb-sc');
subplot(427); plot(x1); title('dsb-sc demodulation,AWGN
  5db');
x2=demod(y_ssb,fc,fs,'amssb');
subplot(428); plot(x2); title('ssb demodulation,AWGN 5
db');
```

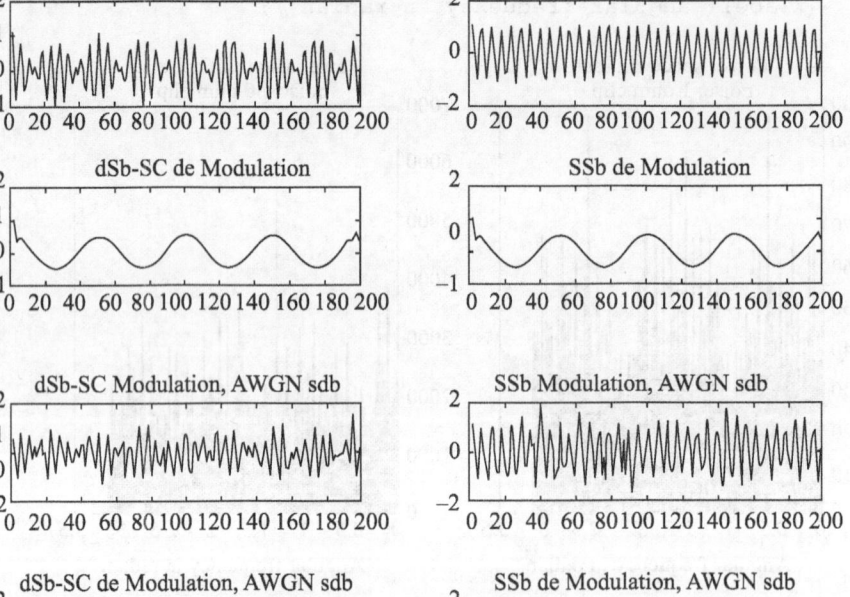

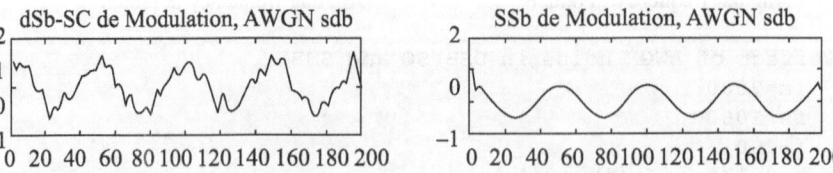

7

Introduction to Digital Communication

LEARNING OBJECTIVES

This chapter will enable the students to
- know what is digital communication
- understand how signal amplitude, frequency, and phase of a carrier are modulated and how they are classified as FSK, ASK, PSK, and QAM
- know the difference between bit rate and baud rate
- appreciate the advantages and disadvantages of various methods of digital modulation

7.1 INTRODUCTION

The term *digital communication* covers a broad range of communication techniques, including digital transmission and digital modulation. In digital transmission, pulses are transferred directly between two or more points in a communication system. It requires a physical connection between the source and destination like a pair of metallic wire, coaxial cable, or an optical fibre. Digital transmission can have analog or digital inputs. However, the analog input should be converted to digital form prior to transmission. Digital modulation is transmittal of digitally modulated analog signal between two or more points. In this type of modulation, the modulated and demodulated signals are digital pulses. However, they are carried through the system on an analog signal called the carrier signal. Digital modulation provides more information capacity, compatibility with digital data services, higher data security, better quality communication, and quicker system availability. Digital modulation schemes have greater capacity to convey large amount of information than analog modulation scheme. We have to face the following constraints while designing a communication system:

(a) Available bandwidth
(b) Permissible power
(c) Inherent noise level of the system

In case of bandwidth, efficiency describes the ability of modulation method to accommodate data within the limited bandwidth. Power efficiency describes the ability of the system to reliably send information at lowest practical power level. As far as the noise is concerned, the system should have a very good noise figure. The parameters to be optimized depend on the demands of a particular system. In the case of digital terrestrial microwave systems, the prime concern is good bandwidth efficiency with low bit error rate. Power is secondary since they have plenty of power available. But in the case of hand-held portable devices, power is the prime concern since they are battery operated. Here we sacrifice some bandwidth efficiency to get cost and power efficiency.

The block diagram of a typical digital communication system is shown in Fig. 7.1. The source is converted to digital form by the analog-to-digital converter. It is passed through a source encoder and then a channel encoder before being modulated. It is then transmitted through a channel. At the receiver, it is demodulated first and then given to the channel decoder. The output of the source decoder gives the digital output. It is then given to a digital-to-analog converter to get back the original signal.

Over the past few years, a major transition has occurred from simple analog modulations like AM, FM, and PM to new digital modulation techniques. Examples of digital modulation include

- digital amplitude modulation
- *I/Q* modulation
- frequency shift keying
- phase shift keying
- minimum shift keying
- quadrature amplitude modulation

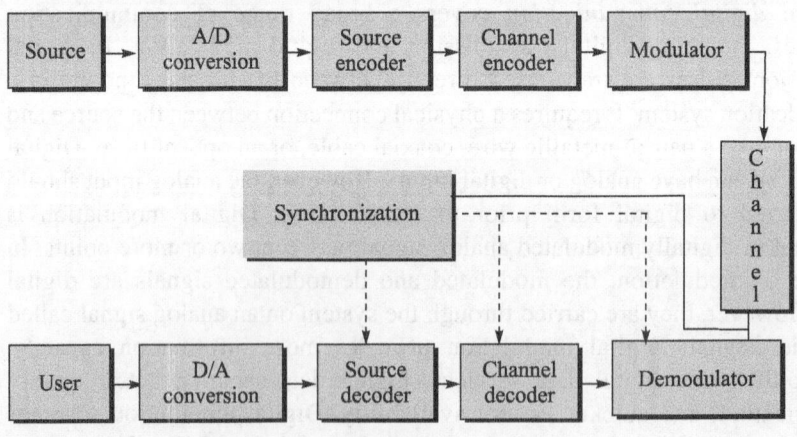

Fig. 7.1 Typical digital communication system

Other more complex systems involve multiplexing. Two principal types of multiplexing or multiple access are TDMA (time division multiple access) and CDMA (code division multiple access). These are two different ways to add diversity to signals allowing different signals to be separated from one another.

Figure 7.2 shows these transitions from analog modulation to digital modulation and to TDMA and CDMA.

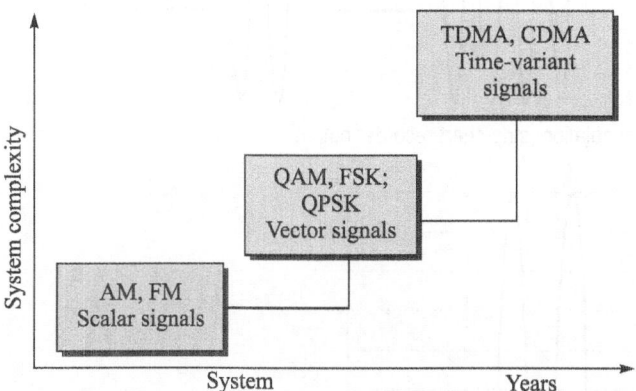

Fig. 7.2 Transition: types of modulation

7.2 DIGITAL AMPLITUDE MODULATION

Digital amplitude modulation is the simplest of digital modulation. It is also called *on-off keying* or *amplitude shift keying* (ASK). This is same as DSBFC except that the modulating signal in this case is a square wave, which is nothing but a binary signal. Mathematically, we can represent this signal by

$$f_{am}(t) = \{1 + e_m(t)\}\left[\frac{E_c}{2}(\cos\omega_c t)\right] \tag{7.1}$$

where $f_{am}(t)$ = digital amplitude modulated wave

$E_c/2$ = unmodulated carrier amplitude

$e_m(t)$ = modulating binary signal in volts

$\omega_c(t)$ = carrier frequency in Hz

since the modulating signal is a digital signal and has either *one* or *minus one* value.

Taking the value as 1, i.e., $e_m(t) = 1$,

$$f_{am}(t) = (1+1)\left[\frac{E_c}{2}\cos\omega_c t\right] = E_c \cos\omega_c t \tag{7.2}$$

and for $e_m(t) = -1$,

$$f_{am}(t) = (1-1)\left[\frac{E_c}{2}\cos\omega_c t\right] = 0 \tag{7.3}$$

Thus, for 100% modulation, $f_{am}(t)$ is $E_c \cos \omega_c t$ or zero, i.e., the carrier is *on* or *off*, and hence the name *on-off* keying. Figure 7.3 shows the waveform of a typical on-off keying. The dotted waveform is the input binary signal and the continuous waveform is the carrier being made on and off. This waveform is the strip chart recorder output. Figure 7.4 shows the waveform for ASK.

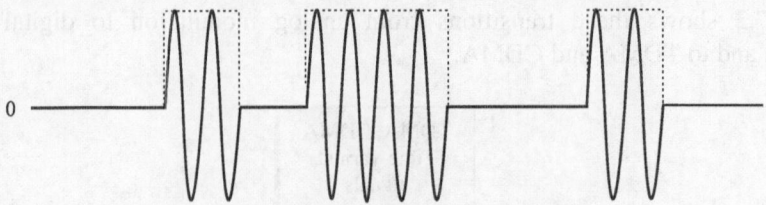

Fig. 7.3 Digital amplitude modulation: strip chart recorder output

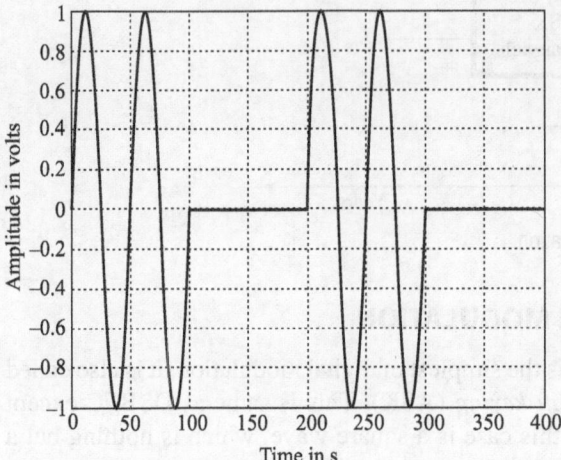

Fig. 7.4 ASK waveform

Example 7.1 Find the baud and the minimum bandwidth necessary to pass a 15 kbps binary signal using amplitude shift keying.

Solution
For ASK, $N = 1$ and hence

$$BW = \frac{15,000}{1} = 15,000$$

$$Baud = \frac{15,000}{1} = 15,000$$

7.3 I/Q MODULATION

In earlier chapters, it was seen that there are only three characteristics of a signal that can be changed over time. They are amplitude, phase, and frequency. But phase and frequency are just different ways to view or measure the same signal change. Amplitude and phase can be modulated simultaneously and separately, but it is difficult to generate and detect such signals. Figure 7.5 shows the signal for amplitude, phase, and both amplitude and phase modulation. It can be seen that the waveform of both amplitude and phase modulation is quite complex. Instead the signal is separated into another set of independent components: the in-phase component I and quadrature component Q. These components are orthogonal and do not interfere with each other. Hence, in digital communication, modulation is often expressed in terms of I and Q.

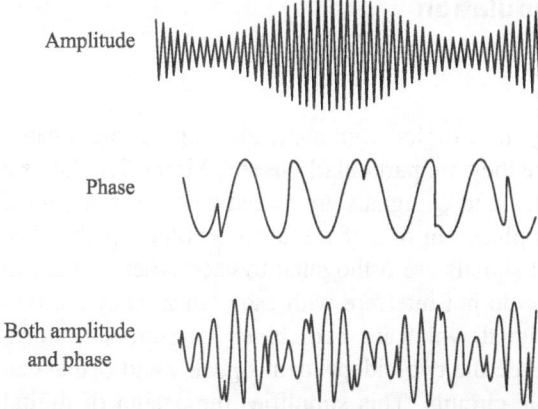

Fig. 7.5   Signal waveform: AM, PM, and both amplitude and phase modulation

7.3.1 The Concept of *I* and *Q* Channels

Signals can be viewed in two forms—rectangular and polar. The signal can be represented in rectangular form by its projection on the *x*- and *y*-axes. In polar form, it is represented by its magnitude and phase angle. The *x*- and *y*-axes are called *I* (the in-phase component) and *Q* (the quadrature component), respectively. These are two canonical ways of representing signal. Figure 7.6 shows these representations. The coefficients s_{11} and s_{12} in Fig. 7.6(a) represent the amplitude of *I* signal and *Q* signal, respectively. The amplitude when plotted on the *x*- and *y*-axes, respectively, as shown in Fig. 7.6(b), gives the signal vector. The angle that signal vector makes with the *x*-axis is its phase. The magnitude of the signal is

$$S = \sqrt{I^2 + Q^2} \tag{7.4}$$

and the phase of the signal is

$$\phi = \tan^{-1}\frac{Q}{I} \tag{7.5}$$

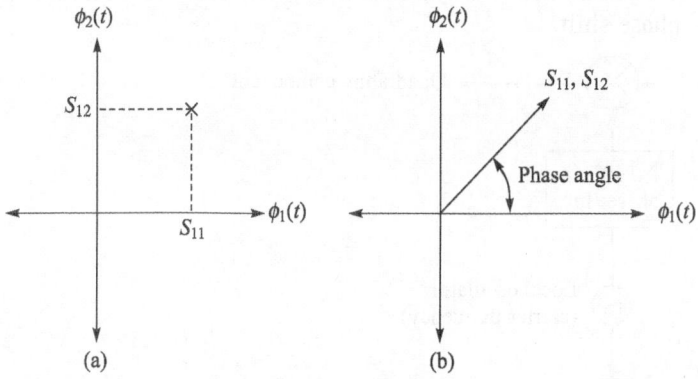

Fig. 7.6   Signal vector plotted on signal scale: (a) *I* and *Q* projections and (b) polar form

7.3.2 Application of I/Q Modulation

First we will discuss I and Q radio transmitter and receiver.

I and Q radio transmitter

The I/Q diagrams mirror the way most digital communication signals are created using an I/Q modulator and hence they are particularly useful. Figure 7.7 shows a typical I and Q radio transmitter. I and Q signals are mixed with the same local oscillator. A 90° phase shifter is placed in one of the local oscillator paths. The original signal and phase shifted signals are orthogonal to each other and are in quadrature. Signals in quadrature do not interfere with each other. They are two independent components of the signal, which are recombined in a summer to form a composite output signal. These are the two independent signals I and Q that can be sent and received with simple circuits. This simplifies the design of digital radios. The main advantage of I/Q modulation is the symmetric ease of combining independent signal components into a single composite signal and later splitting such a composite signal into its independent component parts.

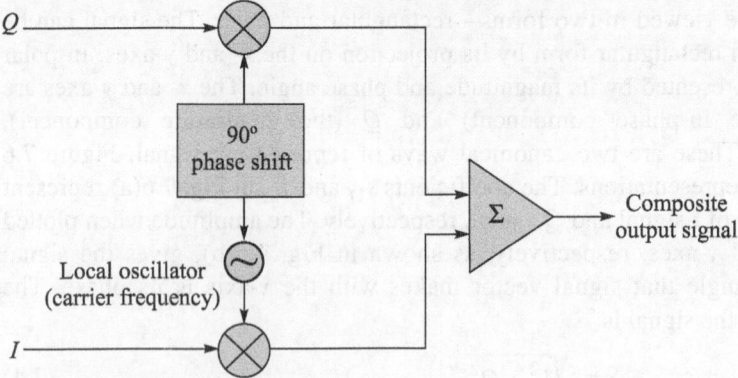

Fig. 7.7 Radio transmitter using I and Q

I and Q radio receiver

The combined I and Q signal from the transmitter arrives at the receiver. Figure 7.8 shows a typical I/Q receiver. The input signal is mixed with the local oscillator signal at the carrier frequency in two ways. One with an arbitrary zero phase and the other with 90° phase shift.

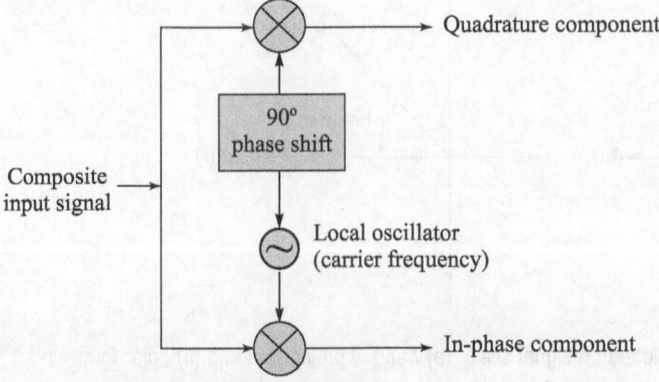

Fig. 7.8 Radio receiver using I and Q

The composite input signal is thus broken into an in-phase I component and a quadrature Q component. Since these components are independent and orthogonal, they can be changed without affecting each other. We cannot normally plot information in a polar format and reinterpret as rectangular values without doing a polar to rectangular conversion. This conversion is exactly what is done by the in-phase and quadrature mixing processes in a digital radio. A local oscillator, a phase shifter, and two mixers can perform the conversion accurately and efficiently.

7.3.3 Need for using *I* and *Q*

What is the need to use I and Q? The answer lies in the fact that digital modulation is easy to accomplish with I/Q modulators. Most digital modulation maps the data to a number of discrete points on the I/Q plane. These are known as constellation points. As the signal moves from one point to another, simultaneous amplitude and phase modulation takes place. This cannot be achieved with an amplitude modulator and phase modulator without getting involved in complexity and difficulty. It is also impossible with a conventional phase modulator. In fact, the signal may circle the origin in one direction forever necessitating infinite phase shifting capability. But simultaneous amplitude and phase modulation is easy with an I/Q modulator. The I and Q control signals are bounded, but infinite phase wrap is possible by properly phasing the I and Q signals.

7.4 SOME IMPORTANT TERMS

7.4.1 Information Capacity, Bits, and Bit Rate

Before going into details of various types of digital modulation, brief explanation of information capacity, bits, and bit rate is required. Information theory is the study of efficient use of bandwidth to transmit information in a communication system. It helps to determine the information capacity of a data communication system. *Information capacity* is a measure of how much information can be propagated through a communication system as a function of bandwidth and transmission time. It also represents the number of independent symbols that can be carried through a system in a given unit of time. Binary digit or bit is the most basic digital symbol used to represent information. Hence, bit rate is a convenient method of expressing the information capacity of a system. Bit rate is simply the number of bits per second transmitted and is expressed as *bits per second* (bps). In 1928, Hartley of Bell Telephone Laboratories developed a useful relationship between information, capacity, bandwidth, and time, which is given below:

$$I \propto B \propto t \qquad\qquad (7.6)$$

where

I = information capacity expressed in bits per second

B = bandwidth in Hz

t = transmission time in seconds

From Eq. (7.6), it can be seen that information capacity is directly proportional to bandwidth and transmission time. If any one of them or both change, a directly proportional change occurs in information capacity. In 1948, Shannon, also of Bell Telephone Laboratories, published a paper relating information capacity of a communication channel to bandwidth and signal-to-noise ratio. The higher the signal-to-noise ratio, the better the performance and higher the information capacity. Mathematically stated, the Shannon limit for information capacity is

$$I = B \log_2\left(1 + \frac{S}{N}\right) = 3.32\, B \log_{10}\left(1 + \frac{S}{N}\right) \tag{7.7}$$

where

I = information capacity expressed in bits per second

B = bandwidth in Hz

$\dfrac{S}{N}$ = signal-to-noise power ratio

7.4.2 M-ary Encoding

This term is derived from the word *binary*. It represents a digit that corresponds to the number of conditions, levels, or combinations possible for a given number of binary variables. This helps to encode a level higher than binary where there are more than two conditions possible. For example, a digital signal with four possible conditions, viz. voltage level, phase, frequency, and power, is an M-ary system where $M = 4$. If there are sixteen possible conditions, then $M = 16$. The number of bits necessary to produce a given number of conditions is expressed as

$$N = \log_2 M \tag{7.8}$$

where

N = number of bits necessary

M = number of conditions, levels, or combination possible
with N bits

The equation can also be written as

$$2^N = M \tag{7.9}$$

7.4.3 Baud and Minimum Bandwidth

Baud rate is often confused with bit rate. Baud like bit rate is a rate of change. However, baud refers to the rate of change of the signal on the transmission medium after encoding and modulation. Bit rate refers to the change of a digital information signal, which is usually binary. Hence, baud rate can be at the most equal to the bit rate but not more. As the Nyquist rate studied in Chapter 5 determines the minimum bandwidth or the minimum Nyquist frequency, which is

$$f_b = 2B \tag{7.10}$$

where f_b is the bit rate in bps and B is the ideal Nyquist bandwidth. However, the actual bandwidth required may depend on several factors, including type of encoding, modulation used, system noise, and required bit error rate. Hence, for a given bandwidth B, the highest bit rate is $2B$. However, if more than two levels are used for signalling, then more than one bit may be transmitted at a time and it is possible to have bit rate that exceeds $2B$. For multilevel signalling, the Nyquist relation for channel capacity is

$$f_b = B \log_2 M \tag{7.11}$$

where

f_b = channel capacity in bps

B = minimum Nyquist bandwidth in Hz

M = the number of discrete signal or voltage levels

From Eq. (7.11), the minimum bandwidth necessary to pass M-ary digitally modulated carrier is given by

$$B = \left(\frac{f_b}{\log_2 M}\right) \tag{7.12}$$

In the above equation, N is substituted for $\log_2 M$ to get

$$B = \frac{f_b}{N} \tag{7.13}$$

where N is the number of bits encoded into each signalling element. If information bits are encoded and then converted to signals with more than two levels, then transmission rate in excess of $2B$ is possible. Also since baud is encoded rate of change and equals the bit rate divided by the number of bits encoded into one signalling element. Thus,

$$\text{baud} = \frac{f_b}{N} \tag{7.14}$$

From Equations (7.13) and (7.14), it is clear that the baud and the ideal minimum Nyquist bandwidth are one and the same and are given by the bit rate divided by the number of bits encoded.

7.5 FREQUENCY SHIFT KEYING

Frequency shift keying (FSK) is another, relatively simple, low performance type of digital modulation. It is a form of constant amplitude angle modulation similar to standard frequency modulation except that the modulating signal is a binary signal that varies between two discrete voltage levels 0 and 1 rather than a continuously changing analog signal. FSK is sometimes called binary FSK (BFSK) or 2FSK. In binary FSK, a 1 is represented by one frequency and a 0 is represented by another frequency.

FSK is used in many applications, including cordless and paging systems. Some of the cordless systems include DECT (digital enhanced cordless telephone) and CT2 (cordless telephone 2).

The general expression for FSK is

$$e_{\text{FSK}}(t) = E_c \cos[2\pi\{f_c + e_m(t)\Delta f\}t] \tag{7.15}$$

where $e_{\text{FSK}}(t)$ = binary FSK waveform

$\quad E_c$ = peak voltage of analog carrier

$\quad f_c$ = analog carrier frequency in Hz

$\quad \Delta f$ = maximum (peak) shift in carrier frequency in Hz

$\quad e_m(t)$ = modulating signal binary input

From the above equation, it is seen that Δf, the peak shift in carrier frequency is proportional to the amplitude of the modulating binary input signal and the direction of the shift is determined by the polarity. When the modulating frequency is a normalized waveform, we get logic $1 = +1$ V and logic $0 = -1$ V. Hence, for logic 1, input $e_m(t) = +1$ and Eq. (7.15) becomes

$$e_m(t) = E_c \cos[2\pi\{f_c + \Delta f\}t]$$

and for logic input, $e_m(t) = -1$ and Eq. (7.15) becomes

$$e_m(t) = E_c \cos[2\pi\{f_c - \Delta f\}t]$$

Thus, in binary FSK, the carrier frequency is shifted up and down in the frequency domain by binary input signal.

The frequency shifts between two frequencies—mark or logic 1 frequency (f_m) and space or logic 0 frequency (f_s). The mark and space frequencies are separated from the carrier frequency by the peak frequency deviation Δf and from each other by $2\Delta f$. Figure 7.9 shows the strip chart recorder output for FSK.

Freq. vs time

One bit per symbol

Fig. 7.9 Strip chart recorder output for FSK

7.5.1 FSK Baud and Bandwidth

Baud rate for the binary FSK is given by making $N = 1$ in Eq. (7.14). With this substitution, the baud rate becomes

$$\text{baud rate} = \frac{f_b}{1} = f_b \qquad (7.16)$$

This shows that in FSK the baud rate equals the bit rate. The minimum bandwidth for FSK is determined by the relation

$$B = 2(\Delta f + f_b) \qquad (7.17)$$

where

B = minimum Nyquist bandwidth in Hz

Δf = frequency deviation ($|f_m - f_s|$) in Hz

f_b = input bit rate

Figure 7.10 shows a waveform for FSK.

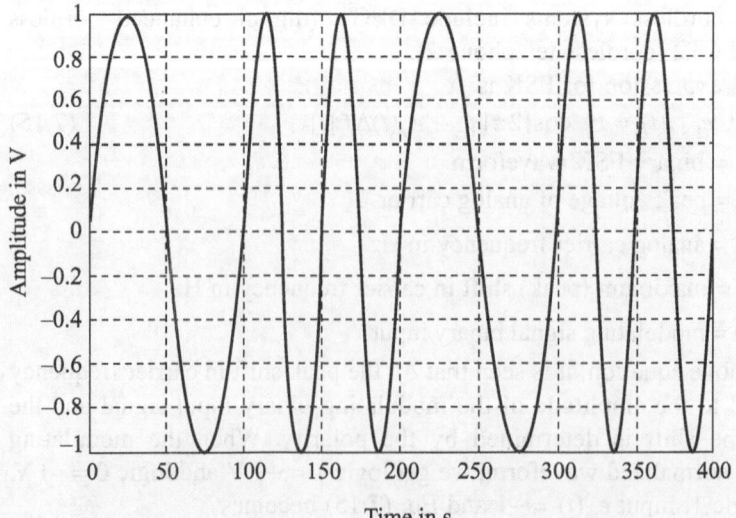

Fig. 7.10 FSK waveform

A binary frequency shift keying (BFSK) signal can be defined as follows:

$$s(t) = E_c \cos 2\pi f_0 t, \quad 0 \le t \le T$$

$$= E_c \cos 2\pi f_1 t, \quad \text{elsewhere}$$

(7.18)

where E_c is a constant (the peak amplitude of the signal), f_0 and f_1 are the transmitted frequencies, and T is the bit duration. The signal has power $P = E_c^2/2$ so that $E_c = \sqrt{2P}$. Equation (7.18) can be written as

$$s(t) = \sqrt{2P} \cos 2\pi f_0 t, \quad 0 \le t \le T$$

$$= \sqrt{2P} \cos 2\pi f_1 t, \quad \text{elsewhere}$$

$$= \sqrt{PT} \sqrt{\frac{2}{T}} \cos 2\pi f_0 t, \quad 0 \le t \le T$$

$$= \sqrt{PT} \sqrt{\frac{2}{T}} \cos 2\pi f_1 t, \quad \text{elsewhere}$$

$$= \sqrt{E} \sqrt{\frac{2}{T}} \cos 2\pi f_0 t, \quad 0 \le t \le T$$

$$= \sqrt{E} \sqrt{\frac{2}{T}} \cos 2\pi f_1 t, \quad \text{elsewhere}$$

(7.19)

where $E = PT$ is the energy contained in a bit duration.

Figure 7.11 shows the BFSK constellation diagram and Figure 7.12 shows the BFSK signal generated by binary sequence 0101001.

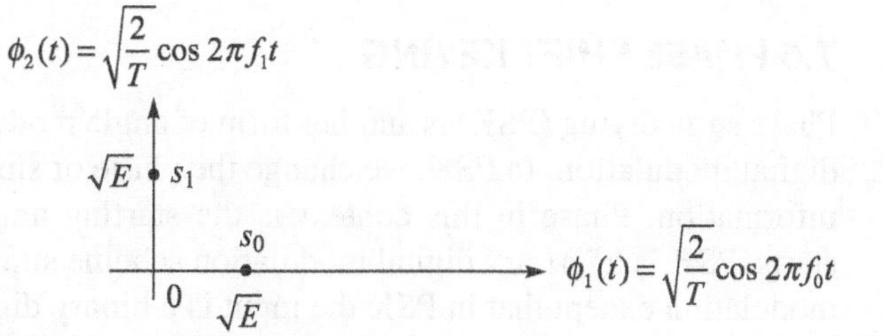

$$\phi_2(t) = \sqrt{\frac{2}{T}} \cos 2\pi f_1 t$$

Fig. 7.11 BFSK signal constellation diagram

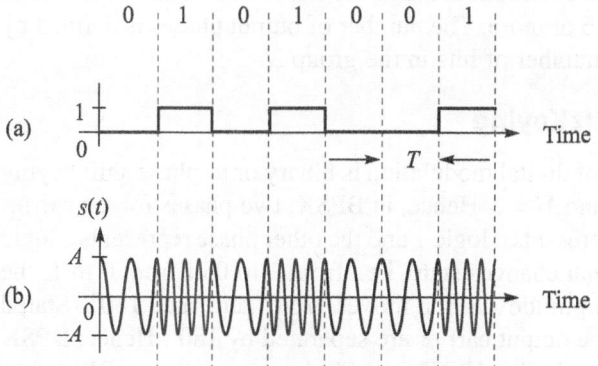

Fig. 7.12 (a) Binary sequence and (b) BFSK signal

Figure 7.13 shows the modulator and coherent demodulator for BFSK signals.

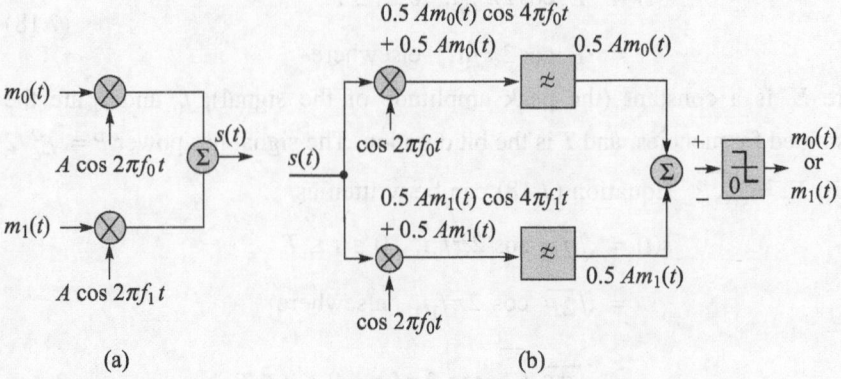

(a)   (b)

Fig. 7.13 (a) BFSK modulator and (b) BFSK coherent demodulator

Example 7.2 Find the peak frequency deviation, minimum bandwidth, and baud rate for a binary FSK signal with a mark frequency of 60 kHz and space frequency of 63 kHz and an input bit rate of 3 kbps.

Solution

Peak frequency deviation $= \left| \dfrac{60\,\text{kHz} - 63\,\text{kHz}}{2} \right| = 1.5$ kHz

Minimum bandwidth, $B = 2(1500 + 3000) = 9$ kHz

Baud rate $= 3000/1 = 3000$

7.6 PHASE SHIFT KEYING

Phase shift keying (PSK) is another form of angle modulated, constant amplitude digital modulation. In PSK, we change the phase of sinusoidal carrier to indicate information. Phase in this context is the starting angle at which the sinusoid starts. PSK is an M-ary digital modulation scheme similar to conventional phase modulation except that in PSK the input is a binary digital signal and there are a limited number of output phases possible. The input binary information is encoded into groups of bits before modulating the carrier. The number of bits in a group ranges from 1 to 16 or more. The number of output phases is defined by M and determined by the number of bits in the group N.

7.6.1 Binary Phase Shift Keying

One of the simplest forms of digital modulation is binary or bi-phase shift keying (BPSK). In BPSK, $N = 1$ and $M = 2$. Hence, in BPSK, two phases for the carrier are possible. One phase represents a logic 1 and the other phase represents a logic 0. As the input digital signal changes state, i.e., from 1 to 0 or from 0 to 1, the phase of the constant amplitude carrier moves from zero and 180°. Stated otherwise, the phases of the output carrier are separated by 180°. Hence, BPSK is also called *phase reversal keying* (*PRK*) and *bi-phase modulation*. BPSK is a form of square wave modulation of a continuous wave signal. Figure 7.14 shows the BPSK carrier. As the figure shows, a logic zero produces a zero phase shift and a logic 1 produces a 180° phase shift. As the binary input shifts from logic 0 to logic 1 and vice versa, the BPSK waveform shifts between 0 and 180°, respectively.

Phase shift represents the change in the state of information. The BPSK signal is special. It lies totally in one axis, the x-axis. It has no y-axis projection. The vector flip-flops on the x-axis depending on the value of the bit. Table 7.1 lists the two symbols and the signals used to represent them. The carrier signal is shown for $f_c = 1$ Hz. The I and Q amplitudes are the x- and y-projections computed by setting $f_c = 0$ and $\sqrt{2E/T} = 1$ (where E is the energy contained in a bit duration and T the bit duration). Then we get $I = 1$ for the first symbol and -1 for the second symbol. Q amplitude is zero for both symbols because sin 0° and sin 180° are both zero.

Table 7.1

Symbol	Bits	S(t)	Phase (degree)	Modulated signal at $f_c = 1$	I	Q
S1	0 0	$\sqrt{\dfrac{2E_s}{T}}\cos(2\pi f_c t + \pi/4)$	45°		1	1
S2	0 1	$\sqrt{\dfrac{2E_s}{T}}\cos(2\pi f_c t + 3\pi/4)$	135°		−1	1

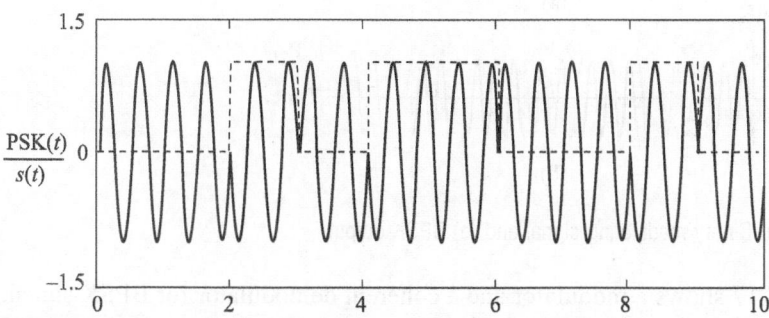

Fig. 7.14 BPSK waveform

The BPSK waveform can be represented as

$$\text{BPSK}(t) = \sin 2\pi ft, \qquad \text{for bit 0}$$

$$= \sin(2\pi ft + \pi), \quad \text{for bit 1}$$

A binary phase shift keying signal can be defined as

$$s(t) = Am(t)\cos 2\pi f_c t, \quad 0 \le t \le T \qquad (7.20)$$

where A is a constant and $m(t) = +1$ or -1, f_c is the carrier frequency, and T is the bit duration.

The signal has a power $P = A^2/2$.

Hence, Eq. (7.20) can be written as

$$s(t) = \pm\sqrt{2P} \cos 2\pi f_c t = \pm\sqrt{PT} \sqrt{\frac{2}{T}} \cos 2\pi f_c t$$

$$= \pm\sqrt{E} \sqrt{\frac{2}{T}} \cos 2\pi f_c t \qquad (7.21)$$

where $E = PT$ is the energy contained in a bit duration. If we take $\phi(t) = \sqrt{2/T}$ $\cos 2\pi f_c t$ as the orthonormal basic function, then the constellation diagram of BPSK signal will be as shown in Fig. 7.15.

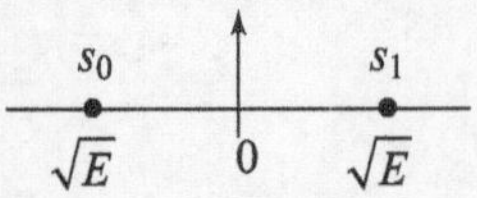

s_0 s_1

$\sqrt{E}$ 0 $\sqrt{E}$

Fig. 7.15 BPSK signal constellation diagram

Figure 7.16 shows the BPSK signal sequence generated by the binary sequence 0101001.

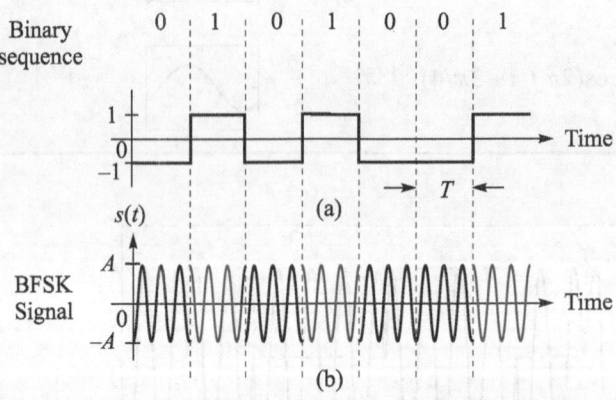

Fig. 7.16 (a) Binary modulating signal and (b) BPSK output

Figure 7.17 shows a modulator and a coherent demodulator for BPSK signal.

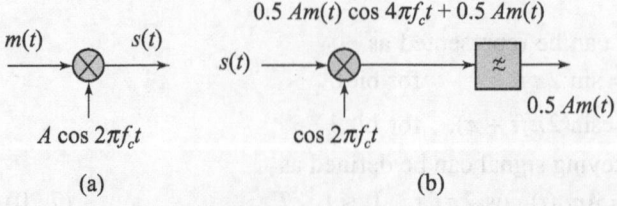

Fig. 7.17 (a) BPSK modulator and (b) coherent demodulator

Example 7.3 For an 8PSK system operating an information bit rate of 36 kbps, find the baud, minimum bandwidth, and bandwidth efficiency.

Solution

$$\text{Baud} = \frac{36,000}{3} = 12,000$$

$$\text{B W} = \frac{36,000}{3} = 12,000$$

Bandwidth efficiency, $\eta = \dfrac{36000}{12000} = 3$ bits/cycle

7.6.2 *M*-ary Phase Shift Keying (MPSK)

When two or more information bits are combined prior to performing PSK modulation, more than two output phases are possible. A general case of *M*-ary phase shift keying (MPSK) will be explained here. An *M*-ary phase shift keying signal can be defined as

$$s(t) = E_c \cos(2\pi f_c t + \theta_i + \theta), \quad 0 \le t \le T$$

$$= 0, \quad \text{elsewhere} \tag{7.22}$$

where
$$\theta_i = \frac{2\pi}{M} i \tag{7.23}$$

for $i = 0, 1, 2, \ldots, M - 1$. Here E_c is the amplitude of the carrier, which is constant, f_c is the carrier frequency, θ is the initial phase angle, and T is the symbol duration. Equation (7.22) can be expanded using trigonometric identities as follows:

$$s(t) = E_c \cos \theta_i \cos(2\pi f_c t + \theta) - E_c \sin \theta_i \sin(2\pi f_c t + \theta) \tag{7.24}$$

The signal has a power $P = \dfrac{E_c^2}{2}$. So, E_c in terms of P is $\sqrt{2P}$. Equation (7.24) can be written as

$$s(t) = \sqrt{PT} \cos \theta_i \sqrt{\frac{2}{T}} \cos(2\pi f_c t + \theta) - \sqrt{PT} \sin \theta_i \sqrt{\frac{2}{T}} \sin(2\pi f_c t + \theta)$$

$$= \sqrt{E} \cos \theta_i \sqrt{\frac{2}{T}} \cos(2\pi f_c t + \theta) - \sqrt{E} \sin \theta_i \sqrt{\frac{2}{T}} \sin(2\pi f_c t + \theta) \tag{7.25}$$

where $E = PT$ is the energy of $s(t)$ contained in a symbol duration for $i = 0, 1, 2, \ldots, M - 1$. It is assumed that the phase angle θ is taken as zero. Taking $\phi_1(t) = \sqrt{2/T} \cos 2\pi f_c t$ and $\phi_2(t) = \sqrt{2/T} \sin 2\pi f_c t$, the signalling constellation diagrams for the MPSK and QPSK are shown in Figures 7.18 and 7.19.

All signal points lie on a circle of radius $\sqrt{E}$.

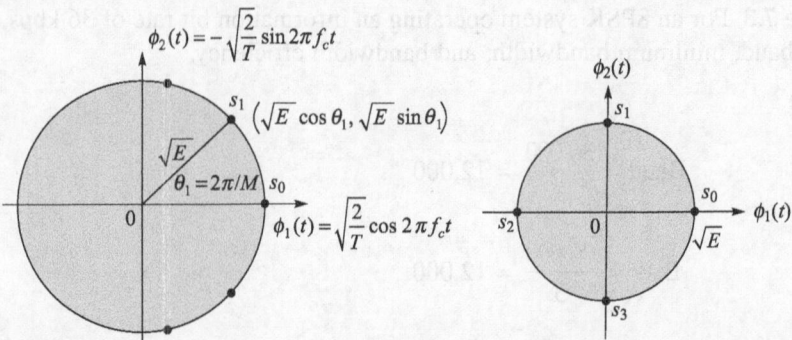

Fig. 7.18 MPSK constellation diagram   **Fig. 7.19** QPSK signal constellation diagram

For the phase to binary vector conversion, the mapping table is as shown in Table 7.2.

Table 7.2 Typical mapping table for MPSK coherent demodulation

θ_i	Natural binary vector
0	0000...00
$\dfrac{2\pi}{M}$	0000... 01
$2\left(\dfrac{2\pi}{M}\right)$	0000... 10
$(M-1)\dfrac{2\pi}{M}$	0000... 11

7.6.3 Quadrature Phase Shift Keying (QPSK)

A QPSK signal is an extension of the BPSK signal. Both of these are a type of M-ary signals discussed earlier. Since dimensionality of a modulation is defined by the number of basic functions used. That makes QPSK a two-dimensional signal. This is not because it sends two bits per symbol, but because it uses two independent signals, sine and cosine, to create the symbols. The modulated signal in polar form is given by

$$s(t) = E_c \, p_s(t) \cos\left(2\pi f_c t + \frac{2\pi i}{M} \right) \tag{7.26}$$

where $i = 0, 1, 2,..., M - 1$, E_c is the amplitude of the carrier, which is constant, f_c is the carrier frequency, and $p_s(t)$ is the pulse shaping function. The initial phase shift θ is assumed to be zero. In digital modulation, the phase of the sinusoid is modified with respect to the received bit. Since a sinusoid can go through maximum of 2π phase change in one period, the maximum phase we can change at any one time is 180°. A variety of PSK modulation can be created by having M quantized levels of 2π. Thus, varying i from 1 to M, the allowed phases are given by

$$\theta_i = \frac{2\pi i}{M} \tag{7.27}$$

where M stands for the order of modulation. $M = 2$ is for BPSK, $M = 4$ is for QPSK, and $M = 8$ for 8PSK, and so on. Figure 7.20 shows three of these modulations and their constellations. These modulations are called rotationally invariant.

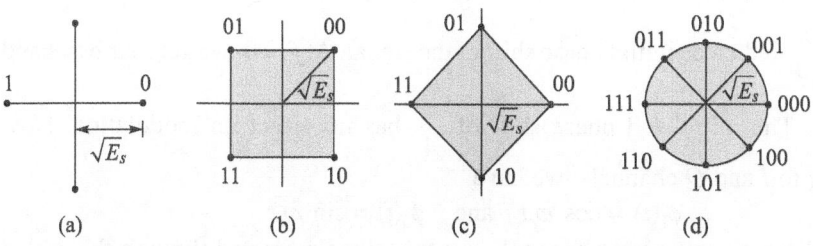

Fig. 7.20 MPSK modulation: (a) BPSK, (b) QPSK, (c) QPSK, and (d) 8PSK

For baseband PSK signals, a square pulse is used. The pulse has an amplitude of A. The energy E in this pulse is equal to the power of the signal times the duration T. With the resistance $R = 1\Omega$,

$$E = \frac{A^2 T}{2}$$

or
$$A = \sqrt{\frac{2E}{T}} \tag{7.28}$$

Substituting this value for $p_s(t)$ in Eq. (7.26), we get

$$s(t) = E_c \sqrt{\frac{2E}{T}} \, \cos\left(2\pi f_c t + \frac{2\pi i}{M}\right) \tag{7.29}$$

Taking E_c inside the square root and letting $E_c^2 \times E = E_s$, Eq. (7.29) becomes

$$s(t) = \sqrt{\frac{2E_s}{T}} \, \cos\left(2\pi f_c t + \frac{2\pi i}{M}\right), \ i = 0, 1, 2, .., M \tag{7.30}$$

In this equation, the amplitude term is constant. The term $2\pi f_c t$ changes with time and the term $\frac{2\pi i}{M}$ changes with information.

Equation (7.30) is plotted as shown in Fig. 7.21.

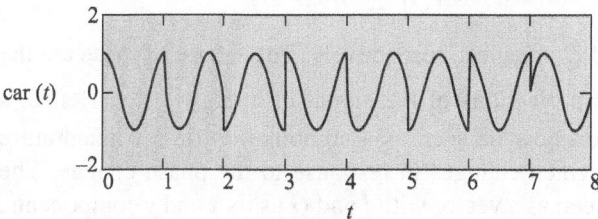

Fig. 7.21 An arbitrary modulated signal

In the plot shown in Fig. 7.21, several phase shifts are seen, some $90°$ as at $t = 1$ and others $180°$ at $t = 6$. A modulated signal for BPSK has only $180°$ phase shifts, whereas QPSK has $90°$ and $180°$ phase shifts. The modulation in Eq. (7.30) can be further expanded using trigonometric identity $\cos (A + B) = \cos A \cos B - \sin A \sin B$ as follows:

$$s(t) = \sqrt{\frac{2E_s}{T}} \left[\cos (2\pi f_c t) \cos \left(\frac{2\pi i}{M} + \frac{\pi}{4} \right) - \sin (2\pi f_c t) \sin \left(\frac{2\pi i}{M} + \frac{\pi}{4} \right) \right] \quad (7.31)$$

where $\frac{\pi}{4}$ indicates initial phase shift of the signal. At $f_c = 0$, we get four baseband signals. The initialized phase shift of $\frac{\pi}{4}$ has no effect on modulation. Now coming to I and Q channels, we have

$$\phi_1(t) = \cos \omega_c t \quad \text{and} \quad \phi_2(t) = \sin \omega_c t$$

As these signals are orthogonal, any two signals created through the scaled versions of these basic signals are also orthogonal. Hence, the scaled I and Q signals are

$$I = \sqrt{\frac{2E_s}{T}} \cos(2\pi f_c t) \quad (7.32a)$$

$$Q = \sqrt{\frac{2E_s}{T}} \sin(2\pi f_c t) \quad (7.32b)$$

They are orthogonal since the basic functions are only multiplied with a constant. Now multiplying them with the angle part of Eq. (7.31) and for $i = 0$, 1, 2, 3 and $M = 4$, we get another set of orthogonal functions as

$$I = \sqrt{\frac{2E_s}{T}} \cos (2\pi f_c t) \left\{ \cos \frac{\pi}{4} \text{ or } \cos \frac{3\pi}{4} \text{ or } \cos \frac{5\pi}{4} \text{ or } \cos \frac{7\pi}{4\pi} \right\} \quad (7.33\text{ a})$$

$$Q = \sqrt{\frac{2E_s}{T}} \sin (2\pi f_c t) \left\{ \sin \frac{\pi}{4} \text{ or } \sin \frac{3\pi}{4} \text{ or } \sin \frac{5\pi}{4} \text{ or } \sin \frac{7\pi}{4} \right\} \quad (7.33\text{ b})$$

The above modulation equation can be expressed as

$$s(t) = \sqrt{\frac{2E_s}{T}} \cos\{\theta (t)\} \cos (2\pi f_c t) - \sqrt{\frac{2E_s}{T}} \sin \{\theta (t)\} \sin (2\pi f_c t) \quad (7.34)$$

This is called the quadrature form of the modulation equation. The two signals are orthogonal. The terms

$$\sqrt{\frac{2E_s}{T}} \cos\{\theta(t)\} \quad \text{and} \quad \sqrt{\frac{2E_s}{T}} \sin\{\theta(t)\}$$

are the amplitudes of I and Q channels, respectively. The values of these are the same as the x- and y-axes projections of the signal of energy $\sqrt{E_s}$. Hence, a phase modulated signal can now be seen as a combination of two quadrature signals, the amplitude of which changes in response to the phase change. The modulating signal can be seen as a vector with I and Q as its x and y components. Since it is not possible to create a signal packet of a particular phase from a free-running sine or cosine wave, the quadrature modulation with I and Q channels come into play. I and Q channels are not just the concepts but also about how modulators are designed. However, the signal created by I and Q channels is not what is transmitted; it is the sum or difference of these two, which is the real modulated signal. There are many ways to map the bits to get the possible phases. But the best way is to see that the adjacent phase means just one bit difference so that when a phase mistake is made and the most likely one is the nearest phase, then only one bit is decoded incorrectly. Grey code satisfies this condition and hence is generally applied in PSK. In QPSK, this can be done perfectly. But in higher-order PSK modulations, this may not be possible. For 8PSK, the codes can be 001, 000, 100, 101, 111, 110, 010, 011.

For QPSK, there are four symbols, each represented by 2 bits. Generally the first is at $45°$ and change the phase by $90°$ each time to get the next symbol. The I and Q values are computed by setting $f_c = 0$ and $\sqrt{\dfrac{2E_s}{T}} = \sqrt{2}$ in Table 7.3.

Figure 7.22 shows a QPSK signal sequence generated by the binary sequence 00, 01, 10, and 11.

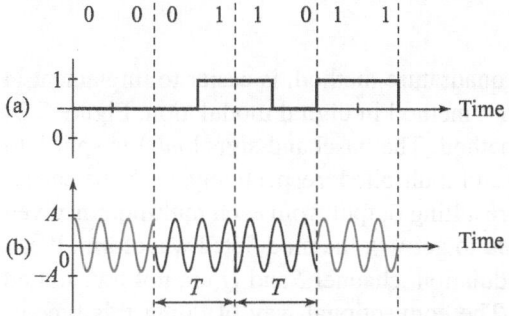

Fig. 7.22 QPSK: (a) binary sequence and (b) QPSK signal

Table 7.3 QPSK mapping

Symbol	Bits	$S(t)$	Phase (degree)	Modulated signal at $f_c = 1$	I	Q
$S1$	0 0	$\sqrt{\dfrac{2E_s}{T}}\cos(2\pi f_c t + \pi/4)$	$45°$		1	1
$S2$	0 1	$\sqrt{\dfrac{2E_s}{T}}\cos(2\pi f_c t + 3\pi/4)$	$135°$		-1	1
$S3$	1 1	$\sqrt{\dfrac{2E_s}{T}}\cos(2\pi f_c t + 5\pi/4)$	$225°$		-1	1
$S4$	1 0	$\sqrt{\dfrac{2E_s}{T}}\cos(2\pi f_c t + 7\pi/4)$	$315°$		1	-1

7.6.4 PSK Modulation

There are two conventional ways of doing PSK modulation. Figure 7.23 shows one of these, which is a straightforward method but requires multiplication and square roots. The information signal $m(t)$ is processed and given to a multiplier. The other input for the multiplier comes from a phase modulator in which the carrier signal f_c is phase modulated. The multiplier output is the modulated signal.

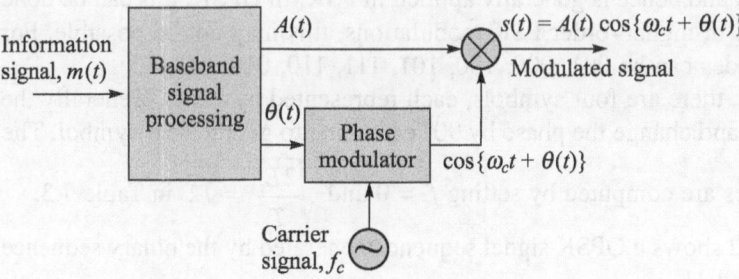

Fig. 7.23 The polar form of modulation

The second method, which is a quadrature method, is easier to implement in hardware and hence the predominant method in digital modulation. Figure 7.24 gives the block diagram for this method. The baseband signal $m(t)$ is split into two signals $x(t)$ and $y(t)$ as shown and multiplied, respectively, with the carrier and the carrier shifted by 90°. The resulting output from each multiplier is given to a summer and algebraically added to get $s(t)$, the modulated waveform. It can be observed that in quadrature modulation, channel I and Q are not transmitted but the real signal is transmitted. The conventional way of doing this type of modulation is called *canonical* form.

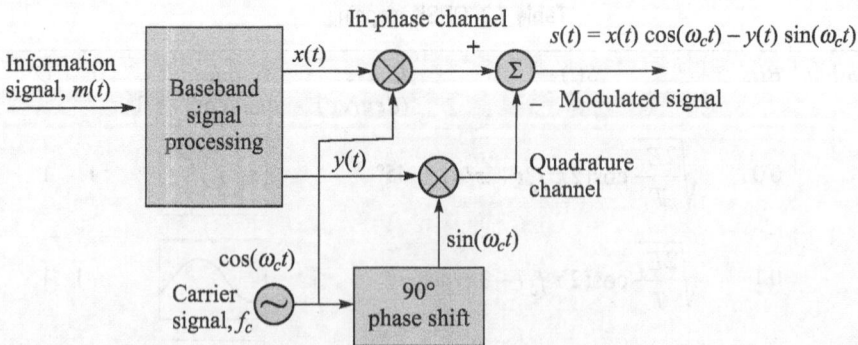

Fig. 7.24 The quadrature form of modulation using I and Q channels

Figure 7.25 shows an MPSK coherent demodulator. The QPSK signal is multiplied with two coherent carriers $\cos 2\pi f_c t$ and $\sin 2\pi f_c t$. The output of the multiplier consists of fundamental and harmonics. The harmonic signals are filtered out by an LPF and given to a phase detector and by the use of mapping table, the message signal is recovered.

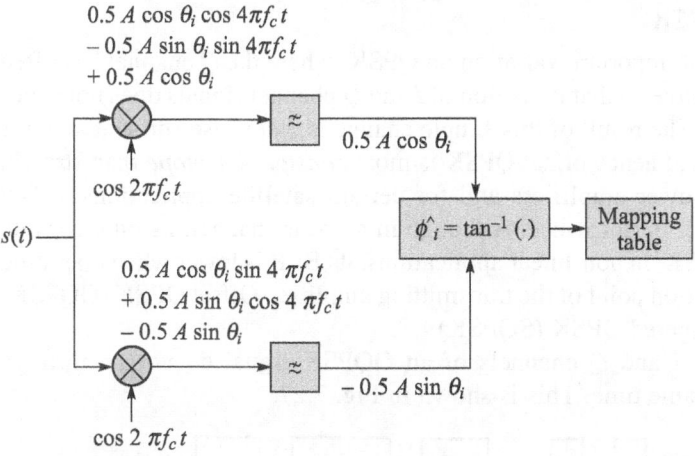

$0.5\,A\cos\theta_i\cos4\pi f_c t$
$-\,0.5\,A\sin\theta_i\sin4\pi f_c t$
$+\,0.5\,A\cos\theta_i$

$\cos2\pi f_c t$

$0.5\,A\cos\theta_i$

$s(t)$

$\hat{\phi}_i=\tan^{-1}(\cdot)$

Mapping table

$0.5\,A\cos\theta_i\sin4\,\pi f_c t$
$+\,0.5\,A\sin\theta_i\cos4\,\pi f_c t$
$-\,0.5\,A\sin\theta_i$

$-\,0.5\,A\sin\theta_i$

$\cos2\,\pi f_c t$

Fig. 7.25   MPSK coherent demodulator

7.6.5 Modulation Index of a QPSK signal

The modulation signal can also be written as

$$s(t) = E_c \cos[\omega_c t + D_p m(t)] \tag{7.35}$$

where $m(t)$ is the information signal, varying between +1 and −1 with a rectangular pulse shape. D_p is called the phase sensitivity. It is equal to the peak phase deviation over one symbol. Its unit is radians per volt. Plotting the above equation, we get the modulated signal as a function of the phase sensitivity.

The phase sensitivity of a modulated signal is related to the traditional modulation index as

$$h = \frac{2\Delta\theta}{\pi} = \frac{2D_p}{\pi\pi} \tag{7.36}$$

For a true carrier suppressed PSK signal, the modulation index is equal to 1 since the peak phase variation is 90°. PSK signals are all carrier suppressed and all have a modulation index of 100%. Figure 7.26 shows a typical PSK signal for various phase sensitivities.

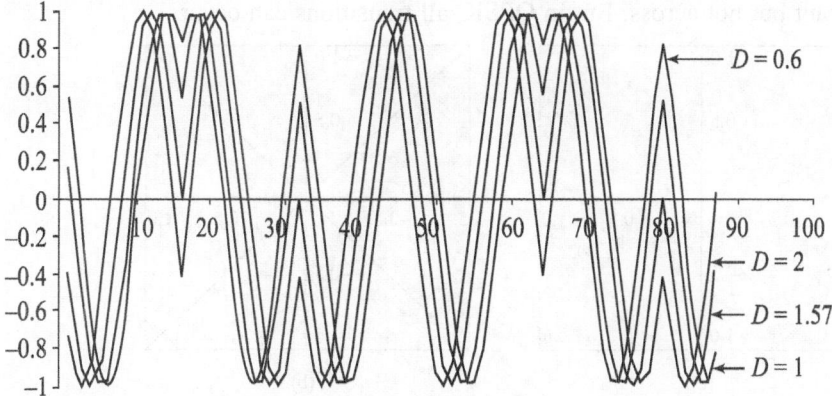

Fig. 7.26   Plot of PSK signal for various phase sensitivities

7.6.6 Offset QPSK

This is a minor but important variation on QPSK, where the Q channel is shifted by half a symbol time so that transition of I and Q channel signals does not occur at the same time. The result of this simple change is that phase shifts at any one time are limited and hence offset QPSK is more *constant-envelope* than straight QPSK. In high-power amplifiers and for certain satellite applications, offset QPSK offers better performance. Although in a linear channel, its bit error rate is the same as QPSK, in non-linear applications, its BER is lower when operating close to the saturation point of the transmitting amplifier. Offset QPSK (OQPSK) is also called staggered QPSK (SQPSK).

Unlike QPSK, I and Q channels of an OQPSK signal do not go through transition at the same time. This is shown in Fig. 7.27.

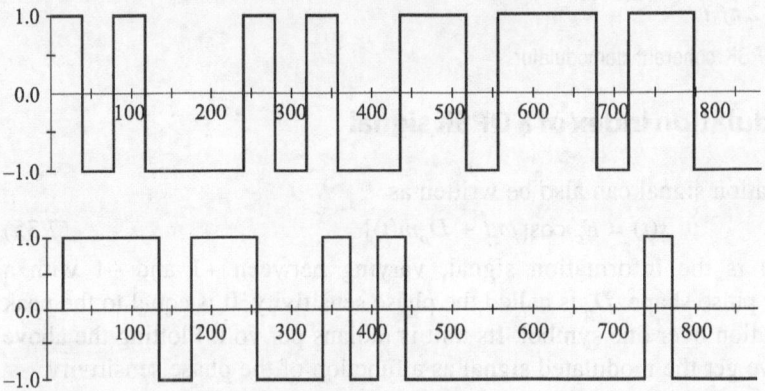

Fig. 7.27 Offset QPSK: I and Q channel mapping

One consequence of this is that when we look at the constellation diagram of the OQPSK as shown in Fig. 7.28, the symbol transitions occur only to neighbours. This means that the transitions are never more than $90°$. At any symbol change, for either I or Q channel, only one axis can change at a time, i.e., either I or Q but not both. Hence, at any transition, only I or Q changes but not both. In the constellation diagram, if the signal was in the right upper quadrant, the next signal can only go to either the lower right quadrant or to upper left quadrant but not across. But in QPSK, all transitions can occur.

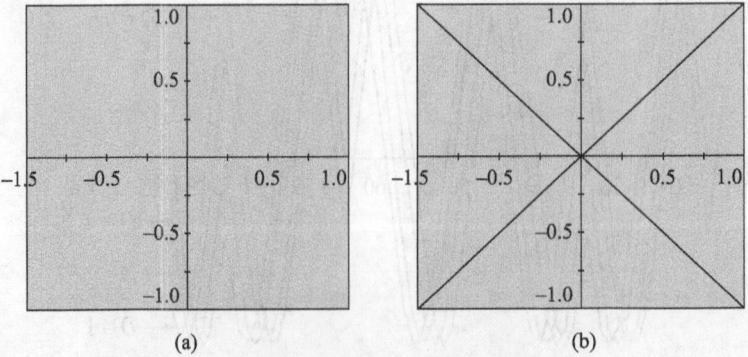

Fig. 7.28 Constellation diagram: (a) OQPSK and (b) QPSK

Figure 7.29 compares the OQPSK signal with a QPSK signal. Note that the OQPSK signal never goes through transitions more than 90°. QPSK, on the other hand, goes through a phase change of 180° for some transitions. The larger transitions are a source of trouble for amplifiers and to be avoided, if possible. In satellite transmission, QPSK reigns supreme, as it is easy to build and-operate. Defence applications often use OQPSK because of its need to use low-power radios and minimum adjacent channel interference issues.

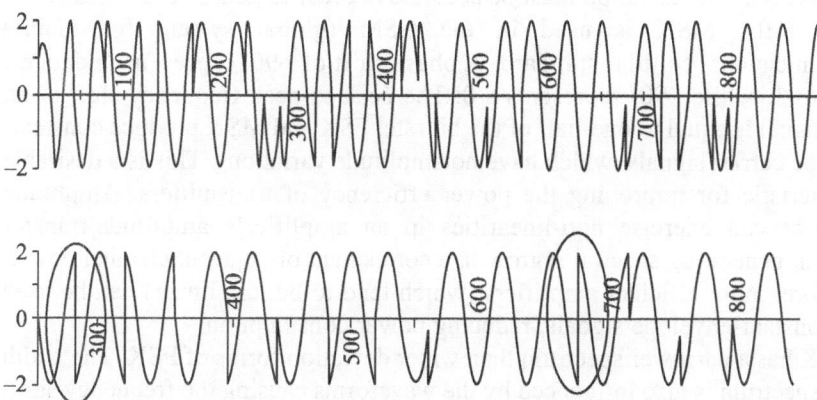

Fig. 7.29   The phase jump at the symbol transition: (a) OQPSK (90° phase shift) and (b) QPSK (180° phase shift)

Figure 7.30 shows how OQPSK differs from QPSK. The Q channel of OQPSK is delayed by a half a symbol time, staggering the two quadrature channels.

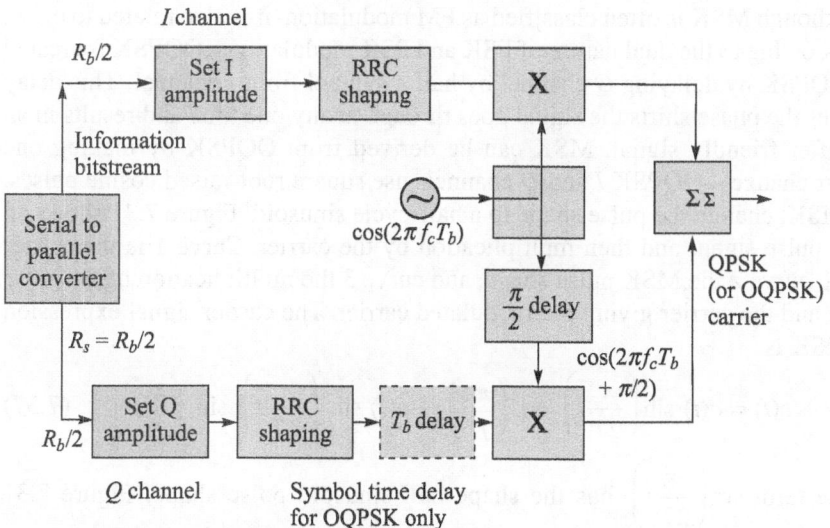

Fig. 7.30   QPSK modified to become OQPS

7.7 MINIMUM SHIFT KEYING

Since a frequency shift produces an advancing or retarding phase, frequency shifts can be detected by sampling phase at each symbol period. Phase shifts of

$(2N + 1)\,\pi/2$ rad are easily detected with an I/Q demodulator. At even numbered symbols, the polarity of the I channel conveys the transmitted data, while at odd numbered symbols the polarity of the Q channel conveys the data. This orthogonality between I and Q simplifies detection algorithms and hence reduces power consumption in a mobile receiver. The minimum frequency shift that yields orthogonality of I and Q is that which results in a phase shift of $\pm \pi/2$ rad per symbol (90° per symbol). FSK with this deviation is called MSK (minimum shift keying). The deviation must be accurate in order to generate repeatable 90° phase shifts. MSK is used in the GSM (global system for mobile communications) cellular standard. A phase shift of +90° represents a data bit equal to 1, while −90° represents a 0. The peak-to-peak frequency shift of an MSK signal is equal to one-half of the bit rate. FSK and MSK produce constant-envelope carrier signals, which have no amplitude variations. This is a desirable characteristic for improving the power efficiency of transmitters. Amplitude variations can exercise non-linearities in an amplifier's amplitude-transfer function, generating spectral regrowth, a component of adjacent channel power. Therefore, more efficient amplifiers (which tend to be less linear) can be used with constant-envelope signals, reducing power consumption.

MSK has a narrower spectrum than wider deviation forms of FSK. The width of the spectrum is also influenced by the waveforms causing the frequency shift. If those waveforms have fast transitions or a high slew rate, then the spectrum of the transmitter will be broad. In practice, the waveforms are filtered with a Gaussian filter, resulting in a narrow spectrum. In addition, the Gaussian filter has no time domain overshoot, which would broaden the spectrum by increasing the peak deviation. MSK with a Gaussian filter is termed GMSK (Gaussian MSK).

Although MSK is often classified as FM modulation, it is also related to offset QPSK owing to the dual nature of FSK and PSK modulations. OQPSK is created from QPSK by delaying Q channel by half a symbol from I channel. This delay reduces the phase shifts the signal goes through at any one time and results in an amplifier friendly signal. MSK can be derived from OQPSK by making one further change—OQPSK I and Q channels use square root-raised cosine pulses. For MSK, change the pulse shape to a half-cycle sinusoid. Figure 7.31 shows an MSK pulse signal and then multiplication by the carrier. Curve 1 is the carrier signal, curve 2 the MSK pulse shape, and curve 3 the multiplication of the pulse shape and the carrier giving the modulated carrier. The carrier signal expression for MSK is

$$c(t) = a(t) \sin\left(\frac{\pi}{2T}t\right) \cos\left(\frac{\pi}{T}t\right) + a(t) \sin\left(\frac{\pi}{2T}t\right) \sin\left(\frac{\pi}{T}t\right) \qquad (7.37)$$

The term $\sin\left(\dfrac{\pi}{2T}t\right)$ has the shape half-sinusoid pulse shape. Figure 7.31 shows how MSK pulses look compared to QPSK square pulses. Remember that in QPSK, the square pulse itself equates to a discrete phase.

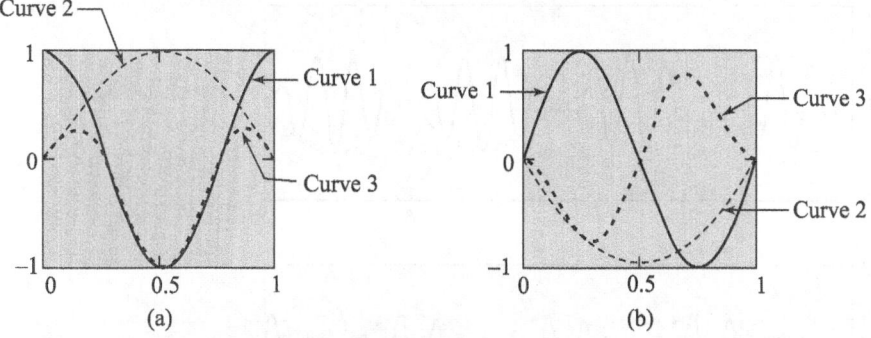

Fig. 7.31 MSK pulse shaping–half sine wave (positive for bit 1 and negative for bit 0): pulse and carrier for (a) a 1 bit and (b) a 0 bit

In MSK, the shape is continuously changing and so there is no discrete jump in the modulated signal at the symbol edge as there is QPSK. For this reason, the modulated signal in Fig. 7.32 has no discontinuities as compared to MPSK signals.

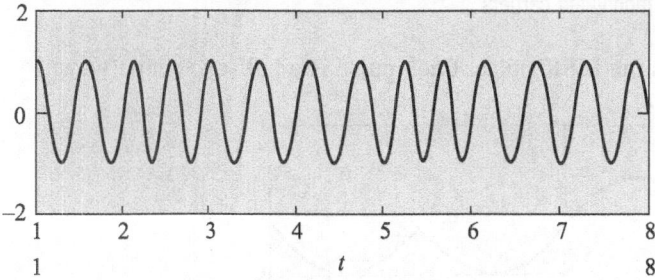

Fig. 7.32 MSK modulated carrier

The dashed line is the QPSK I and Q channel symbols and the solid lines show how these have been shaped by the half sine wave. The I and Q channels are computed by

$$MSKI(t) = QPSKI(t)\left|\sin\frac{\pi t}{2T}\right| \tag{7.38a}$$

$$MSKQ(t) = QPSKQ(t)\left|\sin\frac{\{\pi(t+0.5T)\}}{2T}\right| \tag{7.38b}$$

The I and Q channels are then multiplied by the carrier, cosine for the I channel and sine for the Q channel. Now Eq. (7.38) becomes

$$MSKCI(t) = QPSKI(t)\left|\sin\frac{\pi t}{2T}\right|\cos\frac{\pi t}{T}$$

$$MSKCQ(t) = QPSKQ(t)\left|\sin\frac{\{\pi(t+0.5T)\}}{2T}\right|\sin\frac{\pi t}{T} \tag{7.39}$$

Note that the period of pulse shape is twice that of the symbol rate.

Adding I and Q components gives the MSK carrier of Fig. 7.33. Compare this carrier to a QPSK carrier. This one has much smoother phase shifts at the symbol boundaries. This results in lower side lobes, which is an advantageous property for wireless signals since it results in less adjacent signal interference.

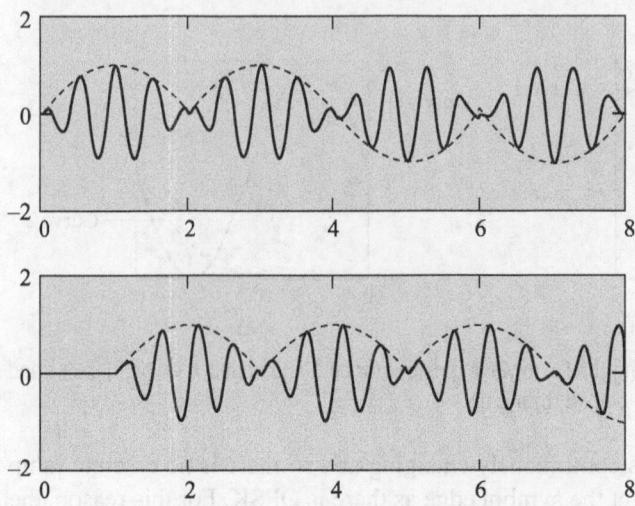

Fig. 7.33 MSK *I* and Q modulated carriers

Figure 7.34 shows the MSK pulse. Each pulse is a half-cycle sine wave.

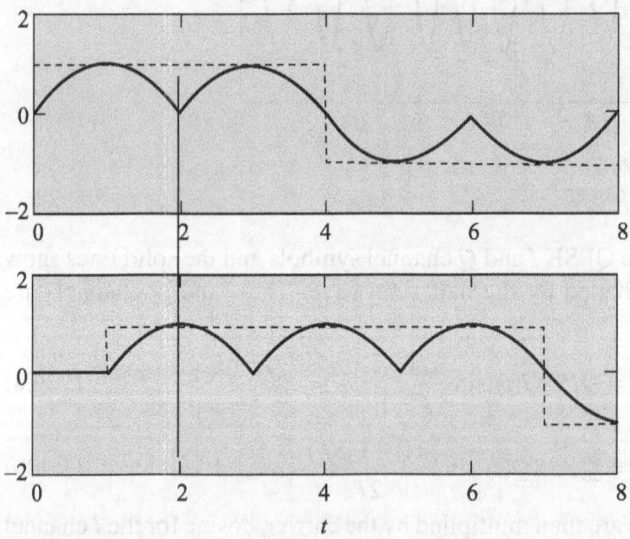

Fig. 7.34 MSK pulse (Each pulse is a half-cycle sine wave.)

Figure 7.35 shows the modification made to the QPSK modulator to create the MSK signal. Only the pulse shaping has been changed. The half-cycle time shift of the OQPSK stays.

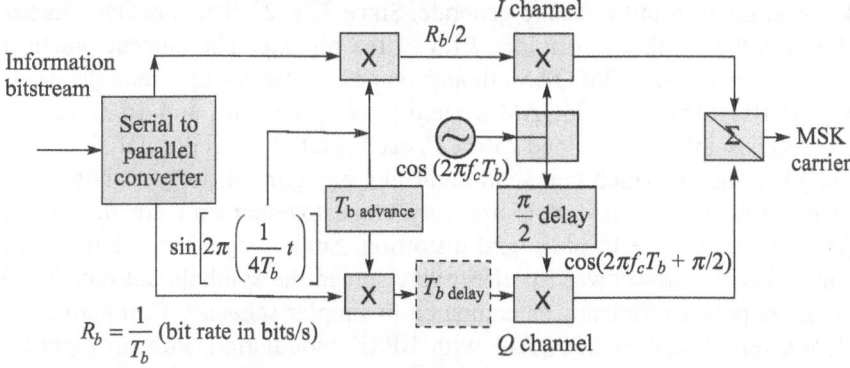

Fig. 7.35   MSK modulator

How MSK differs from QPSK   MSK is generally considered an FSK modulation but it is exactly the same as OQPSK except that it uses a half-sinusoid for pulse shaping instead of root-raised cosine pulses.

7.8 QUADRATURE AMPLITUDE MODULATION (QAM)

Another member of the digital modulation family is quadrature amplitude modulation (QAM). It is used in applications like microwave digital radio, DVB-C (digital video broadcasting – cable), and modems. QAM is a form of digital modulation similar to PSK except that the digital information is contained in both amplitude and phase of the transmitted carrier. In QAM, amplitude and phase shift keying are combined in such a way that the position of the signalling elements on constellation diagrams is optimized to achieve the greatest distance between the elements. This reduces the likelihood of one element being misinterpreted as another element. The degree of bandwidth compression we obtain from QAM is the same as it is with PSK, i.e., $B = \dfrac{f_b}{N}$, where f_b is the channel capacity in bps and N is the number of bits encoded into each signalling element. QAM has the advantage over PSK for phase impairment, which is inherent in all communication systems.

7.8.1 Types of QAM

There are different types of QAM. First we will briefly discuss 16 state quadrature amplitude modulation (16QAM). In 16QAM, there are four I values and four Q values. This results in a total of 16 possible states for the signal. It can go through transition from one state to any other state at every symbol time. Since $16 = 2^4$, four bits per symbol can be sent. This consists of two bits for I and two bits for Q. The symbol rate is one-fourth of the bit rate. So, this modulation format produces a more spectrally efficient transmission. It is more efficient than BPSK, QPSK, or 8PSK. Note that QPSK is the same as 4QAM. Another variation is 32QAM. In this case, there are six I values and six Q values resulting in a total of 36 possible states ($6 \times 6 = 36$). This is too many states for a power of two (the closest power of two is 32). So, the four corner symbol states, which take the most power to transmit, are omitted. This reduces the amount of peak power the transmitter has to generate. Since $32 = 2^5$, there are five bits per symbol and the symbol rate is one-fifth of the bit rate. The current practical limits are approximately 256QAM, though work is underway to extend the limits to 512 or 1024QAM. A 256QAM system uses 16 I values and 16 Q values, giving 256 possible states. Since $256 = 2^8$, each symbol can represent eight bits. A 256QAM signal, which can send eight bits per symbol, is spectrally very efficient. However, the symbols are very close together and are thus more subjected to errors due to noise and distortion. Such a signal may have to be transmitted with extra power (to effectively spread the symbols out more) and this reduces power efficiency as compared to simpler schemes. Comparing the bandwidth efficiency of 256QAM with BPSK modulation with an eight-bit sampler sampling at 10 kHz for voice, BPSK uses 80 kilo symbols per second sending 1 bit per symbol. Whereas a system using 256QAM sends eight bits per symbol and so the symbol rate would be 10 kilo symbols per second. A 256QAM system enables the same amount of information to be sent as BPSK using only one-eighth of the bandwidth. It is eight times more bandwidth efficient. However, there is a trade-off. The system becomes more complex and is more susceptible to errors caused by noise and distortion. Error rates of higher-order QAM systems like this degrade more rapidly than QPSK as noise or interference is introduced. A measure of this degradation would be a higher bit error rate (BER). In any digital modulation system, if the input signal is distorted or severely attenuated the receiver will eventually lose symbol clock completely. If the receiver can no longer recover the symbol clock, it cannot demodulate the signal or recover any information. With less degradation, the symbol clock can be recovered, but if it is noisy and the symbol locations themselves are noisy, it is difficult to recover the symbol clock. In some cases, a symbol will fall far enough away from its intended position that it will cross over to an adjacent position. The I and Q level detectors used in the demodulator would misinterpret such a symbol as being in the wrong location, causing bit errors. QPSK is not as efficient, but the states are much farther apart and the system can tolerate a lot more noise before suffering symbol errors. QPSK has no intermediate states between the four corner symbol locations and so there is less opportunity for the demodulator to misinterpret symbols. QPSK requires less transmitter power than QAM to achieve the same bit error rate. Various QAMs can be called M-ary QAM.

M-ary quadrature amplitude modulation (MQAM)

An M-ary quadrature amplitude modulation (MQAM) signal can be defined by

$$s(t) = E_i \cos(2\pi f_c t + \theta_i), \quad 0 \le t \le T$$
$$= 0, \quad \text{elsewhere} \tag{7.40}$$

Equation (7.40) can be expanded further as follows:

$$s(t) = E_i \cos \theta_i \cos 2\pi f_c t - E_i \sin \theta_i \sin 2\pi f_c t, \quad 0 \le t \le T$$
$$= 0, \quad \text{elsewhere} \tag{7.41}$$

for $i = 0, 1, 2,..., M - 1$, E_i is the amplitude, f_c is the carrier frequency, θ_i is the phase angle, and T is the symbol duration. It has a power $P_i = E_i^2/2$ and hence $E_i = \sqrt{2P_i}$. Substituting this value of E_i in Eq. (7.41), $s(t)$ becomes

$$s(t) = \sqrt{P_i T} \sqrt{\frac{2}{T}} \cos \theta_i \cos 2\pi f_c t - \sqrt{P_i T} \sin \theta_i \sqrt{\frac{2}{T}} \sin 2\pi f_c t$$

$$= \sqrt{E} \cos q_i \sqrt{\frac{2}{T}} \cos 2p f_c t - \sqrt{E} \sin q_i \sqrt{\frac{2}{T}} \sin 2p f_c t \qquad (7.42)$$

where $E = P_i T$ is the energy of $s(t)$ contained in a symbol duration for $i = 0$, 1, 2, ..., $M-1$.

Let $\phi_1(t) = \sqrt{\frac{2}{T}} \cos 2\pi f_c t$ and $\phi_2(t) = \sqrt{\frac{2}{T}} \sin 2\pi f_c t$ be the orthonormal basic functions. The applicable signal constellation diagrams of 4QAM signals, 16QAM signals, and 32QAM are shown in Figures 7.36 and 7.37(a) and (b), respectively.

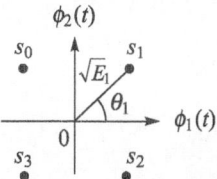

Fig. 7.36 Signal constellation diagram of 4QAM

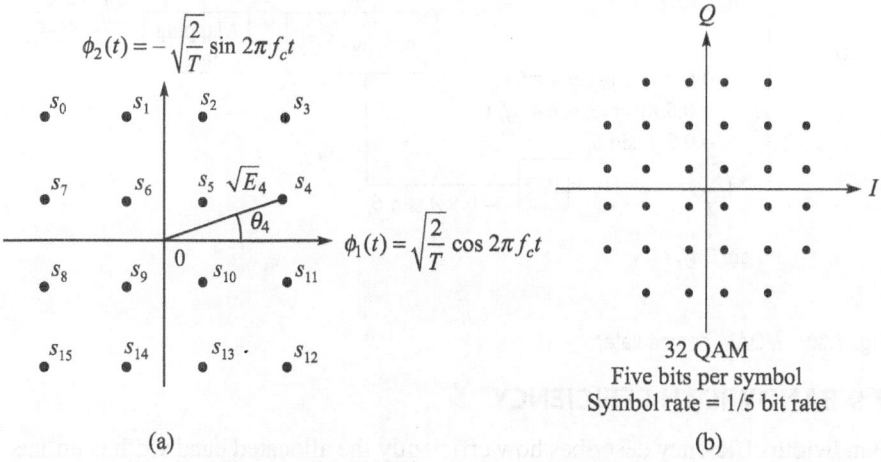

32 QAM
Five bits per symbol
Symbol rate = 1/5 bit rate

(a)   (b)

Fig. 7.37 Signal constellation diagrams of (a) 16QAM and (b) 32QAM

Figure 7.38 is the block diagram of an MQAM modulator. The binary sequence is given to the serial-to-parallel converter, whose output is bits. This is given to the PSK modulator to generate two orthogonal signals, which in turn is given to a multiplier. The other input to the multiplier is the carrier signal. The output from each multiplier is added in a summer to produce the required QAM signal.

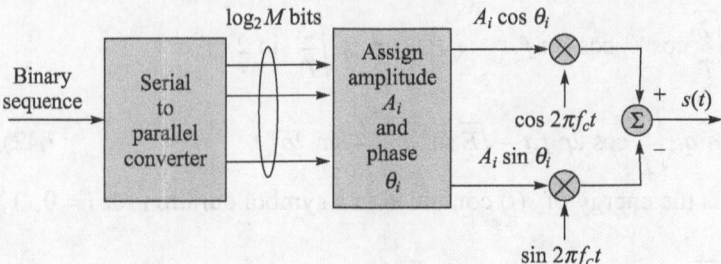

Fig. 7.38 MQAM modulator

Figure 7.39 shows a MQAM demodulator. The incoming QAM signal is given to two multipliers. The other input to the multipliers is the carrier and its quadrature signal. The output of the product detector is passed through low pass filter and given to a phase detector and to a mapping table. The output from the table is the binary sequence.

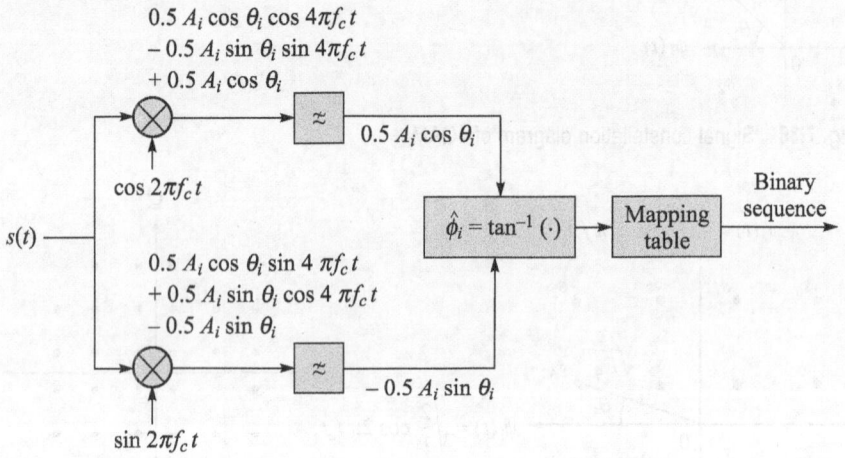

Fig. 7.39 MQAM demodulator

7.9 BANDWIDTH EFFICIENCY

Bandwidth efficiency describes how efficiently the allocated bandwidth is utilized or the ability of a modulation scheme to accommodate data within a limited bandwidth. Stated otherwise, it is often used to compare the performance of one digital modulation technique to another, i.e., *bandwidth efficiency* can be defined as a ratio of the transmission bit rate to the minimum bandwidth required for a particular modulation scheme. It is generally normalized to a 1 Hz bandwidth. Hence, it indicates the number of bits that can be propagated through a transmission medium for each hertz of bandwidth. Mathematically, it can be represented as

$$BW\eta = \frac{\text{Transmission bit rate (bps)}}{\text{Minimum bandwidth (Hz)}} = \frac{\text{bps}}{\text{Hz}} = \frac{\text{bits}}{\text{cycle}} \qquad (7.43)$$

where $BW\eta$ is the bandwidth efficiency. It can also be expressed as a percentage by multiplying $BW\eta$ by 100.

7.9.1 Comparison of Modulation Methods

Table 7.4 shows the theoretical bandwidth efficiency limits for the main modulation types. Note that these figures cannot actually be achieved in practice since they require perfect modulators, demodulators, filter, and transmission paths.

Table 7.4

Modulation format	Theoretical bandwidth efficiency limits
MSK	1 bps/Hz
BPSK	1 bps/Hz
QPSK	2 bps/Hz
8PSK	3 bps/Hz
16 QAM	4 bps/Hz
32 QAM	5 bps/Hz
64 QAM	6 bps/Hz
256 QAM	8 bps/Hz

If the filter is perfect (rectangular in the frequency domain), then the occupied bandwidth could be made equal to the symbol rate. Techniques for maximizing spectral efficiency include the following:

- Relate the data rate to the frequency shift (as in GSM).
- Use pre-modulation filtering to reduce the occupied bandwidth. Use raised cosine filters as used in digital cellular systems and personal handy phones (PHS).
- Restrict the types of transitions.

Table 7.5 gives the summary of various modulation types regarding encoding scheme, possible outputs, minimum bandwidth, baud rate, and bandwidth efficiency.

Table 7.5 ASK, FSK, PSK, and QAM summary

Modulation	Encoding scheme	Possible outputs	Minimum bandwidth	Baud rate	Bandwidth efficiency
ASK	Single bit	2	f_b	f_b	1
FSK	Single bit	2	f_b	f_b	1
BPSK	Single bit	2	f_b	f_b	1
QPSK	Dibits	4	$f_b/2$	$f_b/2$	2
8PSK	Tribits	8	$f_b/3$	$f_b/3$	3
8QAM	Tribits	8	$f_b/3$	$f_b/3$	3
16PSK	Quadbits	16	$f_b/4$	$f_b/4$	4
16QAM	Quadbits	16	$f_b/4$	$f_b/4$	4
32PSK	Five bits	32	$f_b/5$	$f_b/5$	5
32QAM	Five bits	32	$f_b/5$	$f_b/5$	5

7.9.2 Effects of Going Through the Origin

In a QPSK signal, where the normalized value changes from 1, 1 to −1, −1 and this change is simultaneous from I and Q values of +1 to I and Q values of −1, the signal trajectory goes through the origin (the I/Q value of 0, 0). The origin represents 0 carrier magnitude. A value of 0 magnitude indicates that the carrier amplitude is 0 for a moment. Not all transitions in QPSK result in a trajectory that goes through the origin. If I changes value but Q does not (or vice versa), the carrier amplitude changes a little, but it does not go through zero. Therefore, some symbol transitions will result in a small amplitude variation, while others will result in a very large amplitude variation. The clock recovery circuit in the receiver must deal with this amplitude variation uncertainty if it uses amplitude variations to align the receiver clock with the transmitter clock. Spectral re-growth does not automatically result from these trajectories that pass through or near the origin. If the amplifier and associated circuits are perfectly linear, the spectrum (spectral occupancy or occupied bandwidth) will be unchanged. The problem lies in non-linearities in the circuits. A signal which changes amplitude over a very large range will exercise these non-linearities to the fullest extent. These non-linearities will cause distortion products. In continuously modulated systems, they will cause *spectral re-growth* or wider modulation sidebands (a phenomenon related to intermodulation distortion). Another term that is sometimes used in this context is *spectral splatter*. However, this is a term that is more correctly used in association with the increase in the bandwidth of a signal caused by pulsing on and off.

7.10 DIGITAL MODULATION TYPES

The modulation types outlined in sections so far form the building blocks for many systems. There are three main variations of these basic building blocks that are used in communications systems:

- *I/Q* offset modulation
- Differential modulation
- Constant-envelope modulation

7.10.1 I/Q Offset Modulation

The first variation is offset modulation. One example of this is offset QPSK (OQPSK). This is used in the cellular CDMA (code division multiple access) system for the reverse (mobile to base) link.

In QPSK, the I and Q bitstreams are switched at the same time. The symbol clocks or the I and Q digital signal clocks are synchronized. In offset QPSK (OQPSK), the I and Q bitstreams are offset in their relative alignment by one bit period (one-half of a symbol period). This is shown in Figure 7.40. Since the transitions of I and Q are offset, at any given time only one of the two bit-treams can change values. This creates a dramatically different constellation, even though there are still just two I/Q values. This has power efficiency advantages. In OQPSK, the signal trajectories are modified by the symbol clock offset so that the carrier amplitude does not go through or near zero (the centre of the constellation). The spectral efficiency is the same with two I states and two Q states. The reduced amplitude variations (perhaps 3 dB for OQPSK versus 30 to 40 dB for QPSK) allow a more power-efficient, less linear RF power amplifier to be used.

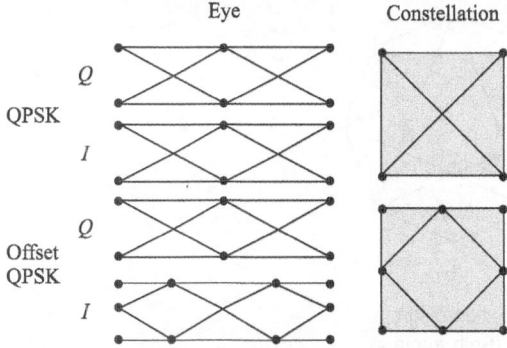

Fig. 7.40   *I/Q offset modulation*

7.10.2 Differential Modulation

The second variation is differential modulation as used in differential QPSK (DQPSK) and differential 16QAM (D16QAM). Differential means that the information is not carried by the absolute state, but it is carried by the transition between the states. In some cases, there are also restrictions on allowable transitions. This occurs in $\pi/4$ DQPSK, where the carrier trajectory does not go through the origin. A DQPSK transmission system can go through transition from any symbol position to any other symbol position. The $\pi/4$ DQPSK modulation format is widely used in many applications including the following:

- Cellular:
  NADC-IS-54 (North American digital cellular)
  PDC (pacific digital cellular)
- Cordless:
  PHS (personal handy phone system)
- Trunked radio:
  TETRA (trans-European trunked radio)

The $\pi/4$ DQPSK modulation format uses two QPSK constellations offset by 45° ($\pi/4$ rad). Transitions must occur from one constellation to the other. This guarantees that there is always a change in phase at each symbol, making clock recovery easier. The data is encoded in the magnitude and direction of the phase shift, not in the absolute position on the constellation. One advantage of $\pi/4$ DQPSK is that the signal trajectory does not pass through the origin, thus simplifying transmitter design. Another advantage is that $\pi/4$ DQPSK with root raised cosine filtering has better spectral efficiency than GMSK, the other common cellular modulation type. Figure 7.41 shows the constellation diagram of differential modulation.

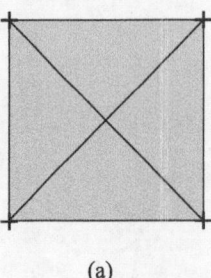 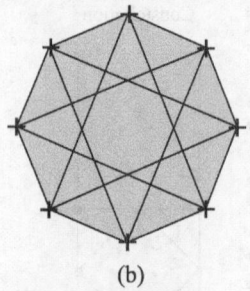

(a)   (b)

Fig. 7.41   Differential modulation—constellation diagram:
(a) QPSK and (b) π/4 DQPSK (Both formats
use 2 bits/symbol.)

7.10.3 Constant-Amplitude Modulation

The third variation is constant-envelope modulation. GSM uses a variation of constant-amplitude modulation format called 0.3 GMSK (Gaussian minimum shift keying). In constant-envelope modulation, the amplitude of the carrier is constant, regardless of the variation in the modulating signal. It is a power-efficient scheme that allows efficient class C amplifiers to be used without introducing degradation in the spectral occupancy of the transmitted signal. However, constant-envelope modulation techniques occupy a larger bandwidth than schemes, which are linear. In linear schemes, the amplitude of the transmitted signal varies with the modulating digital signal as in the BPSK or QPSK. In systems where bandwidth efficiency is more important than power efficiency, constant-envelope modulation is not generally employed. MSK is a special type of FSK where the peak-to-peak frequency deviation is equal to half the bit rate. GMSK is a derivative of MSK where the bandwidth required is further reduced by passing the modulating waveform through a Gaussian filter. The Gaussian filter minimizes the instantaneous frequency variations over time. GMSK is a spectrally efficient modulation scheme and is particularly useful in mobile radio systems. It has a constant envelope, spectral efficiency, good BER performance, and is self-synchronizing. Figure 7.42 shows the difference between QPSK and MSK.

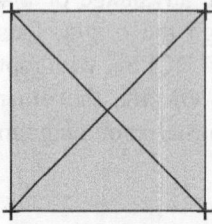 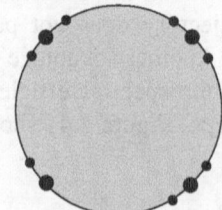

Amplitude (envelope) varies   Amplitude (envelope) does
form zero to nominal value.   not vary at all.

(a)   (b)

Fig. 7.42   Constant-amplitude modulation: (a) QPSK and (b) MSK (GSM)

7.11 SPECTRAL EFFICIENCY VERSUS POWER CONSUMPTION

As with any natural resource, it makes no sense to waste the RF spectrum by using channel bands that are too wide. Therefore, narrower filters are used to reduce the occupied bandwidth of the transmission. Narrower filters with sufficient accuracy and repeatability are more difficult to build. Smaller values of alpha increase ISI (intersymbol interference), because more symbols can contribute. This tightens the requirements on clock accuracy. These narrower filters also result in more overshoot and, therefore, more peak carrier power. The power amplifier must then accommodate the higher peak power without distortion.

The bigger amplifier causes more heat and electrical interference to be produced since the RF current in the power amplifier will interfere with other circuits. Larger and heavier batteries will be required. The alternative is to have shorter talk time and smaller batteries. Constant-envelope modulation, as used in GMSK, can use class C amplifiers, which are most efficient. In summary, spectral efficiency is highly desirable, but there are penalties in cost, size, weight, complexity, talk time, and reliability.

7.12 TIME AND FREQUENCY DOMAIN VIEW OF DIGITALLY MODULATED SIGNAL

There are a number of different ways to view a signal. The simplified example is an RF pager signal at a centre frequency of 930.004 MHz. The pager uses two-level FSK and the carrier shifts back and forth between two frequencies that are 8 kHz apart (930.000 MHz and 930.008 MHz). This frequency spacing is small in proportion to the centre frequency of 930.004 MHz. This is shown in Fig. 7.43(a) and zoomed in (b). The difference in period between a signal at 930 MHz and one at 930 MHz plus 8 kHz is very small. Even with a high-performance oscilloscope, using the latest in high-speed digital techniques, the change in period cannot be observed or measured. In a pager receiver, the signals are first downconverted to an IF or baseband frequency. In this example, the 930.004 MHz FSK modulated signal is mixed with another signal at 930.002 MHz. The FSK modulation causes the transmitted signal to switch between 930.000 MHz and 930.008 MHz. The result is a baseband signal that alternates between two frequencies, –2 kHz and +6 kHz.

The demodulated signal shifts between –2 kHz and +6 kHz. The difference can be easily detected.

This is sometimes referred to as *zoom* time or IF time. To be more specific, it is a band converted signal at IF or baseband. IF time is important as it is how the signal looks in the IF portion of a receiver. The frequency domain representation is shown in Fig. 7.43(c). Most pagers use a two-level, frequency shift keying (FSK) scheme. FSK is used in this instance because it is less affected by multipath propagation, attenuation, and interference, common in urban environment. It is possible to demodulate it even deep inside modern steel/concrete buildings, where attenuation, noise, and interference would otherwise make reliable demodulation difficult.

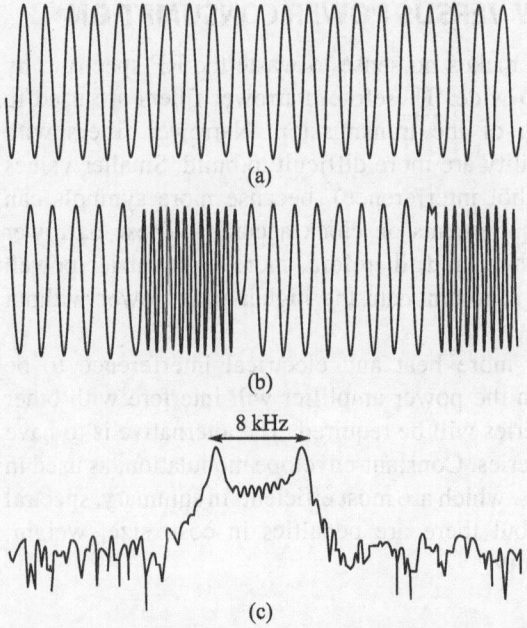

Fig. 7.43 Time and frequency domain diagram of an FSK signal:
(a) time domain baseband, (b) time domain zoom,
and (c) frequency domain narrowband

7.12.1 Power and Frequency View

There are many different ways of looking at a digitally modulated signal. To examine how transmitters turn on and off, a power versus time measurement is very useful for examining the power level changes involved in pulsed or bursted carriers as shown in Fig. 7.44. For example, very fast power changes will result in frequency spreading or spectral regrowth. This is also known as frequency *splatter*. Very slow power changes waste valuable transmit time, as the transmitter cannot send data when it is not fully on. Turning on too slowly can also cause high bit error rates at the beginning of the burst. In addition, peak and average power levels must be well understood, since asking for excessive power from an amplifier can lead to compression or clipping. These phenomena distort the modulated signal and usually lead to spectral regrowth as well.

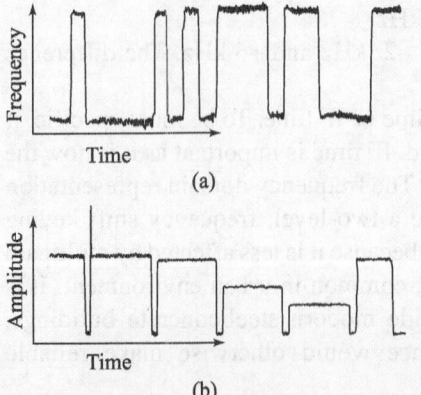

Fig. 7.44 Power and frequency view: (a) frequency vs time and (b) power vs time

7.13 DIGITAL TRANSMITTERS AND RECEIVERS

Figure 7.45 is a simplified block diagram of a digital communication transmitter. It begins and ends with an analog signal. The first step is to convert a continuous analog signal to a discrete digital bitstream. This is called *digitization*. The next step is to add voice coding for data compression. Then some channel coding is added. Channel coding encodes the data in such a way as to minimize the effects of noise and interference in the communication channel. Channel coding adds extra bits to the input data stream and removes redundant ones. Those extra bits are used for error correction or sometimes to send training sequences for identification or equalization. This can make synchronization (or finding the symbol clock) easier for the receiver. The symbol clock represents the frequency and exact timing of the transmission of the individual symbols. At the symbol clock transitions, the transmitted carrier is at the correct I/Q (or magnitude/ phase) value to represent a specific symbol (a specific point in the constellation). Then the values (I/Q or magnitude/phase) of the transmitted carrier are changed to represent another symbol. The interval between these two times is the symbol clock period. The reciprocal of this is the symbol clock frequency. The symbol clock phase is correct when the symbol clock is aligned with the optimum instant(s) to detect the symbols. The next step in the transmitter is filtering. Filtering is essential for good bandwidth efficiency. Without filtering, signals would have very fast transitions between states and, therefore, very wide frequency spectra—much wider than is needed for the purpose of sending information. A single filter is shown for simplicity, but in reality there are two filters, one each for the I and Q channels. This creates a compact and spectrally efficient signal that can be placed on a carrier. The output from the channel coder is then fed into the modulator. Since there are independent I and Q components in the radio, half of the information can be sent on I and the other half on Q. This is one reason digital radios work well with this type of digital signal. The I and Q components are separate. The rest of the transmitter looks similar to a typical RF transmitter or microwave transmitter/receiver pair. The signal is converted up to a higher intermediate frequency (IF), and then further up converted to a higher radio frequency (RF). Any undesirable signals that were produced by the upconversion are then filtered out.

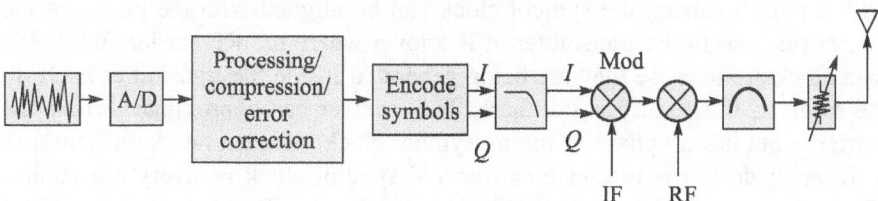

Fig. 7.45 A typical digital transmitter

7.13.1 Digital Receiver

The receiver shown in Fig. 7.46 is similar to the transmitter but in reverse. It is more complex to design. The incoming (RF) signal is first downconverted to (IF) and demodulated. The ability to demodulate the signal is hampered by factors

such as atmospheric noise, competing signals, and multipath or fading. Generally, demodulation involves the following stages:
- Carrier frequency recovery (carrier lock)
- Symbol clock recovery (symbol lock)
- Signal decomposition to I and Q componentsdetermining I and Q values for each symbol (slicing)
- Decoding and de-interleaving
- Expansion to original bitstream
- Digital to analog conversion, if required

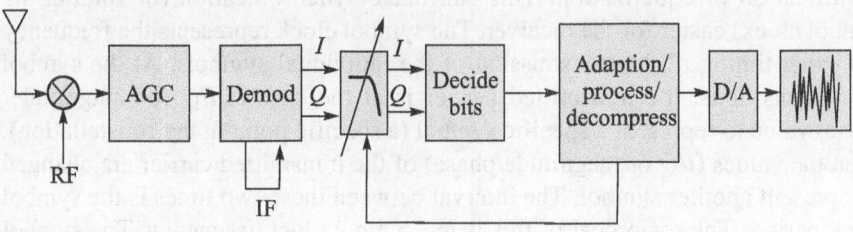

Fig. 7.46 A typical digital receiver

In more and more systems, however, the signal starts out digital and stays digital. It is never analog in the sense of a continuous analog signal like audio. The main difference between the transmitter and the receiver is the issue of carrier and clock (or symbol) recovery. Both the symbol clock frequency and phase (or timing) must be correct in the receiver in order to demodulate the bits successfully and recover the transmitted information. A symbol clock could be at the right frequency but at the wrong phase. If the symbol clock was aligned with the transitions between the symbols rather than the symbols themselves, demodulation would be unsuccessful. Symbol clocks are usually fixed in frequency and this frequency is accurately known by both the transmitter and receiver. The difficulty is to get them both aligned in phase or timing. There are a variety of techniques and most systems employ two or more. If the signal amplitude varies during modulation, a receiver can measure the variations. The transmitter can send a specific synchronization signal or a predetermined bit sequence such as 10101010101010 to 'train' the receiver's clock. In systems with a pulsed carrier, the symbol clock can be aligned with the power of the carrier turn on. In the transmitter, it is known where the RF carrier and digital data clock are because they are being generated inside the transmitter itself. In the receiver, this luxury is not there. The receiver can approximate where the carrier is but has no phase or timing symbol clock information. A difficult task in receiver design is to create carrier and symbol clock recovery algorithms. That task can be made easier by the channel coding performed in the transmitter.

SUMMARY

- Digital modulation provides more information capacity, higher data security, better quality communication, and quicker system availability.

- Examples of digital modulation are digital amplitude modulation, I/Q modulation, ASK, FSK, PSK, QPSK, OQPS, MPSK, and QAM.
- Other more complex systems involve multiplexing. Two common types are TDMA (time division multiple access) and CDMA (code division multiple access) system.
- Digital amplitude modulation is the simplest of digital modulation. It is sometimes called ASK (amplitude shift keying) or on-off keying.
- In I/Q modulation, the signal is separated into in-phase component I and quadrature phase component Q.
- MSK has a narrower spectrum than wider deviation forms of FSK.
- MSK is often classified as FM modulation. It is also related to offset QPSK owing to the dual nature of FSK and PSK modulations.
- Another and most used digital member of the digital modulation family is quadrature amplitude modulation (QAM). It is used in applications like microwave digital radio, DVB-C (digital video broadcasting – cable), and modems.
- This modulation format produces a more spectrally efficient transmission. Hence, it is more efficient than BPSK, QPSK, or 8PSK.
- Different types of QAM are 4QAM, 16QAM, 32QAM, and MQAM.

IMPORTANT FORMULAE

- ASK signal:

$$f_{am}(t) = \{1 + e_m(t)\} \left[\frac{E_c}{2} \cos \omega_c t \right]$$

For $e_m(t) = 1$,

$$f_{am}(t) = (1 + 1) \left[\frac{E_c}{2} \cos \omega_c t \right]$$

$$= E_c \cos \omega_c t$$

For $e_m(t) = -1$,

$$f_{am}(t) = (1 - 1) \left[\frac{E_c}{2} \cos \omega_c t \right]$$

$$= 0$$

- The I and Q signals:

Magnitude, $S = \sqrt{I^2 + Q^2}$

Phase, $\phi = \tan^{-1} \dfrac{Q}{I}$

- Shannon limit for information capacity is

$$I = B \log_2 \left(1 + \frac{S}{N}\right) = 3.32 \, B \log_{10} \left(1 + \frac{S}{N}\right)$$

- Nyquist relation for channel capacity is
$$f_b = B \log_2 M$$
- Minimum bandwidth for M-ary digitally modulated signal:
$$B = \left(\frac{f_b}{\log_2 M} \right) = \frac{f_b}{N} \quad \text{where } N = \log_2 M$$

$$\text{baud} = \frac{f_b}{N}$$

- FSK signal:
$$e_{\text{FSK}}(t) = E_c \cos[2\pi \{f_c + e_m(t)\Delta f\} t]$$
For $e_m(t) = 1$,
$$e_{\text{FSK}}(t) = E_c \cos[2\pi \{f_c + \Delta f\} t]$$
For $e_m(t) = -1$,
$$e_{\text{FSK}}(t) = E_c \cos[2\pi \{f_c - \Delta f\} t]$$
- Minimum bandwidth for FSK:
$$B = 2 (\Delta f + f_b)$$
where f_b = input bit rate
- BFSK signal:
$$s(t) = E_c \cos 2\pi f_0 t, \quad \text{for } 0 \le t \le T$$
$$= E_c \cos 2\pi f_1 t, \quad \text{elsewhere}$$
- BPSK signal:
$$s(t) = Am(t) \cos 2\pi f_c t \quad \text{for} \quad 0 \le t \le T$$
where T is the bit duration
- M-ary phase shift keying signal:
$$s(t) = E_c \cos (2\pi f_c t + \theta_i + \theta), \quad 0 \le t \le T$$
$$= 0, \quad \text{elsewhere}$$
where $\theta_i = \dfrac{2\pi}{M} i$ and $i = 0, 1, 2, 3, ..., M - 1$
- QPSK signal:
$$s(t) = E_c p_s(t) \cos \left(2\pi f_c t + \frac{2\pi i}{M} \right)$$
where $i = 0, 1, 2, 3, ..., M - 1$
- In quadrature form, QPSK signal:
$$s(t) = \sqrt{\frac{2E_s}{T}} \cos \{\theta(t)\} \cos (2\pi f_c t) - \sqrt{\frac{2E_s}{T}} \sin \{\theta(t)\} \sin (2\pi f_c t)$$
- Modulation index of a QPSK signal:
$$h = \frac{2\Delta\theta}{\pi} = \frac{2D_p}{\pi\pi}$$
where D_p is the phase sensitivity

- MSK carrier:

$$c(t) = a(t) \sin\left(\frac{\pi}{2T}t\right) \cos\left(\frac{\pi}{T}t\right) + a(t) \sin\left(\frac{\pi}{2T}t\right) \sin\left(\frac{\pi}{T}t\right)$$

- MQAM signal:

$$s(t) = E_i \cos\theta_i \cos2\pi f_c t - E_i \sin\theta_i \sin2\pi f_c t, \quad 0 \le t \le T$$
$$= 0, \text{ elsewhere}$$

ADDITIONAL EXAMPLES

1. Find out the minimum bandwidth required to pass a 20 kbps binary signal using amplitude shift keying. What is the baud rate of the channel?
   Solution

   Bandwidth = 20,000/1 = 20,000 Hz

   Baud rate = bit rate = 20,000

2. For an FSK signal with a mark frequency of 59 Hz and space frequency of 61 Hz, with an input rate of 5kbps, find the peak frequency deviation, minimum bandwidth, and baud rate.
   Solution

   Peak frequency deviation: $\left|\dfrac{59 \text{ kHz} - 61 \text{ kHz}}{2}\right| = 1 \text{ kHz}$

   Minimum bandwidth = 2(1000 + 5000) = 12 kHz

   Baud = 5000/1 = 5000

3. An 8-PSK system operates with a bit rate of 36 kbps. Find out the minimum bandwidth, baud rate, and bandwidth efficiency.
   Solution

   Minimum bandwidth = 36,000/3 = 12,000

   Baud rate = 36,000/3 = 12,000

   Bandwidth efficiency = 36,000/12,000

   = 3 bits per second per cycle of bandwidth

4. An 8-PSK modulator with an input data rate f_d equal to 5 Mbps uses a carrier frequency of 50 MHz. Find out the minimum double sided Nyquist bandwidth and the baud.
   Solution

   The bit rate in the I, Q, and C channels is equal to one-third of the input bit rate, i.e.,

   $$f_{bI} = f_{BQ} = f_{bC} = \frac{5}{3} = 1.66 \text{ Mbps}$$

   Baud = 1.66 Mbps

   Hence, the fastest rate of change and highest fundamental frequency component presented to either balanced modulator is

   $$f_a = \frac{f_{bC}}{2} = \frac{f_{bQ}}{2} = \frac{f_{bI}}{2} = \frac{1.66}{2} = 0.83 \text{ Mbps}$$

The output wave from the modulator is

$$(\sin 2\pi f_a t)(\sin 2\pi f_c t) = \frac{1}{2}\cos 2\pi (f_c - f_a)t - \frac{1}{2}\cos 2\pi (f_c + f_a)t$$

$$= \frac{1}{2}\cos 2\pi[(50 - 0.83)\text{MHz}] - \frac{1}{2}\cos 2\pi[(50 + 0.83)\text{MHz}]$$

$$= \frac{1}{2}\cos 2\pi (49.17\text{MHz}) - \frac{1}{2}\cos 2\pi (50.83\text{MHz})$$

The minimum Nyquist bandwidth = 50.83 – 49.17 = 1.66 MHz

REVIEW QUESTIONS

1. Explain briefly the different types of digital modulation.
2. What are the constraints we face while designing a communication system?
3. Explain with a neat block diagram, a typical digital communication system.
4. Explain in detail the principle of I/Q modulation and concept of I and Q channels.
5. Explain with a neat block diagram, a typical radio transmitter and receiver employing I/Q modulation technique.
6. Write a brief note on information capacity, bits, and bit rate.
7. Explain with block diagrams, BFSK type of modulator and demodulator.
8. Explain in detail the difference between phase shift keying and binary phase shift keying.
9. What is M-ary phase shift keying? Derive a mathematical expression for it.
10. With the help of constellation diagrams, compare MPSK and FSK.
11. Draw the constellation diagram for the following systems:
    (a) BPSK
    (b) QPSK
    (c) 8PSK.
12. Explain with block diagram, the two conventional ways of doing PSK modulation.
13. Show the block diagram that modifies QPSK to OQPSK.
14. What do you mean by minimum shift keying? Explain in detail the MSK modulator with a block diagram.
15. What is QAM? Explain in detail various types of QAM.
16. Give the signal constellation diagrams for 16QAM and 32QAM.
17. Give a summary of various types of modulation.
18. What do you mean by I/Q offset modulation? With the help of constellation diagram, compare it with QPSK.
19. What is differential modulation? Explain it with constellation diagram.
20. With the help of constellation diagram, explain the constant amplitude modulation.

MATLAB EXAMPLES

1. ```
%program for amplitude shift keying%
   clc; clear all; close all;
   s=[1 0 1 0];
   f1=20;
   a=length(s);
   for i=1:a
       f=f1*s(1,i);
       for t=(i-1)*100+1:i*100
           x(t)=sin(2*pi*f*t/1000);
       end
   end
   plot(x);
   xlabel('time in secs');
   ylabel('amplitude in volts');
   title('ASK');
   grid on;
```

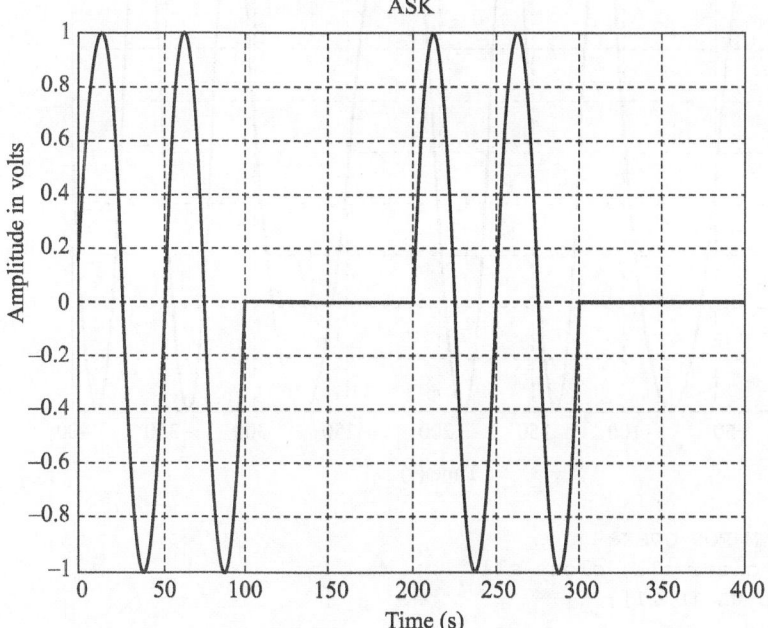

2. ```
%Program for frequency shift keying%
   clc; clear all; close all;
   s=[1 0 1 0];
   f1=10;
   f2=20;
   a=length(s);
   for i=1:a
       if s(1,i)==1
           freq=f1*s(1,i);
           for t=(i-1)*100+1:i*100
               x(t)=sin(2*pi*freq*t/1000);
           end
```

```
            elseif s(1,i)==0
                b=(2*s(1,i))+1;
                freq=f2*b;
                for t=(i-1)*100+1:i*100
                x(t)=sin(2*pi*freq*t/1000);
            end
        end
    end
    plot(x);
    xlabel('time in secs');
    ylabel('amplitude in volts');
    title('FSK');
    grid on;
```

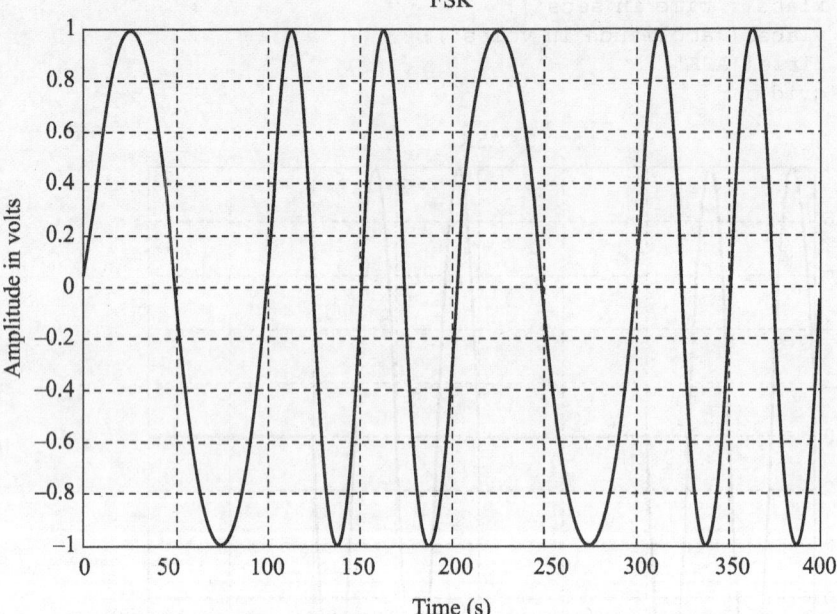

```
3.%PROGRAM FOR QPSK%
    clc; clear all; close all;
    s=[00 01 10 11];
    f1=10;
    a=length(s);
    for i=1:a
        if s(1,i)==00
            for t=(i-1)*100+1:i*100
                x(t)=sin(2*pi*f1*t/1000);
            end
        elseif s(1,i)==01
            for t=(i-1)*100+1:i*100
                x(t)=sin((2*pi*f1*t/1000)+(0.5*pi));
            end
        elseif s(1,i)==10
```

```
            for t=(i-1)*100+1:i*100
                 x(t)=sin((2*pi*f1*t/1000)+(pi));
            end
        elseif s(1,i)==11
            for t=(i-1)*100+1:i*100
                 x(t)=sin((2*pi*f1*t/1000)+(1.5*pi));
            end
        end
    end
    plot(x);
    xlabel('time in secs');
    ylabel('amplitude in volts');
    title('QPSK');
    grid on;
```

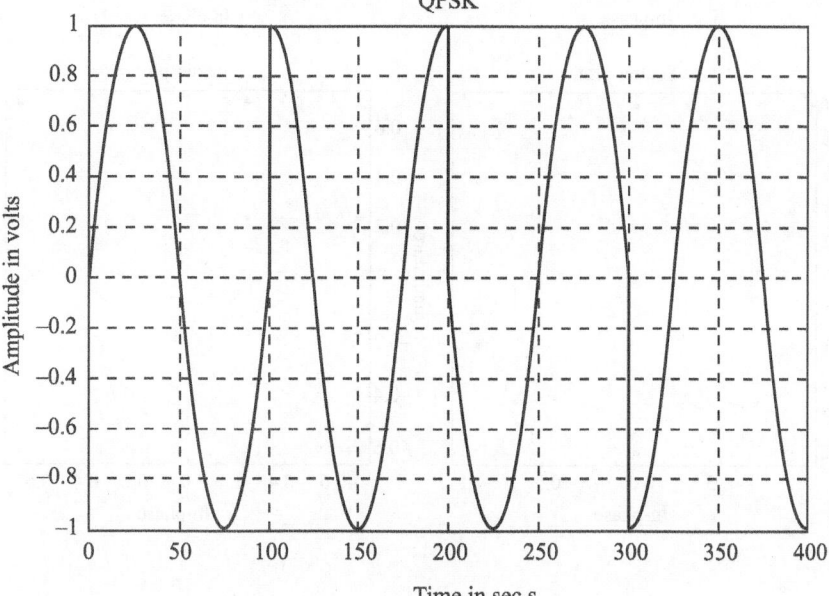

```
4.%program for M-PSK AND M-QASK%
    M = 4; x= [0:M-1];
    scatterplot(pskmod(x,M));

    y = qammod(x,M);
    k = modnorm(y,'peakpow',1);
    y = k*y; scatterplot(y);

    M = 16; x = [0:M-1];
    scatterplot(pskmod(x,M));

    y = qammod(x,M);
    k = modnorm(y,'peakpow',1);
    y = k*y; scatterplot(y);
```

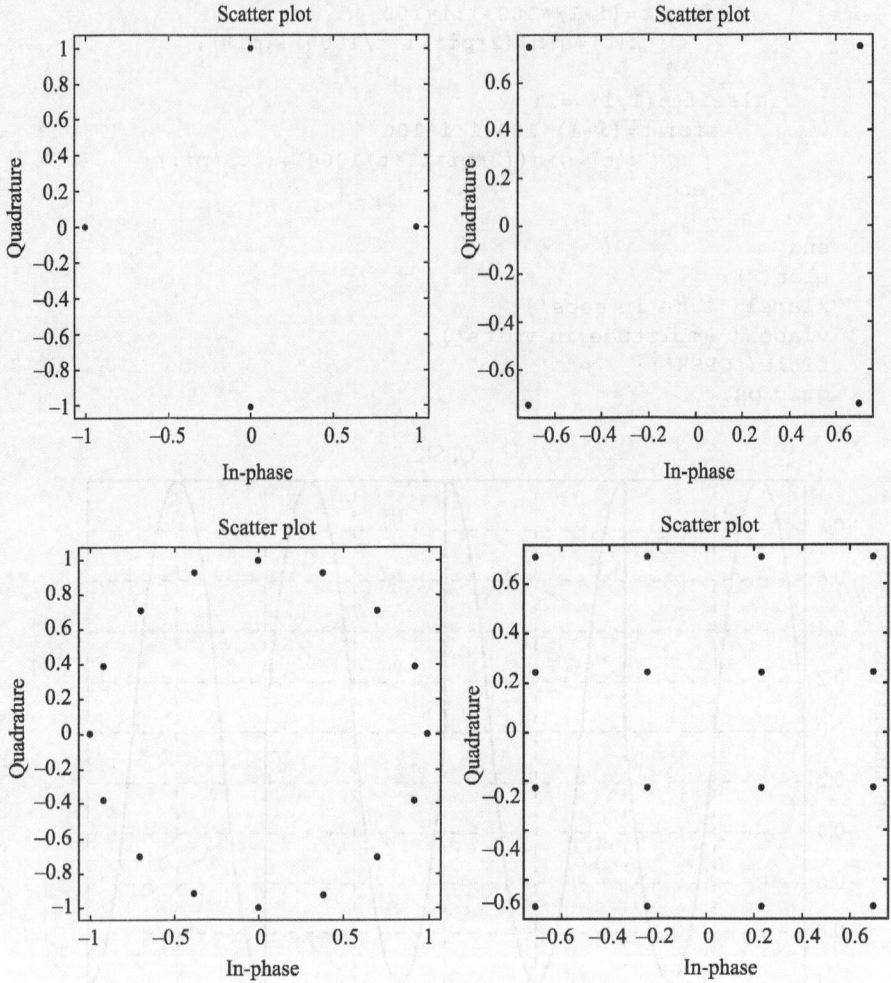

8

Information Theory

LEARNING OBJECTIVES

This chapter will enable the students to become familiar with
- Measure of information
- Entropy
- Source coding theorem and algorithm
- Channel capacity and Gaussian channel capacity
- Shannon's theorem, channel coding theorem, and information capacity theorem
- Rate distortion
- Automatic repeat request (ARQ) and its performance
- Error-free communication over noisy channel
- Channel capacity of continuous channel
- Application of information theory

8.1 INTRODUCTION

Information theory is a branch of mathematical theory of probability and mathematical statistics that quantifies the concept of information. Its applications include information, entropy, communication system, cryptography data transmission, rate distortion theory, data compression, error correction, and related topics.

Shannon, who has been called the father of information theory, was working on the problem of efficiently transmitting information over a noisy communication channel. He employed ergodic theory and probability to study the statistical characteristic of communication systems and thus formulated information theory. Information theory is now used in most of the disciplines like physics, medicine, and computer science apart from telecommunications.

Shannon's theory for the first time considered communication as a mathematical problem in statistics and helped communication engineers to determine the capacity of communication channel in terms of bits.

The transmission part of the theory is not concerned with the meaning of the message conveyed, though the other part of information theory deals with content through lossy compression of messages subjected to fidelity criteria.

Both the parts of information theory are joined together and mutually justified by the information transmission theorems or source channel separation theorems that justify the use of bits for information.

8.2 MEASURE OF INFORMATION

To have a quantitative measure of information, we share with a basic model of information source and define its information content to satisfy some intuitive properties. The source is assumed to be discrete one. If it happens to be a continuous one, we can always convert it to a discrete one by sampling. The output of this source is revealed. Let S_1 be the most likely output and S_N be the least likely output. Let us see which output conveys more information (S_1 or S_N), i.e., which is the most probable or the least probable output. S_N reveals the most information or stated otherwise it follows that a rational measure of information for an information source should be a decreasing function of the probability of that output. Also a small change in the probability of a certain output should not drastically change the information delivered by that output. Hence, the information measure should be a continuous function of the probability of the source output.

Let the information about the output a_k can be subdivided into two independent parts called a_{k1} and a_{k2}, i.e.,

$$X_k = (X_{k1}, X_{k2}), a_k = (a_{k1}, a_{k2}) \tag{8.1}$$

and $\qquad P(X = a_k) = P(X_{k1} = a_{k1}) \, P(Xk_2 = ak_2) \tag{8.2}$

This is possible only when both the components are independent. Since the components are independent, revealing the information about one component does not provide any information about the other component. Hence, the amount of information provided by revealing a_k is the sum of two information components obtained by revealing a_{k1} and a_{k2}. With the above discussion, we can conclude that the amount of information revealed about an output a_k with probability P_k must satisfy the following conditions:

(a) The information of the output a_k depends on the probability of a_k and not on the value of a_k. We denote this function as $I(P_k)$ and call it *self-information*.
(b) Self-information is a continuous function of P_k, i.e., $I(.)$ is a continuous function.
(c) Self-function is a decreasing function of this argument.
(d) If $P_k = P_{k1}P_{k2}$, then
$$I(P_k) = I(P_{k1}) + I(P_{k2}) \tag{8.3}$$

The only function that satisfies all these properties is the logarithmic function, i.e., $I(x) = -\log(x)$. Information is measured in units that depend on the base of the logarithm. If the base is 2, the information is expressed in bits per sample.

Now the information revealed about each source output a_k is defined as the self-information of that output, which is given by $-\log(P_k)$. We can define the information content of the source as the weighted average of the self-information of all source outputs. This is due to the fact that the various source outputs appear with their corresponding probabilities. Hence, the information revealed by an unidentified source output is the weighted average of the self-information of various source outputs, i.e.,

$$\sum_{k=1}^{N} P_k I(P_k) = \sum_{k=1}^{N} -k_i \log k_i \tag{8.4}$$

The information content of the information source is known as the *entropy* of the source and is denoted by $H(x)$. Suppose we have M different and independent messages $m_1, m_2, m_3, \cdots$ with probabilities of occurrence $P_1, P_2, P_3, \cdots$ and suppose further during a long period of transmission a sequence of N messages have been generated. Then if N is very large, we can expect that in N messages sequence, we transmitted $P_1 N$ messages of m_1, $P_2 N$ messages of m_2, and so on and the total information content in such a sequence will be

$$I_{\text{total}} = P_1 N \log \frac{1}{P_1} + P_2 N \log \frac{1}{P_2} + \cdots \tag{8.5}$$

The average information per message interval represented by symbol H will then be

$$H = \frac{I_{\text{total}}}{N} = p_1 \log \frac{1}{p_1} + p_2 \log \frac{1}{p_2} + \cdots$$

$$= \sum_{k=1}^{M} p_k \log \frac{1}{p_k} \tag{8.6}$$

This average information is also called entropy. When there is only a single possible message, i.e., $p_k = 1$, the receipt of that message conveys no information. At the other extreme as $p_k \rightarrow 0$, $I_k \rightarrow \infty$. However, since

$$\lim_{p \rightarrow 0} P \log \frac{1}{p} = 0 \tag{8.7}$$

the average information associated with an extremely unlikely message as well as an extremely likely message is zero.

Now let us consider the case of two messages with probabilities p and $(1 - p)$. The average information per message is

$$H = p \log_2 \frac{1}{p} + (1 - p) \log \frac{1}{1 - p} \tag{8.8}$$

A plot of H as a function of P is shown in Fig. 8.1. It is clearly seen from the figure that H is zero both at $P = 0$ and $P = 1$. Maximum value of H can be located by setting $\frac{dH}{dP} = 0$. For this condition, the value of P works out to be $\frac{1}{2}$, i.e., when two messages are equally likely, the corresponding H is

$$H_{\text{max}} = \frac{1}{2} \log_2 2 + \frac{1}{2} \log_2 2 = \log_2 2 = 1 \tag{8.9}$$

or $\qquad H_{\max} = 1$ bit/message

When there are m messages, H becomes maximum when all the messages are equally likely. In this case, each message has probability $P = \dfrac{1}{M}$ and

$$H_{\max} = \sum \frac{1}{M} \log_2 m = \log_2 m \qquad (8.10)$$

The function H denoted in Fig. 8.1 is known as the *binary entropy function Hb(P)*. From the plot of the binary entropy function, we can see that the entropy of the binary memoryless source is zero when either $p = 0$ or $p = 1$. These two cases correspond to the time when the source generates all zero or all ones, i.e., in both cases, the source is deterministic and completely predictable. On the other hand, entropy is maximized with a maximum equal to 1, when the two source outputs are equally probable and each has probability of 1/2. This is the case when the output is least predictable and hence its entropy is maximized. Hence, it is clearly seen that H depends on the probabilities of messages and entropy as a measure of the information content

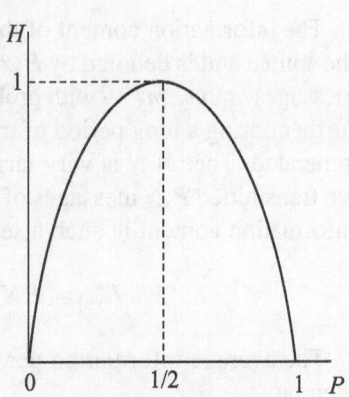

Fig. 8.1 Average information H: two messages plotted as a function of the probability of one of the messages

of a source or measure of uncertainty of it. For discrete signal, the probability density function (PDF) is defined as the probability mass function (PMF).

8.3 JOINT AND CONDITIONAL ENTROPY

When we are dealing with two or more random variables, we can introduce joint and conditional entropy like the way joint and conditional probabilities are introduced. This concept comes into picture when we deal with sources with memory.

8.3.1 Joint Entropy

The joint entropy of two discrete random variables (X, Y) is

$$H(X, Y) = -\sum_{x,y} p(x, y) \log p(x, y) \qquad (8.11)$$

For the case of n random variables $X = (x_1, x_2, \cdots, x_n)$, we have

$$H(X) = -\sum_{x_1, x_2, x_n} p(x_1, x_2, \cdots, x_n) \log p(x_1, x_2, \cdots, x_n) \qquad (8.12)$$

i.e., the joint entropy can be considered as the entropy of a vector valued random variable.

The conditional entropy of the random variable X, given the random variable Y, can be defined by noting that if $Y = y$, then PMF of the random variable X will be $P(X|Y)$ and the corresponding entropy is

$$H(X|Y=y) = \sum_x p(x \mid y) \log p(x \mid y) \tag{8.13}$$

which is nothing but the uncertainty in X when Y is certain. This is known as conditional entropy, which is defined below.

8.3.2 Conditional Entropy

Conditional entropy of the random variable X, given the random variable Y, is defined by

$$H(X|Y) = -\sum_{x,y} p(x,y) \log p(x \mid y) \tag{8.14}$$

In general, we have

$$H(X_n \mid X_1, \cdots, X_{n-1}) = -\sum_{x_1, \cdots, x_n} p(x_1, x_2, \cdots, x_n) \log p(x_n \mid x_1, \cdots, x_{n-1}) \tag{8.15}$$

From the definition of the joint entropy of two random variables, we can infer that the information content of the pair (X, Y) is equal to the information of Y plus the information content of X after Y is known. In other words, it states that the same information is transferred by either revealing the pair (X, Y) or by first revealing Y and then revealing the remaining information in X. We can generalize this relation to the case of n random variables to indicate the chain rule for entropy as follows:

$$H(X) = H(X_1) + H(X_2 \mid X_1) + \cdots + H(X_n \mid X_1, X_2, \cdots, X_{n-1}) \tag{8.16}$$

If the random variables $(X_1, X_2, \cdots, X_n)$ are independent, then the relation reduces to

$$H(X) = \sum_{i=1}^{n} H(X_i) \tag{8.17}$$

If the random variable X_n denotes the output of a discrete source at time n, then $H(X_2|X_1)$ denotes the new information provided by the source output X_2 to someone who already knows the source output X_1. Similarly, $H(X_n|X_1 X_2 \cdots X_{n-1})$ denotes the new information in X_n for an observer who has observed $(X_1, X_2, \cdots, X_{n-1})$. The limit of this conditional entropy as n tends to infinity is known as the *entropy rate* of the random process.

8.3.3 Entropy Rate

The entropy rate of a stationary discrete time random process is defined by

$$H = \lim_{n \to \infty} H(X_n \mid X_1, X_2, \cdots, X_{n-1}) \tag{8.18}$$

The term *stationary* implies the existence of the limit. Alternative definition of the entropy rate for sources with memory is given by

$$H = \lim_{n \to \infty} \frac{1}{n} H(X_1, X_2, \cdots, X_n) \tag{8.19}$$

Entropy rate is basically a measure of the uncertainty per output symbol of the source.

8.3.4 Mutual Information

We have seen that for discrete random variables, $H(X|Y)$ denotes the entropy of the random variable X after the random variable Y is known. Hence, if the entropy of the random variable X is $H(X)$, then $H(X) - H(X|Y)$ denotes the amount of uncertainty of X that has been removed by revealing random variable Y. In other words, $H(X) - H(X|Y)$ is the amount of information provided by the random variable Y about random variable X. This quantity is very much useful in both source and channel coding and is called the *mutual information* between the two random variables.

8.4 DIFFERENTIAL ENTROPY

This term is defined for discrete time continuous alphabet source whose outputs are real numbers. This term, however, does not have the intuitive interpretation of entropy as the uncertainty in source output. An infinite number of bits per source output are required to reconstruct the output of a continuous source reliably. This is due to the fact that the binary expansion of a real number has infinitely many bits.

Differential entropy of a continuous random variable X with the probability density function $f_x(x)$ is denoted by $h(x)$ and defined by

$$h(x) = -\int_{-\infty}^{\infty} f_x(x) \log f_x(x) dx \qquad (8.20)$$

where $0 \log 0 = 0$

Extension of the definition of differential entropy to joint random variables and conditional differential entropy is straightforward.

For two random variables,

$$h(X,Y) = \int_{-\infty}^{\infty} \int_{-\infty}^{\infty} f(x, y) \log f(x, y) dx dy \qquad (8.21)$$

and $\qquad h(X|Y) = h(X,Y) - h(Y) \qquad (8.22)$

The mutual information between two continuous random variables X and Y is given by

$$I(X; Y) = h(Y) - h(Y|X) = h(X) - h(X|Y) \qquad (8.23)$$

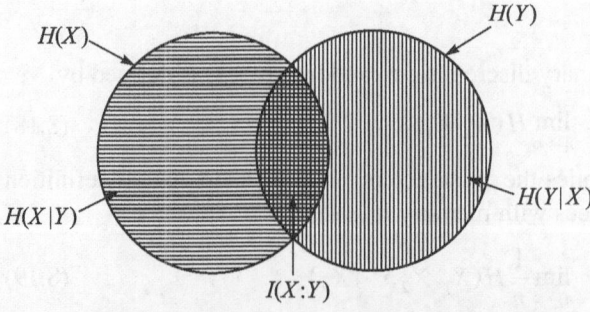

Fig. 8.2 Entropy, conditional entropy, and mutual information

The differential entropy does not have the intuitive interpretation of discrete source entropy. But the mutual information of continuous variables has the same interpretation as the discrete random variables, i.e., the information provided by one random variable about the other random variable. Figure 8.2 shows the entropy, conditional entropy, and mutual information.

8.4.1 Information Rate

Let r message/s be the rate of generation of messages by the source of the message. Then the information rate is defined as

$$R = rH = \text{average number of bits of information/s}$$

Information rate can be better understood by the following example.

Let an analog signal be bandlimited to B Hz and sampled at the Nyquist rate and the samples are quantized to 4 levels. The four quantized levels are independent and occur with probabilities $P_1 = P_3 = \dfrac{1}{4}, P_2 = \dfrac{1}{8}, P_4 = \dfrac{3}{8}$. Then the average information H is

$$H = P_1 \log \frac{1}{P_1} + P_2 \log \frac{1}{P_2} + P_3 \log \frac{1}{P_3} + P_4 \log \frac{1}{P_4}$$

$$= \frac{1}{4}\log_2 4 + \frac{1}{8}\log_2 8 + \frac{1}{4}\log_2 4 + \frac{3}{8}\log_2 \frac{8}{3}$$

$$= \frac{1}{2} + \frac{3}{8} + \frac{1}{2} + 1$$

$$= 2 + \frac{3}{8} = 2.375$$

$$= 2.375 \text{ bits/message}$$

The information rate is $R = rH = 2B(2.375) = 4.75B$ bits/s. We can choose to transmit the messages by binary PCM. Each message is identified by a binary code as indicated.

Message	Probability	Binary code
Q_1	1/4	00
Q_2	1/8	01
Q_3	1/4	10
Q_4	3/8	11

The messages are sampled at Nyquist rate and $2B$ messages/s are transmitted. Since each message requires 2 binits, we will be transmitting $4B$ binits per second. As each binit can convey one bit of information, with $4B$ binits/s, we will be able to transmit $4B$ bits of information/s. But in the above example, we are transmitting only $4.75B$ bits/s and not taking the full advantage of the ability of the binary PCM to convey information. One method to circumvent this is to select different quantization levels such that each level is equally likely. In this case, we find that the average information per message is

$$H = 4\left(\frac{1}{4}\log_2 4\right) = 2 \text{ bits/message}$$

$$R = rH = 2B(2) = 4B \text{ bits/s}$$

If it is not convenient or appropriate to change the messages, we can seek an alternative coding scheme in which on the average the number of bits per message is less than 2.

8.4.2 Source Coding to Increase Average Information Per Bit

Let us consider that we have $M = 2^N$ messages coded into N bits. If the messages are equally likely, then the average information per message interval is $H = N$. Since there are N bits in the message, the average information carried by an individual bit is $H/N = 1$ bit. But if the messages are equally unlikely, then H is less than 1 bit of information. This situation can be avoided and improved by using a code in which not all messages are encoded into the same number of bits. We have to use fewer number of bits in code word for more likely message. This is what was followed in Morse code. The alphabet e, which is frequently used is indicated by a single dot, whereas the less frequently occurring alphabets like v are represented by a long sequence of dots and dashes.

8.5 THE SOURCE CODING THEOREM

An important problem in communication is the efficient representation of data generated by a discrete source. The way by which this representation is accomplished is called *source encoding*. *Source encoder* performs this representation. Knowledge of the statistics of source helps us to make the source encoder efficient, i.e., if some source symbols are known to be more probable than others then, this feature can be made use of in the generation of source code by assigning short codes to frequent source symbols and long codes to less frequent symbols. As we have seen earlier the Morse code is an example of this variable length code. An efficient source code has to satisfy two functional requirements:

(a) binary format used for encoding
(b) source code be uniquely decodable so that original source sequence can be reconstructed faithfully from the encoded binary sequence.

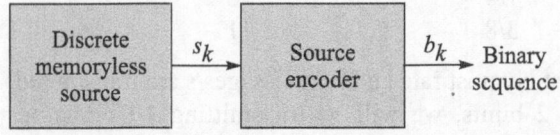

Fig. 8.3 Source encoding

Let us consider a scheme shown in Fig. 8.3 in which a discrete memoryless source whose output S_k is converted by the source encoder into block of 0's and 1's denoted by b_k. It is assumed that the source has an alphabet with k different symbols and the kth symbol s_k occurs with probability p_k, $k = 0, 1, 2, \cdots, (K-1)$. If l_k is the length measured in bits of code word assigned to symbol S_k, then we define the average code word length $\bar{L}$ of the source encoder as

$$\overline{L} = \sum_{k=0}^{k-1} p_k l_k \qquad (8.24)$$

Here $\overline{L}$ represents the average number of bits per source symbol used in the source encoding process. If $L_{\min}$ is the minimum possible value of $\overline{L}$, then coding efficiency is defined as

$$\eta = \frac{L_{\min}}{L} \qquad (8.25)$$

Since $\overline{L} \ge L_{\min}, \eta \le 1$. For a source coder to be efficient, η should approach unity. The value of $L_{\min}$ is determined by Shannon's first theorem called the source coding theorem. The theorem is as follows:

For a memoryless source with entropy $H(X)$, the average code word length $\overline{L}$ for any distortionless source encoding scheme is bounded as

$$\overline{L} \ge H(X) \qquad (8.26)$$

Hence, according to the source coding theorem, the entropy $H(X)$ represents a fundamental limit on the average number of bits per source symbol necessary to represent a discrete memoryless source in that it can be made small but no smaller than the entropy $H(X)$. Hence, with $L_{\min} = H(X)$, the efficency of encoder in terms of entropy $H(X)$ is

$$\eta = \frac{H(X)}{L} \qquad (8.27)$$

The source coding theorem is one of the three fundamental theorems proposed by Shannon. This theorem establishes a fundamental limit on the rate at which the output of an information source can be compressed without causing a large error probability. As we know that the entropy of the source is a measure of its uncertainty or, in other words, the information content of the source, the entropy of the source plays a major role in the context of the source coding theorem.

Let us assume that we observe outputs of length n of a DMS, where n is very large. Then according to the law of large numbers, there is a high probability (which goes to 1 as $n \to \infty$) that letter a_1 is repeated approximately np_1 times, a_2 is repeated approximately np_2 times, $\cdots$, and letter a_N is repeated approximately np_n times. This means that when n is large enough with probability approaching 1, every sequence from the source has the same composition and, therefore, the same probability. The sequence that has this structure is called a typical sequence. Since the assumed source is memoryless, the probability of a typical sequence is given by

$$P(X = x) = \prod_{i=1}^{N} p_i^{np_i}$$

$$= \prod_{i=1}^{N} 2^{np_i \log p_i}$$

$$= 2^n \sum_{i=1}^{N} p_i \log p_i$$

$$= 2^{-nH(X)} \qquad (8.28)$$

Hence, for large n, almost all the output sequences of length n of the source are equally probable with probability $\approx 2^{-nH(x)}$. These are called typical sequences. On the other hand, the probability of the set of non-typical sequences is negligible.

As the probability of the typical sequence is almost 1 and each typical sequence has a probability of almost $2^{-nH(X)}$, the total number of typical sequences is almost $2^{nH(X)}$. Hence, although a source of alphabet size N can produce N^n sequence of length n, the effective number of outputs is $2^{nH(X)}$. The meaning of effective number of outputs is that nothing is lost by neglecting the other outputs and the probability of having lost anything goes to zero as n goes to infinity.

Figure 8.4 shows the set of typical and non-typical sequences indicating the property that is quite important. It tells us that for all practical purposes, only typical sequences can be considered rather than the set of all possible outputs of the source. If n is chosen large enough, the error introduced in ignoring non-typical sequence can be made smaller. This is the philosophy of *data compression*, the practice of representing the output of the source with a smaller number of sequences than the number of the outputs that the source really produces.

Set of typical sequences with $\approx 2^{nH(X)}$ elements

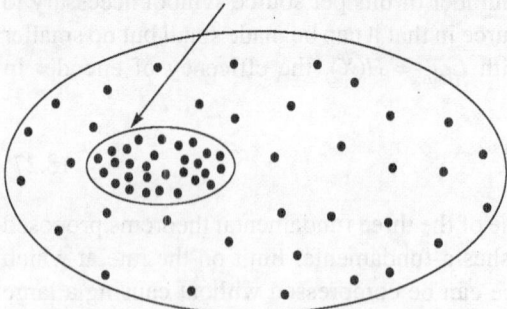

Fig. 8.4 Set of typical and non-typical sequences

Since the error cannot be zero in practical cases, the error can be assumed to be an arbitrary positive number. It can be assumed as minimum as possible. Hence, we can only represent the typical source output without introducing considerable error. Since the total number of sequences is roughly $2^{nH(X)}$, we need $nH(X)$ bits to represent them. These bits are used to represent source outputs of length n. Hence, on the average, any source output requires $H(X)$ bits for an essentially error-free representation. This, once again justifies the notion of entropy as the amount of information per source output. When the source is discrete and memoryless, it can be represented by an independent and identically distributed random variables. Such a source can be compressed only if its probability mass function (PMF) is not uniform. If X is uniformly distributed, then $H(X) = \log N$. Hence, $2^{nH(X)} = 2^{n \log N} = N^n$. This means that the effective number of source outputs of length n is equal to the total number of source outputs and no compression is possible.

Now let us consider the case where source has memory, i.e., the outputs of the source are not independent and previous outputs reveal some information about the future ones. This results in decreasing the rate at which fresh information is produced as more and more source outputs are revealed. This is what happens in English text, which shows a lot of dependency between letters and words. For example 'q' is always followed by a 'u' and a single letter between the spaces is either 'I' or 'a'. The entropy per letter for a large text of English is approximately the limit of $H(X_n|X_1, X_2, \cdots, X_{n-1})$ as n becomes large. In general for stationary sources, the entropy rate has the same significance as the entropy for the case of memoryless sources and defines the number of effective source outputs for any n that is large enough, i.e., 2^{nH}, where H is the entropy/rate.

Statistical models of printed English show that entropy rate converges very quickly. For $n = 10$, we are very close to the limit. For a memoryless source model with $n = 1$, we have $H = 4.03$ bits/letter. Once the memory starts increasing, the size of the space over which conditional probabilities are computed increases rapidly and it is not easy to find the conditional probabilities required to compute the entropy rate. Conditional probabilities for English language estimated to be about 1.3 bits/letter, considering only the 26 alphabets and the space mark. Hence, the source coding theorem can be stated as follows: A source with entropy H can be encoded with an arbitrarily small error probability at any rate R (bit/source output) as long as $R > H$. Stated otherwise, if $R < H$, the error probability will be bounded away from zero. Independent of the complexity of the encoder and decoder employed, this theorem talks about the rate of transmission of information over a communication channel. The communication channel is the one that introduces noise and also limits the bandwidth. The theorem states that in principle, it is possible to find a communication channel that will carry information with a small probability of error, provided the information rate R is less than or equal to c, the channel capacity. In other words, given a source of M, equally likely messages with $M \gg 1$, which is generating information at a rate R and if a channel is there with a capacity c and if $R \le C$, there exists a coding technique such that the output of the source may be transmitted over the channel with a probability of error very much less in the received message.

8.5.1 Source Coding Algorithm

We have seen that H, the entropy of a source, gives limitation on the rate at which source can be compressed for reliable reconstruction, i.e., for the rate above entropy, it is possible to design a code with an error probability as small as required. But at the rate below entropy, such a code does not exist. Shannon's theorem does not provide specific algorithm to design codes approaching close to the entropy of the source. But there are algorithms to design codes that are close to entropy bound.

8.6 DATA COMPACTION

The signals generated by any physical source in their natural form contain a significant amount of information that is redundant. Hence, the transmission of the redundant data is wasteful of primary communication resources. For efficient signal transmission, the redundant information should be removed from the signal prior to transmission. At the same time, care has to be taken so that removing redundant information in no way should affect the loss of information. This removal is generally performed on a signal in digital form, in which case we refer to it as *data compaction* or lossless data compression. The code resulting from such an operation provides the representation of the source output that is not only efficient in terms of average number of bits per symbol but also exact in reconstructing the original signal with no loss of information. The fundamental limit on the removal of redundancy from the data is dictated by the entropy of the source. The data compaction is achieved by assigning short description to the most frequent outcomes of the source output and large description to the less frequent ones. Some source coding schemes are discussed below for data compaction. First we will describe a type of source code known as a *prefix code*, which is not only decodable but also offers the possibility of realizing an average code word length that can be made arbitrarily close to the source entropy.

8.7 PREFIX CODING

Let us consider a discrete memoryless source of alphabet $(s_0, s_1, \cdots, s_{k-1})$ and with probability $(p_0, p_1, \cdots, p_{k-1})$. A source code representing the output of this source has to be of practical use and the code has to be uniquely decodable. This restriction ensures that for each finite sequence symbol emitted by the source, the corresponding sequence of code words is different from the sequence of code words corresponding to any other source sequence. Specific interest is in a special class of code satisfying a restriction known as the *prefix condition*.

Let the code word assigned to source symbol s_k be denoted by $m_{k1}, m_{k2}, \cdots, m_{kn}$, where the individual elements $m_{k1}, m_{k2}, \cdots, m_{kn}$ are '0's and '1's and n is the code word length. The initial part of the code word is represented by the elements $m_{k1}, m_{k2}, \cdots, m_{kn}$ for $i \leq n$. Any sequence made up of the initial part of the code word is called *prefix code word*. Thus, a prefix code is defined as a code in which no code word is the prefix of any code word. The following example will explain the prefix code.

Let us consider the three source codes shown in Table 8.1. Code 1 is not a prefix code since bit 1, the code word for s_0, is a prefix of 11, the code word for s_2. Similarly, the bit 0, the code word for s_1, is a prefix of 01, the code word for s_3.

Table 8.1 Definition of prefix code: an illustration

Source symbol	Probability of occurrence	Code 1	Code 2	Code 3
s_0	0.5	1	1	1
s_1	0.25	0	10	01
s_2	0.125	11	101	001
s_3	0.125	01	1010	0000

Similarly, we can show that code 2 is not a prefix code, but code 3 is. To decode the sequence of the code word generated from a prefix source code, the source decoder simply starts at the beginning of the sequence and decodes one code word at a time. It sets up what is equivalent to a decision tree, which is a graphical portrayal of the code words in the particular code word. This is illustrated in Fig. 8.5. The figure depicts the decision tree corresponding to code 3 in Table 8.1. The tree has an initial state and four terminal states corresponding to the source symbol s_0, s_1, s_2, and s_3. The decoder always starts at the initial state. The first received bit moves the decoder to the terminal state s_0 if it is 1 or to a second decision point if it is 0. Then the second bit moves the decoder one step further down the tree either to terminal state s_1 if it is a 1 or else to the third decision point if it is 0 and so on.

Once each terminal state emits its symbol, the decoder is reset to its initial state. It can be seen that each bit in the received encoded sequence is examined only once. For example, the encoded sequence 0010110000⋯ is decoded as the source sequence s_2, s_1, s_0, s_3,⋯.

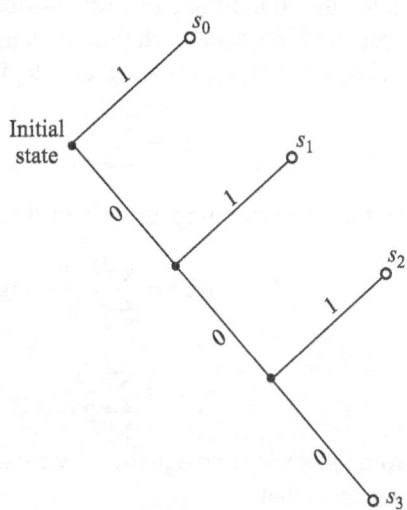

Fig. 8.5   Decision tree for code 3 of Table 8.1

The prefix code is always uniquely decoded. This is an important property. But the converse is not necessarily true. For example, code 2 in Table 8.1 does not satisfy prefix condition, but it is uniquely decodable since the bit 1 indicates the beginning of each code word in the code. Also if a prefix code has been constructed for a discrete memoryless source with source alphabet $(s_0, s_1, \cdots, s_{k-1})$ and source probability $(p_0, p_1, \cdots, p_{k-1})$ and the code word for symbol s_k has length l_k, where $k = 0,1, \cdots, K-1$, then the code word lengths of the code always satisfy a certain inequality, known as *Kraft–McMillan inequality*, given by

$$\sum_{k=0}^{K-1} 2^{l_k} \leq 1 \tag{8.29}$$

where 2 indicates the base, which is the base for binary. But this inequality does not tell us that a source code is a prefix code. It is rather merely a condition on the code word length of the code and not on the code word itself. Prefix codes are distinguished from other uniquely decodable codes by the fact that the end of a code word is always recognizable. Hence, the decoding of a prefix code can be accomplished as soon as the binary sequence representing a source symbol is fully received. Due to this reason, prefix codes are sometimes referred to as *instantaneous codes*.

Let $H(\Upsilon)$ be the entropy of memoryless source, then a prefix code can be constructed with an average code word length $\bar{L}$, which is bounded as follows:

$$H(\Upsilon) \leq \bar{L} < H(\Upsilon) + 1 \tag{8.30}$$

The LHS of the above equation is satisfied with equality under the condition that symbol s_k is emitted by the source with probability $p_k = 2^{-l_k}$, where l_k is the length of the code word assigned to source symbol s_k. Then

$$\sum_{k=0}^{K-1} 2^{-l_k} = \sum_{k=0}^{K-1} p_k = 1 \tag{8.31}$$

Under this condition, the Kraft–McMillan inequality tells us that we can construct a prefix code such that the length of the code word assigned to source symbol S_k is $-\log_2 P_k$. For such a code, the average code word length is

$$\bar{L} = \sum_{k=0}^{K-1} \frac{l_k}{2^{l_k}} \tag{8.32}$$

and the corresponding entropy of the source is

$$H(\Upsilon) = \sum_{k=0}^{K-1} \left(\frac{1}{2^{l_k}}\right) \log_2(2^{l_k}) \tag{8.33}$$

$$= \sum_{k=0}^{K-1} \frac{l_k}{2^{l_k}} \tag{8.34}$$

From the above two equations, we can observe that prefix code is matched to the source in that

$$\bar{L} = H(\Upsilon) \tag{8.35}$$

Use of extended code matches the prefix code to an arbitrary discrete memoryless source. Let $\bar{L}_n$ denote the average code word length of the extended prefix code. For a uniquely decodable code, $\bar{L}_n$ is the smallest possible and from Eq.(8.30), we can infer that

$$H(\Upsilon^n) \le \bar{L}_n < H(\Upsilon^n) + 1 \tag{8.36}$$

Substituting equation $H(\Upsilon^n) = nH(\Upsilon)$ (which states that the entropy of the extended source is equal to n times $H(\Upsilon)$, the entropy of the original source) for an extended source into Eq. (8.36), we get

$$nH(\Upsilon) \le \bar{L}_n < nH(\Upsilon) + 1$$

or

$$H(\Upsilon) = \frac{\bar{L}_n}{n} < H(\Upsilon) + \frac{1}{n} \tag{8.37}$$

In the limit as $n \to \infty$, the lower and upper bounds in the above equation converge to

$$\lim_{n \to \infty} \frac{1}{n} \bar{L}_n = H(\Upsilon) \tag{8.38}$$

Hence, we can state that by making the order of n of an extended prefix source encoder large enough, we can make the code faithfully represent the discrete memoryless source Υ as closely as desired. Stated otherwise, the length of an extended prefix code can be made as small as the entropy of the source provided the extended source has a high enough order, in accordance with the source coding theorem. But we have to pay the price for decreasing the average code word length, i.e., increased decoding complexity. This is brought out by the high order of the extended prefix code.

8.8 SHANNON–FANO CODING

This is one of the methods to generate a more efficient code. An example explains this code as follows. Let us consider that there are eight possible messages m_1 through m_8 with probabilities 1/2, 1/8, 1/8, 1/16, 1/16, 1/16, 1/32, and 1/32. The messages have been ordered in order of decreasing probability as in Fig. 8.6. In column 1, we divide the messages into two partitions such that the sum of the probabilities of each group is the same.

Message	Probability	I	II	III	IV	V	No of bits/message
m_1	1/2	0					1
m_2	1/8	1	0	0			3
m_3	1/8	1	0	1			3
m_4	1/16	1	1	0	0		4
m_5	1/16	1	1	0	1		4
m_6	1/16	1	1	1	0		4
m_7	1/32	1	1	1	1	0	5
m_8	1/32	1	1	1	1	1	5

Fig. 8.6 Shannon–Fano algorithm

Thus, m_1 in one partition has the probability 1/2 and sum of probabilities of all other messages in the second partition is also 1/2. We assign the bit 0 to the messages in one partition and assign the bit 1 to all the messages in the other partition. This method of dividing groups in the same partition into two partitions each with equal sum of probabilities is continued until each message finds itself singled out in a partition. At each partition, one group has a 0 assigned while a 1 is assigned to the other. In the example considered, five partitions are required. Collecting the 0's and 1's for each message m_1 to m_8, we find that m_1 is represented by the single bit 0, m_2 is represented by the three bit code 100 and so on with m_8 having the five bit code word 11111.

With these codes, the average information per message interval is

$$H = \sum_{i=1}^{\infty} p_i \log_2 \frac{1}{p_i} = \frac{1}{2}\log_2 2 + 2 \times \frac{1}{8}\log_2 2^3 + 3 \times \frac{1}{16}\log_2 2^4 + 2 \times \frac{1}{32}\log_2 2^5$$

$$= 2\frac{5}{16}$$

The average number of bits per message

$$= 1\left(\frac{1}{2}\right) + 3\left(\frac{1}{8}\right) + 3\left(\frac{1}{8}\right) + 4\left(\frac{1}{16}\right) + 4\left(\frac{1}{16}\right) + 4\left(\frac{1}{16}\right) + 5\left(\frac{1}{32}\right) + 5\left(\frac{1}{32}\right)$$

$$= 2\frac{5}{16}$$

If we have not used this method of coding, we would have sent 3 bits per message instead of $2\frac{5}{16}$. This method of coding has saved about 30%. This saving in the average number of bits per message can be put to good advantage in a number of ways. We can reduce the bandwidth by giving more time per bit and also avoiding noise to some extent. If we decide to fix the bit rate, the bits not used for transmission can be used for providing error detection, correction, or both. In this example, we selected probabilities in such a manner that at each partitioning, it is possible to arrange the sum of probabilities in each group exactly equal. If it is not possible this way, then we resort to satisfying the equal probability condition as nearly as possible. In this case, code efficiency will be reduced.

8.9 THE HUFFMAN SOURCE CODING ALGORITHM

The Huffman coding in general performs better than Shannon–Fano coding and for a source of given entropy, gives minimum average word length and hence called *optimum code*. In Huffman coding, fixed length blocks of the source output are mapped to variable length binary blocks called *fixed to variable length coding*. This makes the more frequently occurring fixed length sequences to shorter binary sequences and the less frequently occurring sequences to larger binary sequences, thus achieving good compression ratio.

In Huffman coding, we choose a code word length such that more probable sequences have shorter code words. If it is possible to map each source output of probability P_i to a code word of length approximately $\log 1/P_i$ and at the same time ensure unique decodability, then we can achieve an average code word length of approximately $\sum_i p_i \log 1/P_i$, which is equal to $H(X)$.

Huffman codes are uniquely decodable instantaneous codes with a minimum average code word length, which is optional. The meaning of optional is that among all codes that satisfy the prefix conditions, Huffman codes have the minimum code word length.

8.9.1 Huffman Coding Algorithm

The flow chart for this algorithm is shown in Fig. 8.7. The steps are as follows:

(a) Source outputs are sorted in decreasing order of their probabilities.
(b) The two least probable outputs are merged into a single output whose probability is the sum of the corresponding probabilities.
(c) If the number of remaining outputs is 2, then go to step 'd', otherwise go to step b.
(d) Assign arbitrarily 0 and 1 as code word for the two remaining outputs.
(e) If the output is the result of the merger of the two outputs in a preceding step, append the current code word with 0 and 1 to obtain the code word for the preceding outputs and then repeat step 'e'. If no output is preceded by another output in a preceding step, then stop.

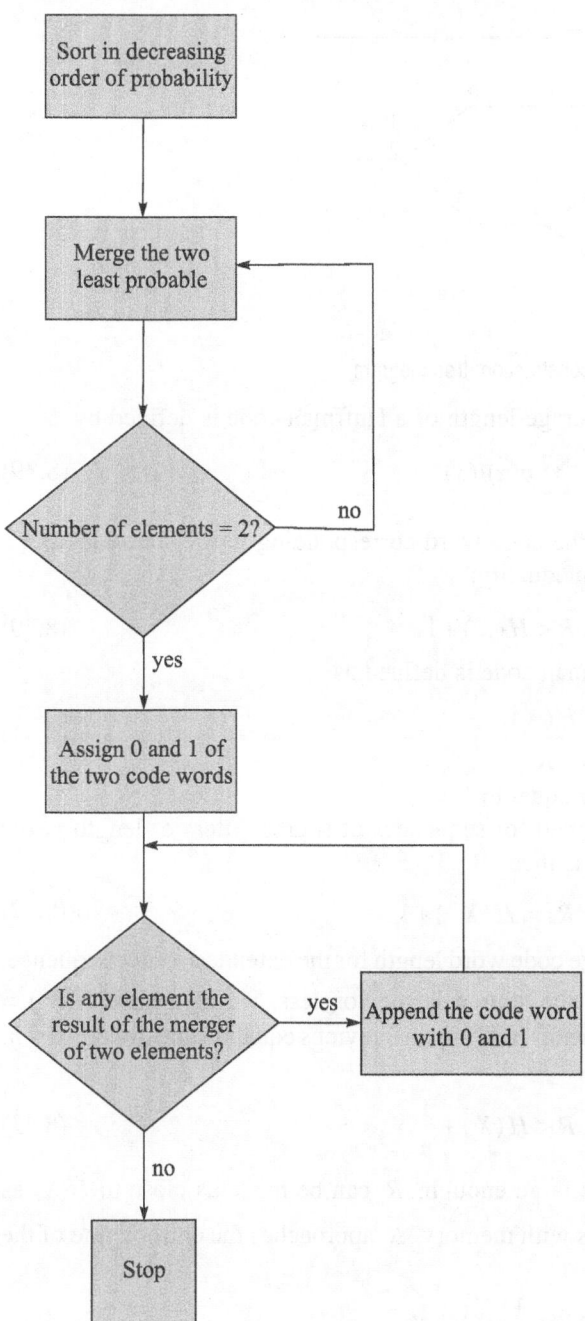

Fig. 8.7   Huffman coding algorithm: flow chart

Let us design the Huffman code for the source with probabilities 1/2, 1/4, 1/8, 1/16, and 1/16. The tree diagram shown in Fig. 8.8 gives the steps for code construction and the resulting code words.

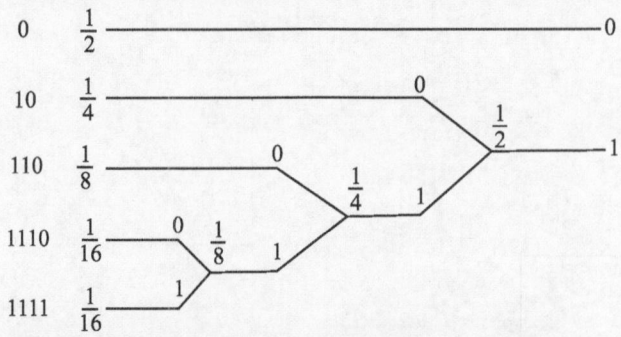

Fig. 8.8 Design steps for code construction: tree diagram

We can show that the average length of a Huffman code is defined by

$$\overline{R} = \sum_{x \in \mathbb{F}} p(x)l(x) \tag{8.39}$$

where $l(x)$ is the length of the code word corresponding to the source output x and satisfies the following inequality:

$$H(X) \le \overline{R} < H(X) + 1 \tag{8.40}$$

The efficiency of a Huffman code is defined as

$$\eta = \frac{H(X)}{\overline{R}} \tag{8.41}$$

which is always less than or equal to 1.

If Huffman code is designed for sequences of source letters of length n, the nth extension of the source is then

$$H(X^n) \le \overline{R}_n < H(X^n) + 1 \tag{8.42}$$

where $\overline{R}_n$ denotes the average code word length for the extended source sequence. Therefore, $\overline{R} = (1/n)\overline{R}_n$. If the source is memoryless, we also have $H(X^n) = nH(X)$. Substituting this value of $H(X^n)$ in the previous equation and dividing by n, we get

$$H(X) \le \overline{R} < H(X) + \frac{1}{n} \tag{8.43}$$

Hence, for any n that is large enough, $\overline{R}$ can be made as close to $H(X)$ as desired. For discrete sources with memory, $\overline{R}$ approaches the entropy rate of the source as defined by

$$H = \lim_{n \to \infty} \frac{1}{n} H(X_1, X_2, \cdots, X_n)$$

8.10 LEMPEL–ZIV SOURCE CODING ALGORITHM

The only drawback of the Huffman code is that it requires knowledge of a probabilistic model of the source. But in practice, source statistics are not always known. Also in modelling text, we find that a storage requirement prevents the Huffman code from capturing the higher-order relationship between words and phrases, thereby compromising the efficiency of the code. These practical limitations are overcome in Lempel–Ziv algorithm, which is intrinsically adaptive and simpler to implement than Huffman coding.

This algorithm belongs to the class of universal source coding algorithm and is independent of the source statistics. This algorithm is a variable to fixed length coding scheme, i.e., any sequence of source outputs is uniquely parsed into phrases of varying lengths and these phrases are encoded using code words of equal length. Parsing is done by identifying phrases of the smallest length that have not appeared so far and the parser observes the source output. As long as the new source output sequence after the last phrase coincides with one of the existing phrases, no new phrase is introduced and another letter from the source is considered. As soon as the new output sequence is different from the previous phrases, it is recognized as a new phrase and encoded. The encoding scheme is as follows: The new phrase is the concatenation of a previous phrase and a new source output. To encode it the binary expansion of the previous phase and the new bits are concatenated.

The encoding in this algorithm is accomplished by parsing the source data stream into segments that are the shortest subsequences not encountered previously. Let us consider an example:

The input sequence is specified as follows:

10010011011101100

We can assume that binary symbols 0 and 1 are already stored in that order in the code book.

We parse the sequence in the following phrases:

0,1,10,01,00,11,011,101,100

It can be seen that all the phrases are different and each phrase is one of the previous phrases concatenated with a new source output. The number of phrases is 9. Hence, for each phrase, we need 4 bits plus an extra bit to represent the new source output. The preceding sequence is encoded by

00000, 00001, 00100, 00011, 00010, 00101, 01001, 00111, 00110

This representation cannot be called a data compression since a sequence of length 17 has been mapped into a sequence of length 45. But as the length of the original sequence is increased, the compression role of this algorithm becomes more apparent. It can be proved that for a stationary and ergotic source, as the length of the sequence increases, the number of bits in the compressed sequence approaches $nH(X)$, where $H(X)$ is the entropy rate of the source.

Let us see in detail how the encoding is done for the example given.

Input binary sequence: 10010011011101100

1. Since it is assumed, symbols 0 and 1 are already stored in that order. Hence, we write

   Subsequences stored: 0, 1

   Data to be parsed: 10010011011101100

   The encoding process begins at left with symbols 0 and 1 already stored, the shortest subsequence of the data encountered for the first time and not encountered before is 10. Hence, the new subsequences stored become and data to be parsed will exclude this, i.e.,

   Subsequences stored: 0,1,10

   Data to be parsed: 010011011101100

2. The second shortest subsequence not seen before is 01 and now we have

  Subsequences stored: 0,1,10, 01

  Data to be parsed: 0011011101100

3. The next shortest subsequence not encountered previously is 00 and now we have

  Subsequences stored: 0,1,10, 01, 00

  Data to be parsed: 11011101100

4. The next shortest subsequence not encountered previously is 11, and hence we have

  Subsequences stored: 0,1,10,01,00,11

  Data to be parsed: 011101100

5. The next shortest subsequences not encountered previously is 011, and hence we have

  Subsequences stored: 0,1,10,01,00,11,011

  Data to be parsed: 101100

6. The next shortest subsequences not encountered previously is 101, and hence we have

  Subsequences stored: 0,1,10,01,00,11,011,101

  Data to be parsed: 100

7. The last shortest subsequence not encountered previously is 100.

The given data stream has been completely parsed. Now we will take up the actual encoding.

Numerical position:	1	2	3	4	5	6	7	8	9
Subsequences:	0	1	10	01	00	11	011	101	100
Numerical representation:			21	12	11	22	42	32	31
Binary encoding:			0100	0011	0010	0101	1001	0111	0110

The binary encoding is done as discussed below.

From the numerical representation of subsequences, the binary equivalent for the first digit is taken followed by the subsequence for the last digit taken as the numerical position (that is the bit value for this numerical position). Since the first two numerical positions are already stored before parsing, their binary equivalent is taken as 000 followed by the subsequences, respectively. Thus, the numeric position is encoded as 0000 and 0001.

The last symbol of each subsequence in the code book is called *innovation symbol* since its appendage to a particular subsequence distinguishes it from all previous subsequences stored, i.e., the last bit of each uniform block of bits in the binary encoded representation of data stream represents the innovation symbol for the particular subsequence under consideration. The remaining bits provide the equivalent binary representation of the *pointer* to the root subsequence that matches the one in question except for the innovation symbol.

The decoder uses the pointer to identify the subsequence and then appends the innovation symbol. Let us take, for example, the binary encoded block 1001 in position 7. In this data, the last bit 1 is innovation symbol. The first three bits give the binary representation of the numeric position, which is in this case 4(100). This points to the subsequence 01 append to this 01 the innovative symbol 1. Then the decoded value is 011, which is correct for the position 7.

From the example given above, it can be seen that in contrast to Huffman coding, the Lempel–Ziv algorithm uses fixed length codes to represent a variable number of source symbols. Thus, the Lempel–Ziv code is suitable for synchronous transmission. Generally, fixed blocks of 12 bits long are used implying a code book of 4096 entries.

The Lempel–Ziv algorithm is now the standard algorithm for file compression. When it is applied to ordinary English text, it achieves a compaction of about 55% compared to 45% achieved by Huffman coding. This is due to the fact that Huffman coding does not take advantage of the intercharacter redundancies of the language.

One problem with the Lempel–Ziv algorithm is how the number of phrases should be chosen. In the example considered, we have chosen 9 phrases, which leaves us 4 bits to represent each phrase. However, any fixed number of phrases will eventually become too small and overflow would occur. If we have to continue coding the source for additional input letters, it will not be possible to more number of new phrases. In our example, we have 9 phrases, which are represented by 4 bits. Hence, we can add 7 more phrases but not more than that. To overcome this problem, the encoder and decoder must purge the older elements, which are not useful anymore, and substitute new elements for them. The purging method should be adopted by both encoder and decoder.

8.11 CAPACITY OF GAUSSIAN CHANNEL

For the channel in which noise is Gaussian, a theorem complementary to Shannon's theorem applies. This theorem is known as Shannon–Hartley theorem. The theorem is stated as follows.

The channel capacity of a white bandlimited Gaussian channel is

$$C = B \log_2\left(1 + \frac{S}{N}\right) \text{ bits} \tag{8.43}$$

where B is the channel bandwidth and S/N is the signal-to-noise ratio within the channel bandwidth, i.e., $N = \eta B$ and $\eta/2$, the two-sided power spectral density. Since channel encountered in most of the physical systems generally is approximately Gaussian and also the results obtained for a Gaussian channel often provide a lower bound on the performance of a system operation over a non-Gaussian channel. Hence, if a particular encoder is used for a Gaussian channel resulting in an error probability P_e, then for a non-Gaussian channel its encoder decoder can be designed so that P_e for this channel will be smaller.

Let us assume that the messages are represented by fixed voltage levels for the purpose of transmission over the channel. As the source generates one message after another in sequence the waveform of the transmitter is as shown in Fig. 8.9. The received signal is accompanied by the noise whose root mean square voltage is σ. The levels have been separated by an interval $\lambda\sigma$, where λ is a number presumed large enough to allow recognition of individual levels with an acceptable probability of error.

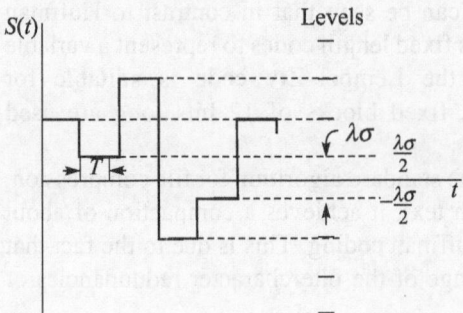

Fig. 8.9 Sequence of waveform of transmitter

If the levels are assumed to be even, the levels are located at voltages $\pm\lambda\sigma/2$, $\pm3\lambda\sigma/2$, etc. If there are to be M possible messages, then there must be M levels. It is then assumed that messages and hence the levels occur with equal likelihood. In that case, the average signal power is given as

$$S = \frac{2}{M}\left[\left(\frac{\lambda\sigma}{2}\right)^2 + \left(\frac{3\lambda\sigma}{2}\right)^2 + \cdots + \left(\frac{(M-1)\lambda\sigma}{2}\right)^2\right] \quad (8.44)$$

$$= \frac{M^2-1}{12}(\lambda\sigma)^2 \quad (8.45)$$

From this equation, we can find M as

$$M = \left(1 + \frac{12S}{\lambda^2\sigma^2}\right)^{1/2} = \left(1 + \frac{12}{\lambda^2}\frac{S}{N}\right)^{1/2} \quad (8.46)$$

where $N = \sigma^2$ is the noise power. Each message is equally likely and, therefore, conveys an average amount of information.

$$H = \log_2 M = \log_2\left(1 + \frac{12}{\lambda^2}\frac{S}{N}\right)^{1/2} \quad (8.47)$$

$$= \frac{1}{2}\log_2\left(1 + \frac{12}{\lambda^2}\frac{S}{N}\right) \text{ bits/message} \quad (8.48)$$

To find out the information rate of the signal waveform $S(t)$ shown in Fig. 8.9, we have to estimate how many messages per unit time may be carried by this signal. The interval T is to be estimated for each message to allow transmitted levels to be recognized individually at the receiver although the bandwidth B of the channel is limited. Since there is an abrupt transition from one level to another at regular interval and this signal when applied to a low-pass filter (LPF), the rise time T of the LPF should be such that it must reproduce the signal in a satisfactory manner, since rise time $\gamma = 0.35/B$. If we get $T = \gamma$, then we may be able to distinguish levels reliably. Since $T = \gamma = 0.35/B$ message rate is $r = 1/T = B/0.35 = 2.85B$, which is better than Nyquist sampling rate $2B$. Since the transmission of any of M messages is equally likely, $H = \log_2 M$ and the channel is transferring information at a rate $R = rH$. Since we have assumed that all precautions are taken to ensure transmission with acceptable probability of error $R = C$, then the channel capacity works out to be as below. Combining the two equations and taking the rate r as the Nyquist rate, i.e., $r = 2B$, we get

$$C = R = rH = 2BH$$

$$C = 2B\,\frac{1}{2}\log_2\left(1 + \frac{12}{\lambda^2}\frac{S}{N}\right) \qquad (8.49)$$

For this equation to satisfy Shannon–Hartley's theorem, $12/\lambda^2 = 1$ or $\lambda^2 = \sqrt{12}$, $\lambda = 2\sqrt{3} \approx 3.5$. If we substitute this value of λ^2 in the equation for C, we get the Shannon–Hartely equation as

$$C = \log_2\left(1 + \frac{S}{N}\right)\ \text{bits/s} \qquad (8.50)$$

Our assumption is that Eq. (8.50) specifies the rate at which information may be transmitted with small error. Shannon–Hartley theorem also contemplates that with adequate care and transmission technique, transmission at channel capacity is possible with arbitrarily small error.

8.11.1 Bandwidth *S/N* Trade-off

Noiseless channel with $S/N = \infty$ indicates a Gaussian channel with an infinite capacity. But as the channel capacity increases, it does not become infinite. As the bandwidth becomes infinite, the noise power also gets increased. Hence, for a fixed signal power and in the presence of white Gaussian noise, the channel capacity approaches an upper limit with increasing bandwidth. We can calculate this limit as follows.

Substituting $N = \eta B$ in the Shannon–Hartley equation, we get

$$C = B\log_2\left(1 + \frac{S}{\eta B}\right) = \frac{S}{\eta}\frac{\eta B}{S}\log_2\left(1 + \frac{S}{\eta B}\right) \qquad (8.51)$$

or $\qquad\qquad C = \frac{S}{\eta}\log_2\left(1 + \frac{S}{\eta B}\right)^{\frac{\eta B}{S}} \qquad (8.52)$

Let $\dfrac{S}{\eta B} = x$, then since $\lim\limits_{x \to 0}(1 + x)^{\frac{1}{x}} = e$,

$$C_\infty = \lim_{B \to \infty} C = \frac{S}{\eta}\log_2 e = 1.44\frac{S}{\eta} \qquad (8.53)$$

This shows that we can trade-off bandwidth for S/N ratio and vice versa. If $S/N = 7$, $B = 4$ kHz, we find $C = 12 \times 10^3$ bits/s. If S/N is made 15 and B decreased to 3 kHz, the channel capacity remains the same with 3 kHz bandwidth and the noise power will be 3/4 as large as with 4 kHz. Thus, the signal power has to be increased by the factor $(3/4) \times (15/7) = 1.6$. Hence, 25% reduction bandwidth requires a 60% increase in signal power.

8.12 DISCRETE MEMORYLESS CHANNEL

So far we have been discussing about discrete memoryless source responsible for information generation. Let us consider now the information transmission with particular emphasis on reliability. For this, we consider a discrete memoryless channel.

This is a statistical model with an input X and an output Y that is noisy version of X; both X and Y are considered random variables. At each instant of time, the channel accepts an input symbol selected from an alphabet α and in response to this it gives an output symbol Y from an alphabet ζ. When both of the alphabets α and ζ have finite sizes, the channel is said to be discrete. It is said to be memoryless when the current input symbol depends only on the current input symbol and not on any of the previous ones.

A discrete memoryless channel is shown in Fig. 8.10. The channel is described in terms of an input alphabet:

$$\alpha = (x_0, x_1, \cdots, x_{j-1}) \qquad (8.54)$$

and an output alphabet

$$\zeta = (y_0, y_1, \cdots, y_{k-1}) \qquad (8.55)$$

and a set of transition probabilities

$$p(y_k \mid x_j) = P(Y = y_k \mid X = x_j) \qquad (8.56)$$

for all j and k. Also we have $0 \le p(y_k \mid x_j) \le 1$ for all j and k.

The input alphabet α and output alphabet ζ need not be of the same size. Sometimes, in channel coding, the size k of the output alphabet ζ may be larger than the size j of the input alphabet α, i.e., $k \ge j$. On the other hand, we may have a situation in which the channel gives out the same symbol when either of the two input symbols is sent, in which case we have $k \le j$. Generally, a

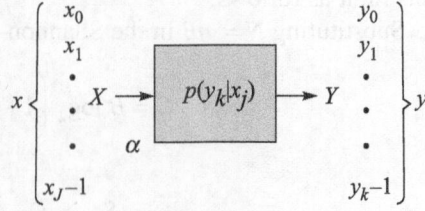

Fig. 8.10 Discrete memoryless channel

discrete memoryless channel is arranged in the form of a matrix and various transition probabilities of the channel are its elements. The matrix P is given as

$$
P = \begin{bmatrix}
p(y_0 \mid x_0) & p(y_1 \mid x_0) & \cdots & p(y_{k-1} \mid x_0) \\
p(y_0 \mid x_1) & p(y_1 \mid x_1) & \cdots & p(y_{k-1} \mid x_1) \\
\vdots & & & \vdots \\
p(y_0 \mid x_{j-1}) & p(y_1 \mid x_{j-1}) & \cdots & p(y_{k-1} \mid x_{j-1})
\end{bmatrix}
\tag{8.57}
$$

The j and k matrix P is called the *channel matrix* or *transition matrix*. Each row of the channel matrix P corresponds to a fixed channel input and each column of the matrix corresponds to a fixed channel output. Also it can be observed that the fundamental property of the channel matrix P is that the sum of the elements along any row of the matrix is always equal to one, i.e.,

$$
\sum_{k=0}^{K-1} p(y_k \mid x_j) = 1 \text{ for all } j
\tag{8.58}
$$

Now if the inputs to a discrete memoryless channel are selected according to the probability distribution $\{p(x_j), j = 0,1,\cdots, J-1\}$. In other words, the event that the channel input $X = x_j$ occurs with probability

$$
p(x_j) = P(X = x_j) \text{ for } j = 0,1, \cdots, j-1
\tag{8.59}
$$

Once we have specified the random variable X denoting the channel input, we can now specify the second random variable Y denoting the channel output. The joint probability distribution of the random variable X and Y is given by

$$
\begin{aligned}
p(x_j, y_k) &= P(X = x_j, Y = y_k) \\
&= P(Y = y_k \mid X = x_j) P(X = x_j) \\
&= p(y_k \mid x_j) p(x_j)
\end{aligned}
\tag{8.60}
$$

The marginal probability distribution of the output random variable Y is obtained by averaging out the dependence of $p(x_j, y_k)$ on x_j as shown by

$$
\begin{aligned}
p(y_k) &= P(y = y_k) \\
&= \sum_{j=0}^{J-1} P(Y = y_k \mid X = x_j) P(X = x_j) \\
&= \sum_{j=0}^{J-1} p(y_k \mid x_j) p(x_j) \text{ for } k = 0, 1, \cdots, K-1
\end{aligned}
\tag{8.61}
$$

The probabilities $p(x_j)$ for $j = 0, 1, \cdots, J-1$ are known as *a priori* probabilities of the various input symbols. The above-mentioned equation suggests that if we are given the input a priori probability $p(x_j)$ and the channel matrix $p(y_k \mid x_j)$, then we may calculate the probabilities of the various output symbols, $p(y_k)$.

8.13 MODELLING OF COMMUNICATION CHANNELS

Communication channel is any medium over which information can be transmitted or in which information can be stored. Examples of communication channels are open wire transmission line, coaxial cable, fibre optic cable, free space, etc. All these channels accept signals at their inputs and deliver signals at their outputs. Hence, each communication channel is characterized by a relation between input and output.

Many factors cause the output of a communication channel different from its input. These factors include attenuation, non-linearities, bandwidth limitation, multipath propagation, and noise. These factors make the relationship between the output and the input a complex one. Also the noise introduced in the channel and fading contribute to this complexity. Since noise is a statistical phenomenon, the relation between the output and the input is generally stochastic.

The channels encountered in practice are continuous waveform channels, i.e., they accept continuous time waveforms as their input and produce continuous waveforms as outputs. But the bandwidths of the practical channels are always limited. Using sampling, we can convert these channels to a discrete time channel, where both the inputs and outputs are discrete time signals.

If the values that the input and output variables can take are finite or countably infinite, the channel is called a discrete channel. Binary input and binary output channel is an example of a discrete channel. In general, a discrete channel is defined by $æ$ (the input alphabet), ζ (the output alphabet), and $p(y \mid x)$. The conditional PMF of the output sequence gives the input sequence. Figure 8.11 gives a schematic representation of a discrete channel.

The output y_i does not only depend on the input at the same time x_i but also on the previous and future inputs or even on previous and future inputs in storage channels. Hence, channels can have memory. However, if a discrete channel does not have memory, it is called *discrete memoryless channel* (DMC). For such a channel for any $y \in \zeta^n$ and $x \in æ^n$, we have

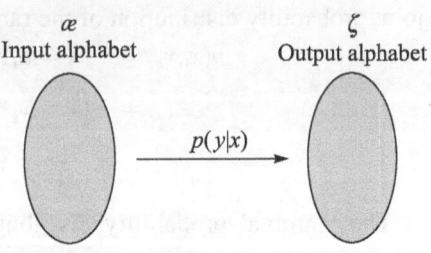

Fig. 8.11 A discrete channel

$$P(y \mid x) = \prod_{i=1}^{n} p\left(y_i \mid x_i\right) \qquad (8.62)$$

All channel models we will be considering are memoryless. A case of a discrete memoryless channel is the *binary symmetric channel* (BSC). It is shown in Fig. 8.12. In binary symmetric channel $\mathbb{C} = P(0 \mid 1) = P(1 \mid 0)$ is called crossover probability. The binary symmetric channel is of great theoretical interest and practical importance.

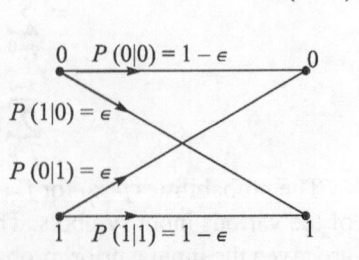

Fig. 8.12 Binary symmetric channel

It is a special case of the discrete memoryless channel with $j = k = 2$. The channel has two input and two output symbols. The input symbols are $x_0 = 0$, $x_1 = 1$ and the output symbols are $y_0 = 0$, $y_1 = 1$. The channel is symmetric because the probability of receiving a 1 if a 0 is sent is the same as the probability of receiving a 0 if a 1 is sent. This conditional probability of error is denoted by p. Figure 8.12 gives the transition probability diagram of binary symmetric channel. The conditional probabilities of error p_{10} and p_{01} can be related to the transition probability diagram. For the case when the binary symbols 0 and 1 are equiprobable, it was shown that the optimized value of these two error probabilities are equal, i.e.,

$$p_{10} = P\big(y = 1 \,|\, x = 0\big) \tag{8.63}$$

and
$$p_{01} = P\big(y = 0 \,|\, x = 1\big) \tag{8.64}$$

For the PCM receiver, $p_{10} = p_{01} = p$

8.14 CHANNEL CAPACITY

It has been shown that $H(X)$ defines a fundamental limit on the rate at which a discrete source can be encoded without errors in its reconstruction. For information transmission over communication channel, there also exists similar fundamental limit. The main objective when transmitting information over any communication channel is reliability, which is a measure of probability of receiving the information correctly at the receiver. But the presence of noise can change the probability on the wrong side. But as per information theory there is a possibility of reliable transmission even over a noisy channel as long as the transmission rate is less than the channel capacity c. This result was shown by Shannon in the year 1948 and is known as *noisy channel coding theorem*. This theorem states that the basic limitation that noise causes in a communication channel is not on the reliability of communication but on the speed of communication.

Figure 8.13 is a discrete memoryless channel with four inputs and outputs. If the receiver receives a, it will not know whether a or d is transmitted since both the inputs are connected to the output marked a and when it receives b, it does not know whether a or b is transmitted and, similarly, for inputs c and d. Hence, there always exists a possibility of error. If the transmitter uses only letters a and c, there will be no ambiguity, i.e., if the receiver receives a or b, it knows that that a was transmitted and if it receives c or d, it knows that c was transmitted. Thus, one can conclude that the two symbols a and c can be transmitted over this channel without error.

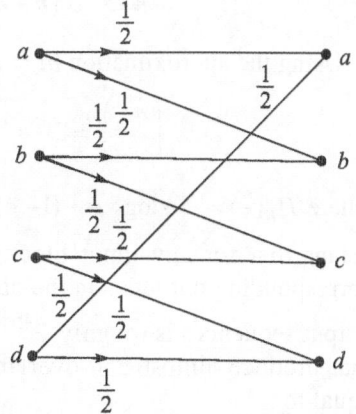

Fig. 8.13 Discrete memoryless channel with four inputs and four outputs

Thus, to avoid the error, only a subset of possible inputs is used over the channel, i.e., using a smaller subset of the possible inputs reduces the number of possible inputs but with a price to be paid for reliable communication. This is the gist of noisy channel coding theorem. This theorem stats that using only those inputs whose corresponding possible outputs are disjoint, causes no ambiguity as to what message was transmitted. But we have to ensure that the chosen inputs are far apart such that their images under the channel operations are non-overlapping. But in practice, it is difficult to avoid non-overlapping outputs. This is mostly the case with all channels. For binary symmetric channel, this argument should not be applied to the channel itself but to the extension channel. The nth extension of a channel with input and output alphabets æ and ζ and conditional probabilities $P(y|x)$ is a channel with input and output alphabet $æ^n$ and ζ^n and

conditional probability $P(y|x\ ) = \displaystyle\prod_{i=1}^{n} p\left(y_i \mid x_i\right)$. The nth extension of a binary

symmetric channel takes binary blocks of length n as its input and its output. This channel is shown in Fig. 8.14. By law of large numbers, if n is large enough and a binary sequence of length n is transmitted over the channel, the output will disagree with the input with high probability at $n\epsilon$ positions.

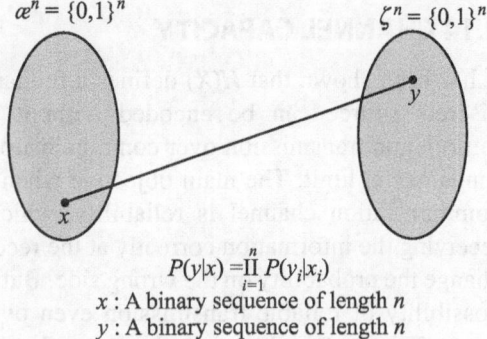

$$P(y|x) = \prod_{i=1}^{n} P(y_i|x_i)$$
x : A binary sequence of length n
y : A binary sequence of length n

Fig. 8.14 A binary symmetric channel: the nth extension

The number of possible sequences that disagree with a sequence of length n at $n\epsilon$ positions is given by

$$\binom{n}{n\,\epsilon} = \frac{n!}{(n-n\,\epsilon)!\,n\,\epsilon!} \tag{8.65}$$

Using the approximation $n! = n^n e^{-n}\sqrt{2\pi n}$, we get

$$\binom{n}{n\,\epsilon} \simeq 2^{nH_b(\epsilon)} \tag{8.66}$$

where $H_b(\epsilon) = -\,\epsilon\log_2\,\epsilon - (1-\,\epsilon)\log_2(1-\,\epsilon)$ is the binary entropy function. This means that for any input block, there are roughly $2^{nH_b(\epsilon)}$ highly probable corresponding outputs. On the other hand, the total number of highly probable output sequences is roughly $2^{nH(Y)}$. The maximum number of input sequences that produce almost non-overlapping output sequence, therefore, is at most equal to

$$M = \frac{2^{nH(Y)}}{2^{nH_b(\epsilon)}} = 2^{n\{H(Y)-H_b(\epsilon)\}} \tag{8.67}$$

i.e., in n uses of this channel, we can transmit at most $\log_2 M = n\{H(Y) - H_b(\epsilon)\}$ bits and the transmission rate per channel use is

$$R = \frac{\log_2 M}{n} = H(Y) - H_b(\epsilon) \qquad (8.68)$$

Figure 8.15 represents this binary symmetric channel. In the above-mentioned relation $R = H(Y) - H_b(\epsilon)$, ϵ depends on the channel and one cannot control it. However, the probability distribution of the random variable Y depends both on the input distribution $p(x)$ and the channel property characterized by ϵ. However, to maximize the transmission rate over the channel, we have to choose a $p(x)$ that maximizes $H(Y)$. If X is a uniformly distributed random variable such that $p(X = 0) = P(x = 1) = 0.5$, then $H(Y)$ will be maximized at one and we obtain

$$R = 1 - H_b(\epsilon) \qquad (8.69)$$

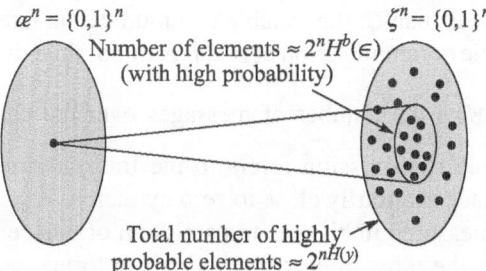

Fig. 8.15 Binary symmetric channel

It can be proved that this is the maximum rate at which reliable transmission over the binary symmetric channel is possible. The reliable transmission means that the error probability tends to infinity. Figure 8.16 gives the plot of channel capacity in this case. The graph is plotted with capacity C as a function of ϵ. It can be seen from the figure that for both $\epsilon = 0$ and $\epsilon = 1$, $C = 1$.

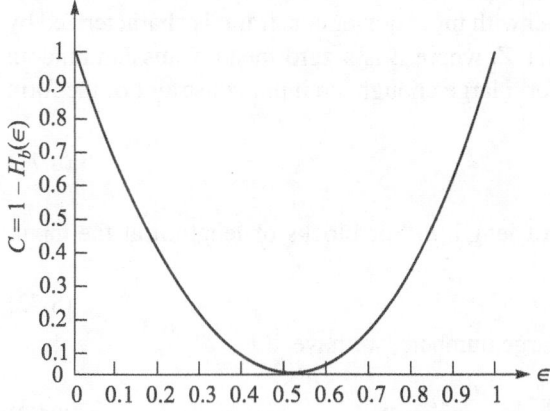

Fig. 8.16 The capacity of binary symmetric channel

This means that a channel that always flips the input is as good as the channel that transmits the input with no errors. The worst case happens when the channel flips the input with probability 1/2.

Channel capacity, denoted by C, is defined as the maximum rate at which we can communicate over a discrete memoryless channel and still make the error probability approach zero as the code block length increases.

8.15 NOISY CHANNEL CODING THEOREM

This theorem gives the capacity of a general discrete memoryless channel. This capacity is given by

$$C = \max_{p(x)} I(X;Y) \tag{8.70}$$

where $I(X;Y)$ is the mutual information between X and Y, the channel input and output. If the transmission rate R is less than C, the reliable communication rate R is possible and if $R > C$, then reliable communication at rate R is impossible. In this theorem R is $\dfrac{\log_2 M}{n}$, where M is the number of messages over the nth extension of the channel and reliable transmission refers to the transmission where the error probability can be made arbitrarily close to zero by increasing n. Both the rate R and capacity C are measured in bits per transmission or bits per channel use. This theorem is one of the most important results in information theory and gives a fundamental limit on the possibility of reliable communication over a noisy channel. Regardless of all other properties, any communication channel is characterized by a number called capacity, which determines how much information can be transmitted over it. Hence, comparing the capacities of channels is enough to compare them from an information transmission point of view.

8.16 GAUSSIAN CHANNEL CAPACITY

A discrete time Gaussian channel with input power constraint is characterized by the input-output relation $Y = X + Z$, where Z is a zero mean. Gaussian random variable with variance P_N and for n large enough, an input constraint of the form

$$\frac{1}{n}\sum_{i=1}^{n} x_i^2 \leq P \tag{8.71}$$

applies to any input sequence of length n. For blocks of length n at the input, output, and noise, we have

$$y = x + z \tag{8.72}$$

In n is large, by the law of large numbers, we have

$$\frac{1}{n}\sum_{i=1}^{n} z_i^2 = \frac{1}{n}\sum_{i=1}^{n} |y_i - x_i|^2 \leq P_N \tag{8.73}$$

or $\qquad |y-x|^2 \le nP_N$ (8.74)

i.e., as probability approaches 1 (as n increases), y will be located in an n-dimensional sphere that has a radius $\sqrt{nP_N}$ and is centred at x. But due to power constraints of p on the input and the independence of the input and noise, the output power is the sum of the input power and the noise power, i.e.,

$$\frac{1}{n}\sum_{i=1}^{n} y_i^2 \le P + P_N$$ (8.75)

or $\qquad |y|^2 \le n(P + P_N)$ (8.76)

This implies that output sequences will be inside an n-dimensional hypersphere of radius $\sqrt{n(P+P_N)}$ and centred at the origin.

Figure 8.17 shows the sequence in the output space. Let us see how many x sequences can we transmit over this channel such that the hyperspheres corresponding to these sequences do not overlap in the output space. If this condition is satisfied, then the input sequences can be decoded reliably. Let us find out how many hyperspheres of radius $\sqrt{nP_N}$ can be packed in a hypersphere of radius $\sqrt{n(P_N+P)}$.

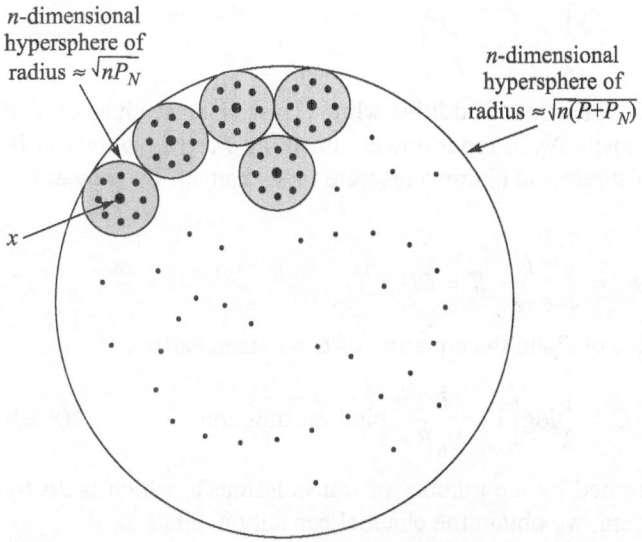

n-dimensional hypersphere of radius $\approx \sqrt{nP_N}$

n-dimensional hypersphere of radius $\approx \sqrt{n(P+P_N)}$

x

Fig. 8.17 Gaussian channel output sequence with power constraint

It can be seen that it is roughly the ratio of the volumes of an ordinary three-dimensional sphere. The volume is $(4/3)\pi R^3$ for a two-dimensional case and the value is πR^2. Hence, the volume of n-dimensional hypersphere is proportional to R^n and is

$$V_n = K_n R^n$$ (8.77)

where R denotes the radius and K_n is independent of R. The number of messages that can be reliably transmitted over this channel is equal to

$$M = \frac{K_n \left\{ n\left(P_N + P\right)\right\}^{n/2}}{K_n \left(nP_N\right)^{n/2}}$$

$$= \left(\frac{P_N + P}{P_N}\right)^{\frac{n}{2}}$$

$$= \left(1 + \frac{P}{P_N}\right)^{\frac{n}{2}} \tag{8.78}$$

Hence, the capacity of discrete time additive white Gaussian noise channel with input power constraint P is given by

$$C = \frac{1}{n} \log M$$

$$= \frac{1}{n} \frac{n}{2} \log\left(1 + \frac{P}{P_N}\right)$$

$$= \frac{1}{2} \log\left(1 + \frac{P}{P_N}\right) \tag{8.79}$$

For a continuous time bandlimited additive white Gaussian noise channel with noise power spectral density $N_0/2$, input power constraint P, and bandwidth B, we can sample at Nyquist rate and obtain a discrete time channel. The power per sample will be

$$P_N = \int_{-B}^{B} \frac{N_0}{2} \, df = BN_0$$

Substituting this value of P_N in the equation of C obtained earlier,

$$C = \frac{1}{2} \log\left(1 + \frac{P}{N_0 B}\right) \text{ bits/transmission} \tag{8.80}$$

If this result is multiplied by the number of transmissions/s, which is $2B$ by Nyquist sampling theorem, we obtain the channel capacity in bits/s as

$$C = B \log\left(1 + \frac{P}{N_0 B}\right) \text{ bits/s} \tag{8.81}$$

This is called Shannon's formula for the capacity of a bandlimited additive white Gaussian noise (AWGN) channel.

8.17 BOUNDS ON COMMUNICATION

From the expression for channel capacity for bandlimited additive white Gaussian noise channel $C = B\log\left(1 + \dfrac{P}{N_oB}\right)$, it is clear that the basic factors that determine the channel capacity are the channel bandwidth B, the noise power spectrum N_o, and the signal power P. A trade-off exists between P and B in the sense that we can compensate for the other. The channel capacity can be increased by increasing input signal power. This is because when we have more power, we can choose a larger number of input levels that are far apart. Hence, we can send more information bits per transmission. But the increase in capacity as a function of power is logarithmic and slow. This is because if we are transmitting with a certain number of input levels that are Δ apart to allow a certain level of immunity against noise and want to increase the number of input levels, we have to introduce new levels with amplitudes higher than the existing level, which requires a lot more power. The capacity of the channel can be increased to any value by increasing the input power. The effect of channel bandwidth is quite different. Increasing B has two contrasting effects. On a higher bandwidth channel, we can transmit more samples/s and hence increase the transmission rate. On the other hand, a higher bandwidth means higher input noise power to the receiver and this reduces its performance. This can be seen clearly from the two B appearing in the relation that describes the channel capacity. Let the bandwidth B tends to infinity. By L' Hospital's rule, we can obtain

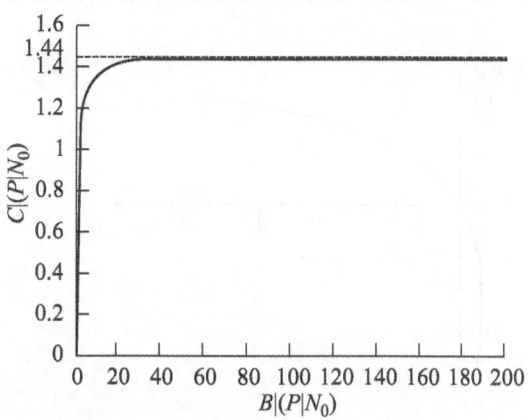

Fig. 8.18 Channel capacity vs bandwidth

$$\lim_{B\to\infty} B\log\left(1 + \frac{P}{N_oB}\right) + \frac{P}{N_o}\log e = 1.44\frac{P}{N_o} \qquad (8.82)$$

i.e., contrary to the power case, by increasing the bandwidth alone, we cannot increase the capacity to any desired value. Figure 8.18 shows C plotted against bandwidth for a practical system $R < C$. If an AWGN channel is employed, we have

$$R < B\log\left(1 + \frac{P}{N_oB}\right) \qquad (8.83)$$

By dividing both sides by B and defining $r = \dfrac{R}{B}$, the spectral bit rate or band-width efficiency

$$r < \log\left(1 + \frac{P}{N_o B}\right) \tag{8.84}$$

If E_b is the energy per bit, then $E_b = \frac{P}{R}$. By substituting this value in the previous relation, we get

$$r < \log\left(1 + r\frac{E_b}{N_o}\right) \tag{8.85}$$

or equivalently, $\quad \dfrac{E_b}{N_o} > \dfrac{2^r - 1}{r} \tag{8.86}$

This relation is shown in Fig. 8.19. The above-mentioned equation gives the relation between the two parameters that are important in a communication system. These parameters are r, the spectral bit rate, which is a measure of bandwidth efficiency of a communication system, and E_b/N_o (the SNR/bit), which is a measure of the power efficiency of a system. The higher the value of r, the higher the efficient bandwidth of the system. With a lower value of E_b/N_o to achieve a certain error probability, the system is more efficient.

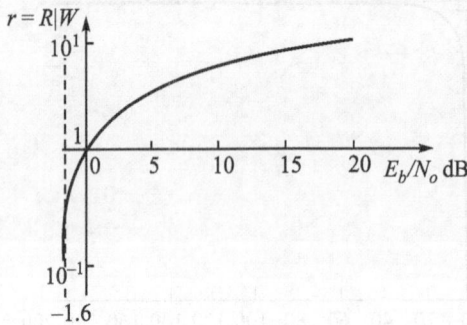

Fig. 8.19 SNR/bit vs spectral bit rate for an optimal system

The curve defined by

$$r = \log\left(1 + r\frac{E_b}{N_o}\right) \tag{8.87}$$

divides the plane into two regions. Reliable communication is possible in the region below the curve and in the region above the curve it is not possible. The performance of any communication system can be denoted by a point in this plane. The closer the point to this curve, the closer the performance of the system is to that of an optimal system. It can be observed from the curve that as $r \to 0$,

$$\frac{E_b}{N_0} = \log 2 = 0.693 \approx 1.6 \text{ dB} \tag{8.88}$$

This is an absolute minimum for reliable communication. Stated otherwise for reliable communication, we must have

$$\frac{E_b}{N_0} > 0.693 \tag{8.89}$$

In Fig. 8.19 when $r \ll 1$, the bandwidth is large and the main concern is limitation of power. This is generally referred to as *power limited case*. Orthogonal, biorthogonal, and simplex signalling schemes with high dimensionality can be used in these cases. The case where $r \gg 1$ happens when the bandwidth of the channel is small and is referred to as *bandwidth limited case*. A few examples are signalling schemes with crowded constellations like PAM, QAM, and PSK.

As we have seen earlier in this chapter, the fundamental limit of coding of information source is expressed in terms of its entropy. Entropy gives the lower bound on the rate of the codes that are capable of reproducing the source with no error. If we want to transmit a source S reliably via a channel with capacity C, then $H(S) < C$. This relation defines the fundamental limit on the transmission of information with the assumption that for each source output, one transmission over the channel is possible.

8.18 INFORMATION CAPACITY OF COLOURED NOISY CHANNEL

The information capacity theorem discussed earlier applies to a bandlimited white noise channel. This theorem is extended to the more general case of a non-white or coloured noise channel. Let us consider the channel model shown in Fig. 8.20(a). The transfer function of the channel is $H(f)$. The channel noise is $n(t)$, which is shown to be additive at the output and is modelled as the sample function of a stationary Gaussian process of zero mean and power spectral densities $S_N(f)$.

We have to find the input ensemble, described by the power spectral density $S_x(f)$, that maximizes the mutual information between the channel output $y(t)$ and the channel input $x(t)$ subject to the constraint that the average power of $x(t)$ is fixed at a constant value P. Also the optimum information capacity of the channel has to be found out. This is a constrained optimization problem. This problem is solved as follows. Since the channel is linear, the model shown in Fig. 8.20(a) can be replaced by an equivalent model shown in Fig. 8.20(b). The spectral characteristics of the signal plus noise measured at the channel output for the both models shown are equal, provided the power spectral density of the noise $n'(t)$ in Fig. 8.20(b) is defined in terms of the power spectral density of the noise in Fig. 8.20(a) as

$$S_N(f) = \frac{S_N(f)}{|H(f)|^2} \tag{8.90}$$

where $H(f)$ is the magnitude response of the channel.

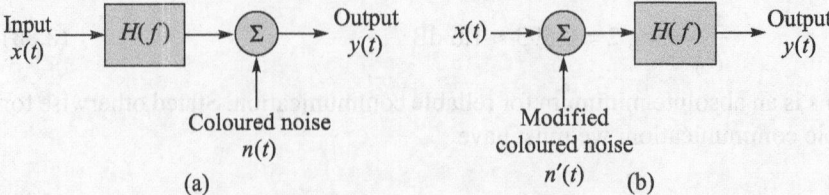

Fig. 8.20 (a) Bandlimited, power limited noisy channel and (b) equivalent model of the channel

To simplify the analysis, the channel is divided into a large number of adjoining frequency slots as shown in Fig. 8.21. The approximation will be better if we make the incremental frequency interval Δf of each subchannel smaller. Thus, the original model of Fig. 8.20(a) is replaced by the parallel combination of a finite number of subchannels N, each of which is corrupted essentially by *bandlimited white Gaussian Noise*. The kth subchannel in the approximation to the model of Fig. 8.20(b) is described by

$$y_k(t) = x_k(t) + n_k(t), \quad k = 1, 2, \cdots, N \tag{8.91}$$

The average power of the signal component $x_k(t)$ is

$$P_k = S_X(f_k)\,\Delta f, \quad k = 1, 2, \cdots, N \tag{8.92}$$

where $S_X(f_k)$ is the power spectral density of the input signal evaluated at frequency $f = f_k$. The variance of the noise component $n_k(t)$ is

$$\sigma^2_k = \frac{S_N(f_k)}{\left|H(f_k)\right|^2}\,\Delta f, \quad k = 1, 2, \cdots, N \tag{8.93}$$

where $S_N(f_k)$ and $|H(f_k)|$ are the noise spectral density and the channel's magnitude response evaluated at the frequency f_k, respectively. The information capacity of the kth subchannel is

$$C_k = \frac{1}{2}\Delta f \log_2\!\left(1 + \frac{P_k}{\sigma^2_k}\right), \quad k = 1, 2, \cdots, N \tag{8.94}$$

where the factor 1/2 accounts for the fact that Δf applies to both positive and negative frequencies. All the N subchannels are independent of one another. So, the total capacity of the overall channel is approximately given by the summation

$$C = \sum_{k=1}^{N} C_k$$

$$= \frac{1}{2}\sum_{k=1}^{N} \Delta f \log_2\!\left(1 + \frac{P_k}{\sigma^2_k}\right) \tag{8.95}$$

To maximize the overall information capacity C subject to the constraint,

$$\sum_{k=1}^{N} P_k = P = \text{constant} \tag{8.96}$$

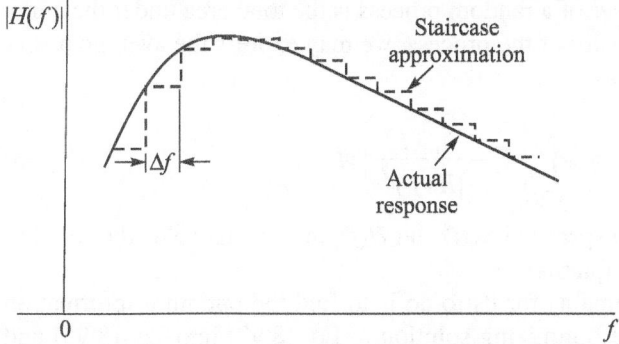

Fig. 8.21 Positive frequency response of an arbitrary magnitude response |H(f)|: staircase approximation

The method of Lagrange multiplier is used to solve a constrained optimization problem. We first define an objective function that incorporates both the information capacity and the constraint as shown by

$$J = \frac{1}{2} \sum_{k=1}^{N} \Delta f \log_2 \left(1 + \frac{P_k}{\sigma_k^2} \right) + \lambda \left(P - \sum_{k=1}^{N} P_k \right) \qquad (8.97)$$

where λ is Lagrange multiplier. Now to differentiate the objective function J with respect to P_k and setting the result equal to zero, we obtain

$$\frac{\Delta f \log_2 e}{P_K + \sigma_k^2} - \lambda = 0 \qquad (8.98)$$

To satisfy this optimizing solution, we impose the following requirement:

$$P_k + \sigma_k^2 = K \Delta f \quad \text{for } k = 1, 2, \cdots, N \qquad (8.99)$$

where K is a constant that is the same for all k. The constant K is chosen to satisfy the average power constraint. Inserting the defining values of Equations (8.92) and (8.93) in the optimizing condition of Eq. (8.99) and simplifying and rearranging terms, we get

$$S_X(f_k) = K - \frac{S_N(f_k)}{|H(f_k)|^2}, \quad k = 1, 2, \cdots, N \qquad (8.100)$$

Let $\Im_A$ denote the frequency range for which the constant K satisfies the condition

$$K \geq \frac{S_N(f)}{|H(f)|^2} \qquad (8.101)$$

then as the incremental frequency interval Δf is allowed to approach zero and the number of the subchannels N goes to infinity, we may use Eq. (8.100) to formally state that the power spectral density of the input ensemble that achieves the optimum information capacity is a non-negative quantity defined by

$$S_X(f) = \begin{bmatrix} K - \dfrac{S_N(f)}{|H(f)|^2}, & \text{for } f \in \Im_A \\[2ex] 0, & \text{otherwise} \end{bmatrix} \qquad (8.102)$$

Since the average power of a random process is the total area under the curve of the power spectral density of the process, we may express the average power of the channel input $x(t)$ as

$$P = \int_{f \in \mathfrak{I}_A} \left(K - \frac{S_N(f)}{|H(f)|^2} \right) df \tag{8.103}$$

For a prescribed P and specified $S_N(f)$ and $H(f)$, the constant K is the solution to the above-mentioned equation.

The only thing that remains for us to do is to find the optimum information capacity. Substituting the optimizing solution to Eq. (8.99) into Eq. (8.95) and then using the defining values of Equations (8.92) and (8.93), we obtain

$$C = \frac{1}{2} \sum_{k=1}^{N} \Delta f \log_2 \left(K \frac{|H(f_k)|^2}{S_N(f_k)} \right) \tag{8.104}$$

When the incremental frequency interval Δf approaches zero, this equation takes the limiting form

$$C = \frac{1}{2} \int_{-\infty}^{\infty} \log_2 \left(K \frac{|H(f)|^2}{S_N(f)} \right) df \tag{8.105}$$

where the constant K is chosen as the solution to Eq. (8.103) for a prescribed input signal power P.

8.19 RATE DISTORTION THEORY

The source coding theorem discussed earlier was for discrete memoryless source according to which the average code word length must be at least as large as the source entropy for perfect coding. But in many practical situations, there are constraints that force the coding to be imperfect, thereby resulting in unavoidable distortion. In other words, the constraints imposed by communication channel may place an upper limit on the permissible code rate and, therefore, average code word length assigned to the information source. On the other hand, in another example, the information source may have a continuous amplitude as in the case of speech and in this case it is required that quantization of the amplitude of each sample generated by the source is code word of finite length as in the case of pulse code modulation. This type of coding is called *source coding with a fidelity criterion* and the branch of information theory that deals with it is called *rate distortion theory*. This theory finds application in two types of situations.

(a) Source coding where the permitted coding alphabet cannot exactly represent the information source, in which case we are forced to do lossy data compression.

(b) The rate at which information is transmitted exceeding the channel capacity. Hence, rate distortion theory can be viewed as a natural extension of Shannon's coding theorem.

8.19.1 Rate Distortion Function

Let us consider a discrete memoryless source defined by $X:\{x_i | i = 1, 2, \cdots, M\}$, which consists of a set of statistically independent symbols together with associated symbol probabilities $\{p_i | i = 1, 2, \cdots, M\}$. Let R be the average code rate in bits per code word. The representation code words are taken from another alphabet $Y: \{y_j | j = 1, 2, \cdots, N\}$. The source coding theorem states that this second alphabet provides a perfect representation of the source provided $R > H$, where H is the source entropy. If one is forced to have $R < H$, then there is an unavoidable distortion and hence loss of information.

Let $p(x_i, y_j)$ denote the joint probability of occurrence of source symbol x_i and representation symbol y_j. From the probability theory, we have

$$p(x_i, y_J) = p(y_J \mid x_i) p(x_i) \tag{8.106}$$

where $p(y_j | x_i)$ is a transition probability. Let $d(x_i, y_j)$ denote a measure of the cost incurred in representing the source symbol x_i by the symbol y_j. The quantity $d(x_i, y_j)$ is referred to as a *single-letter distortion measure*. The statistical average of $d(x_i, y_j)$ over all possible source symbols and representation symbol is given by

$$\bar{d} = \sum_{i=1}^{M} \sum_{j=1}^{N} p(x_i) p(y_j \mid x_i) d(x_i, y_j) \tag{8.107}$$

The average distortion $\bar{d}$ is a non-negative continuous function of the transition probabilities $p(y_j \mid x_i)$ that are determined by source encoder–decoder pair.

A conditional probability assignment $p(y_j \mid x_i)$ is said to be D admissible if and only if the average distortion $\bar{d}$ is less than or equal to some acceptable value D. The set of all D admissible conditional probability assignments is denoted by

$$P_D = \left\{ p(y_j \mid x_i) : \bar{d} \leq D \right\} \tag{8.108}$$

For each set of transition probabilities, we have mutual information

$$I(X ; Y) = \sum_{i=1}^{M} \sum_{j=1}^{N} p(x_i) p(y_j \mid x_i) \log \left(\frac{p(y_j \mid x_i)}{p(y_j)} \right) \tag{8.109}$$

A rate distortion function $R(D)$ is defined as the smallest coding rate possible for which the average distortion is guaranteed not to exceed D. Let P_D denote the set to which the conditional probability $p(y_j \mid x_i)$ belongs for a prescribed D. Then for a fixed D,

$$R(D) = \min_{p(y_j \mid x_i) \in P_D} I(X;Y) \tag{8.110}$$

subject to the constraint

$$\sum_{i=1}^{N} p(y_j \mid x_i) = 1 \text{ for } i = 1, 2, \cdots, M \tag{8.111}$$

The rate distortion function $R(D)$ is measured usually in bits. Thus, the base to the logarithm is 2. We can expect the distortion D to decrease as the rate distortion function $R(D)$ is increased. Conversely, if we can tolerate a large distortion D, then a smaller rate of transmission is allowed.

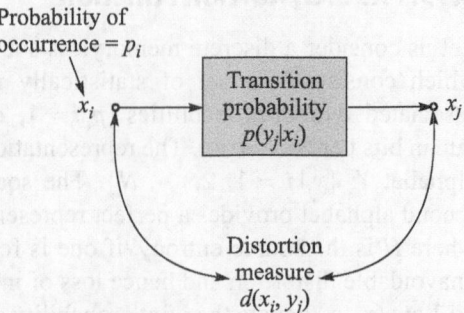

Fig. 8.22 Rate distortion theory

Figure 8.22 summarizes the main parameters of a rate distortion theory. Given the source symbol $\{x_i\}$ and their probabilities $\{p_i\}$ and given a definition of the single-letter distortion measure $d(x_i, y_j)$, the calculation of the rate distortion function $R(D)$ involves finding the conditional probability assignment $p(y_j, x_i)$ subject to certain constraints imposed on $p(y_j|x_i)$. This is a variation problem, the solution to which is not straightforward.

8.20 DATA COMPRESSION

The rate distortion discussed above leads to the idea of data compression, which involves a purposeful or unavoidable reduction in information content of data from a continuous or discrete source. The data compressor or signal compressor supplies a code with the least number of symbols for the representation of the source output, subject to permissible or acceptable distortion. Thus, the data compression retains the essential information content of the source output by altering the fine details or blurring it in a deliberate and controlled manner. This results in reducing the source entropy, which means some loss of information.

In the case of discrete source, data compression is used to encode the source output at a rate smaller than the source entropy. By doing so, the source coding theorem is violated, which means that the exact reproduction of the original data is no longer possible.

For a continuous source, the entropy is infinite and hence a signal compression code must always be used to encode the source output at a finite rate. Hence, it is impossible to digitally encode an analog with a finite number of bits without producing some distortion. This is the case with PCM where we have quantization error due to analog to digital conversion process. Hence, quantizer can also be viewed as a data compressor or signal compressor. Scalar quantizer deals with samples of analog signals one at a time. Each sample is converted into a quantized value with conversion being independent from sample to sample. Scalar quantizer is rather a simple data compressor.

Vector quantizers use blocks of consecutive samples of the source output to form vectors, each of which is treated as a single entity. The vector is encoded by comparing it with a code book consisting of a set of stored reference vectors known as *code vectors* or *patterns*. Each pattern in the code book is used to represent input vectors that are identified by the encoder to be similar to a particular pattern subject to the maximization of an appropriate fidelity criterion.

The encoding process in a vector may thus be viewed as a pattern matching operation.

Let N be the number of vectors in the code book, k be the dimension of each vector, i.e., the number of samples in each pattern, and r be the coded transmission rate in bits per sample. The relation between r, N, and k is given by

$$r = \frac{\log_2 N}{k} \tag{8.112}$$

Assuming that the size of the code book is quite large, the signal to quantization noise ratio (SNR) for the vector quantizer is given by

$$10 \log_{10} (\text{SNR}) = 6\left(\frac{\log_2 N}{k}\right) + C_k dB \tag{8.113}$$

where C_k is a constant expressed in dB that depends on the dimension of k. As per the above-mentioned equation, the SNR for a vector quantizer increases approximately at the rate of $6/k$ dB for each doubling of the code book. Stated otherwise, the SNR increases by 6 dB per unit increase in rate as in the standard PCM using a uniform scalar quantizier. The advantage of the vector quantizer over the scalar quantizer is that its constant term C_k has higher value since the vector quantizer optimally exploits the correlation among the samples constituting the vector. The constant C_k increases with the dimension k, approaching the ultimate rate distortion limit for a given source of information. The improvement in SNR is allowed at the cost of increasing encoding complexity, which grows exponentially with the dimension k for a specific rate r. This is one of the limitations of vector quantization. But with the advent of VLSI technology, complex signal processor can be used for vector quantization.

8.21 AUTOMATIC REPEAT REQUEST

The primary concern in any data transmission is error-free transmission. But as it is practically impossible to achieve, at least one must try to control transmission errors. We have studied so far different coding schemes, which are called feed forward error correction (FEC). FEC relies on the control use of redundancy in the transmitted code word for both detection and correction of errors occurring during the course of transmission over a noisy channel. Irrespective of whether the decoding of the received code word is successful, no further processing is performed at the receiver. FEC depends on the coding to allow error correction. The limitation of FEC is that when the errors are too much, the code will not be effective. Also to achieve low error rates, it is necessary to add a relatively large number of redundant bits. This results in the efficiency of the code being low. Also good code words involve long code word whose processing requires complex and expensive hardware. Hence, channel coding technique suitable for FEC requires only a one-way link between the transmitter and the receiver. There is also an alternative scheme, called automatic repeat request (ARQ), which is employed only when extremely low error rates are required. The main difference between FEC and ARQ is that the receiver is not called upon to correct but only detect errors. When an error is detected, the receiver signals back to the transmitter requesting it to retransmit the word once again. Thus, ARQ uses redundancy only for the purpose of error detection. Since the receiver requests a repeat transmission of the corrupted code word, there must be necessarily a return path or feedback channel. As such ARQ can be used only on half-duplex or full-duplex links. In a half-duplex link, data transmission over the link can be made in either direction but not simultaneously. If one wants data transmission on both sides simultaneously, then the link must be duplex.

8.21.1 Stop and Wait System

A half-duplex link uses the simplest ARQ scheme known as *stop and wait strategy*. In this method, a block of message is encoded into a code word and transmitted over the channel. The transmitter then stops and waits for feedback from the receiver. The feedback signal can be an acknowledgement for a correct receipt of code word or a repeat request for the code word because of an error in decoding. If a repeat request comes from the receiver, the transmitter resends the code word in question before moving on to the next block of words. The idling problem in stop and wait ARQ results in reduced data throughout. Figure 8.23(a) shows the stop and wait ARQ. The transmitter sends a code word to the receiver during the time T_w. The receiver on receipt of the transmitted code word checks for any error and if no error is found then sends the positive acknowledgement signal. Upon receipt of the ACK signal, the transmitter sends the next word. If the receiver does detect an error after the receipt of a code word, it returns a negative acknowledgement (NAK) signal to the transmitter. Then the transmitter sends back the same code word once again and then waits for ACK or NAK response before undertaking further transmission. The time between end of transmission of one word and the start of transmission of next word is T_1. This system is used in IBM's binary synchronous communication (BISYNC) protocol.

8.21.2 Continuous ARQ with Pull Back

This method uses a full-duplex link, thereby permitting the receiver to send a feedback signal while the transmitter is engaged in sending code words over the forward channel. The transmitter continues to send code words one after another, until it receives a request from the receiver on the feedback channel for retransmission. At this point, the transmitter stops, pulls back to the particular code word that was not decoded properly by the receiver, and retransmits the complete sequence of code words starting with the corrupted one. This method is a significant improvement over the stop and wait system. This scheme is shown in Fig. 8.23(b).

8.21.3 Continuous ARQ with Selective Repeat

In this refined version, data throughput is improved further by only transmitting the code word that was received with detected errors. Thus, this method eliminates the need for retransmitting the successfully received code word starting from the corrupted code word. The selective ARQ has the highest transmission efficiency of the three systems but it is costly to implement. This scheme is shown in Fig. 8.23(c).

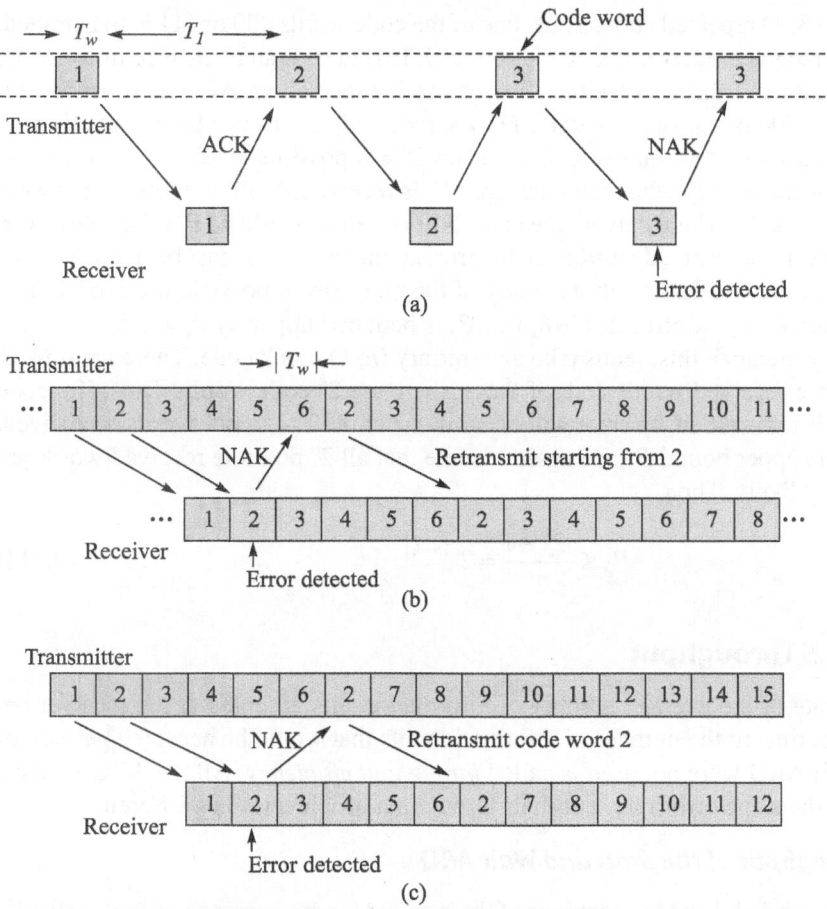

Fig. 8.23 (a) Stop and wait system, (b) continuous ARQ with pull back, and (c) continuous ARQ with selective repeat

The three types of **ARQ** offer trade-off of their own between the need for a half-duplex or full-duplex link and the requirement for efficient use of communication resources. The assumptions for all the three methods are:

(a) Error detection, which makes the design of decoder relatively simple and

(b) Noiseless feedback channel, which is not a severe restriction because the rate of information flow over the feedback channel is typically quite low.

For this reason, ARQ is widely used in computer communication.

8.21.4 Performance of ARQ Systpems

The performance of an ARQ system is measured in two ways, by the probability of error and by the transmission efficiency.

Probability of error

Block codes are used for error detection. Whenever a transmission error occurs and detected, an NAK is returned and the transmitter repeats the message. Let us consider an example. The messages 0 and 1 are encoded prior to transmission using (3, 1) repeated code. Thus, one of the code words 000 or 111 is transmitted. If the receiver receives 001, 010, 100, 110, 101, or 011, an NAK is returned to the transmitter. If, however, either 000 or 111 is received an ACK is returned. But when a 000 is transmitted and a 111 are received, an error will be made. Hence, for a message 000 transmitted, there are $2^3 = 8$ possible received words and an error is made only when the message 111 is received. If all eight messages were equally likely to be received, the probability of error would be $P_e = 1/8$. However, in order for a received word to be in error, all the three bits must be in error. Since such an event is less likely than any of the other seven possible received words, the probability of error is $< 1/8$, i.e., P_e is bounded upper by $P_e = 1/8$.

To generalize this, let us take an arbitrary (n, k) block code. There are now 2^n possible received words and, of these, there are 2^k code words. Thus, if a code word is transmitted, an error will occur if one of the $2^k - 1$ code words is received. For the upper bound P_e, we again assume that all 2^n possible received words are equally likely. Then

$$P_e \leq \frac{2^k - 1}{2^n} \simeq 2^{-(n-k)} \tag{8.114}$$

8.21.5 Throughput

The rate of the average number of information bits accepted at the receiver per unit of time to the number of information bits that would be accepted per unit of time if ARQ were not used is called *throughput efficiency*. All the ARQ systems yield the same error rate, but their throughput efficiencies are different.

Throughput of the Stop and Wait ARQ

Let the probability of acceptance of the receiver for any message on any particular transmission be P_A, then the probability that only a single transmission is all that needed for acceptance is P_A. The probability that two transmissions will be required is $(1 - P_A)P_A$, i.e., the product of the probability $(1 - P_A)$ that the first transmission was rejected and the probability P_A that it was accepted on the second try. The average number of transmissions required for acceptance of a single word is the sum of the products of the number of transmissions j and the probability of requiring j transmissions $P_A(1 - P_A)^{j-1}$. Thus,

$$\overline{N}_{sw} = 1P_A + 2P_A(1 - P_A) + 3P_A(1 - P_A)^2 + \cdots$$

$$= \frac{1}{P_A} \tag{8.115}$$

From Fig. 8.23, it can be seen that the total time required for a single attempt to get the receiver to accept a word is $T_w + T_I$. Hence, on average, the time required to transmit one word is

$$\overline{T}_{sw} = \frac{T_w + T_I}{P_A} \tag{8.116}$$

If ARQ is not used and no coding bits were added to the k information bits, the time needed to transmit the k bits would be

$$T_k = \frac{k}{n}T_w \qquad (8.117)$$

Hence, the throughput efficiency of the stop and wait ARQ system is

$$\eta_{S\&W} = \frac{T_k}{T_{sw}} = \frac{\dfrac{k}{n}T_w}{\dfrac{T_w + T_I}{P_A}} \qquad (8.118)$$

$$\eta_{S\&W} = \frac{k}{n}\frac{P_A}{1 + \dfrac{T_I}{T_w}} \qquad (8.119)$$

Throughput of continuous ARQ pull back

In this method, when an error is detected, once again that particular code and that of the $N - 1$ words that followed is retransmitted. So, the retransmission involves N words. Thus, if a word is received in error, N words are retransmitted and the total number of words transmitted is $N + 1$. If the same word is again in error, the N words are repeated once again. Similar to stop and wait ARQ analysis, the average number of word transmissions required for the acceptance of a single word is

$$N_{Gpull} = 1 \times P_A + (N+1)P_A(1 - P_A) + 2(N+1)P_A(1 - P_A)^2 + \cdots$$

$$= 1 + \frac{N(1 - P_A)}{P_A} \qquad (8.120)$$

Hence, the corresponding time to transmit one word is

$$\bar{T}_{Gpull} = T_w\left(1 + \frac{N(1 - P_A)}{P_A}\right) \qquad (8.121)$$

$$\eta_{Gpull} = \frac{T_k}{T_{Gpull}} = \frac{k}{n}\frac{1}{1 + \dfrac{N(1 - P_A)}{P_A}} \qquad (8.122)$$

Throughput of continuous ARQ with selective repeat

In this method, the mean time for transmission of a word $\bar{T}_{SR}$ is calculated as in the case of stop and wait SRQ except that T_I is made zero. Hence,

$$\bar{T}_{SR} = \frac{T_w}{P_A} \qquad (8.123)$$

$$\eta_{SR} = \frac{T_k}{T_{SR}} = \frac{k}{n}P_A \qquad (8.124)$$

8.22 ERROR-FREE COMMUNICATION OVER NOISY CHANNEL

We have seen that messages of a source with entropy $H(m)$ can be encoded by using an average $H(m)$ digits per message. This encoding has zero redundancy. When this message is transmitted over a noisy channel, there are chances of receiving information erroneously. Error-free communication over a channel is practically impossible when messages are encoded with zero redundancy. To combat noise, we have to necessarily go in for redundancy. The simple example is the introduction of parity bit in a serial transmission. This is called *single parity check code*. In this code, an extra bit is added to each code word to ensure that the total number of 1's in the resulting code word is always even or odd depending on the parity chosen. If single error occurs in the received code word, the parity is violated and the receiver may request a retransmission. But this single parity bit check fails when there is a change of 0 to 1 and 1 to 0 in the transmitted bit. Hence, there are complex coding procedures, which can correct up to n digits.

The addition of extra digit increases the average word length to $H(m) + 1$. Then the efficiency of transmission becomes < 1, since the redundancy bit (the parity bit) does not convey any data. The efficiency with one bit parity works out to be $\eta = H(m)/[H(m) + 1]$ and the redundancy is $1 - \eta = 1/[H(m) + 1]$. Thus, the addition of an extra check digit increases redundancy but it also helps combat noise. Increasing redundancy improves the immunity against channel noise. It was shown by Shannon that error-free communication can be achieved by adding sufficient redundancy. For example, if we have a binary symmetric channel (BSC) with an error probability P_e, then for error-free communication over this channel messages from a source with entropy $H(m)$ must be encoded by binary code with a word length of at least $H(m)/C_s$, where

$$C_s = 1 - \left[P_e \log \frac{1}{P_e} + \left(1 - P_e\right) \log \frac{1}{1 - P_e} \right] \qquad (8.125)$$

where the parameter C_s ($C_s < 1$) is called the channel capacity. The efficiency of these codes is never greater than C_s. For example, if a binary channel has $C_s = 0.4$, a code that can achieve error-free communication must have at least $2.5H(m)$ binary digits per message, which is two and a half times as many digits as required for coding without redundancy. This means that there are $1.5H(m)$ redundant digits per message. Hence, on the average, for every 2.5 digits, transmitted one digit is the information digit and 1.5 digits are redundant, which can be in the form of parity bits check digits, etc. The redundancy of $1 - C_s = (1 - 0.4) = 0.6$.

The error probability of binary signalling P_e varies as e^{-kEb} and hence to make $P_e \rightarrow 0$ either $S_i \rightarrow 8$ or $R_b \rightarrow 0$ because S_i must be finite $P_e \rightarrow 0$ only if $R_b \rightarrow 0$. But according to Shannon, it is not really necessary to let $R_b \rightarrow 0$ for error-free communication, what is required is to hold R_b below C, the channel capacity per second, $C = 2BC_s$. To find this discrepancy, we must investigate in detail the role of redundancy in error-free communication.

Let us consider a simple method of reducing P_e by repeating a given digit as odd number of times. For example, we can transmit 0 and 1 as 000 and 111. The receiver uses the majority role to make decision, i.e., if at least two out of three digits are 1, the decision is 1 and if at least two out of three digits are 0, the decision is 0. Hence, even if one out of three digits is in error, the information received is error free. But this scheme will fail if two out of three digits are in error. In order to correct two errors, we need five repetitions. Thus, to improve P_e, one must have redundancy.

All possible eight combinations of three binary digits are shown as vertices of a cube in Fig. 8.24. For convenience, the binary sequences are mapped as shown. If two binary sequences of the same length differ in j places (j digits), then the Hamming distance between the sequences is considered to be j. Thus, the Hamming distance between 000 and 010 or 001 and 101 is 1, and between 000 and 111 is 3. In case of three repetitions, we transmit binary 1 by 111 and binary 0 by 000. The Hamming distance between these sequences is 3. It can be observed that out of eight possible vertices, only two of the vertices 000 and 111 are for transmitted messages. At the receiver, however, because of channel noise any one of the eight sequences may be received. The majority decision role can be interpreted as a rule that decides in favour of the message (000 or 111) that is at the closest Hamming distance from the received sequence. Sequences 000, 001, 010, and 100 are within 1 unit of Hamming distance from 000 but are at least 2 units from 111. Hence, when any of these four sequences are received, our decession is binary 0. Similarly, when any one of these sequences 110, 111, 011, or 101 is received, the decision binary is 1.

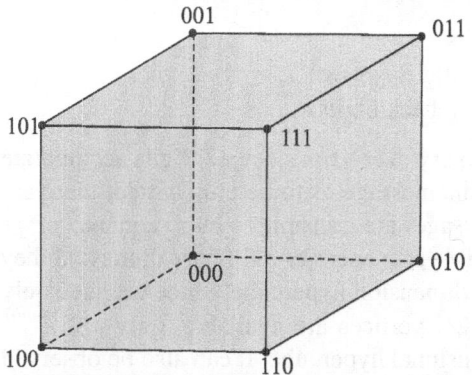

Fig. 8.24   Hamming space: three dimensional

It can be seen that error probability is reduced since we have used only two out of eight possible vertices, which are separated by three Hamming units. If we draw a Hamming sphere of unit radius around each of these two vertices (000 and 111), the two Hamming spheres are not a true geometrical hypersphere because the Hamming distance is not a true geometrical distance. For example, sequences 001, 010, and 100 lie as a Hamming sphere centred at 111 and having radius two. Channel noise can cause a distance between the received sequence and the transmitted sequence and as long as this distance is equal to or less than one unit, we can still detect the message without error. In a similar way, the case of five repetitions can be represented by a hypercube of five dimensions. The transmitted sequences 00000 and 11111 occupy two vertices separated by five units and the

Hamming spheres of two unit radius drawn around each of these two vertices would not be overlapping. Even if channel noise causes two errors, in this case, we can still detect the message correctly. Hence, the reason for the reduction in error probability is that we have not used all the available vertices for messages. If we have occupied all the available vertices for messages, which are the case of not having any redundancy or repetition, then if channel noise caused an error, the received sequence would occupy a vertex assigned to another transmitted sequence and it is certain that a wrong decision is made. But when we leave the neighbouring vertices of the transmitted sequence occupied, then we are able to detect the sequence correctly despite channel noise. The smaller the fraction of vertices used, the smaller the error probability. It is the redundancy that makes it possible to have unoccupied vertices. When the number of repetitions n continues to increase, P_e gets reduced along with R_b by the factor n. But no matter how large we have n, the error probability never becomes zero. The reason for this is that we are adding redundant digits to each information digit.

Using redundant digits not to check on only one transmitted digit but rather a block of digits may be efficient. In this method, suppose we have to transmit m binary information digit/s, then over a period of T s, we have a block of mT binary information digits. If to this block of information digits, we add $(q - m)T$ check digits, which are redundant digits, i.e., $(q - m)$ redundant digits/s, then we need to transmit qT $(q > m)$ digits for every mT information digits. Hence, over a T s interval, we have

mT	information digits
qT	total transmitted digits $(q > m)$
$(q - m)T$	redundant digits (check digits)

Thus, instead of transmitting one binary every $1/m$ s, let mT digits accumulate over a period of T s. If we consider this message as to be transmitted, there are total of 2^{mT} such messages. These messages are transmitted by a sequence of qT binary digits. There are in all 2^{qT} possible sequences of qT binary digits and they can be represented as vertices of a qT dimension hypercube. Since we have only 2^{mT} messages to be transmitted and 2^{qT} vertices are available, only a $2^{-(q-m)T}$ fraction of the vertices of the qT dimensional hypercube. It can also be observed that we have reduced the transmission rate by a factor of m/q. This rate reduction factor is independent of T. The fraction of vertices occupied by transmitted message is $2^{-(q-m)T}$ and can be made as small as possible by simply increasing T. In the limit as $T \to \infty$, the occupancy factor approaches zero and this will make the error probability also tend to zero and we can have the possibility of error-free transmission and hence error-free communication.

Let us now find the optimum rate of reduction ratio m/q for error-free transmission. Increasing T increases the length of the transmitted sequence. If P_e is the digit error probability, then it can be seen from the relative frequency definition or the law of large numbers that as $T \to \infty$, the total number of digits in error in a sequence qT digits as $q \to 8$ is exactly qTP_e. Hence, the received sequences will be at a Hamming distance of qTP_e from the transmitted sequence. Hence, for error-free communication, we must leave all the vertices unoccupied within a sphere of radius qTP_e drawn around each of 2^{mT} occupied vertices, i.e., we must be able to pack 2^{mT} non-overlapping spheres each of radius qTP_e into the Hamming space of qT dimension. This means that for a given q, m cannot be increased beyond some limit without causing overlap in the sphere and the consequent failure of the scheme. Shannon's theorem states that for this scheme to work, m/q should be less than the channel capacity C, which is a function of the channel noise and the signal power $\dfrac{m}{q} < C$.

But such error-free communication is not practical. In the example considered so far, we have accumulated the information digits for T s before encoding them, and for error-free communication we have to wait until eternity before we start encoding since for error-free transmission, T tends to ∞. Hence, there will be an infinite delay at the transmitter and an additional delay of the same amount at the receiver. Also the equipment needed for storage, encoding, and decoding, sequence of infinite digits would be enormous. Hence, we can conclude that in practice error-free communication can never be achieved. Shannon's result indicates the upper limit on the rate of error-free communication that can be achieved on a channel. It also indicates the way to reduce the error probability with only a small reduction in the rate of transmission of information digits. Hence, one has to strike a balance between error-free communication with infinite delay and virtually error-free communication with a finite delay.

8.23 CHANNEL CAPACITY OF CONTINUOUS CHANNEL

Continuous channel deals with analog signals. Analog data deals with continuous random variable. It is possible to extend the definition of entropy for continuous random variable. This definition can be accepted only if we show that it has the meaningful interpretation as uncertainty. Now for a discrete system x taking on discrete values $x_1, x_2, \cdots, x_n$ with probabilities $P(x_1), P(x_2), \cdots, P(x_n)$, the entropy $H(x)$ is defined as

$$H(x) = \sum_{i=1}^{n} P(x_i) \log P(x_i) \tag{8.126}$$

We can generalize this expression for continuous random variable by using the integral instead of discrete summation. Then in this case, $H(x)$ becomes

$$H(x) = \int_{-\infty}^{\infty} p(x) \log \frac{1}{p(x)} dx \tag{8.127}$$

This equation is indeed a meaningful definition of entropy for continuous random variable. A random variable x takes a value in the range $n \Delta x$, $(n + 1) \Delta x$ with probability $p(n \Delta x) \Delta x$ in the limit as $\Delta x \to 0$. The error in the approximation will vanish in the limit as $\Delta x \to 0$. Hence, $H(x)$, the entropy of a continuous random variable x, is given by

$$H(x) = \lim_{\Delta x \to 0} \sum_{n} p(n\Delta x) \Delta x \log \frac{1}{p(n\Delta x)\Delta x}$$

$$= \lim_{\Delta x \to 0} \left[\sum_n p(n\Delta x)\Delta x \log \frac{1}{p(n\Delta x)} - \sum_n p(n\Delta x)\Delta x \log \Delta x \right]$$

$$= \int_{-\infty}^{\infty} p(x)\log \frac{1}{p(x)} dx - \lim_{\Delta x \to 0} \log \Delta x \int_{-\infty}^{\infty} p(x)dx$$

$$= \int_{-\infty}^{\infty} p(x)\log \frac{1}{p(x)} dx - \lim_{\Delta x \to 0} \log \Delta x \qquad (8.128)$$

In the limit as $\Delta x \to 0$, $\log \Delta x \to -\infty$.

Hence, it appears that entropy of a continuous random variable is infinite.

This is quite true since the magnitude of uncertainty associated with a continuous random variable is infinite. Also a continuous random variable assumes infinite number of values and hence the uncertainty is of the order of infinity. This does not mean that there is no meaningful definition of entropy for a continuous random variable. The first term in the equation serves as meaningful measure of entropy of a continuous random variable x. We consider $\int p(x)\log \left(\frac{1}{p(x)} \right) dx$ as relative entropy with $-\log \Delta x$ as a reference. The information transmitted over a channel is actually the difference between the two terms $H(x)$ and $H(x|y)$. If we have a common reference for both $H(x)$ and $H(x|y)$, the difference $H(x) - H(x|y)$ will be the same as the difference between their relative entropies. Hence, we can justify the first term in Eq. (8.128) as the differential entropy of x. But we must always remember that this is a relative entropy and not the absolute entropy. Hence, we can define $H(x)$, the differential entropy of a continuous random variable x, as

$$H(x) = \int_{-\infty}^{\infty} p(x)\log \frac{1}{p(x)} dx \text{ bits} \qquad (8.129)$$

or
$$H(x) = -\int_{-\infty}^{\infty} p(x)\log p(x)dx \text{ bits} \qquad (8.130)$$

Although $H(x)$ is the differential (relative) entropy of x, it is called the entropy of random variable x.

8.24 AN OPTIMUM MODULATION SYSTEM: AN APPLICATION OF INFORMATION THEORY

Consider a generalized communication system shown in Fig. 8.25. The baseband signal $m(t)$ of bandwidth f_M is modulated onto a carrier and is then transmitted over a channel of bandwidth B. Noise is added on the channel and when the modulated signal arrives at the receiver, it has a signal-to-noise ratio S_i/N_i at the receiver input. N_i is the noise power at the receiver input in the bandwidth B. N_i is related to bandwidth as $N_i = \eta B$, where η is constant. The output signal obtained after demodulation is $\hat{m}(t) = m(t) + n_o(t)$, where $n_o(t)$ is the noise accompanying the output signal. The output signal to noise ratio is S_o/N_o and the output waveform $m(t)$ is bandlimited to f_M.

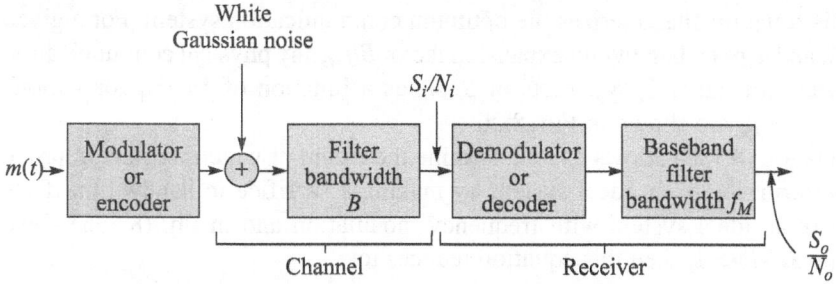

Fig. 8.25 Generalized communication system

The maximum rate at which information may be arriving at the receiver, according to Shannon–Hartley theorem, is

$$C_i = B \log_2\left(1 + \frac{S_i}{N_i}\right) \tag{8.131}$$

While the maximum rate at which information may be issuing from the receiver is

$$C_o = f_M \log_2\left(1 + \frac{S_o}{N_o}\right) \tag{8.132}$$

An ideal modulation is the one in which $C_o = C_i$, i.e., the rate of information output of the receiver is equal to the rate of information input to the receiver. This means that no information is lost as the information passes through the receiver nor is any information accumulated in the receiver. We assume that this feature of the receiver persists at all rates of information flow up to the maximum as determined by Shannon–Hartley theorem. Solving for $\dfrac{S_o}{N_o}$ and setting $C_i = C_o$, we get

$$\frac{S_o}{N_o} = \left(1 + \frac{S_i}{N_i}\right)^{B/f_M} - 1 \tag{8.133}$$

In the example shown, although we limit the baseband signal to f_M, the modulated signal occupies a larger bandwidth B. The ratio B/f_M is called bandwidth expansion factor, the most important parameter of the system. Since $N_i = \eta B$,

$$\frac{S_i}{N_i} = \frac{S_i}{\eta B} = \frac{f_M}{B}\frac{S_i}{\eta f_M} \tag{8.134}$$

Now Eq. (8.133) can be rewritten as

$$\frac{S_o}{N_o} = \left(1 + \frac{f_M}{B}\frac{S_i}{\eta f_M}\right)^{B/f_M} - 1 \tag{8.135}$$

This equation characterizes the optimum communication system. For a given $S_i/\eta f_M$ and a given bandwidth expansion factor B/f_M, any physical communication will yield a smaller S_o/N_o. Plots of S_o/N_o as a function of $S_i/\eta f_M$ for various values of B/f_M are shown in Fig. 8.26.

Equation (8.135) allows us to determine the extent of which we may improve the performance of an ideal system by making a sacrifice in bandwidth. If we compare an ideal system with frequency modulation and in Eq. (8.135) if we assume $S_i/N_i \gg 1$, then this equation reduces to

$$\frac{S_o}{N_o} = \left(\frac{f_M}{B}\frac{S_i}{\eta f_M}\right)^{B/f_M} \tag{8.136}$$

For a wideband FM system,

$$\frac{S_o}{N_o} = \frac{3}{4}\left(\frac{B}{f_M}\right)^2 \frac{S_i}{\eta f_M} \tag{8.137}$$

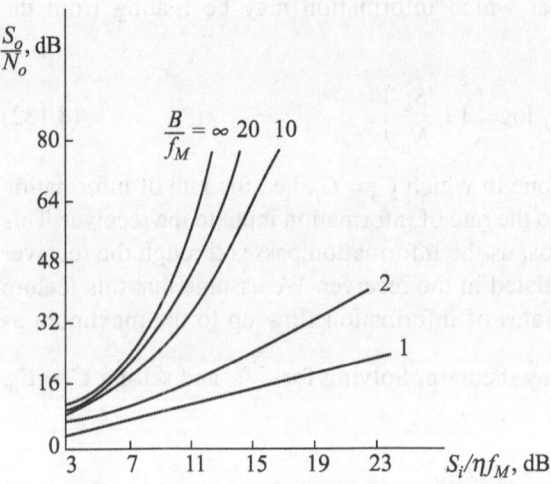

Fig. 8.26 *S/N* ratio characteristic: optimum communication system

Hence, it can be seen that for an ideal system, the performance increases exponentially with bandwidth expansion and the performance of an FM system increases only with the square of the bandwidth expansion factor. Thus, for $B/f_M > 1$, the performance of an ideal system increases much more rapidly with increase in $S_i/\eta f_M$ than that of an FM system.

8.24.1 A Comparison of AM System with an Optimum System

(i) For an SSB system, where $B/f_M = 1$,

$$\frac{S_o}{N_o} = \frac{S_i}{\eta f_M} \tag{8.138}$$

(ii) For a DSBSC synchronous detection, $B = 2f_M$,

$$\frac{S_o}{N_o} = \frac{S_i}{\eta f_M} \tag{8.139}$$

(iii) For square-law detection $\overline{m^2(t)} \ll 1, B = 2f_M$,

$$\frac{S_o}{N_o} = \overline{m^2(t)} \frac{S_i}{\eta f_M} \frac{1}{1 + 3/4 S_i / \eta f_M} \tag{8.140}$$

(iv) Linear envelope detection $\overline{m^2(t)} \ll 1, B = 2f_M$,

Above threshold, $\dfrac{S_o}{N_o} = \overline{m^2(t)} \dfrac{S_i}{\eta f_M}$ $\tag{8.141}$

Below threshold, $\dfrac{S_o}{N_o} = \dfrac{\overline{m^2(t)}}{1.1} \left(\dfrac{S_i}{\eta f_M} \right)^2$ $\tag{8.142}$

Now we compare an ideal system with various systems with the same bandwidth expansion factor.

For an SSB since $B/f_M = 1$ and we set $B/f_M = 1$ in Eq. (8.135), which applies to the ideal system. Then this equation reduces to $S_o/N_o = S_i/\eta f_M$, which is the same expression we get for an SSB system [Eq. (8.138)]. Hence, SSB comes closer to ideal system. The plots of S_o/N_o vs S_i/N_i for $B/f_M = 1$ and for $B/f_M = 2$ are shown in Fig. 8.27. It can be observed that the plot for $B/f_M = 1$ applies both to ideal system and SSB. The plot $B/f_M = 1$ applies for DSBSC also. However, in DSB since $B/f_M = 2$, if DSB were an optimum system, the plot corresponding to $B/f_M = 2$ would apply. DSB, therefore, is not an optimum system. For example, if $S_i/\eta f_M$ is chosen to be 20 dB, then output S_o/N_o is also 20 dB. If, however, DSB were optimum, the output S/N ratio would be larger by 14 dB.

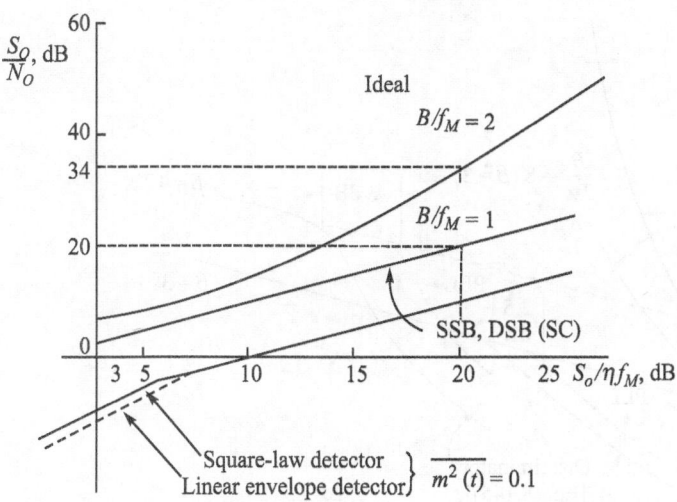

Fig. 8.27   A comparison of AM systems

The curves for asynchronous square law and linear envelope detector are given in Fig. 8.27 for $m^2(t) = 0.1(-10\ \text{dB})$. The performance of these systems is seen to be poor in comparison with the optimum system for $\dfrac{B}{f_M} = 2$.

8.24.2 Comparison of FM Systems

The output SNR for an FM system for sinusoidal modulation is

$$\frac{S_o}{N_o} = \frac{3}{2}\beta^2 \frac{S_i}{\eta f_M} \tag{8.143}$$

where $\beta = \Delta f / f_M$ and Δf is the frequency deviation. To relate β and bandwidth expansion factor from Carson's rule, where $B = 2(\beta + 1)f_M$,

$$\frac{B}{f_M} = 2(\beta + 1) \tag{8.144}$$

The expression for the output S/N ratio of the discriminator, including the effect of sinusoidal modulation and valid for both above and below threshold, is

$$\frac{S_o}{N_o} = \frac{(3/2)\beta^2\left(S_i/\eta f_M\right)}{1 + (12\beta/\pi)\left(S_i/\eta f_M\right)\exp[-\left(f_M/B\right)\left(S_i/\eta f_M\right)]} \tag{8.145}$$

This equation is plotted in Fig. 8.28 for $\beta = 3$ and 12. The output SNR characteristics of the optimal demodulator and second-order PLL are sketched on the same set of axes. It can be observed that the PLL results in 3 dB threshold extension for $\beta = 12$ and a 2 dB threshold extension when $\beta = 3$. The performance of an FM system falls substantially short of ideal system.

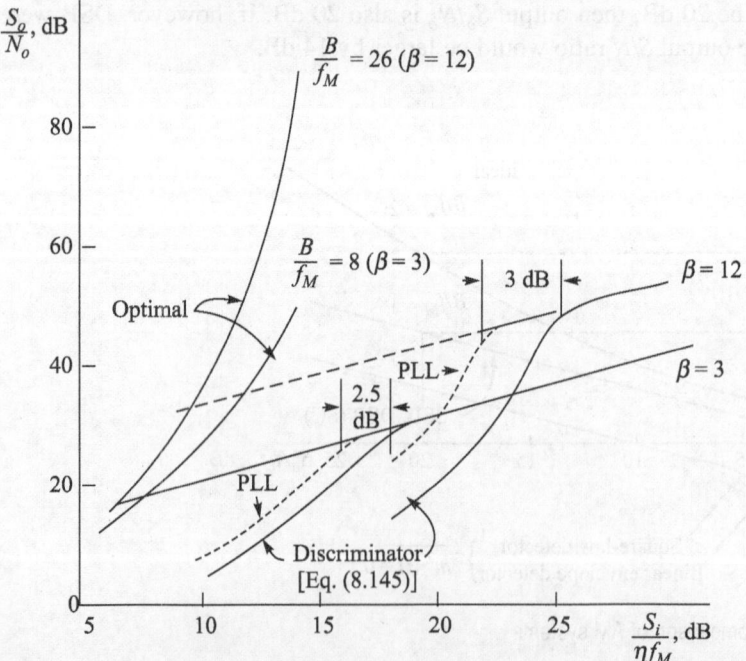

Fig. 8.28 FM demodulators: comparison including the effect of sinusoidal modulation

8.24.3 Comparison of PCM and FM

If we modulate a binary PCM signal onto PSK or FSK carrier, the threshold of FSK will be 2.2 dB greater than the threshold using PSK.

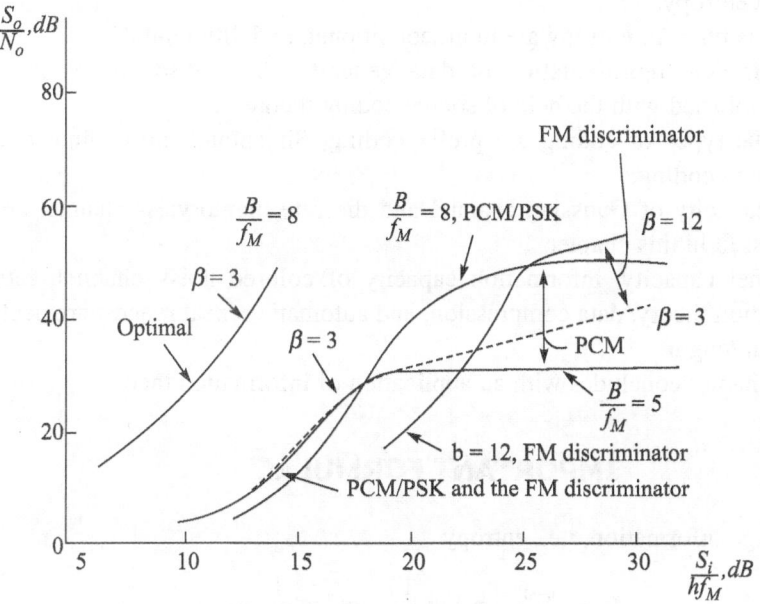

Fig. 8.29   PCM/PSK and the FM: comparison using optimal system and the discriminator demodulation

Figure 8.29 compares PCM/PSK, the FM discriminator, and the optimal demodulator. Let us consider the two FM discriminator characteristics with $\beta = 12$ and $\beta = 3$. The threshold occurs at 25 dB for $\beta = 12$ and at 18 dB and for $\beta = 3$. To compare this with PCM, we consider both threshold and the bandwidth necessary for transmission (bandwidth expansion factor). We observe that bandwidth expansion factor $\beta/f_M = 8$ in PCM, which corresponds to $\beta = 3$, results in an output SNR, which is approximately equal to that of the discriminator operating at $\beta = 12$, $B/f_M = 2(12 + 1) = 26$.

Thus, to obtain an output SNR of 48 dB requires $B/f_M = 26$, when using a discriminator and $B/f_M = 8$ when employing PCM/PSK. Hence, 3.25 times more bandwidth is required for the discriminator. If an output SNR of 28 dB is required, then the FM discriminator with $\beta = 3$ and $B/f_M = 8$ can be employed. However, the same results can be obtained by using PCM-PSK unit with $B/f_M = 5$. In this case, the FM discriminator requires only 1.6 times more bandwidth. Hence, the improvement of PCM over the discriminator increases with increased bandwidth expansion. Comparing the optimal system with PCM for $B/f_M = 8$ indicates that the PCM system operating at threshold requires an S/N ratio that is 11 dB greater than required by the optimal demodulation threshold.

SUMMARY

- Information theory is a subject that quantifies the concept of information.
- The average information per message is represented by the symbol H and is called entropy.
- Various types of entropy are joint, conditional, and differential.
- An efficient representation of data generated by a discrete source is accomplished with the help of source coding theorem.
- Various types of coding are prefix coding, Shannon–Fano coding, and Huffman coding.
- The capacity of Gaussian channel and discrete memoryless channel are discussed in this chapter.
- Channel capacity, information capacity of colored noisy channel, rate distortion theory, data compression, and automatic repeat request are dealt with at length.
- The chapter concludes with an application of information theory.

IMPORTANT FORMULAE

- Average information, i.e., entropy

$$H = \frac{I_{\text{total}}}{N} = p_1 \log \frac{1}{p_1} + p_2 \log \frac{1}{p_2} + \cdots$$

$$= \sum_{k=1}^{M} p_k \log \frac{1}{p_k}$$

- For two messages with probabilities p and $(1-p)$,

$$H = p \log_2 \frac{1}{p} + (1-p) \log \frac{1}{1-p}$$

- For two messages with equal probability 1/2,

$$H_{\text{max}} = \frac{1}{2} \log_2 2 + \frac{1}{2} \log_2 2 = \log_2 2 = 1$$

- For m messages with equal probability $1/M$,

$$H_{\text{max}} = \sum \frac{1}{M} \log_2 m = \log_2 m$$

- Joint entropy, $H(X) = - \sum_{x_1, x_2, x_n} p(x_1, x_2, \cdots, x_n) \log p(x_1, x_2, \cdots, x_n)$

- Conditional entropy,

$$H(X_n \mid X_1, \cdots, X_{n-1}) = - \sum_{x_1 \cdots x_n} p(x_1, x_2, \cdots, x_n) \log p(x_n \mid x_1, \cdots, x_{n-1})$$

- Entropy rate, $H = \lim_{n \to \infty} H(X_n \mid X_1, X_2 \cdots X_{n-1}) \dfrac{2}{M}$

- Differential entropy, $h(x) = -\int_{-\infty}^{\infty} f_x(x) \log f_x(x) dx$
- For two variables, differential entropy,

$$h(X,Y) = \int_{-\infty}^{\infty} \int_{-\infty}^{\infty} f(x,y) \log f(x,y) \, dxdy$$

- Kraft–Mcmillan inequality for average code length for prefix code is

$$\bar{L} = \sum_{k=0}^{K-1} \frac{l_k}{2^{l_k}}$$

and the corresponding entropy of the source is

$$H(Y) = \sum_{k=0}^{K-1} \left(\frac{1}{2^{l_k}}\right) \log_2\left(2^{l_k}\right) = \sum_{K=0}^{K-1} \frac{l_k}{2^{l_k}}$$

- Capacity of Gaussian channel, $C = B \log_2\left(1 + \dfrac{S}{N}\right)$ bits

Its average signal power, $S = \dfrac{2}{M}\left[\left(\dfrac{\lambda\sigma}{2}\right)^2 + \left(\dfrac{3\lambda\sigma}{2}\right)^2 + \cdots + \left(\dfrac{(M-1)\lambda\sigma}{2}\right)^2\right]$

$$= \frac{M^2-1}{12}(\lambda\sigma)^2$$

$$M = \left(1 + \frac{12S}{\lambda^2\sigma^2}\right)^{1/2} = \left(1 + \frac{12}{\lambda^2}\frac{S}{N}\right)^{1/2}$$

and

$$H = \log_2 M = \log_2\left(1 + \frac{12}{\lambda^2}\frac{S}{N}\right)^{1/2}$$

$$= \frac{1}{2}\log_2\left(1 + \frac{12}{\lambda^2}\frac{S}{N}\right) \text{ bits/message}$$

- The Shannon–Hartley equation, $C = \log_2\left(1 + \dfrac{S}{N}\right)$ bits/s

- The capacity of discrete time additive white Gaussian noise channel with input power constraint p is

$$C = \frac{1}{n}\log M$$

$$= \frac{1}{n}\frac{n}{2}\log\left(1 + \frac{P}{P_N}\right)$$

$$= \frac{1}{2}\log\left(1 + \frac{P}{P_N}\right)$$

- For error-free communication over noisy channel, the channel capacity

$$C_S = 1 - \left[P_e \log\frac{1}{P_e} + (1 - P_e)\log\frac{1}{1 - P_e}\right]$$

ADDITIONAL EXAMPLES

1. An analog signal is bandlimited to B Hz, sampled at the Nyquist rate, and the samples are quantized into 4 levels. The quantization levels $Q_1, Q_2, Q_3,$ and Q_4 are assumed independent and occur with probabilities $p_1 = p_2 = p_4 = \dfrac{1}{8}$ and $p_3 = \dfrac{5}{8}$. Find the information rate of the source.

Solution

The average information H is

$$p_1 \log_2 \frac{1}{p_1} + p_2 \log_2 \frac{1}{p_2} + p_3 \log_2 \frac{1}{p_3} + p_4 \log_2 \frac{1}{p_4} = \frac{1}{8} \log_2 8 + \frac{1}{8} \log_2 8$$

$$+ \frac{5}{8} \log_2 \frac{8}{5} + \frac{1}{8} \log_2 8 = 1.55 \text{ bits/message}$$

The information rate R is $R = rH = 2B(1.55) = 3.1B$ bits/s

2. The prabilities of five source messages are $m_1 = 0.2$, $m_2 = 0.3$, $m_3 = 0.2$, $m_4 = 0.15$, and $m_5 = 0.15$. Find the coding efficiecy for (a) Shannon–Fano coding and (b) Huffman coding.

Solution

$$\text{Average information} = 2 \times 0.2 \log_2 \left(\frac{1}{0.2}\right) + 0.3 \log_2 \left(\frac{1}{0.3}\right) + 2 \times 0.15 \log_2 \left(\frac{1}{0.15}\right)$$

$$= 0.4 \log_2 5 + 0.3 \log_2 3.33 + 0.3 \log_2 6.66$$

$$= 2.28 \text{ bits}$$

(a) Shannon–fano coding:

					coding
m_1	0.2	0	0		00
m_2	0.3	0	1		01
m_3	0.2	1	0		10
m_4	0.15	1	1	0	110
m_5	0.15	1	1	1	111

$$\text{Average code word length} = (0.2 \times 2) + (0.3 \times 2) + (0.2 \times 2) + (2 \times 0.15 \times 3)$$

$$= 2.3 \text{ bits}$$

$$\text{Coding efficiency} = \frac{2.28}{2.3} = 0.9913 = 99.13\%$$

(b) Huffman coding:

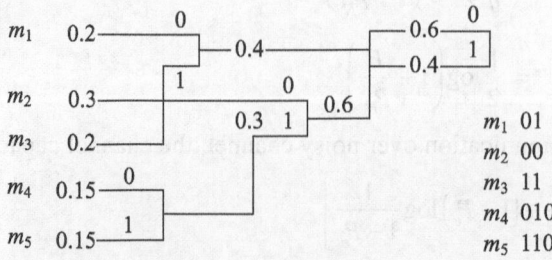

Average code word length $= (0.2 \times 2) + (0.3 \times 2) + (0.2 \times 2) + (2 \times 0.15 \times 3)$

$$= 0.4 + 0.6 + 0.4 + 0.9$$

$$= 2.3 \text{ bits}$$

Coding efficiency $= \dfrac{2.28}{2.3} = 0.9913 = 99.13\%$

In the above case, both codes give the same efficiency.

3. Find capacity of Gausian channel of bandwidth 4 kHz with noise PSD 10^{-9} W/Hz when signal energy is (a) 0.1 J and (b) 0.001 J. (c) How does the channel capacity change in (b) if the bandwidth is increased to 10 kHz?

Solution

(a)

$$C = 4000 \times \log_2\left(1 + \frac{0.1}{2 \times 10^{-9} \times 4000}\right) = 5.44 \times 10^4 \text{ bits/s}$$

(b) Similarly,

$$C = 4000 \times \log_2\left(1 + \frac{0.001}{2 \times 10^{-9} \times 4000}\right) = 2.79 \times 10^4 \text{ bits/s}$$

(c) Here, $B = 10,000$. Then

$$C = 10000 \times \log_2\left(1 + \frac{0.001}{2 \times 10^{-9} \times 10000}\right) = 5.67 \times 10^4 \text{ bits/s}$$

Note that a 100 times fall in energy is more than compensated by increasing bandwidth by 2.5 times.

4. Show that $H(X, Y) = H(X/Y) + H(Y)$.

Solution

$$H(X, Y) = \sum_{i=1}^{m}\sum_{j=1}^{n} p(x_i, y_j)\log_2 \frac{1}{p(x_i, y_j)}$$

or $\qquad H(X, Y) = -\sum_{i=1}^{m}\sum_{j=1}^{n} p(x_i, y_j)\log_2[p(x_i/y_j)p(y_j)]$

since $\qquad p(x_i, y_j) = p(x_i/y_j)p(y_j)$

or $\qquad H(X,Y) = -\sum_{i=1}^{m}\sum_{j=1}^{n} p(x_i, y_j)\log_2 p(x_i/y_j) - \sum_{i=1}^{m}\sum_{j=1}^{n} p(x_i, y_j)\log_2 p(y_j)$

or $\qquad H(X,Y) = H(X/Y) - \sum_{j=1}^{n}\left[\sum_{i=1}^{m} p(x_i, y_j)\right]\log_2 p(y_j)$

or $\qquad H(X,Y) = H(X/Y) - \sum_{j=1}^{n} p(y_j)\log_2 p(y_j)$

(as joint problem is summed over all inputs)

or $\qquad H(X,Y) = H(X/Y) + H(Y)$

5. Compare throughput efficiencies assuming that $T_w = 10$ μs, a BCH (1023, 973) code is used, $P_A = 0.99$, and $T_I = 40$ μs.

Solution

Assume that $N = 4$ so that the retransmission time in the go-back-N system is the same as the idle time in the stop-and-wait system. We then find

$$\eta_{S\&W} = \frac{k}{n} \frac{P_A}{1 + \dfrac{T_I}{T_W}} = \frac{973}{1023} \frac{0.99}{1 + 4} = 0.188$$

$$\eta_{SGpull} = \frac{k}{n} \frac{1}{1 + \dfrac{N(1 - P_A)}{P_A}}$$

$$= \frac{973}{1023} \frac{1}{1 + \dfrac{4(0.01)}{0.99}} = 0.915$$

and $$\eta_{SR} = \frac{k}{n} P_A = \frac{973}{1023}(0.99) = 0.942$$

Thus, there is a significant improvement obtained by using the go-back-N algorithm rather than the stop-and-wait algorithm. However, the improvement made in using the selective repeat algorithm is often not deemed worth the additional complexity.

6. A source with the bandwidth 4000 Hz is sampled at the Nyquist rate. Assuming that the resulting sequence can be approximately modelled by a DMS with alphabet $A = \{-2, -1, 0, 1, 2\}$ and with corresponding probabilities $\left\{\dfrac{1}{2}, \dfrac{1}{4}, \dfrac{1}{8}, \dfrac{1}{16}, \dfrac{1}{16}\right\}$, determine the rate of the source in bits/s.

Solution
We have

$$H(X) = \frac{1}{2}\log 2 + \frac{1}{4}\log 4 + \frac{1}{8}\log 8 + 2 \times \frac{1}{16}\log 16 = \frac{15}{8} \text{ bits/sample}$$

Since we have $2 \times 4000 = 8000$ samples/s, the source produces information at a rate of $8000 \times \dfrac{15}{8} = 15{,}000$ bits/s.

7. A discrete memoryless source has an alphabet of size N and the source outputs are equiprobable (each having a probability of $1/N$). Find the entropy of this source.

Solution

We have $$H(X) = -\sum_{i=1}^{N} \frac{1}{N} \log \frac{1}{N}$$

$$= \log N$$

8. Let X and Y be binary random variables with $P(X = 0, Y = 0) = \dfrac{1}{3}$, $P(X = 1, Y = 0) = \dfrac{1}{3}$, and $P(X = 0, Y = 1) = \dfrac{1}{3}$. Find $I(X,Y)$ in this case.

Solution

We have $\quad P(X = 0) = P(Y = 0) = \dfrac{2}{3}$

Therefore, $\quad H(X) = H(Y) = H_b\left(\dfrac{2}{3}\right) = 0.919$

On the other hand, the (X, Y) pair is a random vector uniformly distributed on three values: $(0, 0)$, $(1, 0)$, and $(0, 1)$.

Therefore, $\quad H(X, Y) = \log 3 = 1.585$

From this, we have

$$H(X|Y) = H(X, Y) - H(Y) = 1.585 - 0.919 = 0.666$$

and $\quad\quad\quad I(X, Y) = H(X) - H(X|Y) = 0.919 - 0.666 = 0.253$

9. Determine the differential entropy of a random variable X uniformly distributed on $[0, a]$.

Solution

From the definition of differential entropy,

$$h(X) = -\int_0^a \frac{1}{a} \log \frac{1}{a}\, dx = \log a$$

Clearly, for $a < 1$, we have $h(X) < 0$, which is in contrast to the non-negativity of the entropy of discrete sources. Also, for $a = 1$, $h(X) = 0$ without X being deterministic. This is again in contrast to the entropy properties of discrete sources.

10. Find the capacity of a telephone channel with bandwidth $W = 3000$ Hz and signal-to-noise ratio of 39 dB.

Solution

The signal-to-noise ratio of 39 dB is equivalent to 7943. Using Shannon's relation, we have

$$C = 3000 \log (1 + 7943) \approx 38{,}867 \text{ bits/s}$$

PROBLEMS

1. A discrete memoryless source has an alphabet $\{a_1, a_2, a_3, a_4, a_5, a_6\}$ with corresponding probabilities $\{0.1, 0.2, 0.3, 0.05, 0.15, 0.2\}$. Find the entropy of this source. Compare this entropy with the entropy of a uniformly distributed source with the same alphabet.

2. Let random variable X be the output of a discrete memoryless source that is uniformly distributed with size N. Find the entropy.

3. An information source can be modelled as a bandlimited process with a bandwidth of 6000 Hz. This process is sampled at a rate higher than the Nyquist rate to provide a guard band of 2000 Hz. We observe that the resulting samples take values in the set $A = \{-4, -3, -1, 2, 4, 7\}$ with probabilities 0.2, 0.1, 0.15, 0.05, 0.3, 0.2. What is the entropy of the discrete time source in bits per output (sample)? What is the information generated by this source in bits/s?

4. A memoryless source has the alphabet $A = \{-5, -3, -1, 0, 1, 3, 5\}$ with the corresponding probabilities $\{0.05, 0.1, 0.1, 0.15, 0.05, 0.25, 0.3\}$.
   (a) Find the entropy of the source.
   (b) Assume that the source is quantized according to the quantization rule
   $$\{q(-5) = q(-3) = -4,$$
   $$\{q(-1) = q(0) = q(1) = 0$$
   $$\{q(3) = q(5) = 4$$
   Find the entropy of the quantized source.

5. An analog signal bandlimited to 4 kHz is sampled at the Nyquist rate. The samples are quantized into eight levels. These levels are assumed to be independent and occur with probabilities $p_1 = p_3 = p_5 = 1/16$, $p_2 = p_4 = p_6 = 1/8$ and $p_7 = p_8 = 7/32$. Find the information rate of the source.

6. An event has two possible outcomes with the three cases of probabilities as follows:

   Case 1: $p_1 = 0.02, p_2 = 0.98$
   Case 2: $p_1 = 0.45$, $p_2 = 0.55$
   Case 3: $p_1 = 0.5, p_2 = 0.5$
   Find out the entropy in each case. What do you infer?

7. Given $\left\{\dfrac{1}{2}, \dfrac{1}{4}, \dfrac{1}{8}, ..., \dfrac{1}{2^{n-1}}, \dfrac{1}{2^{n-1}}\right\}$. Show that the average code word length for such a source is equal to the source entropy.

8. Show that $\{01, 100, 101, 1110, 1111, 0011, 0001\}$ cannot be a Huffman code for any source probability distribution.

9. A discrete memoryless information source is described by the alphabet $Æ = \{x_1, x_2, x_3, x_4, x_5, x_6\}$ with probabilities $\{1/32, 1/8, 1/2, 1/16, 1/32, 1/4\}$, respectively. Design a Huffman code for this source and determine the average code word length of the Huffman.

10. Find the average code length, entropy, code efficiency, and redundancy for six messages with probabilities 0.25, 0.12, 0.15, 0.08, 0.30, 0.10. Use Huffman coding.

11. Messages $Q_1, Q_2, Q_3, \cdots, Q_m$ have probabilities $p_1 + p_2 + p_3 + \cdots + p_m$ of occurring.
   (a) Write an expression for H.
   (b) If $M = 3$, write H in terms of p_1 and p_2, where $p_1 + p_2 + p_3 = 1$.

12. Six messages with probabilities 0.15, 0.1, 0.3, 0.08, 0.25, 0.12 form a zero memory source. Find the 4-ary Huffman code. Find out its average word length, the efficiency, and the redundancy.

13. Two messages are emitted by a zero memory source and their probabilities are 0.6 and 0.4. Find the optimum Huffman binary code. Find also its second-order extensions. What is the code efficiency in each case?
14. Consider five messages given by the probabilities 1/2,1/4,1/8,1/16,1/16.
    (a) Calculate H.
    (b) Use the Shannon–Fano algorithm to develop an efficient code and, for that code, calculate the average number of bits/message. Compare with H.
15. For a binary symmetric channel, find the channel capacity for $p = 0.7$ and $p = 0.4$.

Answers to problems

1. 2.408 bits/message
2. $\log_2 N$
3. 38.528 kbits/s
4. 2.53bits/message, 1.407 bits/message
5. 18.8 kbits/s
6. 0.14, 0.99, 1. Case 1: uncertainty less, Case 2: uncertainty more, Case 3: uncertainty maximum
8. Not starting with 0
9. 1.937 binary digits
10. 2.45 binary digits, 2.418 bits, 0.976 and .024
11. $H = p_1 \log_2 \dfrac{1}{p_1} + p_2 \log_2 \dfrac{1}{p_2} + \cdots + p_m \log_2 \dfrac{1}{p_m}$

$H = p_1 \log_2 \dfrac{1}{p_1} + p_2 \log_2 \dfrac{1}{p_2} + (1 - p_1 - p_2) \log \dfrac{1}{(1 - p_1 - p_2)}$

12. 1.3(4-ary digit), 1.209, 0.93 and 0.07
13. 0.973 bit and 0.953
14. 1.87 and 1.875. Since H and the average number of bits is more or less same, each bit carries one bit of information, which is the maximum information that can be conveyed by a bit.
15. 0.1557 bit/message and 0.0259 bit/message

MATLAB EXAMPLE

```
1. %program in Huffman encoding and decoding%
   symbols= [1:5];
   p= [.4 .15 .15 .15 .15];
   [dict,avglen]= huffmandict(symbols,p)

   avginfo=0;
   for i=1:length(p)
       avginfo=avginfo+p(i)*log2(1/p(i));

   end
   avginfo
```

```
coding_efficiency=avginfo*100/avglen
sig=randsrc(1,100,[symbols;p]);
code = huffmanenco(sig,dict);
decoded=huffmandeco(code,dict);
isequal(sig,decoded)

dict =
   [1]        [ 1]
   [2]        [1x3 double]
   [3]        [1x3 double]
   [4]        [1x3 double]
   [5]        [1x3 double]
avglen =

       2.2000

avginfo =

       2.1710

coding_efficiency =

       98.6796

ans =

       1
```

Introduction to Probability, Random Variable, and Random Processes

LEARNING OBJECTIVES

This chapter will enable the students to know

- what is probability and what is needed for understanding random variable and random process
- random variable and its properties
- that when a signal is transmitted through a communication channel, the signal received has two imperfections—deterministic like non-linear distortion, cross talk, etc. and non-deterministic like addition of noise in the channel, multi-path fading, etc.
- the requirement for random process to be stationary
- how a random process is described in terms of its mean, correlation, and covariance function
- what are the conditions to be satisfied for a stationary random process to be ergodic
- what happens to a random process when it is passed through an LTI circuit
- the power spectral density of a random process
- the characteristic of Gaussian process

9.1 INTRODUCTION TO PROBABILITY

Probability theory is useful in the solution to an engineering problem. This can be done in two different ways. The first is to suggest a philosophy concerning probability. The second one is to note some of the many different types of situations that arise in normal engineering practice in which the use of probability concepts is indispensable.

Several concepts of probability have evolved over the centuries. Three different approaches are discussed in this chapter.

- The classical approach
- The relative frequency approach
- The axiomatic approach

9.1.1 The Classical Approach

In this approach, which is not experimental, the probability of an event is computed a priori by counting the number of ways N_E that an event E can occur and forming the ratio N_E/N, where N is the number of all possible outcomes. An important notion here is that all outcomes (the number of ways that E can occur) are equally likely. The classical approach suffers from several significant problems. However, in those problems where it is impractical to actually determine the outcome probabilities by experimentation and where equally likely outcomes occur, the classical approach is useful. An example of the classical approach can be constructed by considering balls in an urn. Suppose there are 4 red and 5 green balls in an urn. Furthermore, suppose a ball is drawn at random from the urn. Then, by the classical approach, the probability of obtaining a red ball is 4/9.

9.1.2 The Relative Frequency Approach

The relative frequency approach to define the probability of an event E is to perform an experiment n times. The number of times that E appears is denoted by n_E. Then, the probability of E is given by

$$P[E] = \lim_{x \to \infty} \frac{n_E}{n} \tag{9.1}$$

There are some difficulties with this approach (for example, we cannot do an experiment an infinite number of times). Despite the problems with this notion of probability, the relative frequency concept is essential in applying the probability theory to the physical world.

9.1.3 The Axiomatic Approach

This is a mathematical approach that is followed in most modern textbooks on probability. It is based on a branch of mathematics known as measure theory. The axiomatic approach has the notion of a probability space as its main component. Basically, a probability space consists of

- a sample space, denoted by S
- a collection of events, denoted by F
- a probability measure, denoted by P

Without discussing the measure-theoretic aspects, this approach is employed in this chapter. Before discussing (S, F, P), the elementary set theory is reviewed.

9.2 ELEMENTARY SET THEORY

A set is a collection of objects. These objects are called *elements* of the set. Usually, upper case letters in italics are used to denote sets (e.g. *A, B, C*,...). Lower case letters in italics are used to denote set elements (e.g. *a, b, c*, ...). The notation $a \in A$ ($a \notin A$) denotes that a is (is not) an element of *A*. All sets are considered to be subsets of some universal set, denoted here by *S*. Set *A* is a subset of set *B*, denoted by $A \subset B$, if all elements in *A* are elements of *B*. The empty or null set is the set that contains no elements. It is denoted by ($\varnothing$).

Transitivity property

If $U \subset B$ and $B \subset A$, then $U \subset A$, a result known as the transitivity property.

Set equality

The equality of two sets $B = A$ is equivalent to the requirements that $B \subset A$ and $A \subset B$. Often, two sets are shown to be equal by showing that this requirement holds.

Union

The union of two sets *A* and *B* is a set containing all of the elements of *A* and all of the elements of *B* and no other elements. The union is denoted by $A \cup B$.

The union is commutative, i.e., $A \cup B = B \cup A$

The union is associative, i.e., $(A \cup B) \cup C = A \cup (B \cup C)$

Intersection

The intersection of two sets *A* and *B* is a set consisting of all elements common to both *A* and *B*. It is denoted by $A \cap B$. Figure 9.1 illustrates the concept of intersection.

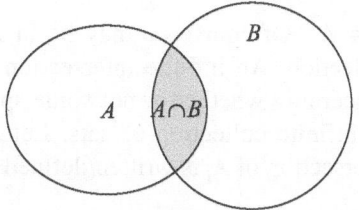

The intersection is commutative, i.e.,

$$A \cap B = B \cap A$$

The intersection is associative, i.e.,

$$(A \cap B) \cap C = A \cap (B \cap C)$$ **Fig. 9.1** The intersection of sets *A* and *B*

The intersection is distributive over unions, i.e.,

$$A \cap B \cup C) = (A \cap B) \cup (A \cap C)$$

Sets *A* and *B* are said to be mutually exclusive or disjoint if they have no common element so that $A \cap B = (\varnothing)$.

Set complementation

The complement of *A* is denoted by $\bar{A}$, and it is the set consisting of all elements of the universal set that are not in *A*. Note that $A \cup \bar{A} = S$ and $A \cap B = (\varnothing)$.

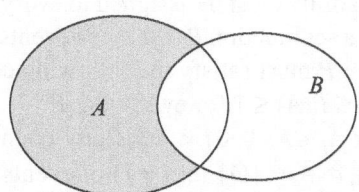

Fig. 9.2 The difference between sets *A* and *B*

Set difference

The difference $A - B$ denotes a set consisting of all elements in A that are not in B. Often, $A - B$ is called the complement of B relative to A. This is shown in Fig. 9.2, where the shaded area is $A - B$.

De Morgan's theorem

According to this theorem, for two sets A and B,

$$\overline{A \cup B} = \bar{A} \cap \bar{B} \text{ and } \overline{A \cap B} = \bar{A} \cup \bar{B} \tag{9.2}$$

If in a set identity, one replaces all sets by their complements, all unions by intersections, and all intersections by unions, the identity is preserved. For example, applying this rule to the set identity $A \cap (B \cup C) = (A \cap B) \cup (A \cap C)$,

we obtain the result $\bar{A} \cup (\bar{B} \cap \bar{C}) = (\bar{A} \cup \bar{B}) \cap (\bar{A} \cup \bar{C})$.

Infinite unions/intersection of sets

An infinite union of sets can be used to formulate questions concerning whether or not some specified item belongs to one or more sets that are part of an infinite collection of sets. Let A_i, $1 \leq i \leq \infty$, be a collection of sets. The union of A_i is written/defined as

$$\bigcup_{i=1}^{\infty} A_i \equiv \{\omega : \omega \in A_n \text{ for some } n, 1 \leq n < \infty\} \tag{9.3}$$

Equivalently, $\omega \in \displaystyle\bigcup_{i=1}^{\infty} A_i$ if and only if there is at least one integer n for which $\omega \in A_n$. Of course, ω may be in more than one set belonging to the infinite collection. An infinite intersection of sets can be used to formulate questions concerning whether or not some specified item belongs to all sets belonging to an infinite collection of sets. Let A_i, $1 \leq i < \infty$, be a collection of sets. The intersection of A_i is written/defined as

$$\bigcap_{i=1}^{\infty} A_i \equiv \{\omega : \omega \in A_n \text{ for all } n, 1 \leq n < \infty\} = \{\omega : \omega \in A_n, 1 \leq n < \infty\} \tag{9.4}$$

Equivalently, $\omega \in \displaystyle\bigcap_{i=1}^{\infty} A_i$ if and only if $\omega \in A_n$ for all n.

9.3 THE AXIOMATIC APPROACH

A probability must be assigned to every event (element of F). To accomplish this, we use a set function P that maps events in F into $[0,1]$, $P : F \rightarrow [0, 1]$. Probability measure P must satisfy the following conditions:

1. $0 \leq P(A) \leq 1$ for every $A \subset F$
2. If $A_i \in F$, $1 \leq i < \infty$, is any countable, mutually exclusive sequence (i.e., $A_i \cap A_j = \{\emptyset\}$ for $i \neq j$) of events, then

$$P\left(\bigcup_{i=1}^{\infty} A_i\right) = \sum_{i=1}^{\infty} P(A_i) \tag{9.5}$$

where P must be countably additive.

3. $\qquad\qquad P(S) = 1 \tag{9.6}$

Conditions 1 through 3 are called the *axioms of probability*. Now, we can say that the permissible set of events F can be any σ-algebra for which there exists a P that satisfies the axioms of probability. As it turns out, for some sample spaces, there are some σ-algebras that cannot serve as a set of permissible events because there is no corresponding P function that satisfies the axioms of probability.

In many problems, it might be desirable to let all subsets of S be events. That is to say, it might be desirable to let *everything* be an event. However, in general, it is not possible to do this because a P function may not exist that satisfies the axioms of probability.

A special case deserves to be mentioned. If S is countable (i.e., there exists a 1-1 correspondence between the elements of S and the integers), then F can be taken as the σ-algebra consisting of all possible subsets of S (i.e., let $F = F_L$, the largest σ-algebra). That is, if S is countable, it is possible to assign probabilities (without violating the axioms) to the elements of F_L in this case. But, in the more general case where S is not countable, to avoid violating the axioms of probability, there may be subsets of S that cannot be events; these subsets must be excluded from F.

As it turns out, if S is the real line (a non-countable sample space), then the σ-algebra F_L of all possible sets (of real numbers) contains too many sets. In this case, it is not possible to obtain a P that satisfies the axioms of probability and F_L cannot serve as the set of events. Instead, for S equal to the real line, the Borel σ-algebra F is usually chosen (it is very common to do this in applications). It is possible to assign probabilities to Borel sets without violating the axioms of probability. The Borel sets are assigned probabilities as discussed below.

Many important applications employ a probability space (S, F, P), where S is the set R of real numbers and $F = B$ is the Borel σ-algebra. The probability measure P is defined in terms of a density function $f(x)$. The density $f(x)$ can be any integrable function that satisfies

1. $f(x)$ 0 for all $x \in S = R$

2. $\displaystyle\int_{-\infty}^{\infty} f(x)dx = 1$

Then, the probability measure P is defined by

$$P(B) = \int_B f(x)dx \text{ for all } B \in F = R \tag{9.7}$$

As defined above, the notion of F, the set of possible events, is abstract. However, in most applications, we encounter only a few general types of F. In most applications, S is either countable, the real line $R = (-\infty, \infty)$, or an interval, i.e., $S = (a, b)$. These cases are discussed briefly in what follows. Many applications involve countable sample spaces. For most of these cases, F is taken as F_L, the set of all possible subsets of S. For events in F_L, probabilities are assigned in an application-specific intuitive manner. On the other hand, many applications use the real line $S = R = (-\infty, \infty)$, a non-countable set. For these cases, it is very common to use an F and P as discussed above. An identical approach is used when S is an interval of the real line.

9.3.1 Implications of the Axioms of Probability

A number of conclusions can be reached from considering the axioms of probability.

1. The probability of the impossible event $[\emptyset]$ is zero.

   Proof: Note that $A \cap [\emptyset] = [\emptyset]$ and $A \cup [\emptyset] = A$

   Hence, $P(A) = P(A \cup [\emptyset]) = P(A) + P(\emptyset)$

   We can conclude from this that $P([\emptyset]) = 0$

2. For any event A, we have $P(A) = 1 - P(\bar{A})$

   Proof: $A \cup \bar{A} = S$ and $A \cap \bar{A} = [\emptyset]$

   Hence,

   $$1 = P(S) = P(A \cap \bar{A}) = P(A) + P(\bar{A})$$

   This result leads to the conclusion that

   $$P(A) = 1 - P(\bar{A}) \tag{9.8}$$

3. For any events A and B, we have

   $$P(A \cup B) = P(A) + P(B) - P(A \cap B) \tag{9.9}$$

   Proof: The two identities

   $$A \cup B = A \cup ((B \cap A) \cup (B \cap \bar{A})) = (A \cup (B \cap A)) \cup (B \cap \bar{A}) = A \cup (B \cap \bar{A})$$

   $$B = B \cap (A \cup \bar{A}) = (A \cap B) \cup (B \cap \bar{A})$$

   lead to

   $$P(A \cup B) = P(A) + P(B \cap \bar{A})$$

   $$P(B) = P(A \cap B) + P(B \cap \bar{A})$$

Subtracting the last two expressions, we obtain the desired result

$$P(A \cup B) = P(A) + P(B) - P(A \cap B)$$

This result is generalized easily to the case of three or more events.

9.4 CONDITIONAL PROBABILITY

The conditional probability of an event A, assuming that an event M has occurred, is

$$P(A|M) = \frac{P(A \cap M)}{P(M)} \tag{9.10}$$

where it is assumed at $P(M) \neq 0$. From Eq.(9.10), we get a useful identity

$$P(A \cap M) = P(A \mid M) P(M)$$

Consider two special cases. The first case is $M \subset A$ so that $A \cap M = M$. For $M \subset A$, we have

$$P\,(A|M) = \frac{P(A \cap M)}{P(M)} = \frac{P(M)}{P(M)} = 1 \tag{9.11}$$

Next, consider the special case $A \subset M$ so that $P(M\,|\,A) = 1$. For this case, we have

$$P\,(A|M) = \frac{P(A \cap M)}{P(M)} = \frac{P(M\,|\,A)}{P(M)}\,P\,(A)$$

or $\qquad \dfrac{1}{P(M)}\,P(A) \geq P(A)$ $\hfill (9.12)$

which is an intuitive result.

Example 9.1 In a fair die experiment, the outcomes are $f_1, f_2, \ldots, f_4, f_6$, the six faces of the die. Let $A = \{f_2\}$ be the event 'a two occurs' and $M = \{f_2, f_4, f_6\}$ be the event 'an even outcome occurs.' Find out $P(A)$, $P(M)$, and $P(A \cap M)$.

Solution

We have $P(A) = 1/6$ and $P(M) = 1/2$

Also $P(A \cap M) = \dfrac{P(A)}{P(M)}$ so that

$$P[\{f_2\}|\{\text{even}\}] = \frac{1/6}{1/2} = \frac{1}{3} > P(f_2) = 1/6$$

Example 9.2 A box contains three white balls w_1, w_2, w_3 and two red balls r_1 and r_2. We remove at random and without replacement two balls in succession. What is the probability that the first removed ball is white and the second is red?

Solution

$P[\{\text{first ball is white}\}] = 3/5$

$P[\{\text{second is red}\}\,|\,\text{first is white}\,] = 1/2$

$P[\{\text{first ball is white}\} \cap \{\text{second ball is red}\}] = P[\{\text{second is red}\,|\,\text{first is white}\}]$
$P[\{\text{first ball is white}\}] = (1/2)\,(3/5) = 3/10$

9.4.1 Total Probability Theorem: Discrete Version

Let $[A_1, A_2, \ldots, A_n]$ be a partition of S. That is

$$\bigcup_{i=1}^{n} A_i = S \text{ and } A_i \cap A_j = [\varnothing] \text{ for } i \neq j \tag{9.13}$$

Let B be an arbitrary event, then

$$P(B) = P[B\,|\,A_1]\,P(A_1) + P[B\,|\,A_2]\,P[A_2] + \cdots + P[B\,|\,A_n]\,P[A_n] \tag{9.14}$$

Proof First note the set identity

$B = B \cap S = B \cap (A_1 \cup A_2 \cup \ldots \cup A_n) = (B \cap A_1) \cup (B \cap A_2) \cup \ldots \cup (B \cap A_n)$

For $i \neq j$, $B \cap A_i$ and $B \cap A_j$ are mutually exclusive. Thus, we have

$$P(B) = P[B \cap A_1] + P[B \cap A_2] + \ldots + P[B \cap A_n]$$

$$= P[B \mid A_1] \, P[A_1] + P[B \mid A_2] + \cdots + P[B \mid A_n] \, P[A_n] \qquad (9.15)$$

This result is known as discrete version of the *total probability theorem* . Let $[A_1, A_2, A_3]$ be a partition of S. Consider the identity

$$P[B] = P[B \mid A_1] \, P[A_1] + P[B \mid A_2] + P[B \mid A_3]$$

This equation is shown graphically in Fig. 9.3.

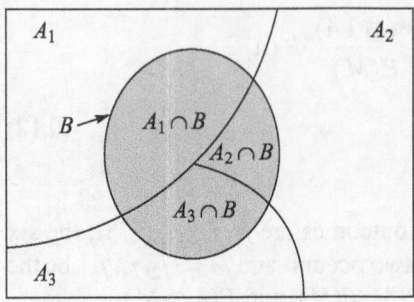

Fig. 9.3 Example of total probability function

9.4.2 Bayes' Theorem

Let $[A_1, A_2, A_3, \ldots, A_n]$ be a partition of S. Since

$$P(A_i \mid B) = P(B \mid A_i) \, \frac{P(A_i)}{P(B)}$$

and

$$P(B) = P[B \mid A_1] \, P[A_1] + P[B \mid A_2] + \cdots + P[B \mid A_n] \, P[A_n]$$

we get

$$P(A_i \mid B) = \frac{P(B \mid A_i) \, P(A_i)}{P[B \mid A_1] P[A_1] + P[B \mid A_2] P[A_2] + \cdots + P[B \mid A_n]} \qquad (9.16)$$

This is called Bayes' theorem. The $P[A_i]$ are called *a priori probabilities* and the $P[A_i \mid B]$ are called *a posteriori probabilities*.

Bayes' theorem provides a method for incorporating experimental observations into the characterization of an event. Both $P[A_i]$ and $P[A_i \mid B]$ characterize events A_i, $1 \le i \le n$. However, $P[A_i|B]$ may be a better and more definitive characterization, especially if B is an event related to the A_i. For example, consider events $A_1 = $ [rain today], $A_2 = $ [no rain today], and $T = 40^\circ$ C. Given the occurrence of T, one would expect $P[A_1|T]$ and $P[A_2|T]$ to more definitvely characterization *rain today* than does $P[A_1]$ and $P[A_2]$.

Example 9.3 We have four boxes. Box 1 contains 2000 components, of which 5% are defective. Box 2 contains 500 components, of which 40% defective. Boxes 3 and 4 contain 1000 components each and 10% are defective in both boxes.

(a) At random, we select one box and remove one component. What is the probability that this component is defective?

(b) We examine a component and find that it is defective. What is the probability that it came from Box 2?

Solution

(a) From the theorem of total probability, we have

$$P[\text{component is defective}] = \sum_{i=1}^{4} P[\text{defective}|\text{box}] \, P[\text{box}_i]$$

$$= (0.05)(0.25) + (0.4)(0.25) + (0.1)(0.25)$$
$$+ (0.1)(0.25)$$

$$= 0.1625$$

(b) By Bayes' law, we have

$$P[\text{box 2} \mid \text{defective}] = \frac{P[\text{defective}|\text{box 2}] \, P[\text{box 2}]}{P[\text{defective}|\text{box 1}] + \cdots + P[\text{defective}|\text{box 4}] \, P[\text{Box 4}]}$$

$$= \frac{(0.4)(0.25)}{0.1625}$$

$$= 0.615$$

9.4.3 Independence

Events A and B are said to be independent if

$$P[A \cap B] = P[A] \, P[B] \tag{9.17}$$

If A and B are independent, then

$$P[A|B] = \frac{P[A \cap B]}{P[B]} = \frac{P[A]P[B]}{P[B]} = P[A] \tag{9.18}$$

Three events A_1, A_2, and A_3 are independent if

(a) $P[A_i \cap A_j] = P[A_i] \, P[A_j]$ for $i \neq j$ and

(b) $P[A_1 \cap A_2 \cap A_3] = P[A_1] \, P[A_2] \, P[A_3]$

Condition a may hold and condition b may not. Likewise, condition b may hold and condition a may not hold. Both are required for the three events to be independent.

Let

$$P[A_1] = P[A_2] = P[A_3] = 1/5$$

$$P[A_1 \cap A_2] = P[A_2 \cap A_3] = P[A_1 \cap A_3] = P[A_1 \cap A_2 \cap A_3] = p$$

If $p = 1/25$, then $P[A_i \cap A_j] = P[A_i] \, P[A_j]$ for $i \neq j$ holds so that requirement a holds.

However, $P[A_1 \cap A_2 \cap A_3] \neq P[A_1] \, P[A_2] \, P[A_3]$ so that requirement b fails.

On the other hand, if $p = 1/125$, then $P[A_1 \cap A_2 \cap A_3] = P[A_1] \, P[A_2] \, P[A_3]$ and requirement b holds.

But $P[A_i \cap A_j] \neq P[A_i] \, P[A_j]$ for $i \neq j$ so that requirement a fails.

More generally, the independence of n events can be defined inductively. We say that n events A_1, A_2, ..., A_n are independent if

(a) all combinations of k, $k < n$, events are independent and

(b) $P[A_1 \cap A_2 \cap ... \cap A_n] = P[A_1] \, P[A_2] \, ... \, P[A_n]$

Starting from $n = 2$, we can use this requirement to generalize independence to an arbitrary but finite number of events.

9.5 RANDOM VARIABLE

Random means a result that obeys laws of chance rather than by any deterministic law. An example is the result of an experiment whose result cannot be predicted. Collective result of a random experiment forms a sample. Collection of outcomes is an event and an event is a subset of sample space.

A random variable is a mapping from the sample space to the set of real numbers. Stated otherwise, a random variable is an assignment of real numbers to the outcome of a random experiment. It is also sometimes called $x = x_i$ as stochastic variable, stochastic function, or random function.

Random variables are denoted by capital letters, e.g., X, Y, Z, etc and individual values of the random variable X are $X(\omega)$. A continuous random variable has its range of value continuous.

9.5.1 Discrete Random Variable

A random variable is discrete if the range of its values is either finite or countably infinite. This range is generally denoted by $\{x_i\}$. We can describe a random variable discrete if there exists a denumerable sequence of distinct numbers x_i such that

$$\sum_i P_x(x_i) = 1 \tag{9.19}$$

Hence, a discrete random variable can assume only certain discrete values, whereas a continuous random variable can assume any value from a continuous interval.

$$\lim_{x \to \infty} F_X(x) = 0$$

In case of two random variables, the joint probability $P_{xy}(x_i, y_j)$ is the possibility that $x = x_i$ and $y = y_j$. In case when the variable x can take values $x_1, x_2, x_3, ..., x_n$ and the variable y can take values $y_1, y_2, y_3, ..., y_m$, then

$$\sum_i \sum_j P_{xy}(x_i, y_j) = 1 \tag{9.20}$$

9.5.2 Cumulative Distribution Function (CDF)

The cumulative distribution function of a random variable X is the probability that X takes a value less than equal to x, i.e.,

$$F_X(x) = P(X \le x) \tag{9.21}$$

and it has the following properties:
- $F_X(x)$ is non-decreasing.
- $0 \le F_X(x) \le 1$
- $F_X(x)$ is continuous from the right, i.e., $\lim_{\varepsilon \to 0} F(x + \varepsilon) = F(x)$
- $P(a < X \le b) = F_x(b) - F_x(a)$

- $P(X = a) = F_X(a) - F_x(\bar{a})$
- $\lim\limits_{x \to \infty} F_X(x) = 0$ and $\lim\limits_{x \to \infty} F_X(x) = 1$

Since in a discrete random variable, the CDF $F_X(x)$ is piecewise continuous, the distribution should look like a staircase denoted by x_i, the points of discontinuity of $F_X(x)$. This is shown in Fig. 9.4.

For discrete random variables, CDF is a staircase function. A random variable can be continuous, discrete, or mixed.

The probability density function (PDF) of a continuous random variable X is defined as the derivative of its CDF and is denoted by $f(x)$.

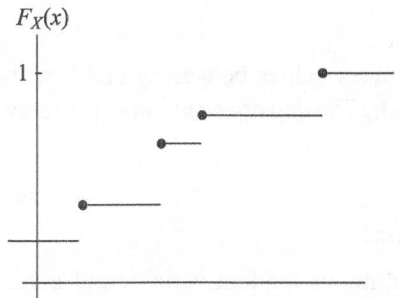

Fig. 9.4 Distribution function of discrete random variable X

$$f(x) = \frac{d}{dx} F_X(x) \tag{9.22}$$

It has got the following properties:

- $f_X(x) \geq 0$

- In general, $P(X \in A) = \int_A f(x)\, dx$

- $F(X) = \int_{-\infty}^{\infty} f_X(u)\, du$

- $\int_a^b f_X(x)\, dx = P(a < X \leq b)$

- $\int_{-\infty}^{\infty} f_X(x)\, dx = 1$

Let us consider a distribution function for a discrete random variable. The discrete steps at each transition point x_1, x_2, x_3, x_4 be k_1, k_2, k_3, k_4, respectively. Then the density function for this discrete random variable will be having discrete amplitudes k_1, k_2, k_3, k_4, respectively. Figures 9.5 and 9.6 illustrate this.

For discrete random variables, a term *probability mass function* (PMF) is defined. It is defined as p_i, where $p_i = P(X = x_i)$. For all i,

$$p_i \geq 0 \text{ and } \sum_i p_i = 1 \tag{9.23}$$

9.5.3 Types of Random Variables

Different types of random variables are used in communications. A few of them will be explained here.

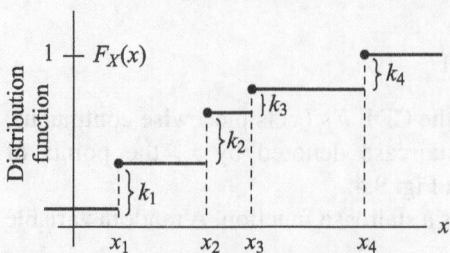

Fig. 9.5 Distribution function for a discrete random variable

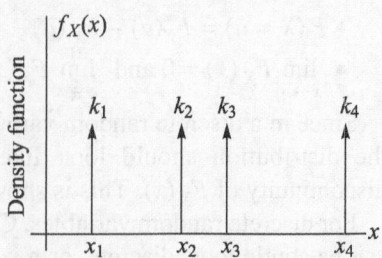

Fig. 9.6 Density function for a discrete random variable

Uniform random variable

This type of variable, which is continuous, takes values between a and b with equal probabilities for intervals of equal length. The density function is given by

$$f_X(x) = \begin{cases} \dfrac{1}{b-a} & a < x < b \\ 0 & \text{otherwise} \end{cases} \tag{9.24}$$

This equation gives the range of this continuous random variable but gives nothing about the expected various values that the random variable can assume. For example, a sinusoidal function with its phase changing at random is modelled as a uniform random variable between 0 and 2π.

Gaussian or normal random variable

The Gaussian random variable is a continuous random variable described by the density function

$$f_X(x) = \frac{1}{\sqrt{2\pi}\sigma} e^{-(x-m)^2/2\sigma^2} \tag{9.25}$$

The two parameters appearing on the expressions are the mean m (which can assume any finite value) and the standard deviation σ (which can assume any finite and positive value). The square of the standard deviation σ^2 is called the *variance*. A Gaussian random variable with mean m and variance σ^2 is denoted by $\mathbb{N}(m, \sigma^2)$. $\mathbb{N}(0, 1)$ is usually called *standard normal*.

This function is the most important and frequently encountered random variable in engineering. In communication, thermal noise (which is one of the major sources of noise) has a Gaussian distribution.

Bernoulli random variable

This discrete random variable takes two values 1 and 0, with probabilities p and $1 - p$. This is a good model for binary data generator. When a binary data is transmitted over a communication channel, there will be some error in the received bits. This error can be modelled by modulo 2 addition of a 1 to the input bit, i.e., a 0 is changed to a 1 and a 1 to a 0. Hence, a Bernoulli random variable can be employed to model the channel errors.

Binomial random variable

This is a discrete random variable giving the number of 1's in a sequence of n independent Bernoulli trials, i.e., it has got a binomial distribution of order n with parameter p and taking the integer values 0, 1, 2,..., n with probabilities

$$P[X = k] = \binom{n}{k} p^k q^{n-k} \quad \text{for} \quad 0 \le k \le n \tag{9.26}$$

Both p and q are known parameters where $p + q = 1$ and

$$\binom{n}{k} = \frac{n!}{k!(n-k)!} \tag{9.27}$$

The binomial random variable X can be represented as $B(n, p)$ and the binomial density function is

$$f_X(X) = \binom{n}{k} p^k q^{n-k} \delta(X - k) \tag{9.28}$$

and the binomial distribution is

$$F_X(X) = \sum_{k=0}^{m_x} \binom{n}{k} p^k q,^{n-k} \quad m_x \le x \le m_x + 1, 0 \le m_x < n - 1$$

$$= 1, \qquad\qquad\qquad x \ge n \tag{9.29}$$

The variable models the number of bits received in error when a sequence of n bits is transmitted over a channel with a bit error probability of p.

Poisson random variable

A random variable X is Poisson with $a > 0$ if it takes on the integer values 0, 1, 2,... with

$$P[X = k] = e^{-a} \frac{a^k}{k!} \delta(x - k) \tag{9.30}$$

and $$F_X(x) = e^{-a} \sum_{k=0}^{mx} \frac{a^k}{k!} \tag{9.31}$$

for $m_x \le x \le m_x + 1$, $m_x = 0, 1, 2, ...$

Rayleigh random variable

A random variable X is Rayleigh distributed with real-valued parameters α, $\alpha > 0$, if it is described by the density

$$f_X[X] = \frac{x}{\alpha^2} \exp\left\{-\frac{1}{2}\left(\frac{x}{\alpha}\right)^2\right\}, \quad x \ge 0$$

$$= 0, \qquad\qquad\qquad x < 0 \tag{9.32}$$

The distribution function for a Rayleigh random variable is

$$F_X[X] = \int_0^x \frac{u}{\alpha^2} \exp\left[\frac{-u^2}{2\alpha^2}\right] du, \quad x \geq 0$$

$$= 0 \tag{9.33}$$

To evaluate Eq. (9.33), change the variable $y = u^2/2\alpha^2$ and $dy = (u/\alpha^2)du$ to obtain

$$F_X[X] = \int_0^{x^2/2\alpha^2} e^{-y} dy = 1 - e^{-x^2/2\alpha^2}, \quad x \geq 0$$

$$= 0, \quad x \leq 0 \tag{9.34}$$

as the distribution function for Rayleigh random variable.

Exponential random variable

A random variable X is exponentially distributed with real-valued parameters λ, $\lambda > 0$, if it is described by the density

$$F_X(X) = \lambda e^{-\lambda x}, \quad x \geq 0$$

$$= 0, \quad x \leq 0 \tag{9.35}$$

The distributed function for an exponential random variable is

$$F_X(X) = \int_0^x \lambda e^{-\lambda y} dy = 1 - e^{-\lambda x}, \quad x \geq 0$$

$$= 0, \quad x \leq 0 \tag{9.36}$$

9.5.4 Functions of a Random Variable

Let X denote a random variable with known density $f_X(X)$ and distribution $F_x(x)$. Let $y = g(x)$ denote a real-valued function of the real variable x. Consider the transformation $Y = g(X)$. This is the transformation of the random variable X into the random variable Y. The random variable $X(\rho)$ is mapping from the sample space into the real line. But so is $g[x(\rho)]$. We are interested in methods for finding the density $f_Y(y)$ and the distribution $F_y(Y)$.

When dealing with $Y = g[x(\rho)]$, there are a few technicalities that should be considered. They are as follows:

1. The domain of g should include range of X.
2. For every y, the set $\{Y = g(X) \leq y\}$ must be an event, i.e., the set $\{\rho \in S : Y(\rho) = g(X(\rho)) \leq y\}$ must be $\Im$, i.e., it must be an event.
3. The events $\{Y = g(X) = \pm \infty\}$ must be assigned a probability of zero.

In practice, the technicalities are assumed to hold, and they do not cause any problem. Define the index set

$$I_Y = \{X : g(x) \leq Y\} \tag{9.37}$$

the composition of which changes with y. The distribution Y can be expressed as

$$F_y(Y) = P[Y \leq y] = P[g(x) \leq y] = P[X \in I_y] \tag{9.38}$$

This provides a practical method for computing the distribution function.

9.5.5 Statistical Averages

The mean, expected value, or expectation of a random variable X is defined as

$$E\ (X) = \int_{-\infty}^{\infty} x f_x(x)dx \qquad (9.39)$$

It is also sometimes denoted by m_x. The expected value is a measure of the average of the values that the random variable takes in a large number of experiments. $E(X)$ is just a real number. The nth moment of a random variable X is defined as

$$m_X^{(n)\ \mathrm{def}} = \int_{-\infty}^{\infty} x^n f_X(x)dx \qquad (9.40)$$

Then the expected value of $Y = g(X)$ is

$$E\{g(X)\} = \int_{-\infty}^{\infty} g(x) f_X(x)dx \qquad (9.41)$$

For discrete random variable, it is

$$E(X) = \sum_i x_i P(X = x_i) \qquad (9.42)$$

and $\qquad E\{g(X)\} = \sum_i g(x_i)P(X = x_i) \qquad (9.43)$

When $g(X) = (X - E(X))^2$, $E(Y)$ is called the variance of X, which gives the measure of the spread of the density function of X. If the variance happens to be small, then the random variable is very much concentrated around its mean and is *less random*. But if the variance is large, then the random variable is highly spread and is less predictable. The variance is denoted by σ_X^2 and its square root is σ_X, which is nothing but *standard variation*. The relation for variance is given by

$$\sigma_X^2 = E(X^2) - (E(X))^2 \qquad (9.44)$$

For any constant c, the following relations hold:

   (a) $E(cX) = c\ E(X)$

   (b) $E(c) = c$

   (c) $E(X + c) = E(X) + c$

The variance has the following properties:

   (a) $\mathrm{Var}(cX) = c^2\ \mathrm{Var}(X)$

   (b) $\mathrm{Var}(C) = 0$

   (c) $\mathrm{Var}(X + c) = \mathrm{Var}(X)$

For other important random variables, the following are the relations for the mean and variance:

Binomial random variable: $E(X) = np$ and $\mathrm{Var}(X) = np(1 - p)$

Uniform random variable: $E(X) = \dfrac{a + b}{2}$ and $\mathrm{Var}(X) = \dfrac{(b - a)^2}{12}$

Bernoulli random variable: $E(X) = p$ and $\text{Var}(X) = p(1-p)$

Gaussian random variable: $E(X) = m$ and $\text{Var}(X) = \sigma^2$

9.5.6 Multiple Random Variables

Let X and Y represent two random variables defined on the same sample space Ω. For these two random variables, the joint *cumulative distribution function* (CDF) can be defined as

$$F_{X,Y}(x, y) = P[\omega \in \Omega : X(\omega) \le x, Y(\omega) \le y] \tag{9.45}$$

This relation can also be expressed as

$$F_{X,Y}(x, y) = P(X \le x, Y \le y)$$

The joint PDF is defined as

$$f_{XY}(x, y) = \frac{\partial^2}{\partial x \partial y} F_{XY}(x, y) \tag{9.46}$$

If the value of the random variable X is equal to x, then the conditional probability density function of the random variable Y is denoted by $f_{Y|X}(y/x)$ and defined as

$$f_{Y|X}(y|x) = \frac{f_{X,Y}(x, y)}{f_X(x)}, \qquad f_Y(x) \ne 0$$
$$= 0, \quad \text{otherwise} \tag{9.47}$$

If the density function after the knowledge of X is the same as the density function before the knowledge of X, then the random variables are said to be *statistically independent*. For statistically independent random variables,

$$f_{X,Y}(x, y) = f_X(x) f_Y(y) \tag{9.48}$$

9.5.7 Multiple Functions of Multiple Random Variables

Let x and Y be two random variables defined by functions

$$Z = g(X, Y) \tag{9.49}$$
$$W = h(X, Y)$$

Then the joint CDF and PDF of Z and W can be obtained directly by applying the definition of CDF, i.e., for all z and w, the set of equations

$$g(x, y) = z \quad \text{and} \quad h(x, y) = w \tag{9.50}$$

has a number of solutions (x_i, y_i) and at these points, the determinant of the Jacobian matrix

$$\begin{bmatrix} \dfrac{\partial z}{\partial x} & \dfrac{\partial z}{\partial y} \\ \dfrac{\partial w}{\partial x} & \dfrac{\partial w}{\partial y} \end{bmatrix} \tag{9.51}$$

is non-zero. Then we have

$$f_{Z,W}(z,w) = \sum_i \frac{f(x_i, y_i)}{|\det J(x_i, y_i)|} \tag{9.52}$$

where $\det J$ denotes the determinant of the matrix J.

9.5.8 Sums of Random Variables

If we have a sequence of random variables $X_1, X_2, X_3, ..., X_n$ with basically the same properties, then their average

$$Y = \sum_{i=1}^{n} X_i$$

is expected to be less random than each X_i. This can be explained with the help of the *central limit theorem* and the *law of large numbers*.

The central limit theorem gives some insight into the distribution of the average apart from giving the convergence of the average to the mean. This theorem states that if X_i's are independent and identically distributed random variables with each of them having a mean m and a variance σ^2, then

$$Y = \frac{1}{n} \sum_{i=1}^{n} X_i$$

converges to $N\left(m, \dfrac{\sigma^2}{n}\right)$.

Thus, the central limit theorem states that the probability density of a sum of N independent random variables tends to approach a Gaussian density as the number N increases. The mean and variance of this Gaussian density are, respectively, the sum of the means and the sum of the variances of the N independent random variables.

The law of large numbers states that if the sequence of random variables X_1, $X_2, X_3, ..., X_n$ is uncorrelated with the same mean m_x and variance $\sigma_X^2 < \infty$, then for any $\epsilon > 0$,

$$\lim_{x \to \infty} P(|Y - m_X| > \epsilon = 0$$

where $$Y = \frac{1}{n} \sum_{i=1}^{n} X_i$$

This implies that the average converges to the expected value.

9.5.9 Jointly Gaussian Random Variables

These are also sometimes referred to as *binormal* random variables. The variables X and Y are distributed according to a joint PDF of the form

$$f_{X,Y}(x, y) = \frac{1}{2\pi\sigma_1\sigma_2\sqrt{1-\rho^2}} \exp\left[-\frac{1}{2(1-\rho^2)}\right.$$

$$\left.\left\{\frac{(x-m_1)^2}{\sigma_1^2} + \frac{(y-m_2)^2}{\sigma_2^2} + \frac{(2\rho(x-m_1)(y-m_2)}{\sigma_1\sigma_2}\right\}\right] \tag{9.53}$$

where m_1, m_2, σ_1^2, and σ_2^2 are the means and variances of X and Y, respectively, and ρ is their correlation coefficient. According to binormal distribution, when two random variables X and Y are distributed, it can be shown that X and Y are normal random variables and the conditional densities $f(x|y)$ and $f(y|x)$ are also Gaussian. This property shows the main difference between the jointly Gaussian random variable and two random variables that each have a Gaussian distribution. This definition can be extended to more random variables. For example, X_1, X_2, and X_3 are jointly Gaussian if any pair of them is jointly Gaussian and the conditional density function of any pair given the third one is also Gaussian.

The main properties of jointly Gaussian random variables are summarized below.

(a) Any set of linear combinations of $(X_1, X_2, X_3, ..., X_n)$ is itself jointly Gaussian. Specifically, any linear combination of X_i's is a Gaussian random variable.

(b) Two uncorrelated jointly Gaussian random variables are independent. Hence, for a jointly Gaussian random variable, independence and uncorrelatedness are equivalent. But this is not true in general for non-jointly Gaussian random variables.

(c) If n random variables are jointly Gaussian, any subset of them is also distributed according to a jointly Gaussian distribution of the appropriate size. In particular, all individual random variables are Gaussian.

(d) Jointly Gaussian random variables are completely characterized by the means of all random variables $m_1, m_2, m_3, ..., m_n$ and the set of all covariance $\text{Cov}(X_i, X_j)$ for all $1 \leq i \leq n$ and $1 \leq j \leq n$. These are called second-order properties, which completely describe the random variables.

Basic properties of joint and marginal cumulative distribution function and probability density function are given below.

(a) $F_X(x) = F_{X,Y}(x, \infty)$

(b) $f_X(x) = \int_{-\infty}^{\infty} f_{X,Y}(x, y)dy$

(c) $f_Y(y) = F_{X,Y}(\infty, y)$

(d) $f_Y(y) = \int_{-\infty}^{\infty} f_{X,Y}(x, y)dx$

(e) $P\{(X, Y) \in A\} = \int\int_{(x,y)\in A} f_{X,Y}(x, y)dxdy$

(f) $F_{X,Y} = \int_{-\infty}^{x} \int_{-\infty}^{y} f_{X,Y}(u, v)dudv$

(g) $\int_{-\infty}^{\infty} \int_{-\infty}^{\infty} f_{X,Y}(x, y)dxdy = 1$

9.6 RANDOM PROCESS

In dealing with signals, a random process is the natural extension of random variable. Generally, the signals we deal with are assumed to be deterministic. But this is not always true. The deterministic assumption on time varying signal is not true and valid. Hence, it is more appropriate to model these signals as random rather than deterministic. A random process can be viewed as a set of possible realization of signal waveforms. The realization is governed by some probabilities law, i.e., the realization of one from the set of possible signals. This is same as the definition for random variable, where a set of possible values is realized by some probability function but the only difference is that we have signals in random process compared to values (numbers) in a random variable.

The random process is a mapping from the sample space into an ensemble of time function known as sample function, i.e., to every sample space $\rho \in S$, there corresponds a function of time, a sample function $X(t, \rho)$. Generally, from the notation, we drop the ρ variable and write just $X(t)$. However, the sample space ρ variable is always there, even if it is not shown explicitly. For a fixed $t = t_0$, the quantity $X(t_0, \rho)$ is a random variable mapping sample space S into the real line. For fixed $\rho_0 \in S$, the quantity $X(t, \rho_0)$ is a well-defined, non-random function of time. Finally, for fixed t_0 and ρ_0, the quantity $X(t_0, \rho_0)$ is a real number.

An example of a random process is the noise waveform. We make repeated measurement of the noise voltage output of a single noise source or make measurement simultaneously of the output of a very large collection of statistically identical noise sources. We call such a collection of noise sources as *ensemble* and individual noise waveforms as *sampled function*. The statistical average is determined from measurements made at some fixed time $t = t_1$ on all the samples functions of ensemble. Hence, to determine $\overline{n^2(t)}$, we measure the voltage $n(t_1)$ of each noise source, square and add the voltages, and divide by the number of sources in the ensemble. The average so determined is the ensemble average of $n^2(t_1)$. Also $n(t_1)$ is a random variable and will have associated with it a probability density function. The ensemble averages will be identical with the statistical averages and may be represented by the same symbols. Hence, the statistical or ensemble average of $n^2(t_1)$ may be written as $E[n^2(t_1)] = \overline{n^2(t_1)}$.

The average determined by measurements on a single sample function at successive times will yield a time average. This is represented as $n^2(t_1)$. In general, time averages and ensemble averages are not the same since the statistical characteristics of the sample functions in the ensemble were changing with time. Such a variation cannot be reflected in measurements made at a fixed time and ensemble average will be different at different times. The random process is called stationary when the statistical characteristics of the sample functions do not change with time. But this does not ensure that ensemble and time averages are the same for it may happen that while each sample function is stationary, the individual sample functions may differ statistically from one another. In such cases, the time average will depend on the particular sample function, which is used to form the average. A random process is referred to as *ergodic* when the ensemble and time averages are identical. An ergodic process is stationary but a stationary process is not ergodic.

9.6.1 Continuous and Discrete Random Processes

For a continuous random process, probabilistic variable ρ takes on a continuum of values. For every fixed value $t = t_0$ of time, $X(t_0, \rho)$ is a continuous random variable. This can be understood with the following example.

Let random variable A be uniform in [0, 1]. The continuous random process is defined as $X(t, \rho) = A(\rho) s(t)$, where $s(t)$ is a unit amplitude square wave with a period T. The sampled function contains periodically spaced in-time jump discontinuities. But the process is continuous.

For a discrete random process, the probabilistic variable ρ takes on only a discrete value. For every fixed value $t = t_0$ of time, $X(t_0, \rho)$ is a discrete random variable. The following example explains this.

Consider a coin being tossed with $S = \{H, T\}$. Then $X(t, H) = \sin t$ and $X(t, T) = \cos t$ define a discrete random process. In this case, the sample functions are continuous functions of time. But the process is discrete.

9.6.2 Distribution and Density Functions

The first-order distribution function is defined as

$$F(x,t) = P[X(t) \le x] \tag{9.54}$$

The first-order density function is defined as

$$f(x,t) = \frac{df(x,t)}{dx} \tag{9.55}$$

These definitions generalize to the nth order case for any given positive integer n. Let $x_1, x_2, x_3, ..., x_n$ denote n realization of variables and let $t_1, t_2, t_3, ..., t_n$ denote n time variables. Then, the nth order distribution function is

$$F(x_1, x_2, x_3,...., x_n, t_1, t_2, t_3,....,t_n) = P[x(t_1) \le x_1, X(t_2) \le x_2, X(t_3) \le x_3,..., X(t_n) \le x_n] \tag{9.56}$$

Similarly, define the nth order density function as

$$F(x_1, x_2, x_3,...., x_n, t_1, t_2, t_3,....,t_n) = \frac{\partial^n F(x_1, x_2, x_3,..., x_n, t_1, t_2, t_3,....,t_n)}{\partial x_1\, \partial x_2\, \partial x_3...\partial x_n} \tag{9.57}$$

In general, a complete statistical description of a random process requires knowledge of all order distribution function.

9.6.3 Stationary Random Process

A process $X(t)$ is said to be stationary if its statistical properties do not change with time. More precisely, process $X(t)$ is stationary if

$$F(x_1, x_2, x_3,...,x_n; t_1, t_2,...,t_n) = F(x_1, x_2, ..., x_n; t_1 + T, t_2 + T, ... t_n + T) \tag{9.58}$$

for all orders n and all time shifts T. Stationarity influences the form of the first- and second-order distribution/density functions. Let $X(t)$ be stationary so that

$$F(x,t) = F(x, t+T) \tag{9.59}$$

for all T. This implies that the first-order distribution function is independent of time. A similar statement can be made concerning the first-order density function. Now, consider the second-order distribution of stationary $X(t)$; for all t, this function has the property

$$F(x_1, x_2; \tau, 0) = F(x_1, x_2, t_2 + \tau, t_2) \tag{9.60}$$

Hence, the second-order distribution function is independent of absolute t, i.e., $F(x_1, x_2; t_1, t_2)$ depends only on the time difference $\tau = t_1 - t_2$. A similar statement can be made concerning the second-order density function. These conditions on $F(X)$ and $F(x_1, x_2; \tau)$ are necessary conditions, but they are not sufficient to imply stationarity. For a given random process, suppose that the first-order distribution/density is independent of time and the second-order distribution/density depends only on the time difference. Based on this knowledge alone, we cannot conclude that $X(t)$ is stationary.

First- and second-order probabilistic averages

First- and second-order statistical averages are useful. The expected value of a general random process $X(t)$ is defined as

$$\eta(t) = E[x(t)] = \int_{-\infty}^{\infty} xf(x,t)dx \tag{9.61}$$

In general, this is a time varying quantity. The expected value is often called first-order statistic since it depends on a first-order density function. The auto-correlation function of $X(t)$ is defined as

$$R(t_1, t_2) = E[X(t_1)X(t_2)] = \int_{-\infty}^{\infty}\int_{-\infty}^{\infty} (x_1 x_2; t_1 t_2)dx_1 dx_2 \tag{9.62}$$

In general, R depends on two time variables, t_1 and t_2. Also, R is an example of a second-order statistic since it depends on a second-order density function. Suppose $X(t)$ is stationary. Then the mean

$$\eta = E[X(t)] = \int_{-\infty}^{\infty} xf(x)dx \tag{9.63}$$

is constant, and the autocorrelation function

$$R(\tau) = E[X(t)X(t+\tau)] = \int_{-\infty}^{\infty}\int_{-\infty}^{\infty} x_1 x_2 f(x_1, x_2; \tau)dx_1 dx_2 \tag{9.64}$$

depends only on the time difference $\tau = t_1 - t_2$ (it does not depend on absolute time). However, the converse is not true, i.e., the conditions ηa is constant and $R(\tau)$ is independent of absolute time do not imply that $X(t)$ is stationary.

Wide sense stationarity (WSS)

A process $X(t)$ is said to be wide sense stationary (WSS) if
   (a) mean $\eta = E[X(t)]$ is constant and
   (b) autocorrelation $R(\tau) = E[X(t)X(t+\tau)]$ depends only on the time difference.
      Note that stationarity implies wide sense stationarity (WSS). However, the converse is not true, i.e., WSS does not imply stationarity.

Ergodic processes

A process is said to be ergodic if all orders of statistical and time averages are interchangeable. The mean, autocorrelation, and other statistics can be computed by using any sample function of the process, i.e.,

$$\eta = E[X(t)] = \int_{-\infty}^{\infty} xf(x)dx = \lim_{T \to \infty} \frac{1}{2T} \int_{-T}^{T} X(t)dt$$

$$R(\tau) = E[X(t)X(t+\tau)] = \int\limits_{-\infty}^{\infty}\int\limits_{-\infty}^{\infty} x_1 x_2 f(x_1 + x_2; \tau) dx_1 dx_2 = \lim_{T\to\infty} \frac{1}{2T} \int_{-T}^{T} X(t)X(t+\tau) d\tau$$

$$(9.65)$$

This idea extends to higher-order averages as well. Since we are averaging over absolute time, the ensemble averages (all orders) cannot depend on absolute time. This requires that the original process must be stationary. That is, ergodicity implies stationarity. However, the converse is not true. There are stationary processes that are not ergodic. The hierarchy of random processes is abstractly illustrated in Fig. 9.7.

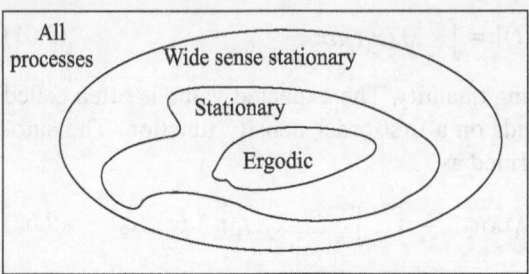

Fig. 9.7   Hierarchy of random process

9.6.4 Multiple Random Processes

When we deal with two or three random processes, multiple random processes arise. When a random process $X(t)$ is passed through a linear time invariant system, for each sample input $x(t; \omega_i)$, the sample function output is defined by $y(t; \omega_i) = x(t; \omega_i)*h(t)$, where $h(t)$ is the impulse response of the system. It can be seen that for each $\omega_i \in \Omega$, we have got two signals $x(t; \omega_i)$ and $y(t; \omega_i)$. Hence, we deal with two random processes, $X(t)$ and $Y(t)$. Two random processes $X(t)$ and $Y(t)$ are said to be independent if for all t_1, t_2, the random variables $X(t_1)$ and $Y(t_2)$ are independent. Similarly, $X(t)$ and $Y(t)$ are uncorrelated if $X(t_1)$ and $Y(t_2)$ are uncorrelated for all t_1, t_2. Sometimes, the dependence between the two random processes also has to be considered.

The cross correlation between two random processes $X(t)$ and $Y(t)$ is defined as

$$R_{XY}(t_1, t_2) = E[X(t_1)Y(t_2)] \qquad (9.66)$$

Hence, it can be inferred that

$$R_{XY}(t_1, t_2) = R_{YX}(t_2, t_1) \qquad (9.67)$$

The concept of stationarity can also be generalized to joint stationarity for the case of two random processes. Two random processes $X(t)$ and $Y(t)$ are jointly wide sense stationary or simply joint stationary if both $X(t)$ and $Y(t)$ are individually stationary and cross correlation $R_{XY}(t_1, t_2)$ depends only on $\tau = t_1 - t_2$.

From the definition and Eq. (9.66), it follows that

$$R_{XY}(\tau) = R_{XY}(-\tau) \qquad (9.68)$$

9.6.5 Bandpass Random Process

If $X(t)$ is bandlimited, it can be expressed in terms of in-phase and in-quadrature component and can be expressed as

$$X(t) = X_c(t)\cos(2\pi f_c t) + X_s(t)\sin(2\pi f_c t) \qquad (9.69)$$

If $X(t)$ is a bandpass signal given to a bandpass filter of same bandwidth, then there will be unaltered $X(t)$. Since multiplication by $\cos(2\pi f_c t)$ at the inputs amounts to shifting the bandpass signal at the output by $\pm f_c(t)$ along the frequency axis, the output signal is passed through a low-pass filter and power spectral density (PSD) for in-phase component is

$$G_{X_c}(f) = G_X(f+f_c) + G_X(f-f_c) \text{ for } |f| \le B \text{ and zero elsewhere}$$

9.6.6 Gaussian Random Process

A random process $X(t)$ is called Gaussian process if

$$X = [X(t_1)X(t_2)X(t_3)...X(t_n)]^T$$

has a jointly multivariate Gaussian density function given by

$$f_x(X) = \frac{1}{(2\pi)^{n/2}|\det C|^{1/2}} e^{(1/2)(x-\mu)^T C^{-1}}(x-\mu) \qquad (9.70)$$

where $x = [x_1, x_2, ..., x_n]$, $\mu = E(X)$, and $C = \begin{pmatrix} C_{11} & \cdots & C_{1n} \\ \vdots & \vdots & \vdots \\ C_{m1} & \cdots & C_{mn} \end{pmatrix}$ such that C_{ij} is a

covariance of $X(t_i)$ and $X(t_j)$.

The element C_{ij} is given by

$$C_{ij} = R_{XX}(t_i, t_j) - \mu_i \mu_j$$

A Gaussian process is completely specified by its set of means and autocorrelations. If the input to the linear system is Gaussian, the output is also Gaussian. If the Gaussian process is strict sense stationary, then it is also wide sense stationary.

White noise

A random process $X(t)$ is called white noise when its PSD, $G(f) = \eta/2$, is constant over entire frequency spectrum and the mean is assumed to be zero. For this, the autocorrelation function is given by $R(\tau) = (\eta/2)\delta(\tau)$. The bandlimited white noise has similar flat spectrum spread over the pass band.

9.6.7 Random Process Through a Linear Time Invariant System

It has been seen already that when a random process passes through a linear time invariant system, the output is also a random process defined on the original probability space. Let a stationary process $X(t)$ be the input to a linear time invariant system with the impulse response $h(t)$ and the output process $Y(t)$ as shown in Fig. 9.8.

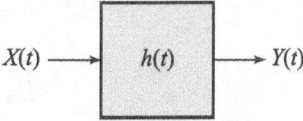

Fig. 9.8 A random process through a linear time invariant system

For such a system, we have to study

(a) under what condition will the output process be stationary,

(b) under what condition will the input and output processes be jointly stationary, and

(c) how we can obtain the mean and autocorrelation of the output process and the cross correlation between the input and output processes.

$$m_y = m_x \int_{-\infty}^{\infty} h(t)dt; \; R_{XY}(\tau) = R_X(\tau)*h(-\tau); \; R_Y(\tau) = R_X(\tau)*h(\tau)h(-\tau) \quad (9.71)$$

The convolution integral is used here to relate the output $Y(t)$ to the input $X(t)$, i.e.,

$$Y(t) = \int_{-\infty}^{\infty} X(\tau)h(t-\tau)d\tau$$

We have

$$E[Y(t)] = E\left[\int_{-\infty}^{\infty} X(\tau)h(t-\tau)d\tau\right] = \int E[X(\tau)h(t-\tau)d\tau = m_x \int_{-\infty}^{\infty} h(u)du \equiv m_y$$

$$(9.72)$$

where $u = t - \tau$. This proves that m_y is independent of t.

The cross correlation function between the output and input is

$$E[X(t_1)Y(t_2)] = E\left[X(t_1)\int_{-\infty}^{\infty} X(s)h(t_2-s)ds\right]$$

$$= \int_{-\infty}^{\infty} E[X(t_1)X(s)]h(t_2-s)ds$$

$$= \int_{-\infty}^{\infty} R_X(t_1-s)h(t_2-s)ds$$

$$= \int_{-\infty}^{\infty} R_X(t_1-t_2-u)h(-u)du$$

$$= \int_{-\infty}^{\infty} R_X(\tau)*h(-\tau) \equiv R_{XY}(\tau)$$

where $u = s - t_2$ and this shows that $R_{xy} = (t_1, t_2)$ depends only on $\tau = t_1 - t_2$.

The autocorrelation function of the output is

$$E[Y(t_1)Y(t_2)] = E\left[\left(\int_{-\infty}^{\infty} X(s)h(t_1-s)ds\right)Y(t_2)\right]$$

$$= \int_{-\infty}^{\infty} R_{XY}(s-t_2)h(t_1-s)ds$$

$$= \int_{-\infty}^{\infty} R_{XY}(u)h(t_1-t_2-u)du$$

where $u = s - t_2$. The above relation can be written as

$$R_{XY}(\tau)*h(\tau) = R_X(\tau)*h(-\tau)*h(\tau) = R_Y(\tau)$$

Hence, the input and output processes are jointly stationary.

9.6.8 Statistical Averages

The random process at any given time defines a random variable at any given set of times defined and a random vector.

This fact enables us to define various statistical averages for the process through statistical averages of the corresponding random variable. Since at any time instant t_0, the random process at that time ($X(t_0)$) is an ordinary variable. It has a density function and its mean and variance can be found out.

Both mean and variance are ordinary deterministic numbers, but they depend on time t_0, i.e., at time t_1, the density function and thus the mean and variance of $X(t_1)$ will generally be different from those of $X(t_0)$.

The mean or expectation of the random process $X(t)$ is a deterministic function of time denoted by $m_x(t)$ that at each time instant t_0 equals the mean of the random variable $X(t_0)$, i.e.,

$$m_x(t) = E[X(t)] \text{ for all } t$$

Since at any time instant t_0, the random variable $X(t_0)$ is well defined with a probability density function $f_{X(t_0)}(X)$,

$$E[X(t_0)] = m_X(t_0) = \int_{-\infty}^{\infty} x f_X(t_0)(x)\, dx \tag{9.73}$$

The autocorrelation function of the random process $X(t)$ is denoted by $R_X(t_1, t_2)$ and is defined by $R_X(t_1, t_2) = E[X(t_1), X(t_2)]$. It is clear from this definition that $R_X(t_1, t_2)$ is a deterministic function of two variables t_1 and t_2 given by

$$R_X(t_1, t_2) = \int_{-\infty}^{\infty} \int_{-\infty}^{\infty} x_1 x_2 f_{X(t_1)X(t_2)}(x_1, x_2)\, dx_1 dx_2 \tag{9.74}$$

9.6.9 Power Spectral Density of Stationary Processes

Spectral characteristics of a random process are determined by the spectrum characteristics of the signals occurring in that process. If these signals vary slowly, then the random process will mostly contain low frequencies and its power will be mostly concentrated at low frequencies. Reverse is the case if the signals change very fast, i.e., random process will mostly contain high frequencies and the power will be concentrated at higher frequencies. Power spectral density or power spectrum of the random process is a useful function that determines the distribution of power of the random process at different frequencies. It is denoted by $S_X(f)$ and gives the strength of the power in the random process as a function of frequency. It has the unit W/Hz.

Wiener–Khinchin theorem gives the relationship between the power spectrum of a random process to its autocorrelation function. This theorem states that for a stationary random process $X(t)$, the power spectral density is the Fourier transform of the autocorrelation function, i.e.,

$$S_X(f) = \Im[R_X(\tau)] \tag{9.75}$$

9.6.10 Power Spectra in LTI System

When a stationary random process with mean m_x and autocorrelation function $R_X(\tau)$ passes through a linear time invariant (LTI) system with the impulse response $h(t)$, the output process will also be stationary with mean

$$m_Y = m_X \int_{-\infty}^{\infty} h(t)\, dt \tag{9.76}$$

and autocorrelation function

$$R_Y(\tau) = R_X(\tau) * h(\tau) * h(-\tau) \tag{9.77}$$

The correlation function of joint stationary processes $X(t)$ and $Y(t)$ is given by

$$R_{XY}(\tau) = R_X(\tau) * h(-\tau) \tag{9.78}$$

Translating these two relations to frequency domain, we get

$$m_Y = m_X H(0) \tag{9.79a}$$

and $\qquad\qquad S_Y(f) = S_X(f)\left|H(f)^2\right| \tag{9.79b}$

Equation (9.79a) states that the mean of the random process is basically its DC value, i.e., the mean value of the response of the system depends only on the value of $H(f)$ at $f = 0$, i.e., at its DC response.

Equation (9.79b) states that when dealing with the power spectrum, the phase of $H(f)$ is irrelevant. Only the magnitude of $H(f)$ affects the output spectrum. This is obvious since the power depends on the amplitude and not on the phase of the signal.

9.6.11 Power Spectral Density of a Sum Process

In the case of communication over a channel in the presence of noise, the noise process is added to the signal process. Let $Z(T) = X(t) + Y(t)$, where $X(t)$ and $Y(t)$ are jointly stationary process. Also $Z(T)$ is a stationary process with

$$R_Z(\tau) = R_X(\tau) + R_Y(\tau) + R_{XY}(\tau) + R_{YX}(\tau) \tag{9.80}$$

Taking Fourier transform on both sides of the above-mentioned equation, we obtain

$$S_Z(f) = S_X(f) + S_Y(f) + S_{XY}(f) + \underbrace{S_{YX}(f)}_{S_{XY}^*(f)}$$

$$= S_X(f) + S_Y(f) + 2\,\mathrm{Re}[S_{XY}(f)] \tag{9.81}$$

The above-mentioned relation shows that the power spectral density of a sum process is the sum of the power spectra of the individual processes with an additional term, which depends on the cross correlation between the two processes. If the two processes $X(t)$ and $Y(t)$ are uncorrelated, then $R_{XY}(\tau) = m_X m_Y$. If one of the processes has zero mean, then $R_{XY}(\tau) = 0$ and

$$S_Z(f) = S_X(f) + S_Y(f) \tag{9.82}$$

9.7 GAUSSIAN PROCESS

A Gaussian process is a stochastic process for which any finite linear combination of samples will be normally distributed. Stated otherwise, any linear functional applied to the sample function X_t will normally give a distributed result. Gaussian process plays an important role in communication systems. Thermal noise, which is the most prevalent noise in communication systems, can be closely modelled by a Gaussian process.

Thermal noise due to thermal agitation of electrons generates a current with a random value. The total current can be thought of as a sum of different current sources. These different current sources can be regarded as random variables. By applying the central limit theorem, which will be discussed later, we can conclude that this total current has Gaussian distribution. This current also forms good models for some information sources as well.

Let a random process $X(t)$ occur over an interval $t = 0$ to $t = T$. If this random process is weighted by some function $g(t)$ and integrated over this interval, we obtain a random variable Y defined by

$$Y = \int_0^T g(t)X(t)dt \tag{9.83}$$

Here Y is referred to as a linear functional of $X(t)$. The distinction between a function and functional is as follows. For example, the sum

$$Y = \sum_{i=1}^N a_i X_i$$

where a'_is are constant and X'_is are random variables, is a linear function of X_i. For each observed set of values for the random variable X_i, there is a corresponding value for the random variable Y. On the other hand, in Eq. (9.83), the value of random variable Y depends on the course of the argument function $g(t)X(t)$ over the entire period from 0 to T. Thus, a functional is a quantity that depends on the entire course of one or more functions rather than on a number of discrete variables. Thus, the domain of a functional is a set or space of admissible functions rather than a region of a coordinate space.

In Eq. (9.83), the weighting function $g(t)$ is such that the mean square value of the random variable Y is finite. If the variable Y is a Gaussian distributed random variable for every $g(t)$ in this class of functions, then the process $X(t)$ is said to be a *Gaussian process*, i.e., the process $X(t)$ is a Gaussian process if every linear functional of $X(t)$ is a Gaussian random variable.

The random variable Y has a Gaussian distribution if its probability density function has the form

$$f_Y(y) = \frac{1}{\sqrt{2\pi}\sigma_\gamma} \exp\left[-\frac{(y - \mu_\gamma)^2}{2\sigma_\gamma^2}\right] \tag{9.84}$$

where μ_γ is the mean and σ_γ^2 is the variance of the random variable Y. When the Gaussian random variable Y is normalized to have a mean μ_γ of zero and a variance σ_γ^2 of one, its probability distribution function becomes

$$f_Y(y) = \frac{1}{\sqrt{2\pi}} \exp\left[-\frac{(y)^2}{2}\right] \tag{9.85}$$

Such a normalized Gaussian distribution is indicated as $\mathbb{N}(0, 1)$.

Gaussian process is very important in the study of communication systems because of its many properties that make analytical result possible and most of the random processes produced by physical phenomena are often such that a Gaussian model is appropriate.

In a Gaussian random process, we look at different instances of time and the resulting random variable will be jointly Gaussian. Now let us see a few definitions of Gaussian process.

- A random process $X(t)$ is Gaussian if for all n and all $t_1, t_2, t_3, ..., t_n$, the random variable $\{X(t_i)\}_{i=1}^n$ has a jointly Gaussian density function. From this definition, it is obvious that at any time instant t_0, the random variable $X(t_0)$ is Gaussian. At any two points t_1, t_2, random variables $\left(X(t_1), X(t_2)\right)$ are distributed according to a two-dimensional jointly Gaussian random variable.

- The random processes $X(t)$ and $Y(t)$ are jointly Gaussian if for all n, m and all $t_1, t_2, t_3, ..., t_n$ and $\tau_1, \tau_2, \tau_3, ...\tau_m$, the random vector $[X(t_1), X(t_2), X(t_3), ..., X(t_n), Y(\tau_1), Y(\tau_2) Y(\tau_3), ..., Y(\tau_m)]$ is distributed according to an $n + m$ dimensional jointly Gaussian distribution. From the above-mentioned definition, it is obvious that if $X(t)$ and $Y(t)$ are jointly Gaussian, then each of them is individually Gaussian. The converse need not be always true, i.e., two individually Gaussian random processes are not always jointly Gaussian.

9.7.1 Central Limit Theorem

To use Gaussian Process as a model for a large number of different physical phenomena, the central limit theorem is used. The theorem is stated as follows.

Let X_i, $i = 1, 2, 3, ..., N$, be a set of random variables that satisfies the following conditions:

- The X_i's are statistically independent.
- The X_i's have the same probability distribution with mean μ_X and variance σ_X^2.

These X_i's are said to constitute a set of independent and identically distributed random variables. These random variables are normalized as

$$Y_i = \frac{1}{\sigma_X}(X_i - \mu_x) \text{ for } i = 1, 2, 3, ..., N$$

so that we have $E[Y_i] = 0$ and $\text{Var}[Y_i] = 1$ and the random variable is defined as

$$V_N = \frac{1}{\sqrt{N\pi}}\sum_{i=1}^{N} Y_i \tag{9.86}$$

The central limit theorem states that the probability distribution of V_N approaches a normalized Gaussian distribution $\mathbb{N}(0, 1)$ in the limit as the number of random variables N approaches infinity. It has to be noted that the central limit theorem gives only the limiting form of the probability distribution of the normalized random variable V_N as N approaches infinity. When N is finite, the Gaussian limit gives a relatively poor approximation for the actual probability distribution V_N even though N may be quite large.

9.7.2 Properties of Gaussian Process

Gaussian and jointly Gaussian random processes have some very important properties that are not shared by other families of random process. These properties are discussed below.

1. If a Gaussian process $X(t)$ is passed through a linear time invariant system, then the output process $Y(t)$ will also be a Gaussian process. This property follows directly from the fact that the linear combination of jointly random variables is itself jointly Gaussian. This property is in contrast to a non-Gaussian process, where knowledge of the statistical properties of the input process does not easily lead to the statistical properties of the output process.

2. If in a set of random variables or samples $[X(t_1), X(t_2), X(t_3), \ldots, X(t_n)]$, obtained by observing a random process $X(t)$ at times $t_1, t_2, t_3, \ldots, t_n$, the process $X(t)$ is Gaussian, then this set of random variables is jointly Gaussian for any n with their n-fold joint probability density function being completely determined by specifying the set of means and the set of covariance functions.

3. If a Gaussian process is stationary, then the process is also strictly stationary.

4. For jointly Gaussian processes, uncorrelatedness and independence are equivalent, i.e., if the Gaussian random variables $[X(t_1), X(t_2), X(t_3), \ldots, X(t_n)]$ are uncorrelated, then they are statistically independent. This in turn means that the joint probability density function of this set of random variables can be expressed as the product of the probability density function of the individual random variables in the set.

SUMMARY

- Probability is a measure of how likely it is that some event will occur, i.e., it is a number expressing the ratio of favourable cases to the whole number of cases possible.
- The fundamental concept in any probabilistic model is the concept of a random experiment whose outcome cannot be predicted, for example flipping a coin.
- Various approaches to calculate probability are the classical approach, relative frequency approach, and axiomatic approach.
- The set theory is used to depict what happens when elements from a group are combined, selected, or excluded.
- Random variable is a mapping of a sample space to the set of real numbers, i.e., it is an assignment of real numbers to the outcomes of a random experiment.
- Important random variables are uniform, Gaussian, Bernoulli, binomial, Poisson, Rayleigh, exponential, and jointly Gaussian random variables.
- Random process is the study of non-deterministic signals, which often occur in communication engineering.
- A collection of statistically identical sources is called an ensemble.
- Important random processes are continuous and discrete, stationary, wide sense stationary, multiple, bandpass, and Gaussian.
- The mean or expectation of a random process is independent of time,

whereas its autocorrelation function depends only on the difference between times.

- Power spectral density of a random process denotes the strength of the power in the random process as a function of frequency.
- Gaussian process is a stochastic process for which any finite linear combination of samples will be normally distributed.
- The central limit theorem indicates that the probability density of a sum of N independent random variables tends to approach a Gaussian density as the number N increases. This theorem can be applied even if the individual random variables are not Gaussian and they are not independent.

IMPORTANT FORMULAE

- In the relative frequency approach:

$$P(E) = \lim_{n \to \infty} \frac{n_E}{n}$$

- In the axiomatic approach:

$$P(B) = \int_B f(x)dx \text{ for } B \in F = R$$

- $P(A \cup B) = P(A) + P(B \cap \overline{A})$

  $P(B) = P(A \cap B) + P(B \cap \overline{A})$

  $P(A \cup B) = P(A) + P(B) - P(A \cap B)$

- Conditional probability:

$$P(A|M) = \frac{P(A \cap M)}{P(M)} \text{ and } P(A \cap M) = P(A|M)P(M)$$

- Total probability theorem–discrete version:

$$P(B) = P[B|A_1]P(A_1) + P[B|A_2]P(A_2) + \cdots + P[B|A_n]P(A_n)$$

- Bayes' theorem:

$$P(A_i|B) = \frac{P(B|A_i)P(A_i)}{P[B|A_1]P[A_1] + P[B|A_2]P[A_2] + \cdots + P[B|A_n]}$$

- Independence:

$$P[A|B] = \frac{P[A \cap B]}{P[B]} = \frac{P[A]}{P[B]} = P[A]$$

- Probability density function (PDF) of a continuous random variable

$$f(x) = \frac{d}{dx} F_X(x)$$

- Properties of PDF:

  (i) $f_X(x) \geq 0$

  (ii) $P(X \in A) = \int_A f(x)dx$ (in general)

(iii) $F(X) = \int_{-\infty}^{x^+} f_X(u)\,du$

(iv) $\int_a^b f_X(x)\,dx = P(a < X \le b)$

(v) $\int_{-\infty}^{\infty} f_X(x)\,dx = 1$

- Uniform random variable:

$$f_X(x) = \begin{cases} \dfrac{1}{b-a}, & a < x < b \\ 0, & \text{otherwise} \end{cases}$$

- Gaussian or normal random variable:

$$f_x(x) = \frac{1}{\sqrt{2\pi}\sigma} e^{-(x-m)^2/2\sigma^2}$$

- Binomial random variable:

$$f_x(X) = \binom{n}{k} p^k q^{n-k} \delta(X-k)$$

- Poisson random variable:

$$F_x(x) = e^{-a} \sum_{k=0}^{mx} \frac{a^k}{k!} \quad m_x \le x \le m_x + 1, m_x = 0,1,2,...$$

- Rayleigh random variable:

$$f_x[X] = \frac{x}{\alpha^2} \exp\left\{-\frac{1}{2}\left(\frac{x}{\alpha}\right)^2\right\} \quad \text{for } x \ge 0$$

- Exponential random variable:

$$f_x(X) = \lambda e^{-\lambda x}, \quad x \ge 0$$
$$= 0, \quad x \le 0$$

- Jointly Gaussian random variable:

$$f_{X,Y}(x,y) = \frac{1}{2\pi\sigma_1\sigma_2\sqrt{1-\rho^2}} \exp\left[-\frac{1}{2(1-\rho^2)}\right]$$

$$\left\{\frac{(x-m_1)^2}{\sigma_1^2} + \frac{(y-m_2)^2}{\sigma_2^2} + \frac{(2\rho(x-m_1)(y-m_2))}{\sigma_1\sigma_2}\right\}$$

- Stationary random process:

$$F(x_1, x_2, x_3, ..., x_n) = t_1 + T, t_2 + T, t_3 + T, ..., t_n + T)$$

- Bandpass random process:

$$X(t) = X_c(t)\cos(2\pi f_c t) + X_s(t)\sin(2\pi f_c t)$$

ADDITIONAL EXAMPLES

1. A box contains five white balls and three green balls. What is the probability that the first removed ball is white and the second is green, assuming that the balls are removed at random and without replacing two balls in succession?

Solution

Probability that the first ball is white, $P(W) = 5/8$
Since a white ball is removed, the remaining balls are four white and three green.
Now, probability that the second ball drawn is green, $P(G) = 3/7$
The desired probability using the multiplication theorem

$$= (5/8)(3/7)$$

$$= 15/56$$

2. There are four boxes containing components. Box 1 contains 5000 components, of which 5% are defective, and Box 2 contains 1000 components, of which 30% are defective. Box 3 contains 1500 components, of which 10% are defective and Box 4 contains 500 components, of which 20% are defective. A box is selected at random and a component is removed. What is the probability that this component is defective?

Solution

From the theorem of total probability, the probability that a component is defective is given by

$$P[\text{component is defective}] = \sum_{i=1}^{4} P[\text{defective}|\text{Box } i]P[\text{Box } i]$$

$$= (0.05)(0.25) + (0.3)(0.25) + (0.1)(0.25) + (0.2)(0.25)$$

$$= 125 + 0.075 + 0.025 + 0.05$$

$$= 0.1625$$

3. In a binary communication channel, '1' is transmitted with probability 0.3 and '0' is transmitted with probability 0.7. The probability of detecting the error when '1' is transmitted is 10^{-3} and the probability of detecting the error when '0' is transmitted is 10^{-7}. Determine the error probability of the channel.

Solution

Given:
Probability that a '1' is transmitted, $P(1) = 0.3$
Probability of detecting the error when '1' is transmitted, $P(e|1) = 10^{-3}$
Probability that a '0' is transmitted, $P(0) = 0.7$
Probability of detecting the error when '0' is transmitted, $P(e|0) = 10^{-7}$

Error probability of the channel $= \sum_{i} P(e, x_i) = P(e, 1) + P(e, 0)$

$$= P(1)P(e|1) + P(0)P(e|0)$$

$$= 0.3(10^{-3}) + 0.7(10^{-0.7})$$

$$= 0.30007 \times 10^{-3}$$

4. Let X is an $N(5,7)$ random variable. Find the density function $Y = -3X + 5$.

Solution

Since Y is a Gaussian random variable with mean

$$m = -3(5) + 5 = -10$$

and the variance

$$\sigma^2 = 4 \times 7 = 28$$

Hence, Y is an $N(-10, 28)$ random variable and

the density function of $Y = f(y) = \dfrac{1}{\sqrt{56\pi}}\, e^{-(y+10)^2/56}$

REVIEW QUESTIONS

1. Explain the three different approaches to the concepts of probability.
2. What are the implications of the axioms of probability?
3. Explain in detail the probability theorem.
4. What is a random variable? Explain continuous and discrete random variables.
5. What do you mean by cumulative density function? Discuss its properties.
6. Explain different types of random variables.
7. Write a brief note on statistical averages.
8. Explain multiple functions of multiple random variables.
9. What are the basic properties of joint and marginal cumulative distribution function and probability density function?
10. What is a random process? Give an example.
11. Describe the continuous and discrete random processes.
12. Explain in detail the stationary random process.
13. Explain in detail the multiple random processes.
14. What are the first- and second-order probabilistic averages?
15. What do you mean by ergodic process? Compare it with wide sense stationary process.
16. Describe a random process when it is passed through a linear time invariant system.
17. Give a brief note on power spectral density of stationary process and sum process.
18. Explain in detail the Gaussian process.
19. What is the central limit theorem? Explain it briefly.
20. Discuss the properties of Gaussian process.

PROBLEMS

1. In a class where foreign languages are offered as elective, 40% of the students take German, 25% take Spanish, and 35% take Russian. The probability of failures in German is 3%, in Spanish 5%, and in French 2%.
   (a) What is the probability that a student has failed in his elective?
   (b) What is the probability that he has taken Spanish?

2. In a binary communication system, the probability of a 0 transmission is 0.6 and the probability of 1 transmission is 0.4. Due to channel noise, the probability of 0 being received in error (i.e., being received as 1) is 0.08 and that of 1 being received in error (i.e., being received as 0) is 0.05.
    (a) What is the probability that the output of this channel is 1?
    (b) Assuming a 0 is observed at the output of this channel, what is the probability that the input to the channel was a 0?
3. A box contains 4 white, 6 red, and 3 green balls. If a ball is drawn at random, what is the probability that it is (a) green and (b) white or red.
4. What is the probability of getting 7 if two dice are thrown?
5. An urn contains 6 white, 4 red, and 3 green balls. If three balls are drawn in succession, what is the probability that they will be of different colours?
6. Find the constant a so that the function

$$f(x) = \begin{cases} a(2x^2 - 3), & 2 < x < 5 \\ 0, & \text{otherwise} \end{cases}$$

is a density function.
7. Let X be $N(-2, 3)$ and Y be $N(-3, 4)$. Determine $f_{XY}(x, y)$ assuming that X and Y are independent.
8. Assuming that X is $N(-1, 2)$ and Y is $N(-2, 3)$ and are independent, determine the covariances of two random variables $R = 2X - Y$ and $S = 2X + Y$.
9. A random variable has an exponential PDF given by $f(x) = 2ae^{-3b|x|}dx$, where a and b are constants. Find the relationship between a and b.
10. When an experiment is conducted 5×10^5 times, the error probability is 10^5. What is the probability that the estimated probability of error does not differ from the error probability by more than 30%?

Answers to problems

1. (a) the probability of a student failure = 0.02540
    (b) the probability that a student has failed in Spanish = 0.5
2. (a) 0.428
    (b) 0.965
3. (a) 0.23
    (b) 0.77
4. 0.166
5. 0.2517
6. 0.014

7. $f(x, y) = \dfrac{1}{4\pi\sqrt{6}} e^{-\frac{(x+2)^2}{8} - \frac{(x+3)^2}{16}}$

8. 5

9. $a = \dfrac{3}{4}b$

10. 22%

Appendix A

MATLAB Exercises

MATLAB is a language that can be used as a powerful tool and can be implemented in the design and analysis of systems in engineering and research. Its flexible program structure helps in developing and analysing various problems encountered. The visualization capability of MATLAB provides a unique insight into the signal character and system response. As with any language, initially the students should try simple examples and as they start getting grip over the language they are encouraged to try more complicated problems. Once mastery over MATLAB is achieved, it becomes very simple for the students to understand the concepts clearly.

Apart from some solved MATLAB exercises with outputs given at the end of most of the chapters, some additional exercises are given here in the form of programs. The students have to key in these programs and then execute and observe the results. With this, they will be able to write on their own MATLAB programs for unsolved problems given at the end of each chapter.

1. **Program to find the plot for $s(t) = A_c[1 + m(t)]$ cos (wct), which is nothing but an AM wave**

```
clc;
clear all;
close all;

fc = 10;
fa = 1;
Ta = 1/fa;
dt = 2*Ta/200;
wc = 2*pi*fc;
wa = 2*pi*fa;

t = 0:dt:2*Ta;

m = cos(wa*t);
m = m(:);
j = sqrt(-1);
g = 1 + m;
phase = exp(j*wc*t);
```

```
g = g(:);
phase = phase(:);
s = real(g.*phase);

subplot(211);
plot(t,m);
xlabel('t');
ylabel('m(t)');

subplot(212);
plot(t,s);
xlabel('t');
ylabel('s(t)');
subplot(111);
```

2. Program to find the average power of a square wave

```
clc;
clear all;
close all;
fprintf('Enter the number of Harmonics to use in the Fourier
Series, N.\n');
N = input('N should be an odd integer: ');
if (rem(N,2) == 0)
    N = N+1;
end;

A = 5;
j = sqrt(-1);

for (n = 1:N)
    c(n) = A/(n*pi*j) * (sin(pi*n/2))^2;
end;

P = (A/2)^2 + 2*sum(c.*conj(c));

fprintf('\n\nThe average normalized power is %G Watts\n',P);
```

3. Program for creating a list of $J0$ (0, Beta) through Jn (n, Beta)')

```
clc;
clear all;
close all;
fprintf('\nThis file creates a list of J0(0,Beta) through
Jn(n,Beta)');
fprintf('\nfor a specified value of Beta\n\n');

Beta = input('Enter the desired value of Beta: ');
n = input('Enter the desired value of n: ');

n = floor(n);
Jn = zeros(n,1);
m = zeros(n,1);
for (i = 0:1:n)
    Jn(i+1) = bessel(i,Beta);
    m(i+1) = i;
```

```
end;
A = [m(:) Jn];

save TABLE5_2.dat A /ascii
fprintf('table5_2.dat','\n\nThe previous columns correspond
to:\n');
fprintf('table5_2.dat','n and Jn(n,Beta) for Beta = %g\n',Beta);
fprintf('\n\nThe output data has been stored in the file
TABLE5_2.dat\n');
```

4. **Program for CDF of Gaussian RV with specified mean and standard deviation**

```
clc;
clear all;

m = input('Enter the desired value of the mean: ');
sigma = input('Enter the desired value of the standard
deviation: ');
sigma = abs(sigma);

x = m-5*sigma:10*sigma/1000:m+5*sigma;

pdf = zeros(length(x),1);
for (i = 1:1:length(x))
   pdf(i) = 1/(sqrt(2*pi)*sigma)*exp(- (x(i) -m)^2/(2*sigma^2));
end;

fprintf('\nSee Window for plot.\n');

plot(x,pdf);
xlabel('x');
ylabel('f(x)');
title('PDF of Guassian R.V. with specified mean and standard
deviation');
grid;
```

5. **Program to determine the number of bits required to represent a character and to calculate the information content of each character assuming that each character is equally likely to be sent**

```
clc;
clear all;
chars = input('Enter the Number of available characters: ');
p = 1/chars;
I = -log(p)/log(2);
fprintf('The information content of each character is %G
(bits)\n',I);
B = input('Enter the Channel bandwidth in Hz: ');
SNRdB = input('Enter the Signal to Noise ratio in dB: ');

SNR = 10^(SNRdB/10);
C = B*log(1 + SNR)/log(2);
fprintf('\n\nThe channel capacity is %G (bits/sec)\n',C);
fprintf('or equivalently %G (chars/sec)\n\n',C/b);
```

```
b = ceil(log(chars)/log(2));
fprintf('\n\nThe number of bits required to represent a
character = %G\n\n',b);
```

6. Program to generate delta function

```
clc;
clear all;
close all;
f = -1:0.01:1;
PSDc = zeros(length(f),1);
for (i = 1:1:length(PSDc))
  PSDc(i) = 4*sqrt(pi)*exp(-(pi*f(i))^2);
end;
PSDd = zeros(length(f),1);
for (i = 1:1:length(PSDd))
  if (f(i) == 0)
    PSDd(i) = 3;
  end;
end;
PSD = PSDc + PSDd;
plot(f,PSD);
xlabel('f');
title('PSD');
Brms = (sqrt(2/7))/pi;
fprintf('\nThe rms bandwidth is %g',Brms);
fprintf (' Hz\n\n');
```

7. Program to find the entropy of a binary source

```
clc;
clear all;
i = 0;
for (loop = 0:0.05:1)
  i = i+1;
  P1(i) = loop;
  if P1(i) == 0 | P1(i) == 1
    H(i) = 0;
  else
    H(i) = -1/log(2) * (P1(i)*log(P1(i)) + (1-P1(i))*log(1-
P1(i)));
  end;
end;

plot(P1,H);
title('Entropy of Binary Source');
xlabel('Probability P1');
```

8. Program for PM and FM generation

```
clc;
clear all;
close all;

t = -4:0.01:4;
fc = 1;
fm = 0.25*fc;
Dp = pi;
```

```
Df = pi;
m = cos(2*pi*fm*t);
theta = Df/(2*pi*fm)*sin(2*pi*fm*t);
sp = cos(2*pi*fc*t + Dp*m);
sf = cos(2*pi*fc*t + theta);

%PLOT_PR(3);
plot(t,m);
title('Sinusoidal message waveform');
xlabel('t');
ylabel('m(t)');
pause;

plot(t,m,'-',t,sp);
title('PM modulator output waveform for sinusoidal input');
xlabel('t');
ylabel('sp(t)');
pause;

plot(t,m,'-',t,sf);
title('FM modulator output waveform for sinusoidal input');
xlabel('t');
ylabel('sf(t)');
```

9. Program for PDF of Gaussian RV with specified mean and standard deviation

```
clc;
clear all;

m = input('Enter the desired value of the mean: ');
sigma = input('Enter the desired value of the standard
deviation: ');
sigma = abs(sigma);

x = m-5*sigma:10*sigma/1000:m+5*sigma;

pdf = zeros(length(x),1);
for (i = 1:1:length(x))
   pdf(i) = 1/(sqrt(2*pi)*sigma)*exp(- (x(i) -m)^2/(2*sigma^2));
end;

fprintf('\nSee Window for plot.\n');

plot(x,pdf);
xlabel('x');
ylabel('f(x)');
title('PDF of Guassian R.V. with specified mean and standard
deviation');
grid;
```

10. Program for generating Gaussian distribution function for different alpha

```
%
% Generating Gaussian distribution function for different
alpha
```

```
x = -5:0.05:5;
y=normpdf(x,0,1); % mean = 0 , std deviation = 1 normpdfin
built function
subplot(131) ; plot(x,y); axis([-5 5 0 1 ]); title ( 'mean=0,
std dev =1');
y = normpdf ( x,0,0.5 );
subplot (132); plot (x,y) , axis([-5 5 0 1]) ; title('mean=0,std
dev =0.5');
y = normpdf  (x ,1,1);
subplot (133) ; plot (x,y) , axis ([-5 5 0 1]) ; title ('mean=1,
std dev=1');
```

11. **Program to show that for a binary source, maximum entropy will occur when the probability of sending a binary 'one' is equal to the probability of sending a 'zero' and finding that maximum entropy**

```
clc;
clear all;

fprintf('Calculating the Entropy for a Binary Source\n');

i = 0;
for (loop = 0:0.02:1)
  i = i+1;
  p(i) = loop;
  if p(i) == 0 | p(i) == 1
    H(i) = 0;
  else
    H(i) = -1/log(2) * (p(i)*log(p(i)) + (1-p(i))*log(1-p(i)));
  end;
end;

[Hmax,pmax] = max(H);
pmax = p(pmax);
fprintf('The maximum entropy is H(p) = %G\n',Hmax);
fprintf('Which occurs for a Probability p = %G\n',pmax);
fprintf('\n\nNow press any key to plot the Entropy as a
function of p\n');
pause

plot(p,H);
title('Entropy of Binary Source');
xlabel('Probability p');
ylabel('H(p)');
```

12. **Program for generation of a square wave of different duty cycle**

```
% Generation of square wave of different duty cycles and
plotting
f =50;fs=1000;t=0:1/fs:0.1 ; % A s in Experiment 1
v1 = square(2*pi*f*t,30);    % duty cycle of 25 percent
v2 = square(2*pi*f*t,70);    % duty cycle of 75 percent
subplot  ( 2, 1, 1) % divides plot window into 2 x 1, select 1,
',' optional
plot(t,v1 ); axis([0 0.1 -1.2 1.2 ]); % As in Experiment    1
title( 'sq. wave duty cycle 30,f = 50Hz,fs = 1000Hz' )
```

```
subplot( 2, 1, 2 ) % divides plot window into 2 x 1 matrix,
2 nd selected
plot ( t, v2 ) ; axis([ 0 0.1 -1.2 1.2 ]); % As in Experiment   1
xlabel('time in second') ; ylabel  ('amplitude') ;
title ('sq.wave duty cycle 70,f = 50 Hz,f s = 1000Hz')
```

13. Program for SSB generation

```
clc;
clear all;
close all;
N = 1024;
fs = 2048;

t = (0:N-1)/fs;
fc = 600; %Carrier frequency !! Limit fc<800 to avoid freqdomain
aliasing
fm1 = 200;
fm2 = 100;
Em1 = 2;
Em2 = 2;

m = Em1*cos(2*pi*fm1*t)+Em2*cos(2*pi*fm2*t); %Message
mh = imag(hilbert(m)); %Hilbert transform of the message
signal

sbu = m.*2.*cos(2*pi*fc*t) -mh.*2.*sin(2*pi*fc*t); %Expression
for USB SSB
sbl = m.*2.*cos(2*pi*fc*t) + mh.*2.*sin(2*pi*fc*t); %Expression
for LSB SSB
SBU = 2/N*abs(fft(sbu));
SBL = 2/N*abs(fft(sbl));

freq = fs * (0 : N/2) / N;
clc;
display('Single SideBand Modulation');
sprintf('Carrier frequency: %d Hz',fc)
sprintf('Message frequency: %d Hz and %d Hz',fm1,fm2)
sprintf('USB spectra at: %d Hz and %d Hz',fc+fm1,fc+fm2)
sprintf('LSB spectra at: %d Hz and %d Hz',fc-fm1,fc-fm2)

close all;
subplot(211);
plot(10*t(1:200),sbu(1:200),'b'); %Time Domain Plot
title('Time Domain Representation');
xlabel('Time'); ylabel('Modulated Signal');

subplot(212);
plot(freq,SBU(1:N/2+1),freq,SBL(1:N/2+1)); %Frequency domain
plot
title('Frequency Domain Representation');
xlabel('Frequency(Hz)'); ylabel('Spectral Magnitude');
legend('USB','LSB');
```

14. Program for generating unit step function

```
%%1. Program for creating a Unit Step Function:
clc;
clear all;
close all;
function y = U_STEP(t,Tbegin);
n = length(t);
i = 1;
j = 0;

temp = 0;
while(temp == 0)
  if (t(i) >= Tbegin)
    temp = 1;
    j = i;
  else
    i = i+1;
    if (i > n)
      temp = 1;
    end;
  end;
end;

y = zeros(n,1);
if (j > 0)
  for (i = j:1:n)
    y(i) = 1;
  end;
end;

n = length(t);
i = 1;
j = 0;

temp = 0;
while(temp == 0)
  if (t(i) >= Tbegin)
    temp = 1;
    j = i;
  else
    i = i+1;
    if (i > n)
      temp = 1;
    end;
  end;
end;

y = zeros(n,1);
if (j > 0)
  for (i = j:1:n)
    y(i) = 1;
  end;
end;
```

Important Mathematical Relations/ Formulae

L'Hopital's rule

If $\lim f(x)/g(x)$ results in the indeterministic form $0/0$ or ∞/∞, then

$$\lim \frac{f(x)}{g(x)} = \lim \frac{\dot{f}(x)}{\dot{g}(x)}$$

Power series

$$e^x = 1 + x + \frac{x^2}{2!} + \frac{x^3}{3!} + \cdots + \frac{x^n}{n!} + \cdots$$

$$\sin x = x - \frac{x^3}{3!} + \frac{x^5}{5!} - \frac{x^7}{7!} + \cdots$$

$$\cos x = 1 - \frac{x^2}{2!} + \frac{x^4}{4!} - \frac{x^6}{6!} + \frac{x^8}{8!} - \cdots$$

$$\tan x = x + \frac{x^3}{3} + \frac{2x^5}{15} - \frac{17x^7}{315} + \cdots \quad x^2 < \frac{\pi^2}{4}$$

$$Q(x) = \frac{e^{-x^2/2}}{x\sqrt{2\pi}}\left(1 - \frac{1}{x^2} + \frac{1\cdot 3}{x^4} - \frac{1\cdot 3\cdot 5}{x^6} + \cdots\right)$$

$$(1 + x)^n = 1 + nx + \frac{n(n-1)}{2!}x^2 + \frac{n(n-1)(n-2)}{3!}x^3 + \cdots + \binom{n}{k}x^k$$
$$+ \cdots + x^n \approx 1 + nx, \quad |x| \ll 1$$

$$\frac{1}{1-x} = 1 + x + x^2 + x^3 + \cdots, \quad |x| < 1$$

Series expansions

Taylor series:

$$f(x) = f(a) + \frac{f'(a)}{1!}(x-a) + \frac{f''(a)}{2!}(x-a)^2 + \cdots$$
$$+ \frac{f^{(n)}(a)}{n!}(x-a)^n + \cdots$$

where $f^{(n)}(a) = \dfrac{d^n f(x)}{dx^n}\bigg|_{x=a}$

Maclaurin seires:

$$f(x) = f(0) + \frac{f'(0)}{1!}x + \frac{f''(0)}{2!}x^2 + \cdots + \frac{f^{(n)}(0)}{n!}x^n + \cdots$$

where $f^{(n)}(0) = \dfrac{d^n f(x)}{dx^n}\bigg|_{x=0}$

Binomial series:

$$(1+x)^n = 1 + nx + \frac{n(n-1)}{2!}x^2 + \cdots, \quad |nx| < 1$$

Exponential series:

$$\exp x = 1 + x + \frac{1}{2!}x^2 + \cdots$$

Logarithmic series:

$$\log(1+x) = x - \frac{1}{2}x^2 + \frac{1}{3}x^3 - \cdots$$

Trigonometric series:

$$\sin x = x - \frac{1}{3!}x^3 + \frac{1}{5!}x^5 - \cdots$$

$$\cos x = 1 - \frac{1}{2!}x^2 + \frac{1}{4!}x^4 - \cdots$$

$$\tan x = x - \frac{1}{3}x^3 + \frac{2}{15}x^5 + \cdots$$

$$\sin^{-1} x = x - \frac{1}{6}x^3 + \frac{3}{40}x^5 + \cdots$$

$$\tan^{-1} x = x - \frac{1}{3}x^3 + \frac{1}{5}x^5 - \cdots, \quad |x| < 1$$

$$\operatorname{sinc} x = 1 - \frac{1}{3!}(\pi x)^2 + \frac{1}{5!}(\pi x)^4 - \cdots$$

Sums

$$\sum_{m=0}^{k} r^m = \frac{r^{k+1} - 1}{r - 1}, \quad r \ne 1$$

$$\sum_{m=M}^{N} r^m = \frac{r^{N+1} - r^M}{r - 1}, \quad r \ne 1$$

$$\sum_{m=0}^{k} \left(\frac{a}{b}\right)^m = \frac{a^{k+1} - b^{k+1}}{b^k (a - b)}, \quad a \ne b$$

Complex numbers

$$e^{\pm j\pi/2} = \pm j$$

$$e^{\pm jn\pi} = \begin{cases} 1, & n \text{ even} \\ -1, & n \text{ odd} \end{cases}$$

$$e^{\pm j\theta} = \cos\theta \pm j \sin\theta$$

$$a + jb = re^{j\theta}, \quad r = \sqrt{a^2 + b^2}, \quad \theta = \tan^{-1}\left(\frac{b}{a}\right)$$

$$(re^{j\theta})^k = r^k e^{jk\theta}$$

$$(r_1 e^{j\theta_1})(r_2 e^{j\theta_2}) = r_1 r_2 e^{j(\theta_1 + \theta_2)}$$

Trigonometric identities

$$e^{\pm jx} = \cos x \pm j \sin x$$

$$\cos x = \frac{1}{2}(e^{jx} + e^{-jx})$$

$$\sin x = \frac{1}{2j}(e^{jx} - e^{-jx})$$

$$\cos\left(x \pm \frac{\pi}{2}\right) = \mp \sin x$$

$$\sin\left(x \pm \frac{\pi}{2}\right) = \pm \cos x$$

$$2 \sin x \cos x = \sin 2x$$

$$\sin^2 x + \cos^2 x = 1$$

$$\cos^2 x - \sin^2 x = \cos 2x$$

$$\cos^2 x = \frac{1}{2}(1 + \cos 2x)$$

$$\sin^2 x = \frac{1}{2}(1 - \cos 2x)$$

$$\cos^3 x = \frac{1}{4} (3 \cos x + \cos 3x)$$

$$\sin^3 x = \frac{1}{4} (3 \sin x - \sin 3x)$$

$$\sin (x \pm y) = \sin x \cos y \pm \cos x \sin y$$

$$\cos (x \pm y) = \cos x \cos y \mp \sin x \sin y$$

$$\tan (x \pm y) = \frac{\tan x \pm \tan y}{1 \mp \tan x \tan y}$$

$$\sin x \sin y = \frac{1}{2} [\cos(x - y) - \cos (x + y)]$$

$$\cos x \cos y = \frac{1}{2} [\cos(x - y) + \cos (x + y)]$$

$$\sin x \cos y = \frac{1}{2} [\sin(x - y) + \sin (x + y)]$$

$$a \cos x + b \sin x = C \cos(x + \theta)$$

$$\text{in which } C = \sqrt{a^2 + b^2} \text{ and } \theta = \tan^{-1} \left(\frac{-b}{a} \right)$$

Indefinite integrals

$$\int u \, dv = uv - \int v \, du$$

$$\int f(x) \, \dot{g}(x) dx = f(x) \, g(x) - \int \dot{f}(x) g(x) \, dx$$

$$\int \sin ax \, dx = -\frac{1}{a} \cos ax, \quad \int \cos ax \, dx = \frac{1}{a} \sin ax$$

$$\int \sin^2 ax \, dx = \frac{x}{2} - \frac{\sin 2ax}{4a}, \quad \int \cos^2 ax \, dx = \frac{x}{2} + \frac{\sin 2ax}{4a}$$

$$\int x \sin ax \, dx = \frac{1}{a^2} (\sin ax - ax \cos ax)$$

$$\int x \cos ax \, dx = \frac{1}{a^2} (\cos ax + ax \sin ax)$$

$$\int x^2 \sin ax \, dx = \frac{1}{a^3} (2ax \sin ax + 2 \cos ax - a^2 x^2 \cos ax)$$

$$\int x^2 \cos ax \, dx = \frac{1}{a^3} (2ax \cos ax - 2 \sin ax + a^2 x^2 \sin ax)$$

$$\int \sin ax \sin bx \, dx = \frac{\sin(a - b)x}{2(a - b)} - \frac{\sin(a + b)x}{2(a + b)}, \quad a^2 \neq b^2$$

$$\int \sin ax \cos bx \, dx = -\left[\frac{\cos(a - b)x}{2(a - b)} + \frac{\cos(a + b)x}{2(a + b)} \right], \quad a^2 \neq b^2$$

$$\int \cos ax \cos bx \, dx = \frac{\sin(a-b)x}{2(a-b)} + \frac{\sin(a+b)x}{2(a+b)}, \quad a^2 \neq b^2$$

$$\int e^{ax} \, dx = \frac{1}{a} e^{ax}$$

$$\int xe^{ax} \, dx = \frac{e^{ax}}{a^2} (ax - 1)$$

$$\int xe^{ax^2} \, dx = \frac{1}{2a} e^{ax^2}$$

$$\int x^2 e^{ax} \, dx = \frac{e^{ax}}{a^3} (a^2 x^2 - 2ax + 2)$$

$$\int e^{ax} \sin bx \, dx = \frac{e^{ax}}{a^2 + b^2} (a \sin bx - b \cos bx)$$

$$\int e^{ax} \cos bx \, dx = \frac{e^{ax}}{a^2 + b^2} (a \cos bx + b \sin bx)$$

$$\int \frac{1}{x^2 + a^2} \, dx = \frac{1}{a} \tan^{-1} \frac{x}{a}$$

$$\int \frac{x}{x^2 + a^2} \, dx = \frac{1}{2} \ln (x^2 + a^2)$$

$$\int \frac{1}{a^2 + b^2 x^2} \, dx = \frac{1}{ab} \tan^{-1} \frac{bx}{a}$$

$$\int \frac{x^2}{a^2 + b^2 x^2} \, dx = \frac{x}{b^2} - \frac{a}{b^3} \tan^{-1} \frac{bx}{a}$$

$$\int \frac{x}{a^2 + b^2 x^2} \, dx = \frac{1}{2b^2} \ln (a^2 + b^2 x^2)$$

Definite integrals

$$\int_0^\infty \frac{x \sin(ax)}{b^2 + x^2} \, dx = \frac{\pi}{2} \exp(-ab), \quad a > 0, b > 0$$

$$\int_0^\infty \frac{\cos(ax)}{b^2 + x^2} \, dx = \frac{\pi}{2b} \exp(-ab), \quad a > 0, b > 0$$

$$\int_0^\infty \frac{\cos(ax)}{(b^2 - x^2)^2} \, dx = \frac{\pi}{4b^3} [\sin(ab) - ab \cos(ab)], \quad a > 0, b > 0$$

$$\int_0^\infty \operatorname{sinc} x \, dx = \int_0^\infty \operatorname{sinc}^2 x \, dx = \frac{1}{2}$$

$$\int_0^\infty e^{-ax^2} \, dx = \frac{1}{2}\sqrt{\frac{\pi}{a}}, \quad a > 0$$

$$\int_0^\infty x^2 \, e^{-ax^2} \, dx = \frac{1}{4a}\sqrt{\frac{\pi}{a}}, \quad a > 0$$

Functions

1. Rectangular function: $\text{rect}(t) = \begin{cases} 1, & -\frac{1}{2} < t < \frac{1}{2} \\ 0, & |t| > \frac{1}{2} \end{cases}$

2. Unit step function: $u(t) = \begin{cases} 1, & t > 0 \\ 0, & t < 0 \end{cases}$

3. Signum function: $\text{sgn}(t) = \begin{cases} 1, & t > 0 \\ 0, & t = 0 \\ -1, & t < 0 \end{cases}$

4. (Dirac) delta function: $\delta(t) = 0, \quad t \neq 0$

$$\int_{-\infty}^\infty \delta(t) \, dt = 1$$

or, equivalently, $\int_{-\infty}^\infty g(t) \, \delta(t - t_0) \, dt = g(t_0)$

5. Sinc function: $\text{sinc}(x) = \dfrac{\sin \pi x}{\pi x}$

6. Sine integral: $\text{si}(u) = \displaystyle\int_0^\infty \frac{\sin x}{x} \, dx$

7. Error function: $\text{erf}(u) = \dfrac{2}{\sqrt{\pi}} \displaystyle\int_0^\infty \exp(-z^2) \, dz$

   Complementary error function: $\text{erfc}(u) = 1 - \text{erf}(u)$

8. Binomial coefficient: $\dbinom{n}{k} = \dfrac{n!}{(n-k)!k!}$

9. Bessel function of the first kind of order n:

$$J_n(x) = \frac{1}{2\pi} \int_{-\infty}^\infty \exp(jx \sin\theta - jn\theta) \, d\theta$$

10. Modified Bessel function of the first kind of zero order:

$$I_0(x) = \frac{1}{2\pi} \int_{-\infty}^\infty \exp(x \cos\theta) \, d\theta$$

11. Confluent hypergeometric function:

$$_1F_1(a; b; x) = 1 + \frac{a}{c}\frac{x}{1!} + \frac{a(a+1)}{b(b+1)}\frac{x^2}{2!} + \cdots$$

Fourier Series Representation and Its Properties

Table A.1   Fourier series representation of a periodic signal of period T_0 ($\omega_0 = 2\pi/T_0$)

Series form	Coefficient computation	Conversion formulae
Trigonometric:	$a_0 = \dfrac{1}{T_0}\displaystyle\int_{T_0} f(t)\,dt$	$a_0 = C_0 = D_0$
$f(t) = a_0 + \displaystyle\sum_{n=1}^{\infty} a_n \cos n\omega_0 t$	$a_n = \dfrac{2}{T_0}\displaystyle\int_{T_0} f(t) \cos n\omega_0 t\,dt$	$a_n - jb_n = C_n e^{j\theta n} = 2D_n$
$\quad + b_n \sin n\omega_0 t$	$b_n = \dfrac{2}{T_0}\displaystyle\int_{T_0} f(t) \sin n\omega_0 t\,dt$	$a_n + jb_n = C_n e^{-j\theta n} = 2D_{-n}$
Compact trigonometric:	$C_0 = a_0$	$C_0 = D_0$
$f(t) = C_0 + \displaystyle\sum_{n=1}^{\infty} C_n \cos(n\omega_0 t + \theta_n)$	$C_n = \sqrt{a_n{}^2 + b_n{}^2}$	$C_n = 2\lvert D_n \rvert, \quad n \ge 1$
	$\theta_n = \tan^{-1}\left(\dfrac{-b_n}{a_n}\right)$	$\theta_n = \angle D_n$
Exponential:		
$f(t) = \displaystyle\sum_{n=-\infty}^{\infty} D_n e^{jn\omega_0 t}$	$D_n = \dfrac{1}{T_0}\displaystyle\int_{T_0} f(t)\, e^{-jn\omega_0 t}\,dt$	

Table A.2 Fourier transforms

No.	$x(t)$	$X(\omega)$	
1	$e^{-at}u(t)$	$\dfrac{1}{a+j\omega}$	$a>0$
2	$e^{at}u(-t)$	$\dfrac{1}{a-j\omega}$	$a>0$
3	$e^{-a\lvert t\rvert}$	$\dfrac{2a}{a^2+\omega^2}$	$a>0$
4	$t\,e^{-at}u(t)$	$\dfrac{1}{(a+j\omega)^2}$	$a>0$
5	$t^n e^{-at}u(t)$	$\dfrac{n!}{(a+j\omega)^{n+1}}$	$a>0$
6	$\delta(t)$	1	
7	1	$2\pi\,\delta(\omega)$	
8	$e^{j\omega_0 t}$	$2\pi\,\delta(\omega-\omega_0)$	
9	$\cos\omega_0 t$	$\pi[\delta(\omega-\omega_0)+\delta(\omega+\omega_0)]$	
10	$\sin\omega_0 t$	$j\pi[\delta(\omega+\omega_0)-\delta(\omega-\omega_0)]$	
11	$u(t)$	$\pi\,\delta(\omega)+\dfrac{1}{j\omega}$	
12	$\operatorname{sgn} t$	$\dfrac{2}{j\omega}$	
13	$\cos\omega_0 t\,u(t)$	$\dfrac{\pi}{2}\,[\delta(\omega-\omega_0)+\delta(\omega+\omega_0)]+\dfrac{j\omega}{\omega_0^2-\omega^2}$	
14	$\sin\omega_0 t\,u(t)$	$\dfrac{\pi}{2j}\,[\delta(\omega-\omega_0)-\delta(\omega+\omega_0)]+\dfrac{\omega_0}{\omega_0^2-\omega^2}$	
15	$e^{-at}\sin\omega_0 t\,u(t)$	$\dfrac{\omega_0}{\left(a+j\omega\right)^2+\omega_0^2}$	$a>0$
16	$e^{-at}\cos\omega_0 t\,u(t)$	$\dfrac{a+j\omega}{\left(a+j\omega\right)^2+\omega_0^2}$	$a>0$
17	$\operatorname{rect}\left(\dfrac{t}{\tau}\right)$	$\tau\operatorname{sinc}\left(\dfrac{\omega\tau}{2}\right)$	
18	$\dfrac{W}{\pi}\operatorname{sinc}(Wt)$	$\operatorname{rect}\left(\dfrac{\omega}{2W}\right)$	
19	$\Delta\left(\dfrac{t}{\tau}\right)$	$\dfrac{\tau}{2}\operatorname{sinc}^2\left(\dfrac{\omega\tau}{4}\right)$	

Contd

Contd

No.	$x(t)$	$X(\omega)$	
20	$\dfrac{W}{2\pi}\ \text{sinc}^2\left(\dfrac{Wt}{2}\right)$	$\Delta\left(\dfrac{\omega}{2W}\right)$	
21	$\displaystyle\sum_{n=-\infty}^{\infty}\delta(t-nT)$	$\omega_0\displaystyle\sum_{n=-\infty}^{\infty}\delta(\omega-n\omega_0)$	$\omega_0=\dfrac{2\pi}{T}$
22	$e^{-t^2/2\sigma^2}$	$\sigma\sqrt{2\pi}\,e^{-\sigma^2\omega^2/2}$	

Appendix D

Miscellaneous

Table A.3   Convolution table

No.	$x_1(t)$	$x_2(t)$	$x_1(t) * x_2(t) = x_2(t) * x_1(t)$
1	$x(t)$	$\delta(t-T)$	$x(t-T)$
2	$e^{\lambda t} u(t)$	$u(t)$	$\dfrac{1-e^{\lambda t}}{-\lambda}\, u(t)$
3	$u(t)$	$u(t)$	$tu(t)$
4	$e^{\lambda_1 t} u(t)$	$e^{\lambda_2 t} u(t)$	$\dfrac{e^{\lambda_1 t} - e^{\lambda_2 t}}{\lambda_1 - \lambda_2}\, u(t), \quad \lambda_1 \neq \lambda_2$
5	$e^{\lambda t} u(t)$	$e^{\lambda t} u(t)$	$te^{\lambda t} u(t)$
6	$te^{\lambda t} u(t)$	$e^{\lambda t} u(t)$	$\dfrac{1}{2}\, t^2 e^{\lambda t} u(t)$
7	$t^N u(t)$	$e^{\lambda t} u(t)$	$\dfrac{N!e^{\lambda t}}{\lambda^{N+1}}\, u(t) - \displaystyle\sum_{k=0}^{N} \dfrac{N!\, t^{N-k}}{\lambda^{k+1}(N-k)!}\, u(t)$
8	$t^M u(t)$	$t^N u(t)$	$\dfrac{M!\, N!}{(M+N+1)!}\, t^{M+N+1} u(t)$
9	$t\, e^{\lambda_1 t} u(t)$	$e^{\lambda_2 t} u(t)$	$\dfrac{e^{\lambda_2 t} - e^{\lambda_1 t} + (\lambda_1 - \lambda_2) te^{\lambda_1 t}}{(\lambda_1 - \lambda_2)^2}\, u(t)$
10	$t^M e^{\lambda t} u(t)$	$t^N e^{\lambda t} u(t)$	$\dfrac{M!\, N!}{(N+M+1)!}\, t^{M+N+1} e^{\lambda t} u(t)$
11	$t^M e^{\lambda_1 t} u(t)$	$t^N e^{\lambda_2 t} u(t)$	$\displaystyle\sum_{k=0}^{M} \dfrac{(-1)^k\, M!\,(N+k)!\, t^{M-k} e^{\lambda_1 t}}{k!\,(M-k)!\,(\lambda_1 - \lambda_2)^{N+k+1}}\, u(t)$
	$\lambda_1 \neq \lambda_2$		$+ \displaystyle\sum_{k=0}^{N} \dfrac{(-1)^k\, N!\,(M+k)!\, t^{N-k} e^{\lambda_2 t}}{k!\,(N-k)!\,(\lambda_2 - \lambda_1)^{M+k+1}}\, u(t)$

Contd

Contd

No.	$x_1(t)$	$x_2(t)$	$x_1(t) * x_2(t) = x_2(t) * x_1(t)$
12	$e^{-\alpha t} \cos{(\beta t + \theta)}\, u(t)$	$e^{\lambda t}\, u(t)$	$\dfrac{\cos(\theta - \phi)e^{\lambda t} - e^{-\alpha t} \cos(\beta t + \theta - \phi)}{\sqrt{(\alpha + \lambda)^2 + \beta^2}}\, u(t)$ $\phi = \tan^{-1}\left[- \beta/(\alpha + \lambda)\right]$
13	$e^{\lambda_1 t}\, u(t)$	$e^{\lambda_2 t}\, u(-t)$	$\dfrac{e^{\lambda_1 t}\, u(t) + e^{\lambda_2 t}\, u(-t)}{\lambda_2 - \lambda_1}$, $\quad \mathrm{Re}\ \lambda_2 > \mathrm{Re}\ \lambda_1$
14	$e^{\lambda_1 t}\, u(-t)$	$e^{\lambda_2 t}\, u(-t)$	$\dfrac{e^{\lambda_1 t} - e^{\lambda_2 t}}{\lambda_2 - \lambda_1}\, u(-t)$

Table A.4 A short table of (unilateral) Laplace transforms

No.	$x(t)$	$X(s)$
1	$\delta(t)$	1
2	$u(t)$	$\dfrac{1}{s}$
3	$tu(t)$	$\dfrac{1}{s^2}$
4	$t^n\, u(t)$	$\dfrac{n!}{s^{n+1}}$
5	$e^{\lambda t}\, u(t)$	$\dfrac{1}{s - \lambda}$
6	$te^{\lambda t}\, u(t)$	$\dfrac{1}{(s - \lambda)^2}$
7	$t^n\, e^{\lambda t}\, u(t)$	$\dfrac{n!}{(s - \lambda)^{n+1}}$
8a	$\cos{bt}\, u(t)$	$\dfrac{s}{s^2 + b^2}$
8b	$\sin{bt}\, u(t)$	$\dfrac{b}{s^2 + b^2}$
9a	$e^{-at} \cos{bt}\, u(t)$	$\dfrac{s + a}{(s + a)^2 + b^2}$
9b	$e^{-at} \sin{bt}\, u(t)$	$\dfrac{b}{(s + a)^2 + b^2}$
10a	$re^{-at} \cos{(bt + \theta)}\, u(t)$	$\dfrac{(r\cos\theta)s + (ar\cos\theta - br\sin\theta)}{s^2 + 2as + (a^2 + b^2)}$
10b	$re^{-at} \cos{(bt + \theta)}\, u(t)$	$\dfrac{0.5re^{j\theta}}{s + a - jb} + \dfrac{0.5re^{-j\theta}}{s + a + jb}$

Contd

Contd

No.	$x(t)$	$X(s)$
10c	$r\,e^{-at}\cos(bt+\theta)\,u(t)$	$\dfrac{As+B}{s^2+2as+c}$
	$r=\sqrt{\dfrac{A^2c+B^2-2ABa}{c-a^2}}$	
	$\theta=\tan^{-1}\left(\dfrac{Aa-B}{A\sqrt{c-a^2}}\right)$	
	$b=\sqrt{c-a^2}$	
10d	$e^{-at}\left[A\cos bt+\dfrac{B-Aa}{b}\sin bt\right]u(t)$	$\dfrac{As+B}{s^2+2as+c}$
	$b=\sqrt{c-a^2}$	

Table A.5 The Laplace transform properties

Operation	$x(t)$	$X(s)$
Addition	$x_1(t)+x_2(t)$	$X_1(s)+X_2(s)$
Scalar multiplication	$kx(t)$	$kX(s)$
Time differentiation	$\dfrac{dx}{dt}$	$sX(s)-x(0^-)$
	$\dfrac{d^2x}{dt^2}$	$s^2X(s)-sx(0^-)-\dot{x}(0^-)$
	$\dfrac{d^3x}{dt^3}$	$s^3X(s)-s^2x(0^-)-s\dot{x}(0^-)-\ddot{x}(0^-)$
	$\dfrac{d^nx}{dt^n}$	$s^nX(s)-\displaystyle\sum_{k=1}^{n}s^{n-k}x^{(k-1)}(0^-)$
Time integration	$\displaystyle\int_{0^-}^{t}x(\tau)d\tau$	$\dfrac{1}{s}X(s)$
	$\displaystyle\int_{-\infty}^{t}x(\tau)d\tau$	$\dfrac{1}{s}X(s)+\dfrac{1}{s}\displaystyle\int_{-\infty}^{0^-}x(t)dt$
Time shifting	$x(t-t_0)\,u(t-t_0)$	$X(s)\,e^{-st_0},\ t_0\geq 0$
Frequency shifting	$x(t)e^{s_0t}$	$X(s-s_0)$
Frequency differentiation	$-tx(t)$	$\dfrac{dX(s)}{ds}$
Frequency integration	$\dfrac{x(t)}{t}$	$\displaystyle\int_{s}^{\infty}X(z)dz$
Scaling	$x(at),\ a\geq 0$	$\dfrac{1}{a}X\left(\dfrac{s}{a}\right)$
Time convolution	$x_1(t)*x_2(t)$	$X_1(s)X_2(s)$

Contd

Contd

Operation	$x(t)$	$X(s)$
Frequency convolution	$x_1(t)\,x_2(t)$	$\dfrac{1}{2\pi j}\,X_1(s) * X_2(s)$
Initial value	$x(0^+)$	$\displaystyle\lim_{s\to\infty} sX(s)\quad (n > m)$
Final value	$x(\infty)$	$\displaystyle\lim_{s\to 0} sX(s)\quad [\text{poles of } s\,X(s) \text{ in LHP}]$

Table A.6   (Unilateral) z-transform pairs

No.	$x[n]$	$X[z]$
1	$\delta[n-n]$	z^{-k}
2	$u[n]$	$\dfrac{z}{z-1}$
3	$nu[n]$	$\dfrac{z}{(z-1)^2}$
4	$n^2 u[n]$	$\dfrac{z(z+1)}{(z-1)^3}$
5	$n^3 u[n]$	$\dfrac{z(z^2+4z+1)}{(z-1)^4}$
6	$\gamma^n\, u[n]$	$\dfrac{z}{z-\gamma}$
7	$\gamma^{n-1}\, u[n-1]$	$\dfrac{1}{z-\gamma}$
8	$n\gamma^n\, u[n]$	$\dfrac{\gamma z}{(z-\gamma)^2}$
9	$n^2\gamma^n\, u[n]$	$\dfrac{\gamma z(z+\gamma)}{(z-\gamma)^3}$
10	$\dfrac{n(n-1)(n-2)\cdots(n-m+1)}{\gamma^m m!}\,\gamma^n\, u[n]$	$\dfrac{z}{(z-\gamma)^{m+1}}$
11a	$|\gamma|^n \cos\beta n\, u[n]$	$\dfrac{z(z-|\gamma|\cos\beta)}{z^2 - (2|\gamma|\cos\beta)z + |\gamma|^2}$
11b	$|\gamma|^n \sin\beta n\, u[n]$	$\dfrac{z|\gamma|\sin\beta}{z^2 - (2|\gamma|\cos\beta)z + |\gamma|^2}$
12a	$r\,|\gamma|^n \cos(\beta n + \theta)\, u[n]$	$\dfrac{rz[z\cos\theta - |\gamma|\cos(\beta-\theta)]}{z^2 - (2\,|\gamma|\cos\beta)z + |\gamma|^2}$
12b	$r\,|\gamma|^n \cos(\beta n + \theta)\, u[n]$ $\gamma = |\gamma|e^{j\beta}$	$\dfrac{(0.5re^{j\theta})z}{z-y} + \dfrac{(0.5re^{-j\theta})z}{z-y^*}$

Contd

Contd

No.	$x[n]$	$X[z]$
12c	$r\|\gamma\|^n \cos(\beta n + \theta)\, u[n]$	$\dfrac{z(Az + B)}{z^2 + 2az + \|\gamma\|^2}$

$$r = \sqrt{\frac{A^2\|\gamma\|^2 + B^2 - 2AaB}{\|\gamma\|^2 - a^2}}$$

$$\beta = \cos^{-1}\frac{-a}{\|\gamma\|}$$

$$\theta = \tan^{-1}\frac{Aa - B}{A\sqrt{\|\gamma\|^2 - a^2}}$$

Table A.7 Selection of ITU voiceband (telephone line) modem standards

	ITU standard*	Type of modulation	Bit rate, b/s	Symbol rate, bauds
(a) Symmetric modems	V.21	Binary FSK	300	300
	V.22 bis	QPSK	1200	600
	V.26	QPSK	2400	1200
	V.27	8-PSK	4800	2400
	V.32	16-QAM	9600	2400
	V.34	1024-QAM	28,800	3429
	V.34 High speed	Nested-constellation of four 960-QAM constellations	33,600	
(b) Asymmetric modems	V.90:			
	Downstream	Digital	56,000	
	Upstream	V.34 High speed	33,600	

*The suffix 'bis' designates the second version of a particular standard.

Table A.8 Useful constants

Physical constants:

Boltzmann's constant	$k = 1.38 \times 10^{-23}$ joule/kelvin
Planck's constant	$h = 6.626 \times 10^{-34}$ joule-second
Electron (fundamental) charge	$q = 1.602 \times 10^{-19}$ coulomb
Speed of light in vacuum	$c = 2.998 \times 10^8$ metres/second
Standard (absolute) temperature	$T_0 = 273$ kelvin
Thermal voltage	$V_T = 0.026$ volt at room temperature
Thermal energy kT at standard temperature	$kT_0 = 3.77 \times 10^{-21}$ joule

One hertz (HZ) = 1 cycle/second; 1 cycle = 2π radians
One watt (W) = 1 joule/second

Mathematical constants:

Base of natural logarithm	$e = 2.7182818$
Logarithm of e to base 2	$\log_2 e = 1.442695$
Logarithm of 2 to base e	$\log 2 = 0.693147$
Logarithm of 2 to base 10	$\log_{10} 2 = 0.30103$
Pi	$\pi = 3.1415927$

Table A.9   Recommended unit prefixes

Multiples and submultiples	Prefixes	Symbols
10^{12}	tera	T
10^{9}	giga	G
10^{6}	mega	M
10^{3}	kilo	K(k)
10^{-3}	milli	m
10^{-6}	micro	μ
10^{-9}	nano	n
10^{-12}	pico	p

Index